SPACE PIONEERS

in their own words

Loretta Hall

Published by Río Grande Books
Los Ranchos, New Mexico

For more information www.SpacePioneerWords.com

 Rio Grande Books, Los Ranchos, New Mexico
www.LPDPress.com

Printed in the U.S.A.
Book design by Paul Rhetts

Library of Congress Cataloging-in-Publication Data

Hall, Loretta, author.
Space pioneers in their own words / by Loretta Hall.
pages cm
Includes index.
ISBN 978-1-936744-27-5 (pbk. : alk. paper)
1. Astronautics--United States--Biography. 2. Pioneers--United States--Biography. 3. Outer space--Exploration--United States. I. Title.
TL789.8.U5H285 2014
629.4092'273--dc23
2014013328

Front cover: Walter Cunningham during the Apollo 7 Mission.
Photo courtesy of NASA and Walter Cunningham.

Acknowledgements

This book would not have been possible without the generous cooperation of Chris Orwoll, George House, and Michael Shinabery at the New Mexico Museum of Space History in Alamogordo. They gave me access to their oral history transcripts and historic photograph collection. I am also grateful to Howard Hallmark, a volunteer at the museum, for writing the foreword for *Space Pioneers: In Their Own Words*.

The Historical Society of New Mexico provided a grant that covered my expenses for an extra trip to Alamogordo so I could select and digitize photographs for this book. The Southwest Oral History Association also provided a grant to cover part of the cost of publishing this book. I am deeply grateful to both organizations.

I appreciate the skill and dedication with which Paul Rhetts and Barbe Awalt at Rio Grande Books published and help with marketing both this book and my previous space history book, *Out of this World: New Mexico's Contributions to Space Travel.*

Finally, I would like to thank my editor, Wayne Grumet. His suggestions improved the quality of this book. I remain solely responsible for any errors, typographical or otherwise, that may have slipped past my final review.

Note to the Reader

Space Pioneers: In Their Own Words consists of annotated excerpts from the oral history collection of the New Mexico Museum of Space History. Men and women interviewed for that collection are not limited to people who worked on space projects in New Mexico; in fact, one of the people featured in this book was a Russian cosmonaut. The ninety oral histories I selected were, I thought, the most informative and relevant to the development of manned spaceflight.

Each excerpt is in the first person, as the interviewee spoke about his or her involvement with space programs. Each excerpt is introduced with a brief description of the individual whose comments follow. Two tiny spacecraft symbols mark the beginning of each excerpt, and three small satellite symbols mark the end of the excerpt.

I edited the excerpts with what I hope was a gentle but firm hand. In order to create a smooth narrative, I sometimes moved paragraphs or sentences around. I omitted material that I thought was redundant or extraneous to the presentation at hand. Because I wanted the excerpts to read smoothly and without undue distractions, I did not mark omissions with ellipses. In a few instances, I realized that an interviewee had misremembered or misstated a fact. To the extent possible, I verified the correct information and substituted it for the incorrect version. When I inserted explanatory words or sentences, I did put them brackets.

Researchers who wish to study the oral histories in detail should contact the New Mexico Museum of Space History (nmspacemuseum.org) to request access to the full transcripts.

For more information on the space research programs at Holloman Air Force Base and White Sands Missile Range, see my previous book *Out of this World: New Mexico's Contributions to Space Travel.*

Enjoy!

Loretta Hall
loretta@authorhall.com

Contents

Foreword

There are no extraordinary men . . . just extraordinary circumstances that ordinary men are forced to deal with.

William Halsey

It is unlikely that any of the men and women who have contributed to the space program have considered themselves as extraordinary. It is more likely that they were merely ordinary men and women doing their job, an ordinary job. But these ordinary men and women doing their ordinary jobs accomplished an extraordinary feat. They set man on the Moon. Some of those ordinary men walked on the Moon.

Loretta Hall in *Out of This World* rose to the extraordinary challenge to chart our trek toward the stars. Perhaps, since the beginning of time we have looked at the stars and dreamed of traveling to them. Now Ms Hall has once more risen to the challenge of charting the space program through the words of many of the participants in that program. None of them thought of themselves as "making history." They were just doing their job.

Ms Hall has done a very good job of taking the words of these space pioneers and weaving them into a narrative of their involvement in the development of the space program. Ms Hall didn't stop with the walks on the Moon. She has also included the space station and the shuttle program, and even a look at the dreams for the future. She has also given a well-balanced presentation of UFOs. You can make your own decision.

These men and women in *Space Pioneers: In Their Own Words* came from all walks of life. Some of them accomplished great things. Some didn't contribute very much. But each contribution was important. Regardless, each contribution is a gem that helps complete the picture. Ms Hall has captured those gems for us. I recommend Ms Hall's *Space Pioneers: In Their Own Words* to enhance your knowledge of Space History.

Howard Hallmark, Ph.D.
Volunteer Transcriber of Oral Histories
New Mexico Museum of Space History

Introduction

> *So much is being lost. One of our prime reasons for being here, as a museum, is to collect and preserve space-related history. If that wasn't gleaned via oral histories, video histories, and artifacts, we wouldn't have it for posterity's sake, for future generations.*
>
> George House, Curator
> New Mexico Museum of Space History

A cube of bronze glass supported at two corners by massive stone pillars, the New Mexico Museum of Space History sits on a ledge part way up a mountain, overlooking Alamogordo. Several steps downhill and across the access roadway sits the museum's Hubbard Space Science Research Building. The simple appearance of the small building belies the treasure trove it contains. While the impressive museum building houses artifacts that visually portray mankind's history of space exploration, the Hubble building protects written documents, photographs, and audio and video recordings of activities that made space travel possible.

An important component of that collection is the oral history program, which is funded by the International Space Hall of Fame Foundation. Its value is explained by museum staff in this statement:

> While the official governmental and private industry reports reflect the results of [space development] activities, such reports are "cut and dried" and reflect the technical aspects of the various programs. The human element is missing from such reports. The New Mexico Museum of Space History Oral History Program is attempting to recapture the history of "the old days" and

> in particular, the activities of the various individuals involved in these momentous events. No activity is considered too small to be recorded and preserved for posterity.

A visit to the museum in southern New Mexico is a rich experience because of the many outdoor and indoor exhibits that depict space research from 1930 to the present day. The documents stored in the Hubble building, however, are available only by arrangement to "legitimate researchers." The museum staff have provided access to the oral history transcripts for the purpose of writing this book, making substantial portions of those valuable recollections fully available to the general public for the first time. They present vivid accounts of daily life as well as the research activities that have taken place during the past eighty years in various phases of the US space program.

Introduction

Since the New Mexico Museum of Space History Oral History Program began in 1986, 140 people have been interviewed. Ninety of them are represented in this book, their stories spanning the time period from the end of World War II through the space shuttle era. As of this writing, nineteen of the space pioneers quoted in this book have been inducted into the International Space Hall of Fame. Many will never be chosen for that distinction, as they were smaller cogs in the gears that propelled space research. Whether individuals became famous or contributed in a nearly anonymous way, their thoughts and experiences offer unique glimpses into the activities that shaped various aspects of America's space programs. As the United States moves into the future of space exploration along with international partners, it is important to look back at the way individuals worked together to produce incredible technological advances over a relatively short time span. Their stories provide insight into the past and inspiration for the future.

These oral history recordings and transcripts, along with other documents and photographs, are valuable assets for this country. They exist because of the vision and dedication of the staff of the museum in Alamogordo, so it is fitting to begin this journey through space history by understanding how the museum came into being.

On July 24, 1973, Dwight Ohlinger told several friends he thought a space hall of fame should be established, and that Alamogordo would be the right place to build it. Ohlinger had moved to that southern New Mexico city in 1950 and served as its mayor in the mid-1960s. His suggestion drew enthusiastic support from the group, and a few days later, New Mexico Governor Bruce King issued a proclamation naming Alamogordo as the site for the International Space Hall of Fame. An outpouring of support arose among the state's cities, counties, legislature, and US Congressional delegation. The New Mexico American Revolution Bicentennial Commission proposed that the Space Hall of Fame and its associated New Mexico Museum of Space History be the centerpiece of the state's celebration of the nation's bicentennial in 1976.

Just over three years after Ohlinger suggested the idea, the International Space Hall of Fame was dedicated in early October 1976. Construction delays kept the facility from being opened to the public until November 23 of that year, but the dedication celebration was a momentous occasion. The formal dedication ceremony also opened a five-day technical conference chaired by Dr. Ernst Steinhoff, one of the

German rocket developers who came to the United States with Wernher von Braun at the end of World War II. The dedication's timing was significant for several reasons. Robert Goddard was born on October 5, 1882. The year 1976 was the fiftieth anniversary of Goddard's first liquid-fueled rocket flight. The Soviet Union launched Sputnik I on October 4, 1957, so the space history museum opened on the first day of the twentieth year of the Space Age.

This introduction sets the stage with recollections from Ohlinger and another individual who were instrumental in establishing the International Space Hall of Fame and the New Mexico Museum of Space History.

Dwight Ohlinger came to Alamogordo as an insurance agent in 1950. One evening in 1973, he was watching television and saw a commercial for the Baseball Hall of Fame. It struck him that there's also a Football Hall of Fame, a Cowboy Hall of Fame, and several others. Why shouldn't there be a Space Hall of Fame? In his May, 1998, oral history interview, he picked up the story:

I was just sitting in my chair, and I looked at that and a thought hit me. I thought, gee, let's get one of those in Alamogordo for the research people that have done all of this. Because actually, all of the history was done here. It was all tried out here before it went back to Cape Canaveral [NASA's manned spaceflight launch facility in Florida]. We did the dangerous work. And I had known quite a few of those people, personally. So I thought this is the biggest natural I ever saw, and it would be a shame if we would let that skip by without taking advantage of it.

I was a good friend of Dan King. He was at the Security Bank at that time. At lunch the next day, I was eating at the Desert Air down there, and Dan and his son were eating lunch. I went over and sat down with them, and told Dan I had a real wild idea. He threw his fork down and said, "That is the most wonderful thing I have ever heard of. We've got to get cracking on this." So we met with the Chamber within two or three days. And, jeez, it was off and running. Then in 1976 this [building] is done.

It's amazing how everybody got on board. I was real proud. It was almost non-political. We've had Republican governors and Democrat governors and everything else, and everybody has always been for us. I guess when I got that idea, I lucked on it. That's the only thing I've seen that is non-political.

We got together with the BLM [Bureau of Land Management] and found some land for [a good price]. Several of us looked at it, and we said, "What a beautiful spot for that [building], like a jewel above the town down there. People driving in or flying in can see that Space Hall.

When we first got going, a lot of those people came to us and said, "Let me help." I think Dr. Stapp [former head of the Aeromedical Field Laboratory at Holloman Air Force Base in Alamogordo] was here from the very first. Dr. Steinhoff [one of the German missile experts who joined the US space program after World War II] was here from the very first. We made many trips together up to Santa Fe and back when we were trying to get this place [established]. It has made life so interesting. I still remember one of the darndest characters I ever met. [Mercury, Gemini, and Apollo astronaut] Gus Grissom came to visit here. We went to church all the time with Major [David] Simons [head of the Space Biology Branch at Holloman]. You know, you mingle with these people and didn't even think anything about it.

Bruce [Governor King] and his wife were here [for the groundbreaking ceremony]. We came out here the first day and flattened a little land, and got some bushes out of the way so you could drive up here and park. We got all set and were going to dedicate the thing, and they got the dynamite and the plunger ready. They said, "Okay, here we go." Everybody is taking a picture and they signaled Bruce to go ahead. He goes *zunk* on the handles and nothing happens. And *zunk* again. Somebody said, "I think we had better fix the plunger." Somebody went and rechecked. I guess there was a loose wire or something and they fixed it up. Then they had a small piece of dynamite that finally went *kerbloom!* and the Space Hall got dedicated.

Aubrey Dunn, Sr., was born in Alamogordo in 1928. He was a US Navy veteran, a sixteen-year New Mexico state senator, and an honorary master sergeant in the US Air Force. Dunn mentioned White Sands Missile Range, which is located about 50 miles west of Alamogordo, on the opposite side of the Tularosa Basin. Established in 1945 as White Sands Proving Ground, the overland rocket testing facility was renamed White Sands Missile Range in 1958.

In 1950, I got a job at Holloman Air Force Base as a civilian budget officer. I was there for about three years, and I got to sign a lot of the orders for Dr. Steinhoff, who went to Germany and brought a lot of the [other German] scientists back over here. He and I got to be friends at that time.

Then I began keeping books at the *Alamogordo News*. I kept all of the books and sold all of the ads. [The paper] was published once a week. In 1950, the population here was right around 6,000. That is when they started all of the work out at White Sands Missile Range. In 1954, I got elected county tax assessor. Alamogordo was just booming. Between 1954 and '55, the town grew immensely. And by 1960 there was a population of 21,600 and something. Most of that growth was between '54 and '58. All of these companies were coming down here to test all of those missiles. They had a lot of computer stuff out here, and the Air Force was testing the Aerobee missile, working with White Sands Missile Range. In town, there was such a crowd here. They were coming back, similar to what it was before the Base ever got here. These guys were in here renting houses. They got $40 a day per diem, which we couldn't imagine, plus their salary. So it made all of the houses sell, and everything moved fast. It was a great growth. We went to five days a week on the paper.

All of this built up in Alamogordo, and then they came up with the Cape [Canaveral]. All of the missile stuff got transferred down there. Where the votes are is where things go. So they took all of this computer stuff from here up to Kirtland [Air Force Base in Albuquerque], up to Sandia [National Laboratories], and just gutted us. In 1959 and '60 we had over 700 FHA and VA foreclosures in our little town.

So then, we really started to figure out all kinds of things we could do to make the town stable and make it grow. We really started being conscious of tourists and everything else we could do to bring it back. We got involved with the Base in every way we could, and we got some more activity, a fighter squadron and a bunch of other things. But the town really became conscious of all of the missiles. And that is where the whole concept came as to what we had going out here. I had seen and been part of all that space program, and everything started right out here. All of these Germans came in here. At one time, we had more doctor's degrees, scientists and physicists, all here. We were second only behind Los Alamos [the town in northern New Mexico where the atomic bomb was first developed] in the whole state. We were also second in per capita income as compared to the whole state, because we had all of these folks in here.

We had a problem. Otero County is a huge piece of property. In 1963, when I really got involved, we had no way to get money for the schools. Here we were with all of this per capita income and all of these scientific people, but we couldn't get money to educate the kids. So we set up a deal to have a four-star general come, Schriever, I believe was his name. We set a big deal out at the Officers' Club. We had about 250 out there, and this guy came. We were trying to be a real advocate of all the things we had at the Missile Range. Ted Bonnell [a local banker] came to me and said, "I want you to get your cowboy chaps on and your boots, and a plaid shirt and cowboy hat. We want to give this guy a Stetson hat. We want you to come in as an old cowboy off the [Missile] Range."

I said, "Well, okay, I'll see what we can do." So I let my beard grow, it was kind of red in those days. It was about three weeks of beard, I guess. I got all of that rig on, and this general got up there to speak. I came in there, had a lariat rope in my hand, and walked out on the stage with my spurs and all. I interrupted this guy while he was speaking. It kind of brought the house down. Nobody knew I was going to do that. I patted him on the back. I had worked up a little skit, and I bragged on the Range. I said, "You've got all of the cameras out here all over this Range. It is the greatest Range in the world. A guy can drive forty miles up there, and he gets out to take a leak in a bush. And by the time he gets back, those guys have already analyzed it and tell him he has a urine problem!" The whole skit was kind of like that. I said, "You know right here where we are, General, we got this headquarters building. It really is sorry, and you guys don't want to put in more money for that." I said, "We roped this jackass one day. Right here where all of these officers are, we had all of these jackasses, kind of like it is now." It kind of made a hit with the town people, and he took it real good. Later, we gave him a Stetson hat. He didn't forget where Holloman was after that little deal. It worked out good.

We had a Chamber of Commerce which was active, and I was active in that. I was co-owner of the newspaper and [the other co-owner] Billy Holder was active in the Chamber of Commerce. One day [in 1973] they called a meeting of a few people, and then later on they called me to come down, and some others. A guy named Dwight Ohlinger came up with the idea to create the International Space Hall of Fame. He was kind of a happy-go-lucky guy. He said, "Well, we ought to have [the hall of fame]." He expressed this in first one committee and then this other committee, and they decided then we were going to have it. We [decided we] ought to get the governor to proclaim this as the International Space Hall of Fame before somebody else gets the idea. So we went back to the newspaper, Billy Holder and I. I got on the phone and called the governor. I said, "Bruce, we want to create an International Space Hall of Fame here in Alamogordo. What do you think?" He said, "Well, all right, Aubrey, what are you going to do?" I said, "I want you to put out one of those little deals that you send out to people that voted for you, that you created today the International Space Hall of Fame. That is what I want you to do." He said, "It will be done." So that was all done in one day, and from then it went to grow. Everybody thought it was the greatest idea.

Chapter 1
A Century of Paradigm Shift

Today, technology changes at a whirlwind pace. No sooner do we buy the latest smartphone or tablet than a new model hits the market with more features, faster processing, and larger memory. Society has become so used to this pace that people forget this is a recent phenomenon. Before the Industrial Revolution some two hundred years ago, societies changed little over a period of centuries or even millennia. Improvements in technology accelerated at increasing rates during the nineteenth and twentieth centuries. Pioneers who settled the American West evolved into pioneers who ventured into outer space.

As glamorous as the term *space pioneer* sounds, many of them came from humble beginnings rooted in simpler technologies. In their oral history interviews, some recalled challenging childhood experiences that helped them develop into hard-working, resourceful adults. Sometimes, serendipity played a role, too. Their stories exemplify how the right mix of curiosity and opportunity (sometimes arising by blind luck) can take people from all kinds of upbringings and parts of the world, and bring them together to the same pivotal point in history.

Much of the early space research was conceived and/or conducted at White Sands Proving Ground/Missile Range and Holloman Air Force Base, which are located on opposite sides of the Tularosa Basin, a remote desert area of southern New Mexico. Consequently, many of the excerpts in this book are from men and women who served at those facilities.

Clyde Tombaugh was born only three years after the Wright brothers flew the first powered airplane. As a self-taught astronomer, he became famous in 1930 for discovering Pluto, the ninth planet in this solar system. (In 2006, Pluto was reclassified as a "dwarf planet," a change Tombaugh's widow said might have disappointed him, although he would have understood the rationale for it.) Two years later, he began his college education at the University of Kansas, where he earned Bachelor's and Master's degrees. In 1946 he was hired as head of optical measurements for the missile testing program at White Sands Proving Ground in New Mexico (see Chapter 3).

I was born in February 1906 on a farm Streator, Illinois, and I lived there until I was sixteen. Then the family moved to southwest Kansas, where I helped my father raise wheat on a big farm. That's where I started making my own telescopes. In the winter seasons, I would work on grinding mirrors for telescopes. The third one was a 9-inch mirror of excellent quality, which I struggled for weeks and weeks to get a good curvature. I made drawings of Mars and Jupiter at the eyepiece of the telescope and sent them to Flagstaff [Arizona] to the Lowell Observatory.

Apparently, they were favorably impressed with the quality of them. They could compare them with the markings on the current photographs of those dates, and they were quite impressed. They became very interested in me as a possibility to run a new photographic camera there. They finally asked if I would care to come out there and work in a cold, unheated dome on a three-month trial basis. Well, I stayed fourteen years. I must have been satisfactory!

I can hardly remember the time that I wasn't interested in astronomy. I remember that, in my early grades, I had a very lively interest in geography. That was the super-interest of my life. Then one day, probably when I was in about the fifth or sixth grade, the thought occurred to me: I wonder what the geography would be like on the other planets. So, that was a natural lead-in to astronomy. I had an uncle who lived nine miles from us, who was an amateur astronomer. He had a 3-inch non-achromatic telescope of poor quality, and my first views of Saturn's rings, Jupiter's moons, and craters on the Moon were through that telescope.

I have it with me today at home. Then in 1920 they got an achromatic telescope from Sears, Roebuck. That was only about a 60-millimeter one. I wanted to have more power to see those things I read about in the books. And so, my family being rather short on money, the only way to solve that problem was to make my own telescopes. That's how I got into making my own telescopes.

Dr. John Paul Stapp played a key role in early space research at Edwards Air Force Base in California, Holloman Air Force Base in New Mexico, and Wright-Patterson Air Force Base in Ohio. He also pioneered restraint systems for automobiles. Dr. Stapp was inducted into the International Space Hall of Fame in 1979.

I was born in Bahia, Brazil, in July 1910, of missionary parents and lived my first twelve years in Brazil. I was taught by my father and mother, who were certified teachers in the state of Texas. My father was president of the Baptist College in Bahia. The college took all grades from sub-kindergarten to high school. On furlough in 1916, I saw America and started to learn English for the first time. My parents had not allowed me to learn or speak English because they wished for me to become bilingual with Portuguese. After a very interesting year where I saw snow for the first time and many other things, we returned to Brazil. Once more on furlough, we came in 1922, and I was left to go to school in San Marcos [Texas] Baptist Academy, where I was registered in 1923.

After graduation from San Marcos Academy in 1925, I earned a Bachelor's degree at Baylor University in Waco [Texas] in 1931, and in 1932 I received my Master's degree. In 1940 I was awarded a Ph.D. in biophysics from the University of Texas. When I first went into Baylor, I was going to be a writer, but they didn't teach any courses in writing. I already knew how to appreciate literature, and I felt taking courses on that was a waste of time. At the same time, I was also going to finish in music. I did complete about three years of music. In the meantime, my second year at Baylor, I started with biological sciences and immediately went into pre-med work because I was quite interested and involved when I began in zoology. In fact, that's what I took my degree in, zoology and chemistry. Likewise, the Master's degree. I was one of those who took on just about everything in sight. I took speech courses and belonged to Alpha Psi Omega (dramatic fraternity) and Little Theater on campus. I was also in the band, and I was a distance runner in cross-country and on the track team. I was able to stay busy from about 5:00 a.m. until 1:00 a.m. every day of the time I was there.

After completing an M.D. from the University of Minnesota and nine months of internship in 1944, I was called to Army active duty. I had been drafted during medical school. I wasn't choosing a military career. In fact, I would have had a fellowship at Mayo [Clinic] if I had not gone into active duty. Having gotten into the military, I had thirty-six months obligatory service from October 4, 1944. Before that time was completed, I found myself involved in a million-dollar-a-year research project which I could not walk out on, and so I stayed.

Colonel Ulysses Hess was chief of staff for personnel at Holloman Air Force Base from 1955 until he retired from the Air Force in 1959. He continued working as a civilian employee at Holloman until 1975, when he became executive director of the International Space Hall of Fame.

I was born in March 1911 at Carterville, Missouri, but very little of it remains. Most of it fell in a hole. Carterville was lead and zinc mining country back in the early 1900s on up. In 1917 and '18, they were digging all those underground drifts everywhere, thirty to forty feet underground. Now, most of these have caved in.

My mother died when I was a year old. My dad died when I was two years old. In 1911 or '12, there wasn't too much child welfare or anything like that, just somebody had to raise you. But as soon as you were a big husky boy, you were out on your own. So, at fifteen years of age, or thereabouts, I was out on my own.

I had a real interesting experience. I was in Ponca City, Oklahoma. No reason at all to be there, except I was there and I was looking for work or something. I was sleeping in the park. I think I was fifteen years of age at the time. Somebody tapped me. "Boy, what're you doing here in the park?" I said, "Obviously, I'm sleeping." He said, "Come, go home with me." He took me home with him, and he took me in as a member of his family. He had a son and a daughter in high school. He said, "What are you gonna do?" And I said, "I'm going to look for work." He said, "I'll help you."

At that time, Ponca City Fire Department kept two great big, husky boys on the payroll, working straight nights. He got me a job on the fire department, working twelve hours a night. That's the way I went through high school, working on the fire department. It paid pretty good money, sixty-five dollars a month, and I slept in the dormitory. I was among the high pay ones in high school.

I remember the first time I got the monthly report card in high school. I had nobody to sign it, so I gave it to the [fire] chief. He said, "I'm not your dad" and handed it back to me and walked off. So I signed my own report card and turned it in. That raised quite a little question to the high school at first, but later on they got used to it because I worked my way through high school that way. Then I got a football scholarship to Tulsa University. But I didn't last very long, because about the first week of practice, I tore up my knee and that lost the scholarship. So, there I was out looking for another job again. I joined the Army.

James J. Orbison was born during World War I. By the end of his career as an electronics engineer, he had worked on manned and unmanned missions to the Moon and Mars.

When I was born in 1915, radio was just barely beginning to be used, and mainly it was used by the Merchant Marine and the Navy. The early equipment was the old spark-gap transmitter. If you've ever heard one, it sounds like lightning striking every time it makes a dot or a dash. Radio as such didn't really begin to appear until in the early twenties. *Electronics*, incidentally, is a term that has come up since World War II, in popular usage. Up until then it was radio, pure and simple. There were a few electronic controls used in vacuum tubes, but really nothing.

For a while, while I was still in high school, I even thought about working for the railroad and becoming a locomotive engineer and going on up into the railroad business. When I was ten, I built my first receiver, using a crystal receiver. I bought the parts at Kress's. Kress's at that time, back in 1925, had a radio counter, just as any other kind of hardware counter, and I began to get interested in radio. Then, when I was in high school, I had a chance to start working at the local theater and became interested in sound projection. From that was where I really began to develop my interest in what we now call electronics.

When World War II started, I went into the Air Force and was fortunate enough to be accepted into the radio school at Scott Field, Illinois. That particular school, before World War II, was actually a college-level course, because we learned all of the engineering that was necessary to build receivers and transmitters and whatever was necessary. I suppose from that, and the fact that I was in the Air Force and working with airborne and ground radio equipment and eventually electronic controls of various kinds and that sort of stuff, it just grew like Topsy. I don't think that I ever really had a definite goal that that was where I wanted to go. It was just that I wanted to learn as much about electronics as I can and work with it.

When I was a young boy right after World War I, I was all wrapped up in these battle stories and all those flying stories. I knew all of the World War I aces and all this kind of stuff. About this time science fiction began to come in, *Amazing Stories*, *Astounding Stories*, and this kind of stuff, and I read those voraciously. My dad was a good, stable farmer, hard-working type of individual, and his imagination didn't even go much into airplanes, much less space science. I was wasting my time, totally wasting my time reading that junk, nothing would ever come out of it. There were several times there when I had to kind of keep my magazines under cover, because if dad saw them he'd jump me again. Of course, I grew up. When I retired from the Air Force, I went out to Goldstone [Deep Space Communications Complex, operated by NASA's Jet Propulsion Laboratory] and got into the space business out there. We had already put Ranger 7 down on the Moon when I went home for a two-week vacation. We were sitting there one Saturday afternoon watching a ball game on the TV, and dad got to laughing. I said, "What's so funny, Dad?"

"Well," he said, "I just got to thinking, seeing you sitting over there. I remember how I used to raise hell with you because you wasted your time reading all that science fiction crap. Now I see you sitting there, and I'm knowing that you're one of the engineers that's going to put the spacecraft on the Moon to get the first close-up pictures of the Moon." He said, "It's just unbelievable to me that you have done that, but I know you've done it. It's just like watching that television game. I know that that game is being played in New York City right now. I don't have any idea how it gets here. I know that if I asked you, you could give me a long, detailed explanation of exactly how that picture gets on that glass screen on that set, but I don't want to know. At my age I don't care about knowing. All I know is that picture is there, I'm watching that game, and I know it's being played in Yankee Stadium in New York City, that's all I need to know about it. I'm perfectly happy." And indirectly, without dad realizing it, he was expressing the feeling of about 95 percent of the human race. They know it's happening, they accept it, they make use of it, but they do not have the remotest idea how it came about and how it's being operated.

Justin (Jed) Durrenberger was born in 1916. He helped develop the Baker Nunn satellite tracking cameras that were placed around the world to track Sputnik and subsequent satellites. He received the Robert Goddard Memorial Award in 1972 for his work in photographic tracking of missiles.

My dad died during the flu epidemic during the First World War. I was two years old, so I don't remember him. My mother had to go to work to keep us going. We moved to Greenwich Village in New York. I did most of my growing up in Greenwich Village about the time that [Charles] Lindbergh crossed the Atlantic to Paris. That began the interest in aviation, flying, and so forth. I hoped to be able to fly until [I found out] how bad my eyes really were. So flying was not for me. That led to a career of taking pictures of aircraft in the 1920s and the early 1930s. Because if you were going to make models of them, and that was the thing people would do in the late 1920s and early 1930s, was making airplane models. And in order to get the detail, you had to have pictures of those airplanes. So that got me started in photography.

My whole career from the 1920s until now into the 2000 era has been all oriented around optics, photography, recording, visualizing. They have changed the name now to *visualizing*. It's no longer *photography*.

I grew up during the Depression, the end of the 1920s and early 1930s. Those were days when there wasn't much money, especially in Greenwich Village, for a mother who went to work and didn't ever remarry. Since I had to go to work while going to school, I began working in the darkroom of a photo finishing plant in a retail photographic store. The owner of the store became an alcoholic, and I ended up pretty well running the business. At rather an early age, I was trying to earn a living.

In 1950 I graduated from what is now New Mexico State University with a Bachelor's degree in mechanical engineering. People asked me why on earth did I want to get a degree in mechanical engineering when everything I did was in optics. I was interested in designing and producing the equipment which is used in that particular type of work. I went out to White Sands and filed my application as a mechanical engineer. In those days, a Bachelor in engineering had the annual salary of $5,000 a year.

The main building at White Sands Missile Range was under construction. The building wasn't finished, and I didn't get a job until March 15, 1951. In the meantime, I had to do something to continue being able to eat and stay in the area. So I was laying cinder blocks. I can't work with leather gloves to set cinder blocks. It turned out that I wore the fingerprints off of my fingers. So I had security problems in getting a job at White Sands because I had no fingerprints.

Despite the fact that Al Metzger did not have a college degree, he belonged to a group known as the Broomstick Scientists. In 1945, as the Nazi government was about to be defeated in World War II, more than a hundred German scientists, engineers, and technicians who had developed the powerful V-2 missile surrendered to the US Army. About half of those so-called "Paperclip Scientists" who were admitted to the United States were sent to live at Fort Bliss at El Paso, Texas, and work at White Sands Proving Ground nearby in New Mexico. Equipped with 300 train car loads of V-2 missile components, launch equipment, and research documentation, they began working with Americans to further develop rocketry, not only as weapons but for scientific research and ultimately for manned spaceflight. The Army employed General Electric as a contractor to oversee the V-2 program. The Broomstick Scientists were US Army recruits with technical backgrounds who took over the V-2 missile program at White Sands in 1951, when the Army's contract with General Electric ended.

I was born in Altoona, Pennsylvania, in June 1918. I graduated from Altoona High School in 1936, and then I went to work for the Pennsylvania Railroad as a machinist apprentice for steam locomotives. My father was a machinist for Pennsylvania Railroad. Being raised in a railroad town with over 30,000 employees, it's the big industry and it is just natural a lot of times to follow in your father's footstep. And that's what I did. I worked with him about six months. He was a hard taskmaster, but he taught me a lot of basic values of work, work ethic, and so forth. They federalized the National Guard. I was in the 110th Infantry, National Guard, in Pennsylvania. When we were federalized right before Pearl Harbor, they put us on active duty for a year. Then Pearl Harbor came along. I was in a muddy field down in Virginia, eating cold beans and hot dogs when the announcement came over the radio. We knew at that time, since we were already federalized, that we were in for the duration, which we were. So we went down for maneuvers in Louisiana until the time came to break us out for different assignments. With my railroad background, I was sent to Camp Shelby, Mississippi, where they were activating the 727th Railway Operating Battalion.

After World War II, I went back to the railroad. I was only back there a couple months and they went on strike. Being married, with a child to feed, I had to find work. After World War II, you were allowed to reenlist without a loss of grade for ninety days. On the eighty-eighth day, I went over to the recruiting office and signed up to go back in.

Edward Dittmer served in the US Army in World War II. After the war, he returned to civilian life for a brief period and then reenlisted in the Army Air Corps. In 1957 he was assigned to Holloman Air Force Base, where he participated in such high-profile programs as Project Manhigh (see Chapter 4) and training chimpanzees for the Mercury Program (see Chapter 7). He was inducted into the International Space Hall of Fame in 2001.

I was born in Laverne, Minnesota, in September 1918. My dad was a harness maker by trade. In 1942 I was up for the draft and classified as 1-A, so I just volunteered. I entered the Army the first of April, April Fool's Day, during World War II. I was assigned to the First Cavalry Division, Fort Bliss, Texas – the horse cavalry at that time. From there, I went to the 713th Combat Engineers and the 770th Railway Operating Battalion and served in the Pacific. After three years and seven months, I had enough points to get out. I enlisted in the Air Force in 1945.

I was a medic during World War II and Vietnam. I'd always had an interest in medicine. In fact, at one time, I had the desire to become an M.D. But money was tight and so I just thought I'd give it a shot at the medical field. I wanted to further my education, and I felt I had a better opportunity in the Air Force than I would on the outside. At that time, things were pretty tough. I had married right after World War II, and there were lots of opportunities for the family to be taken care of, so I decided to go back in. I had a good job at the time I went back in. I was working at the VA hospital in Marion, Indiana. I just got a promotion. They started us out on the mild wards, then you go to the semiviolent and the violent wards. That was quite an experience, too, and some good training. When I left there, I was in the hydrotherapy department, where they bring the violent patients and wrap them in cold sheets and calm them down and then send them back to the wards.

Frank Conway worked on various missile programs at White Sands Proving Ground, beginning in 1950. He was an expert marksman and won New Mexico state championships in high-power rifle outdoors and three-position small-bore rifle indoors.

I was born in February 1919. Of course, the talk at the time when I was growing up was the First World War. It was going to end all wars. That was the way *that* worked out.

My dad joined the Navy Air Corps in 1908, when he already had five years in the Navy. At that time the Navy only had balloons. In 1910 they got their first three airplanes from money appropriated by Congress. The admirals couldn't see any use for airplanes, but Congress could. In 1914, as far as we can tell, my dad was the first aviation chief petty officer in the Navy.

I wanted to work with tools and so on, so I went to Worcester [Massachusetts] Trade High School. I wanted to get into the electrical department because I saw that was a coming thing. You went into the technical school, then the shops themselves for two weeks, and then you went into the vocational part of it. Every day, you had two hours of drafting, for instance. If you were studying the building trades, it was architectural. If you are studying to be a welder or a machinist or something like that, then you were into mechanical drafting. So it was a very outstanding thing.

I went to work at the Worcester Pressed Steel Company. [The airplanes] had, at that time, a big aluminum dome covering the pitch-changing mechanism on the front. We were the only plant in the whole United States that had the capability for making that. So we were running two shifts to make these big aluminum things that covered up the pitch-changing mechanism. And we got another contract from Britain for two million 20-shot clips for the Tommy gun, and 250,000 of the 50-round drums for the Tommy gun.

The company had a privately owned museum. Mr. Higgins had gone into Europe during the Depression and had brought out all kinds of armor and so on. He had probably close to sixty to seventy complete suits of armor. He had another wing, which was Stone Age weapons, another wing which was Bronze Age weapons, and an industrial wing that had mostly major bomber parts. And in the museum was the Yale University Firearms Collection, on loan. Every day I came to work on the second shift. I was night shift electrician. I had to check out the fluorescent lamps. They had just come in at that time in the '30s and they burned out quite often. So I walked really slowly through those 2,300 guns on a daily basis. I studied everything. It was a fabulous collection! Weapons, some were experimental, almost all of the Colts, and inventions, fantastic type of stuff up there. About seventy-five full-stock Kentucky rifles from every famous maker.

One day, I came in and there was gun missing. So I went to the curator, and I said, "Do you have a gun on loan?" And he said, "No." I said, "Well, you've got one missing." He said, "Where?" So we went downstairs and I showed him it was missing. I told him it was a match target Colt Woodsman .22. They called the police, and they had already caught the kids the night before. They broke into a hardware store and got two cartons of ammunition. They were shooting out streetlights with this pistol when the police caught them. But they wouldn't tell them where they got it. Three or four months later, there was another one missing. So, again, I went in and asked about it, and they said, "No, we don't know about it." This time it was a .45, 1938 National Match, and had a firing pin safety in the experimental [model]. Some kids had stolen that too.

So then Mr. Higgins wanted to see me. I went in and reported to him. He said, "I can understand, maybe, that you would see something was missing, but you know exactly what's missing, and there are 2,300 guns down there." I said, "I couldn't do it with all of them, but these particular ones were unusual, so I knew that." He said, "With that interest, anything you want to borrow, take apart, assemble, shoot, anything like that out of the whole collection, just tell my secretary what you are taking." Well, I was in heaven right there. I didn't have to go to church any more. So anyway, now I am the plant gun nut. On Fridays when I came to work, for about a year and a half, I went down into this shop area after I went to the museum and got the Tommy gun out of the museum. Then I would take five rounds, load them into a random selected clip and drum, and shoot a five-shot little burst into a twenty-foot tunnel we had built just for that purpose. So when I came in the Army, I knew a Thompson [machine gun] inside out.

I had signed up for the Army the day after Pearl Harbor, but I couldn't actually raise my right hand until 15 January 1942, because I had to give the plant three weeks' notice and let them try to find somebody else for the second shift. When I came to Aberdeen [Proving Ground in Maryland], I took technical training and basic training all in two months. I was in hog heaven there too, because I had read about Aberdeen my whole life. And now I am going to be working there. They gave me aircraft armament, and among them was the Thompson. I took it apart and put it together, and so on, and showed them. The guns they didn't know about, I told them about them. They had dragged some guns out of storage areas, the 1917 rifle from the First World War, for instance, was being issued as a substitute for the carbine. The carbines were coming, but we could only get 1917s.

Freda Herrera was born in Nogales, Arizona, in 1921. After starting out in the field of nursing, she earned a Master's degree in physiology. She worked at the Aero Medical Laboratory at Wright-Patterson Air Force Base in Ohio and later at the Aeromedical Field Laboratory at Holloman Air Force Base.

My father was plumber. He was known as the utility man at [US Army] Camp Steven D. Little in Nogales, Arizona. My mother, of course, was a housewife. I went to high school at University High School in Los Angeles, California. I graduated from the University of California, San Francisco in the nursing program in December 1944.

I was always very interested in the military, and I was always interested in medicine. Basically, in the '30s, the best advice any of the teachers could give you was that nursing was a good occupation for a woman's employment at that time. I went into the Army Nurse Corps in 1945. I was released from that in December of '46. I went back to college at Berkeley in California, and finally ended up at Earlham College in Indiana, where I procured a Bachelor's degree. At that time, some of the universities didn't consider a Bachelor of Science in Nursing as a college degree, so I majored in biology, assuming that I would proceed into a field of medicine.

In 1959 Dr. William Douglas was selected to be the personal physician of the Mercury Seven astronauts. In his oral history interview, he remembered not only his own childhood experiences, but those of his parents. He was inducted into the International Space Hall of Fame in 1992.

I was born in September 1922, in the little town of Estancia, New Mexico, just southeast of Albuquerque. My father came out here with his father from Clay Center, Kansas. They came out here around the turn of the century sometime, I would guess about 1903, '04, '05, something like that. He and his father came out here in what was called an immigrant train. It was a boxcar. They put their livestock in the boxcar, and grandfather and my father lived also in the boxcar. Dad relates how occasionally the train would stop, go out on the siding, they'd take the stock out to water it and to exercise it, and get back in and proceed on.

What influenced his father to come to Estancia, I don't know, but I do know that they took over a homestead that had been abandoned. I can't really call them homesteaders, because as I understand it, when one homesteaded, one had to spend a certain amount of money or make certain improvements on the property in order for it to be that individual's own, deeded to them. The person who staked out this homestead didn't do all those things, so my grandfather took over that place. He actually dug a cave into the side of a hill and then built a little house on the outside of the cave, where they lived while they constructed a proper house and dug a well.

My father built a windmill made out of wood and all, and he constructed it as a boy in his early teens. When they had the house all completed, the windmill, and the well dug and all that, they wrote back to Kansas, and my grandmother and the younger child of the family, my father's brother, came out. They were dirt-poor farmers and ranchers, and they always were very, very poor people. Finally, they moved into Estancia, where my grandmother taught school. Even when my father was in high school, I think he would go to school for a year and teach for a little bit. Apparently, they didn't have very strict requirements for teaching certificates in those days. But because of his interest in teaching I guess, he went to what was called New Mexico State Normal Institute, which was at Silver City. It's Western New Mexico University now. Because of living out here in the Southwest, he became completely bilingual and had a wonderful command of the Spanish language. And again in his teens, he went up to Tierra Amarilla, which is up in northwestern New Mexico, and taught school to the Spanish-speaking children up there because he could teach in Spanish.

Apparently, there was a smallpox epidemic up there and he caught smallpox. So he wrote his mother, or sent her a telegram, or something – somehow communicated with her – and said, "Please come nurse me, take care of me." The train was running at that time into Chama, I believe, and so he met her with a horse team and wagon at Chama. As he tells the story, she came in at night, and he stayed with the team and wagon back out of the lights. She got off the train, and she obviously was concerned there was nobody there to meet her. After the few spectators had departed, dad drove the team and wagon up, got his mother and drove back to his home. The reason he did all this was that if the people had seen him with these pockmarks on his face, he would have been put in the pest house. That's why he wanted his mother there to take care of him, too. He didn't want to have to stay in the pest house.

My mother was from Mississippi. Her father was a conductor in the Illinois Central Railroad back there, and she came out here about 1920. Her mother's sister lived in Estancia and her husband was a physician. It was thought that mother might have had tuberculosis. She had a chronic cough and so they sent her out West. Again, the interesting thing about how teaching was different in those days – a lady comes out here because of a presumptive diagnosis of tuberculosis and immediately starts teaching school. I don't think she had TB, at least there was no indication of it since then. Her chest X-rays were always normal and she lived to a ripe old age.

Even as a very small child in the first, second, third grades, I had an abiding interest in biology. I used to like to collect bugs. As a matter of fact, some of the kids used to call me "Buggy" because I offered them a penny for every bug they gave me. I was dissecting frogs, I guess, when I was ten or twelve years old. My parents went off to the World's Fair, and I stayed home. The price I extracted from them for staying home, or the bribe that they gave me for staying home, was that they gave me a microscope. I got a Bausch and Lomb New Gem microscope, which I still have, and it's still a nice little thing. So, at any rate, it just continued, and at one time I thought I'd like to be a laboratory technician. That sounded great. And then sometime, I decided I might as well go to medical school and that was it. Ever since I can remember, I've been interested in biological sciences. Medicine was sort of the ultimate in the biological sciences, and that's where I went.

Cynthia Sommers-Guzevich's background is similar to that of the famous Paperclip Scientists, although she did not come to the United States with that group. Her story sheds some light on German and American approaches to missile research in the 1940s. The first paragraph below is from her oral history interview transcript. The subsequent paragraphs are from an interview this author conducted with her in April 2013.

I was born in Munich, Germany, in May 1923. In Munich, I received my education to become an artist. At the beginning of the war, the people that had more than a semester to graduate had to work for War Production. Because I was good with my hands in drawings, I was educated as a draftsman by the German government. I started out with a company that was called Loeffelbein, which was a technical company that worked for War Production, designed speed boats and engines and various parts for wartime use. I worked for Messerschmidt for six months on loan from the company and then was transferred to Linde Air Products, which is now Union Carbide. I liked the company so much and I liked the project so much that I transferred there permanently. Until the end of the war, I worked there, and this is where I became chief of their drafting section and was introduced into the space program, although at that time it was unbeknownst to me that that's what we were working on. Obviously, the objects that were in relationship to the V-1 and the V-2 was so highly secretive that the work was doled out in parts to different firms.

I was drafted into it [working for the Nazi government] when I was not quite nineteen. I studied to be an artist, but nobody was allowed to be anything but [in] the government. I was drafted into work making ammunition in an underground factory. I had a very influential uncle that was a personal war friend of Hitler's. He was never a Nazi, though.

He was smart. He put in the questions to the right people. You had no choice what you wanted to do. But he called me into that office and they said, "What would you *like* to do?" I said, "Something with my hands." They said, "Would you like to be a draftsman?" I said, "Oh, what's that?" They said, "You draw pictures of houses and engines and airplanes." I said, "That sounds interesting."

They assigned me to an office, and there were maybe sixty or seventy women sitting there. At that time, everything had to be done by hand. The engineers would draw the drafting and the girls had to fill out all the writing and all that. I said, "I don't want to do that." I was really arrogant. (Laughs) The man that interviewed me said, "What do you want to do?" I said, "I really don't know. What are those men doing in there?" It was a big office, just men. He said, "Those are engineers. Would you know what to do?" I said, "Show it to me. Maybe I do." So we went in there, and they had all the drawings there. I had no idea what it was. I said, "That's what I want to do." He said, "Do you know anything about engineering?" I said, "No, but I can learn. I'm not stupid."

Two engineers took me under their wings, and they were teaching me like you wouldn't believe. At night, I went to engineering school. They assigned me that. Sometimes I wasn't allowed out of that office for a week. That office was really working on the missile program. But it was so categorized that they knew they were doing something on engines, but they didn't know what for, because that was another office. It was very interesting, very challenging. And I enjoyed it.

In 1948 I married an American and came to the United States. When I finally got my permit to come over to the United States, I had to prove I wasn't a Nazi. I was applying to marry my husband, Mr. Guzevich. They [the American government] found out that I was in that program. When I applied for my citizenship, they said, "You worked for the Nazis." I said, "Everybody did." [They said,] "But we can't find you as a member of the Party." I said, "I'm not a joiner." "But you have to prove that you were not a Party member." I said, "If you can't find my name, then doesn't that prove you something?" It took me one year to convince them. Every month, I had to go before the American military to be questioned. Once, I got down there and they wanted to find out what I was doing [in the German war program]. I said, "I can't tell you." He said, "Why not?" I said, "I raised my right hand and I swore what I would do, I would never reveal to anybody." "But we're Americans."

I said, "They didn't say I can reveal it to Americans." So I had a lot of trouble.

But they brought me over here with my husband. In California, I worked on Travis Air Force Base. Then, somehow, they found out what my office [in Germany] did. So they transferred me to New Mexico and my husband to Holloman, because he was still military. There were some of the [other German] people there, like Dr. Dohm, that I worked with on White Sands [Proving Ground].

William Bowers was another member of the Broomstick Scientists. He recalled how he got a job at Rocketdyne's test lab at Santa Susana Field Laboratory in southern California, testing components for every rocket engine from the Redstone to the space shuttle.

I was born in Clarksburg, West Virginia, in October 1923. My father was a civil engineer at Hope Natural Gas Company in Clarksburg. My dad's engineering got me interested in that from the very beginning. I really became aware when we went to Arkansas in 1928. He went down there on a special program for the gas company, and I got kind of interested in traveling by train at that time. And I never lost my interest in railroads to this day. I graduated [with a Bachelor of Science in mechanical engineering, from West Virginia University in Morgantown] in 1948, and went to work with the Hope Gas Company, where my dad worked.

On June 10, 1952, I went to the Tenth Annual World Science Fiction Convention in Chicago. Now by this time, a lot of my friends had left [for] White Sands, and I was still in Clarksburg. Anyway, on the train I met an employment rep from North American Aviation. He asked me where I was going, and I told him. He said, "Where do you work?" And I told him I worked for the gas company. He said, "If you are interested in science fiction, why in the hell are working for the gas company? We need people like you out there building rocket engines." So that is where I went in 1953, and stayed there at Rockwell for the next thirty-five years. [Rocketdyne and Rockwell merged in 1967.]

After a thirty-two year career in the Air Force, Wayne Mattson retired to Alamogordo in 1990. He served as a volunteer at the New Mexico Museum of Space History until his death in 2012.

I was born in Fond du Lac, Wisconsin, in November 1929. My father drove a bus and when the company was sold out from under him, so to speak, he wound up as a machinist. My father first took me up in an airplane when I was three years old, and he said he had to hang on to me to keep me from walking over the side of an open cockpit job. Ever since that point in time, I was strongly interested in aviation.

I've always had an interest in rocketry. As a young kid, we were always messing around with chemicals. One day when I was about thirteen years old, we inadvertently made our own rocket. We took a glass tube, filled it full of a mixture, set it up at an angle, ignited another mixture of the propellant on the outside, because we didn't know anything about fuses. Lo and behold, it ignited the fuel on the inside of the tube and the tube took off across the room, and we finally succeeded. We were using gunpowder.

Wallace (Wally) Adam, born in 1934, was an Air Force pilot who taught aeronautics at the Air Force Academy in Colorado Springs during the early 1970s. He served as director of the Holloman High Speed Test Track in the late 1970s.

My parents were cattle ranchers in western Nebraska, so I was raised out in the sticks, nowhere. I went to high school in the little town of Alliance, Nebraska, in the middle of the panhandle. Graduated in 1951 from that. I did the cowboy thing for a couple of years and decided that wasn't the way I want to make my living, so I went to the University of Nebraska to an engineering program and graduated in mechanical engineering in 1958.

I wish I could remember the quote, but Edgar Allen Poe has a quote something about those that dream in the daytime getting much more than those that dream only at night. I was always kind of a daydreamer as a youngster. When we were kids growing up and after I got out of high school, and my brother and I were on the ranch down there, just because it was such a beautiful area, we used to get our horses out and ride at night sometimes on a full moon. The stars were so brilliant in that area, where there's no pollution or anything. You could see clear to the beginning of time, almost, at night, you know. You could really see back. And part of my dreaming was, I wonder what it would be like to go to the Moon. That was just a fantasy thing, but it turns out when Sputnik came along, the fantasy wasn't as far-fetched. So when I decided I didn't want to be a cowboy anymore, I said what would it take if I were to pursue that wild-hair dream to make it possible? An engineering education seemed to be the thing, so that's what daydreaming led me to think.

[Note: The Edgar Allan Poe quote, from his short story "Eleanora," is "They who dream by day are cognizant of many things which escape those who dream only by night."]

Story Musgrave was born in 1935. Despite growing up in an abusive home that he said "had no intellectual things" in it, he voraciously pursued formal education. He earned two Bachelor's degrees (one in mathematics and statistics, and one in chemistry), three Master's degrees (in operations analysis, business administration, and computer programming; physiology and biophysics; and literature), and a Doctor of Medicine degree. He flew six missions as a space shuttle astronaut.

I was born in Massachusetts, lived on a thousand-acre dairy farm in Berkshire County with 200 head of Guernsey cattle, sheep, chickens, ducks, horses, every other kind of animal.

The very foundations of my approach to space were born in nature – having our own thousand acres, having a river, having forests. I took to the outside. Even at the age of three, I'd go out in the forest at night. I love the dark. I love night. I was totally at home and intuitive. You couldn't get me lost in a forest even at night. I was really comfortable with that entire world, and at the age of three I didn't know too much. I'm saying three because my memories probably don't go back any further. So from the earliest age, I was immersed in the rivers, the forest, the soil. I'd say that was the major influence which carried on forward into spaceflight. And now I just see the world from a different vantage point, but there's that basic relationship to nature.

I was steam locomotive. When I was a child, diesels had not arrived yet. And so, I could not have envisioned spaceflight. The farm was unbelievably narrow as well. I have no regrets, because everything in my past got me to here, and I love what I have. But you do raise the questions as to how you want to raise children. Farm kids, in general, had an incredibly narrow exposure, especially in the '30s. Nowadays, I think due to the information age, television, and transportation and other things, farm kids do get a broader look at the world than the farm kids of the '30s did. But still, it is not as cosmopolitan as if you were raised in the middle of the city.

Nature became my world of exploration because I never really got outside the county. I never really got over the hill. My education there was principally nature. There was never a discussion in the family of anything of any importance other than getting the hay in. There were no books. There was no homework. There was nothing. I'm out exploring nature because there's nothing else to explore.

The traditions of the late poets of the English Romantics, I relate to them and they strike a chord with me. And it's the tradition in America, especially Whitman, what you call the American transcendentalists, Emerson and Thoreau. That's the basic tradition that strikes a chord with me. Humanity is basically a quest for spirituality, religion, god if you will, through nature, direct revelation. And so, I'd say in terms of my childhood, that was the basic force that really led me into the aviation world, the space world, and my approach to the experience of space.

Chapter 2
Life in the Trenches

Most space pioneers were ordinary people who had the opportunity to work on extraordinary projects. Their oral histories include snapshots of what daily life was like, both on and off the job. Much of the early space research was conceived and/or conducted at White Sands Proving Ground/Missile Range and Holloman Air Force Base, both of which are located in the Tularosa Basin, a remote desert area of southern New Mexico. Consequently, most of the following excerpts are from men and women who served at those facilities.

Wally Adam was commissioned as a second lieutenant in the Air Force in 1958 after participating in the ROTC program during college. In 1976, he began a five-year stint at Holloman Air Force Base.

When we got our introductory briefing at the base, the minister was also relatively new, and I think he defined it fairly well in his story about the Base. He said, "We were coming from Eglin Air Force Base [in Florida] out here. Eglin's nice and green and on the coast, and it's a nice, beautiful area. We were driving through west Texas and, you know, that's just awful looking country. So we drive through Artesia [New Mexico] and we start heading into mountains and, man, it's gorgeous." And the more up they get to the mountains, they come through Cloudcroft and they knew they were getting close to Holloman. "We're getting close, oh this is great!" They came down to High Rolls and all the apple orchards. It was fall time, the apples were hanging on the trees ripe, and he said, "This is just great! Then we rounded that curve and looked out on the Tularosa Basin. Frankly, we about puked!" And that just about defines the initial impression. I was not quite that surprised. But I'll tell you, it only takes about a year to be in this area and you wouldn't leave it. It really grows on you.

Dr. Richard Bick was a medical officer at White Sands Proving Ground from 1956 until 1958, the year the base was renamed White Sands Missile Range. He recounted his experiences in his 2007 oral history interview.

In 1955 I had a letter inviting me to join the Army. There was a thing known as the Doctor Draft Law. I was sent to Fort Sam Houston, where they trained doctors to be military people, which can be somewhat of a comedy. They even had us on a parade ground mixed with dentists and doctors, and when the big old sergeant said "Right Face," half of us turned left. When he said "About Face," some of us went forward. It was kind of a comedy. At that time, it was supposed to be a six weeks indoctrination course. At three weeks, somebody showed up from Washington and said, "We have an acute need to send at least half of you to Europe. If you want to volunteer, here's your chance." I volunteered, but I think the Army made an appropriate decision and declined my voluntary attempt, because I had three children at the time. And it wasn't economically sound, I don't think.

I waited for my assignment, but before that, an interesting thing occurred. A colonel from Washington, in the medical department, came to Fort Sam and stood before us and said, "We do our best to assign you to a pleasant place that you can do your work. If in fact, you are at a place you are not happy with, you can call me directly and not go to your commanding officer," which is a little un-Army-like, I think. Okay.

So, my assignment was Fort Bliss. I was assigned to a little clinic out in the maze of Fort Bliss. To give you an example of why I didn't like that, I went to work, and I sat there with a nurse, in a little mini clinic, and I averaged one patient per day. But three days a month, I had fifty patients.

It was after payday, after Juarez, after gonorrhea. So I took care of those folks, ordered their penicillin, and they all got well. Then there was one patient daily for the next twenty-seven days! Prior to that, I had been private practice for five years and had done a lot of things, and was in tune a little bit to a good family practice scene, with a side specialty in surgery. So I could visualize doing some things. Well, it didn't take me long to contact that colonel.

Then they shipped me to White Sands, and it was great, because that was a real nice place to be. It was like a small town community. There were families and a nice swimming pool, nice weather, and I began to appreciate the desert.

This was 1956. At that time, of course, the big things were Nikes [first successful, guided surface-to-air missle] and smaller missiles. But one little interesting thing happened when I was Fort Bliss. On two occasions, I was on temporary duty, TDY, to Red Canyon. At Red Canyon, I was housed in a little trailer, where I lived with another person. I was the assigned physician at Red Canyon for a week. We flew up from Fort Bliss Military Airport in a little, small airplane. While I was there, there was little or nothing to do, except to appreciate that. All the Nike batteries from around the country were shooting at a little red, radio-controlled airplane.

Several things happened there. One was the sergeants. You know, a sergeant will frequently befriend a doctor, because they may need you a little later. They went out to Trinity Site and brought me back a handful of the frozen sand. And they also got me a bunch of long cacti [ocotillo] from which I made and carved canes, because I didn't have much to do there anyway. But the last incident that I really remember is returning at the end of the week. We were supposed to fly back. The plane was there. There was a horrendous windstorm and dust storm. You couldn't see the top of a nearby hill. It was just horrible! So I drive out to where the airplane was and they said, "Come on, let's go!" I said, "Well, it's howling and you can't even see. Are you guys nuts?" And they said, "We are taking off." I said, "Well, I reserve the right to change the way I get back. I'm going in a car." Of course, they called me everything they could think of, you know.

The second part of the story is, on the flight back, the Plexiglas [window] blew out of the airplane. The pilot is nearing El Paso, at the field down there. He said, "We are coming in to land." They said, "Don't come here. You can't land here. There is no visibility. Go to Las Cruces." They go to Las Cruces, and they said, "Don't land here. There is no visibility. Go somewhere else." Finally, he turned around and called El Paso, and said, "We are running out of gas. We are going to land." And they landed with practically no gas in El Paso. Well, with one more guy in there, guess what [would have happened].

The clinic [at White Sands] was really a dispensary level clinic. When I came there, I think, a lieutenant colonel was in charge. And he was leaving shortly. Then a doctor came. He had a medical MOS [Military Occupation Specialty] that I didn't quite understand. I mean, he was really a clinical practicing doctor, but he was an expert in supply and in setting up clinics and that sort of things. At least, that was his MOS at the time. Also, there was one Navy doctor, and five Army doctors. I was kind of the senior inasmuch as the other four Army doctors had just come out of their internship and they hadn't practiced anywhere.

So when little opportunities came along, I got some of the nice TDYs. During that period the Secretary of the Army came down. I think his name was Brucker [Wilber M. Brucker]. He came down with some brass from Washington. They picked up some officers from White Sands, and we flew in a little DC-9, one of those small, two-motor propeller airplanes, to different sites on the Proving Ground for two days. They set up fancy lunches and had military missile meetings. I was just having fun watching it. Mr. Brucker was dressed in a civilian blue suit, as I remember. The lunches were terrific. The only medical issue I had was one of the colonels was constipated. At any rate, that was a fun thing.

Another thing I thought was really a lot of fun. For the first time since the opening of White Sands, at least that is what they told me, they had a deer hunt open to the public. I was the doctor that went along with that. We drove into the Proving Ground in a place that was nice and hilly, and it was not flat desert. We stayed in the cabin of a former rancher who was now employed as a Range rider. But it was a great experience at his house. He was there, I was there, a couple of sergeants were there, and two or three game wardens. We got there on a Friday. The hunt was on Saturday and Sunday. Friday afternoon, a game warden came in and said, "Well, I am going to go out and find somebody that shot a deer before the season opened and confiscate it." Well, I know damn well he went out and [shot] one, because in about an hour and a half, he came back with a nice one.

They hung it up in a little anteroom, and skinned it off, and let it cool for the night. For the next two days, we had deer steak at least twice, and sometimes three times. Their idea of a deer steak was just to whack off a piece with an axe and cut it real thin and overcook it. But the whole thing was fun. The last afternoon, the rancher said, "Doc, have you got a gun?" I said, "Yes." He said, "Do you have a license?" I said, "Yes." He said, "Well, I'm going to take you deer hunting." So he has me jump in his truck and he drove around. All of a sudden, he came to a place and he stops. I said, "What's this all about?" He said, "There is a big gully there." It was shaped just like a bathtub, and there are trees on all sides of it. He said, "Here is where you are going to get your deer. Just start picking up big rocks and start throwing them in there." Which we did. Pretty soon, about five deer come climbing up the other side, and I fortunately managed to get one.

As far as I know, there has only been one military fatality ever at WSPG. It was a sad event, and it occurred while I was there. [There's a] large concrete blockhouse about a half a mile south of the headquarters building on the road to El Paso. A static test exploded there, and the force from the explosion severely injured a soldier who was observing the firing through the observation port. When they brought him to the dispensary, he was semiconscious and bleeding profusely from multiple lacerations of the face and neck. In addition, he was struggling to breathe due to the bleeding into his throat. His immediate need was an unobstructed airway to supply oxygen to his body. As no tracheotomy tubes were available, we were able to fashion one using available rubber tubing. After I performed the tracheotomy, we inserted the makeshift tube. It functioned well, and we were then able to supply his needed oxygen. While we were working on the airway problem, other clinic members were helping to control the bleeding. He soon was in a reasonably stable condition so we could transfer him to Beaumont Hospital. He arrived safely, but unfortunately expired several hours afterward. It was a shocking event. The entire process in stabilizing this injured soldier was a group effort. I was helped by an expert and dedicated group of clinic staff. Needless to say, we soon stocked tracheotomy tubes in the dispensary.

I didn't see any snakebites while I was there. I was interested in the flora and the fauna. There were guards that drove the roads at night, that road to El Paso, and maybe all over. They would come up on a snake out on the road, because it was warmer. They would catch them, and they would bring them to me in the dispensary. I would kill them and skin them. Everybody wanted a skin. That happened about, oh, six or seven times. And I pickled a lot of tarantulas and even those little vinegaroons, and different bugs. We even had a baby bobcat there once. So when I left I had a lot of specimens, which I gave to the school.

Several of the people quoted in this chapter belonged to the elite group called the Broomstick Scientists. At the beginning of the Korean War, men were being drafted or recalled to active duty in the US military. Those with university degrees in engineering, mathematics, and physics or other sciences were routed to the First Ordnance Guided Missile Support Battalion. Some were assigned to White Sands Proving Ground to work on the extended V-2 missile program or other missile development programs. Others worked on temporary assignment to missile-related contractors like General Electric, Douglas Aircraft, or the Jet Propulsion Laboratory. These young, highly educated servicemen advanced quickly through military ranks, sometimes to the consternation of more traditional Army personnel. In fact, one supervisor once told members of the group that he was not more impressed with them than we would be by a bunch of broomsticks. Thereafter, the group adopted the name "Broomstick Scientists," turning the slur into an emblem of honor.

Darby Chiles was considered one of the Broomstick Scientists, even though he held a desk job at White Sands Proving Ground in the post-Paperclip years. His oral history interview took place in September 2001.

I was born in January 1928, in Hamilton, Ohio, a suburb of Cincinnati. I attended high school at Hamilton High School and college at Miami University, Oxford, Ohio. You have to make a living at something, and I'm not much of a dancer. So I went for industrial engineering. I started to work in 1947 with the Crosley Automobile Company. The Crosley is no longer made, but it was a small automobile with four cylinders, stick shift, kind of like, but of course better than, the Volkswagen Bug. It was a car a little bit ahead of its time, and it was getting 50 miles per gallon in 1940.

On June 25, 1950, the Social Democrats invaded South Korea. On July 1, I had a letter in my hand from the President. The best I can remember, it started out "Greetings." August 1 was the day of our induction physical. As we walked in, they handed us a sheave of papers, but they already were filled out. So we just walked out the other door, and we were in the Army. I am not joking. The physical examination was already filled out when we got there. They took our temperature. I don't know that they read it, but they did take our temperature. And we went home.

On September 1, I was in the Army and was assigned to Basic Training at Fort Knox, in the Third Armored Division. This was a real gung-ho kind of a thing that would have made Georgie Patton proud, with comments like, "We are going to Korea. Nobody is getting out of this outfit. Nobody is leaving Fort Knox." "On January 1, you will be in Korea," and so on. The regular Army kind of thing. You couldn't even leave the company area. You weren't allowed to go anywhere.

They have an induction process. You go through a building where they sort your skills and abilities, they give you an IQ test, and that sort of thing. Near the end, they throw clothes at you. At the very end of the processing line, a corporal noticed my sheet and said, "Hey I see you can type." Well, typing sounded better than a lot of things I could think of in the Army. So, I took a typing test. And at the conclusion of the test, he said, "Oh no, this looks better, because SR615-205-1, Army Special Regulations for Scientific, Technical, and Professional Personnel is a pretty good one." He said, "I am marking your paper at this point. You will have six weeks of Basic Training, and you will be leaving from wherever you are assigned Basic Training to this program." Hey, anything is better than what I could conceive. So I was in the Third Armored six weeks less one day. And after six weeks of the captain and sergeants saying, "Nobody is leaving this outfit," they were a little angry because they got a document from post headquarters, all made out, that transferred me to South Post, Fort Myers in Virginia.

I jumped on the train and went to South Post. This was about October of 1950. I didn't have any choice. Others said they were asked their preference for assignment. I was never asked a preference or given a choice on anything. I received my orders to White Sands Proving Ground, and I said, "Where is that?" No one at Fort Myers knew where White Sands was in October 1950. They said, "Well, go across the street to the Pentagon Travel Office and ask them how to get there." I went to the Pentagon. I am serious. No one at the Pentagon Travel Office knew where White Sands was located. I said, "Well, where do I go?" He said, "Usually we try to send fellows in a group, two, three, or twenty together with meal tickets, that sort of thing. But we can't send you with anyone because we don't know where you are going. We'll send you alone. Here's an allowance for travel by personal car in lieu of railroad ticket and meals."

They gave me so many days to get there, which allowed me several days at home en route. "When you get to Chicago, ask them, they'll know." So when I got to Chicago, nobody knew, except they knew where New Mexico was. "Get on this train to New Mexico and the conductor will know." Well, I got on the train and the conductor sadly shook his head and said, "No, son, I don't know where it is, but there's a new conductor that gets on just as the train enters New Mexico. He will know because he is right there in New Mexico." So he got on the train, and he says "White Sands Proving Ground. No son, but I know where White Sands is." And I said, "Fine, would you help me get off?" He said, "Yes, it is near a town, the stop is called Alamogordo, New Mexico."

I got to Alamogordo about eleven or twelve o'clock at night. The train station was completely deserted. No one was there at that time of the night. But fortunately, after I got off, a few cars back, an Air Force captain got off the train. There were the two of us all alone in the Alamogordo station at midnight. I was a bit concerned because my paper said, "This date be at White Sands." Being a neophyte, I thought you were in the brig if you weren't there at that time. So I explained my dilemma to him during our conversation. He said, "Oh well, I'm going to call for a staff car. Come ride with me. We'll go over to Alamogordo Air Force Base and see if we can't work it out." Where else are you going to go in Alamogordo at midnight? So technically, I was in the Air Force for one day in Alamogordo.

I learned that Signal Corps trucks run between White Sands and Alamogordo frequently, so the next morning I hitchhiked a ride in the back of this two-and-a-half-ton Signal Corps truck and rolled into White Sands a day late. It was still morning but I was there. I was concerned because I am late, I've had it. First Sergeant Chronister said, "Oh yes, we have been expecting you. Here is your pass." What! It was on a Friday morning, so he said, "There is no point of doing anything around here today. Here is your pass. A first class pass, go anywhere you want." Hey, remember, this is just after basic training, where you weren't even allowed to leave the company area. I grabbed the pass. I promptly changed to my civilian clothes and on my first day I was in Las Cruces. I came Friday night and returned again on Sunday.

Monday morning, I got up and went over to the company headquarters, where the first sergeant and the company commander work. I walked in. They said, "Yeah, we've been looking for you, been waiting for you. You go over to the comptroller's office, now! They open at nine o'clock." I showed up promptly at nine o'clock at the comptroller's office. It was a wooden building. The Post engineers and the comptroller shared some space in that little one-story building. So I walked in. The comptroller said, "Oh yes, we have been waiting for you. Your desk is over here." They had my desk ready, waiting for me. By ten o'clock on the morning of the first day, I had been introduced to everyone and was sitting at my desk when he was handing me my first assignment. Understand, no one had asked me for a preference at any time, but there I was assigned the same work I had been doing as a civilian. I could not have been selected and assigned any quicker to work for which I was suited.

Well, things went all right until one day the first sergeant said, "Darby, you are on KP." Okay. Everybody in the Army is on KP, so I went over to Sergeant Mabe's Ptomaine Palace and had the green eggs for breakfast, and spent, oh, I don't know, I spent about a week that day doing the various KP kinds of chores, you know. What I didn't know then was that Colonel Eddy had brought Sergeant Mabe and Frank Woods, the comptroller, and a couple of other people on base with him from Japan, which was his previous assignment. He had done this before at several different bases. I guess that was a typical practice, I don't know. But Sergeant Mabe could do nothing wrong, except prepare food.

December would be the next month. My turn for KP came up again. But I had heard that the Officers' Club had real food, and if you took KP there, you could eat their leftovers. They asked, "Who volunteers?" So I volunteered for KP at the Officers Club. That is the only time I ever asked for anything, any assignment or anything in the Army. So I showed up at the Officers' Club and sure enough, they even had meat to eat! That was really quite a nice experience.

The new administration building was under construction. The move-in date was February 1, 1951. The comptroller's office was the first to move, and we moved in about two weeks before the official date. Well, I liked this because, while I didn't have a private office, I shared my office with a civilian. But in a brand new building, air-conditioned, I mean, I'm coming home! And I also learned very quickly if you are in the headquarters company you don't pull KP. The reason being that you pull Charge of Quarters (CQ) with the OD (Officer of the Day) every two months or so. Well, I mean, I'd heard of people going to heaven, but "Here I am," you know!

To make it all the nicer, the Comptroller Division was all civilian. I was the only military person in the entire division. And not only that, but it was a petticoat division. A young, single guy, I mean, good grief! It doesn't get any better than this.

When it was about the first of December, another fellow came to camp. He drove in with this absolutely gorgeous car, a 1948 Pontiac, and it was a convertible! You know how young fellows are. This was a car! This guy had to lock the doors when he stopped at a stop sign to keep the girls from jumping in. He said, "Let's go down to El Paso." I think there were four of us in the car. I can't remember if we were going to Juarez or somewhere down there, and there's the four of us rolling across the desert. I think we may have been singing in the car. Here came the flashing red light. They pulled Tom over. "Do you know you were going 48 miles an hour in a 45 mile zone?" It was just a couple of miles over the speed limit. What can you say? The cops got you. You have your choice: "Come back in two weeks and go to court, or follow me to the Justice of Peace." There is not much real choice. We followed him to the local Justice of Peace, who happened to be in a very small adobe house. The house measured about 8 feet across, and about that deep, one little front room, and had a small half desk. The Justice of Peace was already sitting there, waiting, when we walked in. Unfortunately, the Justice of Peace didn't speak any English. His little seven-year-old daughter stood beside the desk, and would interpret for us. So they disposed of the case rather quickly and we were on our way.

I was assigned to the 9393 Technical Service Unit. I remember the first day I came in the gate from Route 70. I was a young fellow. I don't remember my exact age, in the twenties. I had lived in Cincinnati, Ohio, and I was very perturbed because the government had imposed on my lifestyle. I mean, this was rutting season [a very active time for deer, and therefore, the best time to choose a good hunting spot], and they've drafted me for some place in far-off New Mexico. My girlfriend wanted letters, but I didn't feel like writing letters that first day. I remember, for the first letter I took a handful of sand, put it in an envelope, and sealed it. And her letter that day was just sand. I thought, how better to describe how I felt than the sand.

My many assignments were management studies of various segments of the Proving Ground. This one study happened to be about the Signal Corps Division. The lieutenant colonel who was the commander of the Signal Corps Division was just a couple of doors down in the administration building, a real nice guy. We had many conversations. He went to Colonel Eddy and said, "I want Darby to come and do a management analysis of the Signal Corps Division." Since it didn't cost anything, they would send me. So okay, here we go. I learned quickly that anyone who is a lieutenant colonel or above is usually a real friendly guy. But the fellows under the rank of major are striving so hard that they're obnoxious. So you just learn how to deal with each of them in that way.

This colonel was a particularly nice guy. We had started getting ready for the management study, reviewing organization charts, relationships, and that sort of thing. Well, the big problem in the Signal Corps Division was a certain regular Army captain. He really thought he was John Wayne, you know, in starched, close-pressed fatigues. So when he heard that this private was coming in to do a management analysis of his operation, he knew that he was also the cause of the problem over there.

He caught me in the administration building one day. We would do a lot of little things to tweak the tail of the lion, you know. Well, I wore argyle sox with my Class A uniform. So I am sitting at my desk with my pant legs pulled up. Captain Barr caught me with argyle sox in my uniform, and he reported me to my company commander for being out of uniform. I was walking down by the PX one day, and Captain McGaffick stopped me. He said, "Son, let me see your sox." I pulled up my pant legs. My sox were red and brown and yellow, argyle you know. And he says, "Well, you are out of uniform, aren't you"? I said, "Yes, sir." What can you say with your pants pulled up? I was one of the few people who got caught with my pants up! He said, "Well, what is your story?" I said, "Well, laundry didn't come back (which was actually true!) and I wore these today." "Well, that is not the thing to wear is it?" "No, sir."

But that was McGaffick. He was a housemother, really. He called Barracks Four the fraternity house, because the first time he came through on a formal inspection, there were so many college pennants and the college dormitory riff-raff hanging on the wall. He says, "What in the hell is this, a fraternity house?" And someone in the back softly said, "Yes, sir." And that was my introduction to the Signal Corps.

Life here wouldn't have been the same without McGaffick, because he always gave us something to tweak. His name was Captain Clyde O. McGaffick. You didn't say, "Captain McGaffick." You said, "Captain Clyde ***O.*** McGaffick." Don't leave out the full name. Well, one of the guys picked up a stray puppy that wandered around the barracks. He ensconced the dog under the barracks. We all fed the dog and took care of it for quite some time. He named the dog Klido. Not Fido, but Klido. So can you imagine these draftees, "Here Klido, here Klido." And we always wondered if he ever put the two together. We were just kind of tweaking him, you know.

There was one management study after another. The reason that they asked for someone from Fort Myers was for performance evaluation. I was in the automobile industry, and performance evaluation was my forte. The automobile industry, in those days, was considered the epitome of efficiency and capitalism at its best. Detroit made the best products and was the most efficient. And so by the time I arrived here, it seemed everybody concerned knew my name.

Well, performance measurement at White Sands, I'm trying to select my words right now. I'll just give you the anecdote that I told the comptroller. He wanted me to do work measurement of the entire base. He said we could get some stopwatches and whatever measurement equipment we needed. I said, "Sir, we don't need stopwatches to measure the work activity at WSPG. We need calendars." We went ahead with it, though, and developed quite a thorough program. Nobody knew what things cost. They had all of these programs, the V-2, the Nike, and others. Nobody had an idea what it cost. They knew the total cost for the whole operation, but nobody knew what the segments cost. And so we developed a standard cost system for cost accounting. Every person that worked here at that time began reporting time and kept track of the number of hours. So within a few weeks, we could tell you how many hours were spent on the V-2 project, how many were spent on housing, how many were spent here, wherever.

I had six whole weeks of Basic Training with the Third Armored Division. Armored carries the carbine because it is smaller and fits in the tank better. If you made a certain score they called you a sharpshooter. They had a tag you pinned on. It didn't do a darn thing for your pay, but they seemed to value badges more than pay anyway. So late during my tour at White Sands, his highness Captain Clyde O. McGaffick was assigned to a re-training job. He then was moved into the administration building. Now Captain Clyde O. McGaffick had the same office down the hall that I had for two years. He called me in one day and said, "Sergeant, I see you qualified as sharpshooter on the carbine at Fort Knox. But you didn't qualify on the M-1 Garand rifle, which is the standard arm of White Sands Proving Ground." And I said, "That is right, sir, I didn't. I have been pretty busy." "Well, you are a noncommissioned officer here at Fort White Sands, and so you should be qualified in the standard arm, the M-1 Garand rifle." "Yes, Sir, whatever you say, Sir."

I got ahead of myself. It must have been during the first week or two that I was assigned to White Sands that I was assigned a rifle. They kept it locked up in the company supply room. They kept all the rifles together in a rack with a lock. The fire extinguishers were locked up in a similar fashion. They issued me an M-1 rifle. I had no idea how to take it apart or to do anything with it. And it was in cosmoline [a brown, wax-like substance used to prevent rust]. Oh, what a mess that was. I said, "What am I supposed to do with this?" They said, "Clean it up. That's yours." I got in trouble because I called it a gun. But it sure looked like a gun to me. "This is your rifle for the full time you are at Fort White Sands."

I'm standing with this gun loaded with cosmoline, and I'm supposed to do something with it without any rags or solvent. I was told, "You have to be resourceful in combat." And both Captain Clyde O. McGaffick and First Sergeant Chronister said, "Go take care of your rifle." I knew it was dry desert here, so you shouldn't have much problem with rust on steel. So I took my gun in the shower with me. I got the water good and hot and soaped up the sucker. I used my toothbrush, you know, like this, back and forth, and got it looking pretty nice, and when it dried I quickly turned it back in to the supply room.

I didn't do anything more about it for some time, until Captain Clyde O. McGaffick said I had to go down to the Range. Well, once I fired the thing I was in trouble. Because my toothbrush wasn't long enough to get all the way down that long tube, you see. So I didn't know what to do. Well, the guy sleeping next to me in the fraternity house was a great big Armenian kid. He was a prince of a fellow. He had taken basic training at Fort Dix in the infantry, and had M-1 experience. So he took pity on me. He said, "Darby, I can do that." He cleaned it in a few seconds, and oiled it and shined it up and I quickly took it over to the company supply room. I guess that is what they do in the infantry. And I said, "Well, I'll be indebted forever." And that was the last I saw of my rifle, excuse me, gun. No, it's rifle isn't it? That's right, yeah it's rifle.

I remember the first sandstorm. I was standing on the second floor looking out of the window of the fraternity house. I don't know where it came from, but here comes an outside toilet, a privy. I had never seen any on the base. But now there's one rolling end over end across the company compound. And then there were other small sheds and oil drums and what-have-you, because it was like a hundred mile an hour wind.

The fraternity house housed all the guys who had different jobs all over the place. Some of them went to work at six o'clock in the morning, and others went at noon, or two. It wasn't a reveille kind of an operation. I was in the administration building, and we went to work at nine o'clock a.m., you know, like civilized people. They tried to get us up at five o'clock. Well, I couldn't see much point of getting up at five o'clock and going over and eating Sergeant Mabe's green eggs, and then go back and lay down for a few more hours before you had something to do, and there wasn't any place to sit down in the barracks. So, hey, I would stay in bed until 8:30, get up and shower and go to the office.

Well, we got in a new sergeant. He told me, "You get up at five o'clock, and you be in that mess hall at six o'clock. I don't care what you do the rest of the day, but that's how we're a-running this here barracks." Well, I did. Next morning I was up and gone. But then the second morning it was five after five, and the third morning, until I was back to a sensible schedule, you know. We went through a few arguments about this. He was accusing me of insubordination and rejection of authority. I guess these are military terms that end up court martial. So he talked to Captain McGaffick about it, and he said, "Yeah. He's done all of those things. He is not much of a soldier." I said, "I'm not any kind of a soldier." So the punishment was, instead of putting me in the brig, I was assigned every Saturday afternoon to scrub the exterior of the barracks with a toothbrush. And I'll bet you a quarter that barracks number four is still whiter than the other three barracks after all of these years. It was something else. But at least it satisfied the situation without too much problem.

They used to have Army trailer court. We had a million dollar view, because the mountains were just a spectacular view. One could just sit in a chair and enjoy watching the young mountains grow. It was that nice. Well, the trailer park wasn't the greatest. This was 1950, you remember. They didn't have house trailers like now. We had a 25-foot Adams Caravan house trailer, and you could park it there for $3.50 a month, which just about matched my salary. It included water, electricity, and sewer connections. Well, Mama lasted about six weeks there. So we moved on to El Paso and bought a little house right outside the entrance to Fort Bliss. I remember it cost sixty-five hundred dollars, and wow, that was quite a jump for us.

We had a 1950 Ford, so I traded the 1950 Ford for a down payment on this house. We had no furniture, of course. So I was living in El Paso but with no car. Well, the Army had an old bus. It was an old ambulance really, but the Army had mounted seats in it. It kind of looked like a bus, but it was designed to carry litters, and they just installed seats in it. We took that rickety bus every day 60 miles up to work and back. So here we are. We've no car, but we have a house and a house trailer, with no furniture. We parked the trailer so the door of the trailer matched the kitchen door and we used the stove and the sink in the trailer. And then stepped right into the house.

We dragged the mattress in and plopped it on the living room floor in front of the fireplace, which was the complete furnishings of the house when we moved in. How romantic can you get! But at that age, you know, a sunrise is romantic. So we enjoyed that a great deal, until I could sell the trailer. I sold the trailer to a draftee, and he hauled the trailer 60 miles back to the same spot and lived in the trailer. We now had some bucks, so we went down and bought some furniture for the house. Then a few months later I was out of the Army and dumped everything. But we had a lot of fun and good experiences.

Frank Conway joined the Army the day after Pearl Harbor. After giving his employer three weeks' notice, he reported for Basic Training at Aberdeen Proving Ground in Maryland in January 1942.

I went to the 8th Armored Division, which was forming at Fort Knox, Kentucky. Very shortly, I ran into my first rocket, the bazooka. Then in 1943, we came out [to White Sands Proving Ground in New Mexico]. We had two hundred .50 calibers on a firing line that was five miles long, and we shot at towed targets and rocket targets. They had big fins on them.

We were probably shooting sixty or seventy a day, and then again at night.

One of the interesting things was, they had a rhyme. Everybody had to learn about leads on targets, and everybody had to learn that little rhyme. "If you think you are ahead, you are a way behind. If you think you are way ahead, you are still behind it," and so on. The public address system down through the main control tower and four auxiliaries of the PA system were hooked in to the airplanes also, and women pilots [were towing targets]. So anyway, it comes on the PA system and this woman says, "Tell that silly son-a-bitch that I am pulling that target, not pushing it." She had taken three rounds through the tail compartment. So that was a little humorous event. I stayed out there for three weeks. And right then and there I decided, I am going to live out there someday.

I was assigned to the Ordnance Board detachment after the war, at Aberdeen [Proving Ground]. Major Keller was one of the very few bomb disposal officers left in 1950. He came into our shop all the time. He was on the Ordnance Board, and there were only a thousand enlisted men left in the whole Ordnance Corps. He said that he and another officer from the Ordnance Board had watched a V-2 take off from White Sands Proving Ground. I said, "I would give three months' pay to watch something like that." Well, this was on a Wednesday. The next day, our sergeant major from the Board came down, and he said, "Hey, Frank, we put a letter in on that thing that you were talking to Keller about." I didn't know what we were talking about at all. Saturday, I was standing tall for the shop inspection. I was responsible for the machine shop, twenty-eight lathes, and all of this kind of stuff. The warrant officer who was our personnel officer came in, and he says, "When do you want to leave?" And I said, "What are you talking about?" He said, "Well, didn't you tell Major Keller you wanted to go to White Sands Proving Ground?" I said, "I want to watch a missile, yeah." He said, "Well, you are transferred, and you can leave whenever you want."

I went home first, up to Massachusetts, to my folks'. I had a Jeep, a 1946 Willys, that in 1947 I bought for $500. I came down through the Blue Ridge Mountains, the Shenandoah Parkway, sleeping on the ground, cooking over a fire, watching the turkeys and the deer. I finally got to Birmingham [Alabama], and it dawned on me, I hadn't had a haircut in three weeks. The barber said, "What do you think of the war?" And I said, "What war?" He said, "You've got Aberdeen Proving Ground on your bumper, don't you?" I said, "Yes, Sir." He said, "Well, what are you?"

I said, "Well, I am a sergeant in the Army." He said, "And you don't know anything about a war?" And I said, "No, Sir." He said, "Well sergeant, North Korea attacked South Korea five days ago." I didn't know anything about it. I never turned on the car radio. I speeded up and I got to White Sands on the fourth of July in 1950.

In the time I was here, they kept switching me all the time, so I would get a little experience on each one of the missiles. Once, when we finished a successful shoot with Nike, they gave all of us the rest of the day off. The enlisted men and the contractors just stayed there. Well, a friend of mine and I went down the hill in uniform fatigues. We were going to hunt jackrabbits. He had a green pickup, and we parked it underneath some trees, screw-bean mesquites or something, to get some shade for it. We were out there walking, and an L-5 [airplane] came over, looking. He didn't see us because we matched the range characteristics. He went off and he made a circle and he came back. Well, all of a sudden the ground shook. We looked back to the launcher, and there was a big cloud of smoke. We said, "Jesus, something blew up on the launch pad!" We were really worried about it, you know. And about that time, we looked over about four or five hundred yards away, and there was another cloud. They had shot a conventional warhead in an HJ [Honest John]. It was classified at that time, that we actually had a conventional warhead. It had landed about five or six hundred yards from us. [The airplane pilot] was looking to see if anybody was in the impact area. So we got the hell out of there, because we weren't supposed to be there in the first place.

Lieutenant Regretto was the pilot. He got into the habit of hunting coyotes from that L-5. They got a trench gun from the MPs, and the guy would sit in behind them with the windows open. When he would get a coyote running out in the desert, he would bank it up so this guy could go out and shoot them. He would throttle it way down, you know, as slow as he could. He was doing it one day, and he hung a wing tip on a barbed wire fence and cartwheeled the L-5. He had some fancy explaining to do. He wasn't looking for missiles that day. I never did get to ride that thing, but I would have loved to have gone coyote hunting with him.

One day in the hangar, we had some stuff that was being taken apart. One of the guys on it was an Air Force staff sergeant. He was taking the regulator and the aluminum can apart. I say "can" because the PSI [pressure] had been up around 4,800 or 5,000 pounds when it was pressurized, and the can had been recovered. He was going to take it apart, and when he took that last bolt out, that rocket took off with just the air pressure alone in it. It went through the hangar. It sailed up in the air, and it went underneath me. I was about three feet in the air when that thing went by. It put a hell of a dent in one of the doors, a big rolling door. The engineers figured out it still had about 2,400 pounds pressure on it when he took that last bolt out. So it was a very effective missile.

We had a pet crow, Jimmy, and we had an IG [inspector general] inspection coming up. This particular guy went through the screens in the barracks and searched around to see if there was something hid behind there. He got a step ladder, and they got out about four hundred brilliantly colored socks that had been hanging out on the line when the guys would wash. Jimmy was hiding them. We never did figure out how he got up into the attic in the barracks. They used to feed us at ten o'clock on Sunday morning. A bunch of us were sitting around waiting, and I came out about the time to witness this thing. Jimmy was lying out there with his wings spread out, on his back. So this guy looked at him, and went over and picked him up, opened up a garbage can and dropped him in it. Everybody hollered at him, "What are you doing!?" He said, "Well, he is dead." We said, "Hell, no. We put beer in his water cup out here. He was just drunk and passed out on us."

At that time, the main gate was about the size of a phone booth. They had a little table in there with a visitor's pen. Jimmy the Crow flew in there, picked up a pencil that was on the pad, and flew out. The best they can tell, about seven or eight minutes later he flew in the door at what we refer to as the El Paso gate, and put the pencil down on the pad. They didn't have one. That is supposed to be the truth.

In my twenty-nine years in the Army, there were probably three or four months that I was unhappy with what I had. And the rest of the time, if I had written my own job description, I couldn't have done it any better.

Don Cooper was seventeen when he joined the Air Force in 1950. He reported to Lackland Air Force Base in San Antonio for initial processing. The following excerpt from his 2007 oral history interview picks up the story there:

We got to Lackland about six o'clock that night, and we got off the train. We were all broke. We didn't have any money, but we didn't need it, really. They opened up the chow hall and made us some sandwiches and everything. They had chocolate pudding. That was my introduction to the Air Force.

Back when we were in high school, we had an old '34 Plymouth. It was great. The only thing is that it didn't have any reverse. They were all forward gears. There were five of us, and when we went anywhere we had to park, we would have to pull it up, put it in neutral, and push it in. We ran that thing for a long time, and then we decided we were going to work on that transmission. So we tore it down, and, man, I knew everything there was to know about a transmission. We put it back together and it worked great. There was one thing. We could only get reverse gear. There were no forward gears now. So we decided to put it back the way it was. We wound up with three forward gears. That was my first experience as a mechanic. Then I said, that's not for me.

But when I went into the Air Force in October 1950, they changed my mind. They took us in for orientation, and they said, "What would you like to do?" I said, "I don't know, maybe clerical work." They said, "Okay, that sounds good." When the orders came out, I was going to stay there and go to Aircraft Engine School. I graduated from Aircraft Engine School, and they sent a bunch of us to Alexandria, Louisiana. The base we went to was Alexandria Municipal Airport. They abandoned the municipal airport, then they leased it to the Army Air Force for a while. Then they took it back and leased it out to a turkey farmer. Then when the Korean Conflict started out, they reopened it as an Air Force base.

We had gone out one Saturday night, and we had a party over at the Pelican Club in Alexandria. When we got back, it was about two or three o'clock in the morning, and we got into the old sack. They rousted us out about four o'clock and said, "You're moving out." "Moving out? Where are we supposed to be going? We haven't packed anything." They said, "Well, your gear is already packed. It is on the plane on the runway waiting for you." So they threw us in a carryall and escorted us up there and threw us on the plane. We hit Holloman just about daylight. This was 1952. I looked out and I saw the walls. I nudged my buddy, and I said, "What did we do last night?" He said "Why?" I said, "I think they sent us to Leavenworth!" He said, "What do you mean?" I said, "Look at those walls out there!" Well, I used a few other expletives, you know. It happened to be the Sacramento and the Organ Mountains. But they looked like walls to me, at that time of morning.

So here we were at Holloman. We were assigned to different units, one went over into sheet metal, and one in the orderly room. I wound up out on the flight line, because there were only two of us that were mechanics. We took care of all the jet aircraft. We had the F-86s out here. We were working in conjunction with White Sands Proving Ground, because there were a lot of missiles being launched, and some of them would go crazy, go wild. If they couldn't destroy them from the ground, it was up to our 86s to take them out of the air.

They sent a crew, I think it was two individuals, back to my hometown, and they did a background check on me. They were there, oh, for eight months doing a background, because we were working in a very, very sensitive area. They issued us a little button. It had color codes, so when you saw someone with that little badge there, and you noticed the code, you knew where they could be and where they couldn't be. I got tangled up with a full-bird colonel because his button said he could not be anywhere except in the general area. I was in a highly restricted area, and he was there. It got to be a tussle. He said, "You don't respect my uniform?" I said, "Yes, Sir, I respect your uniform and I respect your rank, yes, Sir. But you are still in the wrong area." So I called the air police, they are now security police. Two carryalls with four air police came over. He said, "You know, you were a gentleman about this whole thing." I said, "Well, Colonel, I was born and raised as a gentleman. It took an act of Congress to make you one." I had to get in that last lick. He had given me a whole raft of crap.

We were part of that transition part coming out of the Army Air Force into regular Air Force. I was in about four or five months before we finally got the blues [blue uniforms]. When we got the blues, a lot of people took us for bus drivers, especially if you were standing in a bus station.

I lived in the middle of history in my youth. The Fredericksburg area in Virginia is known as the most historic city in the United States, based on everything that happened in the War Between the States, otherwise known as the Civil War. Here you are, playing around Martha Washington's, and Mary Washington's home, John Paul Jones' house, and the [Battles of] Chancellor[sville] and the wilderness, Marye's Heights. You grew up and really didn't pay any attention to it. The same thing happened out here [at Holloman and White Sands]. When I was serving out here, we really didn't plan any history! We were living history, and we didn't even know it.

Aubrey Dunn , Sr., was born in Alamogordo in 1928. After a few years, his family moved closer to Roswell, and then returned to Alamogordo in 1938.

We were here and saw the Air Force come in, and the town flourished a little bit, and started growing in '42. In fact, our house was a two-story house, and one time, we had five couples of GIs what lived with us. [My brother] Bill and I had a room, and then we had bedrooms for five of those couples, and my dad and mother were there. That's the way we had to do it, because there weren't any houses [available]. GIs and officers came and stayed with us. One guy's name was Moran and he was in charge of digging the ditches, and building the roads, and doing all of that stuff out there. Knowing those guys and the conditions they lived and worked under, rattlesnakes, you never...!

Jed Durrenberger had become proficient in photography in the 1920s and 1930s. He worked in missile tracking at White Sands Proving Ground/Missile Range from 1951 until 1980. In 1958, the facility's name was changed from Proving Ground to Missile Range to reflect its primary mission of missile research.

World War II came along, and I applied to the Navy. I went from New York down to where they had started a new, very highly specialized Naval photographic lab outside of Washington, DC. I passed the test for chief petty officer in that particular job, until they gave me my physical exam and found that my eyes were so poor that I wouldn't even qualify to go to Officer Candidate School. So they said, you may want the Navy but the Navy doesn't want you. So I thought I was immune from World War II. But eventually I was drafted. I had, in the meantime, joined the Coast

Guard Auxiliary. Fortunately, I learned enough Morse Code to be able to completely foul up the aptitude test that they gave me to see what I could do in the Army. What they wanted then was code intercept operators. I didn't want to spend the war listening to dah-dit-dit-dah. So I completely fouled that up.

Luckily, the second aptitude was photo lab chief, which I had been. So I spent the war as a photo lab chief. And because of the background, I ended up with a signal intelligence service, which was highly classified. Among the things we used to do was look for German secret messages and break down their codes.

[When I came to White Sands,] I started work with Dr. Clyde Tombaugh. The IGY, the International Geophysical Year, was scheduled to take place in the middle of 1957 until almost the end of 1958. Clyde Tombaugh had retired from White Sands Missile Range to take the solar program looking for extra-terrestrial circulating bodies around the Earth – things that were not in close enough to become meteors, and yet they were too fast to be recorded by normal star-type imaging devices. The United States intended to launch the first [artificial] satellite in 1958. But the Russians beat us to it in October 1957. They launched Sputnik. In order to photograph the satellites when they were actually launched, the Baker Nunn satellite tracking camera was designed for the wide field and the curved surface for the stretching of the film to follow the focus of the large aperture system of lenses.

Twelve of them were being made, and Smithsonian had the almighty budget in those days of exactly one million dollars to build twelve of those Baker Nunn tracking cameras. The first station was on the White Sands side of the San Augustin Pass. The things that were stationed there was the first of the Baker Nunn satellite tracking cameras and two super-wide field cameras that were specifically for the use of the meteor team from Harvard. They were in a building that had two slide-back rooms that could be opened, depending on whichever one was doing what.

When the first camera was ready to be installed at the White Sands Missile Range, the Smithsonian people had nobody to assemble the camera when it was delivered from where it was fabricated in Pasadena, California. So I ended up volunteering and borrowed crews from the Navy machine shops, and vehicles, and we assembled the first of the Baker Nunn satellite tracking cameras here. That was how I became familiar with that particular model camera. One of the embarrassing things, the large 30-inch diameter mirror had Invar rods. Invar is a material that doesn't expand or contract significantly with changes in temperature over a fairly large range. They had micrometers upon top of the Invar rods that went down for support on top of the camera and down to where the big mirror was. That would level and center the mirror so it would be where it was supposed to be. In assembling that, one had to actually lean down through the aperture to reach down to the primary mirror in order to attach the Invar rods to the sockets. I, as an idiot, forgot I had a jeweler's screwdriver in my pocket, and hanging upside down from my waist inside there, the jeweler's screwdriver worked out of my pocket, dropped down and hit the main mirror and made a small chip. But it didn't do any serious damage.

Thomas Foley had a varied career as a mechanical engineer, sometimes "designing things which cost $200 and we made a million of them a year," and other times designing "things that cost a million dollars each, and we made 200 a year." He was one of the Broomstick Scientists, working at White Sands Proving Ground from December 1950 until September 1952.

I was born in Worcester, Massachusetts, in October 1926. I can remember back at the age of four or five, declaring that I wanted to be an inventor. And somehow that happened. I entered Worcester Polytechnic Institute in 1944. I was a little too young to be drafted yet. I stayed right in school through that period, got through school in two years and eight months, nonstop. They kept calling me up for the draft and rejecting me. I was classified 4-F [disqualified for medical reasons]. I wanted to go in the worst way. But my mother encouraged me to stay in school and complete my degree, which I was able to do. I got a Bachelor of Science in mechanical engineering.

After I got my degree, I was interviewed by General Electric Company, and took a job with them in the household refrigerator department, designing parts and assemblies for household refrigerators. It was quite challenging because anything I designed, we made a million of them a year. So they had to be right. One of the things I designed and had patented was the first magnetic door latch, which prevented kiddies from locking themselves in refrigerators. When I was first hired, my position was Engineer for Butter Conditioners, which was the most unexciting, dull thing one could imagine. As I said, when you make a million of them, everything has to be exactly right.

I had just bought my first good car, a used Pontiac convertible, and I was just sitting on top of the world. I can remember hearing the announcement of the Korean War on June 25, 1950, and thinking, huh, that's interesting. I wonder what will happen. Within weeks, I was shipping out as a draftee to Fort Jackson, South Carolina.

It was a strange experience because I knew I had been 4-F. I thought, well, when I get to camp they will know all about me and they will send me back again. They are not going to keep me, surely [not] this emaciated looking young man who is unhealthy. That didn't happen. No doctor even stood in the same room with me until once when I collapsed during training in the heat. But it was a shock to realize I am not going to get out of this place. They are really going to keep me. And I still have that damn car that I'm paying $75 a month for.

I completed Basic Training at Fort Jackson. I think it was thirteen weeks of basic infantry training. That was one of the best things that ever happened to me. First of all, I thought every day, if I ever get out of here I will never complain about anything. And that thought has supported me most of my life. The discipline and the activities were so severe, intended to break people, I guess. They were successful at it. I remain amazed at how they could take a group of hillbillies and engineers and lawyers and other people and point them in one direction and make them all do the same thing. I hope they still have training that is that good.

After basic training I was assigned to, thank God, Fort Myer, Virginia, to be reassigned under the special technical program, SPP (Scientific and Professional Personnel). I applied to go to Fort Belvoir [Virginia], because I thought that was where all the mechanical stuff was. I got shipped to White Sands, which was a pleasant surprise when I got there. I drove out in my $75-a-month Pontiac. I came in early December 1950, and like many of the others, I was both shocked and pleased at the same time. It looked like a far more interesting place to be than many I could think of.

After KP and sweeping out hangars and things, which I did willingly, I was assigned to the Nike program. I worked on Nike erection and handling equipment. We did some design work; we were supervised by

Douglas Aircraft engineers. And I did office work, administration, typing. Not very exciting work. Yet all around me were all of these exciting things happening. One thing that happened to me on the Nike program, I was injured by a Nike missile. I was working on it in the shop and got underneath it and stood up abruptly into the tail fin. I needed six stitches in the top of my head. I was probably one of the first casualties of the Nike missile.

I witnessed some of the Nike launches close up. I was quite impressed. I saw some Honest John firings. I was also underneath that V-2 that blew up one night. In fact, I still had my car and I had driven it out to within, I would say, a thousand feet of the actual missile, and parked, and was sitting on the hood of the car with my friends, watching it. When it took off, the shock waves were such that car went up and down on its springs. And we thought, "Oh my God, this is really exciting." It was much more exciting a few minutes later, when the damn thing [went] down under power. We could hear things thudding into the desert, *zoom, zoom!* Pieces were falling all around us. I think my impression is still there in the sand. There was no place to go and no place to hide. It was a kind of dumb thing to have done, but it happened.

I lived in Barracks Four, which is a barracks with a lot of history for us. We had the bunks upside down [see page 63] on the first floor. There were so many of us that lived there that I don't think the fire marshals would have ever agreed to it. We used to say that New Mexico being the Land of Enchantment, the enchantment blew into the barracks frequently, along with bugs, and spiders, and lizards, and other things. And that required frequent cleaning. But it was a very jovial place. I designed and erected on the end of my bunk a coffee shelf, which was totally forbidden. I managed to hide it for inspections, as it was quickly demountable. And on Sunday mornings, we would gather there with doughnuts and hot coffee, and reflect upon the week.

[Life at White Sands Proving Ground] was varied in the things that we could do. We could mountain climb. We used to hike out into the desert and look for pottery fragments, which were all over the place out there. Probably wouldn't be permitted today. And I, with several friends, entered into an intense study of Spanish. We thoroughly enjoyed that. Spent a lot of time in Mexico. It was very exciting to be able to communicate in the language. I made several trips to Chihuahua and to Mexico City. All of my leave was spent like that. I went with friends. I never went alone, as I recall.

I spent one New Year's Eve, I think in Chihuahua, in a bar full of dancing people, and Mexicans shooting pistols into the air. It was quite exciting. The Spanish that I learned here served me well later, when I accepted an assignment with General Electric Company to participate in a factory start-up in Bogotá, Colombia. My family and I stayed for four years in Colombia, which never would have happened had I not been here.

I still laugh about some of the things that we did here. We, of course, drank a lot. We used to buy 85-cent-a-liter gin in Juarez. Oso Negro gin, great stuff. Of course on forty dollars a month with tax deductions and car payments, we were always short of money. But we compensated. We made our own fun.

Once, on KP, I broke a cup. Dropped it by accident. There were pieces all over the floor. I was seen breaking it, so I couldn't push it some place and pretend it didn't happen. They were going to deduct thirty-five cents from my pay for this. I rebelled about that. I said, "That is not right. That is a pretty low breakage rate for handling that many cups in a day, and I am not going to pay it." Well, I didn't hear anything for a few days, until Lieutenant Thomas spoke to me and said, "Foley, you don't really want to carry this up to a court-martial point, do you?" And I said, "No, I am happy to pay."

But that was one good thing about the way the Army operated. Rules, no exceptions. If you got sick, you checked out of your barracks, went to sickbay. When you got well, you signed back in again, very orderly. The rules were well known. And the key is, no exceptions. That served me well later in management in companies that I worked for.

Security at White Sands was fairly good. In fact, with our interest in Mexico and Mexican music, there was a civilian employee, a lady here on the base, who had a tape recorder. One time, we went to Mexico with her tape recorder to record musicians there. That was successful. It was a fun thing we did. Two years later, after I had been discharged and gotten married to my wife, who was a European war refugee, a knock came on the door of our apartment in Erie, Pennsylvania. The CIC [Counterintelligence Corps] people were there, showed their badges. They said, "We need to talk to you, Mr. Foley." And so they began an interview. My wife was scared to death because she had been interviewed by the Gestapo in the war. So she was, I guess, hiding in another room. She was really frightened that here in America, these people could come and start questioning. The CIC people knew every detail of my life. They knew what dates I had traveled to Mexico City, where I had stayed, who I had traveled with. I believe the inquiry had to do with this lady who, while based at White Sands, had a tape recorder carrying it on and off the base. I think she was quite innocent, but the sequence of events would lead one to be suspicious. I think that was what they were checking up on. They asked me if I knew her. And of course I did. I never heard more about it. But it was interesting to know that our security was that well managed, at least at that point.

James Gimbel was "seventeen plus one day" when he joined the Army in 1959. His first tour of duty was in Germany, where he worked on missile guidance systems.

I came to White Sands Missile Range in 1961. I remember it was in the summer here, and fall in Europe. We landed in El Paso, and the buses at that time had no air conditioners. We were dressed in winter uniform, which you might recall was OD woolen, thick. On the bus out to White Sands, I remember all of us were really complaining to the bus driver.

They put us in Quonset huts with tin roofs, no air conditioner, and there was an inch of sand on the floor. The old mattresses were still rolled up with spiders and snakes in them. I remember they fed us all the 3.2 [low-alcohol] beer we could drink, and we drank a lot of it.

Howard Hallmark enlisted in the Air Force in 1948. He went to Perrin Air Force Base in Texas for Basic Training. Between his assignments at WSPG and Holloman, he spent nineteen months in Guam as a typhoon chaser.

I didn't know that I had any choice [of training] when I went into Basic. A lot of the boys there had letters saying that they were going to this school, or that school. One of the boys had a letter guaranteeing him Weather Observer School. When it came time for us to put down our three choices, I just put down weather observer, and the second one was heavy equipment operator. They said, "The Heavy Equipment Operator School is closed right now." I said, "Well, just weather observer." "Don't have any other choices?" "No, just weather observer." I got it.

I went to Chanute Air Force Base at Rantoul, Illinois, for Weather Observer School, and the following year, I signed up for Rawinsonde School back at Chanute for three months. [Rawinsonde is a gas balloon carrying meteorological instruments and a radar target.] In that we studied upper air sounding. We had a battery-operated instrument that we would put on the balloon, and it would send back temperature and pressure and humidity. With the radar tracker we would get the direction and the angles of the wind. And then we could determine what direction the balloon was going and the speed of the wind. We could determine how high we were from the recorder and also from the tracker.

That information helped us determine what the weather was going to be. We could, on our weather maps, plot the weather, and we could see these fronts coming through in the upper air, as well as surface. It was valuable for the forecaster, and of course the pilots, military and civilian, were very dependent on what the upper air weather was going to be. Of course, in rockets they had to be very careful because they wanted to make sure when that rocket went up, it wasn't going to land on some populated area when it came back down.

In Perrin, just south of the Oklahoma state line, south of the Red River, we had a lot of tornados in the summertime, a lot of rolling thunderstorms. They could be bad. Unfortunately, sometimes we'd lose a cadet in one of those storms. They'd wander off course. When we'd have an accident such as that, they really raked us over with a fine-tooth comb to make sure. They tried their best to find out what was the cause of it. If someone was goofing off, that was going to be exposed. When we'd have an accident, we had two teletype machines. Both machines actually sent and received. One we kept for the record, and every time we came to the end of the roll of paper, we would take that, and date it, time it, initial it, and that roll was put into storage in case something did happen. Of course, all of our weather maps and observation charts were available as well. So they would really rake us over with a fine-tooth comb trying to find out if we had something to do with that pilot error, or whether it was all the cadet's inexperience, or whatever.

After Rawinsonde School, I went out to White Sands Proving Ground. The only Air Force on the post was the Weather Detachment. They had been a mobile unit, and had just changed to permanent status when I went out there around the first of November of '49. It was very primitive. The base was activated in '45. Now they have signs, miles away, "White Sands Missile Range, such and such." Back then there were no signs. When we got down to the turnoff to the Post, there was a little bitty sign saying "This way to White Sands Proving Ground." It was very secretive. It wasn't on the map or anything.

[When Hallmark left WSPG, he served nearly two years overseas.] Around the middle of January 1952, I came into Holloman. Our work at Holloman was just routine upper air weather soundings. We did go over to the launch pad or the blockhouse sometimes when they were going to launch, and take a special wind run for them. They wanted to know the exact winds every ten or twenty seconds up to a certain distance. So we would go over next to the blockhouse and release a pibal [pilot balloon] and take a reading every ten or twenty seconds, up whatever altitude they wanted. They wanted to know exactly what the wind was right there at the launch site before they fired.

Ordinarily, we took two regular Rawinsonde runs each day. But we did run specials for them when they called for it. I don't remember what kind of rockets they were testing. There were several.... Most of them got off the ground, but sometimes they didn't. In fact, I've seen some of them sit there on the launch pad, and big old orange smoke just boil up from the bottom of it, and when it was all over with it would still be there. I remember when President Kennedy announced, very absolutely, "We will put a man on the Moon in this decade." I said to myself, "Mr. President, you don't know anything about rockets." But they did it.

I hadn't been at Holloman very long when the Base weather officer called a standby inspection. I was used to white-collar inspections, so I was busy putting a white collar on my bed. [The "white collar" was the top edge of the white sheet folded down over the blanket.] The other guys, one of them looked around and said, "What are you doing?" "I'm getting ready for inspection." "Oh, we don't do that here. Put that dust cover back on." Well, I didn't know whether to believe him or not, but I did. By noon I knew why we didn't have white-collar inspections. I couldn't even see the high line pole outside of the barracks window. That was my first experience with a dust storm. That night I took a broom and swept off my bed.

I finally gave up trying to keep my car clean. I used to like to keep it spotless, but every time I'd get it all polished up, there would come a dust storm and then the mud storm. It never got to the rainstorm. It was just a mixture of rain and dirt.

Lewie "Tiny" Harris began working at White Sands Proving Ground as a mechanic in 1949. He was responsible for servicing the generators that were scattered throughout the Range.

In the early days, the Range roads, if you could call them roads, were bulldozer trails. A bulldozer would go along and make a road, and then the vehicles traveling across there would smooth it down so you could get over it. Some of them were terribly bad in the winter time. I remember one time that it had been snowing and raining, and snowing and raining for about three days, and we had been uprange. We stayed uprange during the week and worked, and we were headed home on Friday afternoon. We got down approximately forty miles from camp, and we sunk the vehicle clear to the frame, a six-by-six [six-wheel drive vehicle]. It took two weeks to get it out, because when you sink one of those six-by-six's, it's bad! It was down past the running board in the mud. We just stepped right out in the mud and walked away. I sank in almost to my knees.

There was a crippled fellow with me, Lyge Bright, and he had a terrible time getting out. Then it was about a twenty-five mile walk before we were picked up. We walked almost to the Tularosa gate before the MPs found us. Lyge was in pretty bad shape by then. I tried to get him to stop and wait and I'd go on, but he said he was afraid he'd freeze to death. It was pretty cold. So we kept going slowly. We made it, and the MPs came along in a Jeep and picked us up. Then we liked to froze to death until we found some fire at the guard gate. We walked probably about six hours, because the first two and half miles of it was in deep mud. We had a water supply and a little bit of lunch supplies that we had left over from spending a week uprange.

Bill Hermann was a mechanical engineer who came to White Sands Proving Ground in the Broomstick Scientist period. He worked primarily on the Corporal missile.

I came to White Sands in January of 1951. Like many other people, I got off the train with a fellow GI at the Oro Grande station, and it was just slightly before dawn. There was this guy in there with a kerosene lamp in this little shack, and an EE-8 field telephone. He rang up the base, and about a half an hour later an Army truck rumbled in. We threw our duffle bags on the truck and took out for the base. I had never been west of the Mississippi before. It was a totally new experience for me to be out in the desert.

There were a lot of people coming in, and my first job was to double deck all of the bunks. We actually took these bunks apart and turned them over and put the tops together, somehow or other bolted them together. Took the mattress spring and tuned that around and used supports for that and we actually had double deck bunks. I did that for, I don't know how long. I guess it was waiting for a clearance.

One time, we were having an inspection, and a friend of mine called me over to look at his gear that was all laid out in his foot locker. He said, "Do you see anything wrong with this?" I said, "Everything looks regulation to me." And he said, "Look at my combat boots." I looked at them and said, "They look fine to me." And he said, "Oh good, they probably won't catch it then." He happened to be big and wore a size fourteen shoe, and for some reason his combat boots were in the repair shop. He borrowed somebody's size eight. If somebody had said to him, "Soldier, put on those combat boots," it would have broken up everybody.

Ralph Kopplin was assigned to the finance office at Holloman Air Force Base from 1951 until 1953. He sent his recollections to the NMMSH oral history project in written form in 2002.

Upon arrival I was given a "going on active duty" physical. The doctors discovered that I had a hernia that needed to be repaired. Within a week of reporting I was in the operating room and then confined to the hospital for ten days until I could function as an assistant finance officer. I remember the bottle of water on the bedside table as the minerals settled out to the bottom. It seemed there would be half an inch of such deposit in a few hours. These deposits plagued everyone when they bathed – no soap suds nor lather. Detergents had not been introduced to the American market.

In the beginning of my tour, there were only forty-two officers on the base. Much of the work on the Range was performed by civilian contractors. As a result, the additional temporary assignments came around frequently, such as Officer of the Day, auditing the funds at the Officers Club, parade duty and others. One duty I volunteered for and maintained most of my tour was Trial Counsel for Special Courts Martial. Since I was a very junior officer, the law officer had little difficulty finding and assigning a defense lawyer superior to me as required by the Uniform Code of Military Justice, 1951. The majority of cases were for the charge of absence without leave (AWOL). They took minimum time to prepare or conduct. The more unusual cases included disrespect toward a superior non-com, pistol whipping, and arson of the base stockade.

As part of military pay responsibility, Captain Carlson and I, accompanied by two carloads of military police, went into Alamogordo each month to pick up the cash (about $250,000) for payday. Some of the airmen preferred to get paid cash. We met the train from El Paso which carried cash from the federal reserve bank. The depot manager brought the cash into the depot, where the captain signed for it. Then under cover of the airmen, we took it to the car and then to the bank about two blocks away. Again under cover of the airmen, we took the cash into the bank and counted every bill to ensure the correct amount. Then it was return to the car and drive out to the base and the finance office, where the cash was put in the safe and remained under guard until the next morning. The captain and I both wore loaded .45 automatic pistols during the monthly event.

In the BOQ [bachelor officers' quarters] we had quite an assortment of officers. All but one other and myself had served in World War II and were either career military or had been recalled for the Korean conflict. The most veteran were pilots for the drone squadron. This group was a tenant organization assigned to Holloman from a base in Florida. Their mission was to fly B-17 mother ships and then, through a remote-control black box, pick up a B-17 drone and control it at a distance for the purpose of serving as a target for air-to-ground missiles (without warheads).

These missions were conducted early in the morning, and the drone pilots were airborne even earlier, so it seemed they were always around the BOQ early in the afternoon. With time on his hands, one Captain "Dutch" Mitwaulk undertook building a large model airplane. At least four to six months were spent on the project, which was conducted in the day room. The day of launch came. The plane was too big to get out the door. After a few days of remodeling, the plane was taken outside and its engine was started outdoors. It flew well until the landing, when it crash landed, never to be reflown again.

Mitwaulk always had a project going. First was buying a Cadillac. To guard it from sandstorms, Dutch put paste wax on it three weekends in a row. It really gleamed until he got into another interest, deer hunting in the mountains surrounding the base. Regardless of the size of the trail, Dutch took his Cadillac on it in search of the deer. He bagged one deer and had it butchered. He brought it back to the BOQ, where we spent all afternoon and evening boiling the venison along with whatever was left in our food boxes that divided each person's provisions. We threw cabbage, lettuce, radishes, ketchup, and all sorts of things into the pot. That evening we had a great meal and enough for all.

Another experience for Dutch was the finding of a gold mine south of Alamogordo at Oro Grande. Oro Grande was a small, obscure location halfway to El Paso. It seemed to me it was on the White Sands Proving Ground, but apparently was on the eastern edge. Dutch heard that this had been a gold mine location quite a few years before, probably before statehood for New Mexico. He drove his Cadillac up a very old trail toward a rock outcropping. There he was confronted by an old prospector who wielded a rifle and wanted to know what Dutch was doing on his property.

After a few words of explanation, we all felt the "prospector" had felt that he had found a sucker with some money. It was no time until Dutch was buying equipment to establish a water sluice to prospect for gold. The prospector said he knew the gold was there and all he needed to find it was some money for the equipment. When we did see Dutch, he was studying geology books and mining books. This effort went on for about two months until the fun was over and Dutch decided to leave the mine to the prospector.

Golf was a challenge in those days in that area. You could go east up the mountains to Cloudcroft, New Mexico, and play the course Ben Hogan used to enjoy. It was a pretty course in the valley between the mountains. The first hole, the regulars said, was one Hogan used an 8 iron from the tee to reach the green – not very much distance forward, but a long way down. By contrast, the number nine hole was the reverse, straight up it seemed, and hope the ball didn't roll back down. To the west from the base, the other choice was at Las Cruces, New Mexico, on the edge of the Tularosa basin. Sand greens, rough fairways bordered by mesquite trees and sand. It was very primitive and not used very much.

Toward the end of 1952, Major Hanning, the base comptroller, transferred me from assistant finance officer to acting budget and fiscal officer. A veteran civilian had resigned from heading the Budget and Fiscal office and I was seen as the next best bet. The 1955 budget was due in a few months and someone had to fill the slot. I was completely lost at first, but found several key civilian employees and one airman very capable, as they had prepared a budget before. The annual budget request included documentation on every research project being conducted on the base and those that were expected to be started in 1955. We funded for fruit flies, pedigreed dachshund dogs, chimpanzees, and the high speed track, then at Edwards AFB, which was planned to be moved to our base. All these and many more items were to study the effect of speed, force, and cosmic rays on a man who would soon be launched into space. We finally finished the budget, got it printed, and Major Hanning and I hand carried the copies to Research and Development Command Headquarters in Baltimore, Maryland. I only served as a strong back to aid the delivery of the budget.

Wayne Mattson enlisted in the Air Force three days after graduating from high school in 1950. As there was no Air Force Academy at that time, he applied for and was accepted at the Naval Academy. After earning his degree in engineering, he entered flight training and became an Air Force pilot. In 1965 he was assigned to Edwards Air Force Base in California, and a year later he transferred to Holloman AFB as chief of the operations branch of the Target Drone Division. As many of the people involved in rocket and missile launches, Mattson often refers to one of those vehicles as a "bird."

When I first got to Holloman, this base was known as the Air Force Missile Development Center. I was assigned to aircraft and missile tests, and we had three main projects in the organization I was in. One was Firefly. Two was the BQM target drones. And three was sounding rockets in support of ABRES [Advanced Ballistic Reentry System] and in support of NASA for the landing on Mars. Project Firefly involved a bird known as an AQM-91. This was a high-altitude reconnaissance drone. Highly classified at the time. It was secret within a radius of fifty feet. None of us could figure out why it was secret within a radius of fifty feet until the later years. Son of a gun! That was stealth technology we were testing!

Every time a Russian photo-reconnaissance satellite passed over where it could take pictures of Holloman, we knew the transit time. We had either to cover up that bird if it was outside, or to move it into a hangar. We planned our missions around satellite transit times so that there would be no chance of the bad guys finding out about it.

You could program it on a preprogrammed flight path and it would fly above any interceptor or any surface-to-air missile. The bird had a camera and it could resolve a one-square-foot target on the ground. You got some good results with it. The Air Force finally decided we are getting our satellite technology, we've got the SR-71, the U-2s are still flying, so let's do away with this. The birds eventually wound up at the boneyard at Davis-Monthan [AFB in Tucson, Arizona], and there they were deliberately cut up so that no one else could get their hands on them.

This AQM-91 was launched from a C-130 aircraft. I was on temporary duty and just made it back to Alamogordo, and ran into my center commander. He was telling me, "While you were gone, they dropped one." Where'd they drop it? Fortunately, it landed just inside the fence up at Los Alamos. That's good. He said, no it wasn't that good. What happened, there was some guy who was the editor of a little newspaper nearby, who heard about it. He climbed a tree, got a picture of it, and he was going to publish that. The Air Force asked him nicely, "Please don't do that because it is a highly classified project." And his answer was, "I hated the Air Force when I was in it, and anything I can do to embarrass you guys, I'm going to do it." So he published that photograph.

Then later on, on a mission I was flying, we launched out over the Pacific and we would fly with the C-130 basically a straight path back and the drone would fly a zig-zag path and, of course, we had plotting boards and everything else on board with radar on top of the C-130. We could keep track of it. And we also had what we called SIDS, Stellar Inertial Doppler System, on board, which we were developing at the same time. This tied into the tracking computers. In this particular mission, all of a sudden, the AQM-91 drone went into an inadvertent parachute recovery, and Denver center gave our call sign and said, "Looks like your bird's gone into recovery mode." There was a little silence and then, "It looks like your bird's going to come down in the Four Corners area." We confirmed that. We're orbiting, we've got a visual on it now. And he said, "Coming down in the Four Corners area, you don't need to worry about it. You'll get your drone back." Little pause. "In about three to four months, in the form of jewelry!"

It impacted on the Indian reservation, maybe fifty meters or so east of a dirt road in a little clearing, thank heaven. The parachute came down nicely and draped over the top of it. Beautiful. We marked the location and headed for the closest place that would hold a C-130, which happened to be Farmington, New Mexico. So, we come into Farmington and landed, and we've got these strange-looking pylons sticking out on the wing. And as we landed and rolled out, the tower gave our call sign and says, "What are those funny-looking things you've got on the wing?" We couldn't tell them. So we said, "Those are training wheels."

At Farmington, we kicked off the crew from the back end. It was partial Air Force, partial Teledyne-Ryan contractors. They rented automobiles and headed for that spot in the reservation. One of the contractor types leased a chopper and a pilot and headed out there. I gave him the radio from a survival kit so we could communicate, and we took off again in the 130 to orbit the area. I got out there and we couldn't find the parachute.

We had marked the position with SIDS, we had marked it with TACAN [tactical air navigation] and visual references, and we couldn't find that parachute. So finally, we started an expanding square search pattern and found it. The Indians had gotten there. They had the parachute rolled up already and had it in the back of a pickup truck and were about to drive off with it when the reservation police arrived and stopped them. About five to ten minutes later, the helicopter arrived on the scene and we secured it. Everybody was briefed as to what we had, so we said, "If you guys want the parachute, you can have it. Just forget what you saw!" And then, of course, we sent a commendation to the Indian reservation police for the fine work that their people had done. We had to go in with a truck and a crane to get that thing out of there, but we finally got it back to Holloman.

At Holloman, there is a site called ZEL [zero-length launch]. It is a large building, open in front, and out of the back end comes a big tube. That building was used initially for zero-length launches of F-84s and F-100s. We had two launchers that we had installed out there and we would launch the drones off of those. They came pretty close to the edge of the test track facility. In fact, one day a drone screwed up on launch. When you recover a drone, you recover it via parachute. It was put into emergency parachute recovery, and it landed right in the test track commander's parking slot. Fortunately, his car wasn't there.

In 1952, the New Mexico College of Agriculture and Mechanic Arts (now New Mexico State University) established a co-op program that allowed students to study for six-month periods and work the alternating six-month periods at White Sands Proving Ground. Grady Nicholson was one of those co-op students.

Ardmore, Oklahoma, is my hometown and I graduated from high school there. My uncle, who lives in Alamogordo, arranged for me to get on the co-op program at White Sands Missile Range. I came out right from high school in 1954, and I went on that program for the full five years. I think I am one of the few that did the program the way it was designed to do. Most took more than five years to complete their degrees.

It worked out really well for me. In '59 when I graduated, I went to work at White Sands for about six months. Then I went to school to work on my master's at New Mexico State. I got my Master's degree in 1961, and I came to work at Holloman Air Force Base.

In the co-op program, most of the time I was working for an outfit called Missile Geophysics. The job I was involved with was to try to take weather data, wind and whatnot, and predict where the missiles that they were shooting were going to impact, so they could go out and be located. They also placed microphones out in the desert, triangulation techniques, and tried to zero in on the shock wave as the missile came in. What I did a lot of my career was to go out and set up those microphones in the desert and then I would take the wind data. It was very interesting.

Sometimes, if they were having a hard time locating [a missile], we would go out and try to locate it for them. We used to have the Army Jeeps, or the three-quarter ton cargo trucks, like an enlarged Jeep. They all had governors on them. We would sometimes set the governor at a real low speed in the area where we were hunting, and we would get out on the top and let the vehicle drive itself. One time, this vehicle I was in picked out a boondock [a remote, brushy area] and went right up on it and high centered, and we had to dig it out of there. That was embarrassing. That was fun, actually.

Albert Poitra joined the Air Force in December 1948. Six months later, he was assigned to Holloman. In his 1989 oral history interview, he said that when he arrived in Alamogordo, he was "an 18 year old runny-nosed kid not even dry behind the ears, who didn't exactly comprehend the whole scope of things." While he served at Holloman, he had a chance to mature and develop some self-assurance.

Lackland Air Force Base was Basic Training. After completing the Basic Training, we all took tests to see what you were more adapted to. I guess they kind of liked my mechanical skills, that's what I scored highest, so from there I went to Lowry Air Force Base [in Colorado]. I went to remote control school for B-36s. This involved one gunner firing four guns, above and below. So, remote control, he could aim his guns and have firepower of eight, .50 caliber guns, aimed at the same target. Each gun was positioned differently on the fuselage, so all this had to be synchronized to where, regardless of where he aimed the weapons, he wouldn't shoot himself down, or any other aircraft within the vicinity. That's what they sent me to school for and without me knowing it, those things were ready for the scrap heap because the B-52 was on the assembly line already. So, they sent me to Holloman to set up a gunnery situation where they could use the same system.

I think it was right at the middle of June 1949. I got off at the train station at Alamogordo with my duffel bag over my shoulder. The place was kind of a dusty old tank town with no significance. I couldn't understand why I was here, you know, who had blessed me with this place?

I can remember the old tar paper shacks we had for quarters, the old rickety mess hall with the morgue out back. We ate off picnic tables. I think I made $86 a month, got paid once a month. I suppose somebody from the Civil War would probably say he got paid $16, but anyway, that's what my pay was. I had a little girlfriend downtown and that was my primary reason for coming to town very often. I'd go over to their house for supper.

For a long time I was in limbo. I didn't have a job, and the job that I did have they didn't want to do anymore because of the B-52 coming along. So, we checked in at the squadron orderly room, and they said you're gonna be on tumbleweed detail or you're gonna go on KP or you're gonna do this and you're gonna do that. Meaningless jobs, you know. For about six months. I got tired of that after a while. So I met this guy, and I asked him where he was working, and he said he was with missile recovery. Missile recovery, what's that? Well, he says, why don't you go out and talk to the first sergeant, maybe you can get on over there. So I went and talked and they said come over and you can help out.

Going out to the Alkali Flats, there'd be three or four of us in a weapons carrier or a personnel carrier. This one particular time, the missile was supposed to have landed in the Alkali Flats, and we were going out to see how many pieces we could pick up. But it never left the sled because somebody, on their checklist, forgot to pull some pins. But we stayed out there for a day and a half, maybe. Going on the way out there, some of these old ranches out in that area had been abandoned during World War II, and they had some fence posts out there. I was in alert of those because these other guys that I was with didn't realize how cold it could get out there at night. So, I'd holler, "Stop the truck," and they'd say, "What does this guy want to stop the truck for?" I'd jump off and load that old fencepost in there and we would go and I'd get down a little ways and I'd see another one. "Stop the truck," I'd say and load me up another fencepost. After two or three times, these guys was getting pretty well disturbed about it. We just kept on going. Personnel carriers are wide open. If you were lucky, you might have got one out of the motor pool with a canvas top, and the heater in that just can't warm six or eight guys, especially in the middle of the night. But I had my firewood and they didn't realize what I was really thinking about ahead of things. Eating C-rations out in the middle of nowhere, sixty, seventy miles to the nearest situation. I'm starting them a fire about maybe ten, twelve o'clock at night, and everybody tried to huddle around my fire. I said "You were the guys that were complaining about me stopping the truck, and now you're sitting around here soaking up the heat from my fire, *my* fire." And then, eating C-rations. I'm warming mine on the manifold of the truck. I popped the lid a little bit, wired it to the manifold, started up the truck, and in about ten minutes you had warm chow. These guys couldn't understand what was going on, so I guess I was kind of a seasoned troop where they were kind of green in that respect.

We could get out there two or three o'clock in the morning or sometimes stay all night, trying to get this thing ready. We'd eat our C-rations and we'd sit in front of the heater and try to keep warm. I remember wrapping up in a tarpaulin that was laying there, just to stay warm. Couldn't build any fires in this one particular spot. But out there on the Alkali Flats we could build a fire if we had something to build a fire with. A lot of the vehicles were coming back with the boards tore off some of the equipment that they chopped up to burn. The fold-up seats on them weapons carriers had oak planks we were sitting on and the backboard. They chopped them up in the night just to stay warm.

We had to follow the beaten track of most of the vehicles because if you had a heavy rain, it would make pools out in there. The alkali would mix with the rainwater, and it would kind of frost over on the top of the pool, which gave you the impression by looking at it that there's solid alkali right out there that you could land a space shuttle on. But when that water soaked into the ground or evaporated, it would leave this thin layer of alkali on the top. And when that dried out, you had a hollow spot underneath. You could be driving along, if you got off the beaten track, you could lose a whole vehicle in there or maybe even more, maybe lose people in there. So you had to be very careful because the desert out there is only for those that really know.

The support equipment that we had at that time, we're talking 1949, all of this stuff was left over from World War II, whether it be frozen turkeys in the mess hall, C-rations, weapons, vehicles, aircraft, clothing, anything that you want. And we had some support aircraft that were used on different experiments such as the chase planes.

One guy smashed up a Jeep out on one of the launch sites there, chasing jackrabbits. Ran into the only tree on site. There was only one tree there within 30 miles, I guess, and he piles up a Jeep. He had a lot of answering to do. Little funny things like that went on, especially in our idle time, late at night, or whatever. And the things we had to go through in comparison to what the military people have now, hell, they're living in glory. I'm telling you, they really have it nice.

Robert Ritter graduated from Youngstown College in Ohio with a major in mathematics and a minor in physics. He was drafted six months later, in January 1951. After boot camp, he was assigned to White Sands Proving Ground as one of the Broomstick Scientists.

It was my understanding that they had put in a request for six mathematicians. When they didn't get it one month, they put it in another month, and someone was just adding those up. So we ended up with quite a few mathematicians out here. I think it took a little time to figure out what to do with us. I ended up in the department where we kept track of the personnel. It was an administration job, really. It didn't have too much to do with mathematics. But we kept track of the various missiles that were active at that time, and the personnel that were on them. It may have had something to do with how they kept track of what each missile was costing us. We had a large board that had each missile listed on it. Then, underneath it, we had the personnel that were on that [missile assignment] each day. I think the hours then that were applied to each man were somehow applied to those programs in the accounting department.

Our emblem became crossed broomsticks, which were put on our discharge certificates that we all have.

Someone organized the General's Glee Club, which for me was the cultural thing of the base. We got a group going of tenors and basses, a very balanced group. I think we were directed by some professional people. They were good soloists and also excellent directors. We had some very, very good trips going.

I think one of the highlights was when General Houge at Fort Bliss sent his plane to take us down to Fort Sam Houston and tour through the Brooke Army Medical Center. Oh, this was going to be something. We were going to be down there in San Antonio, and we'll get this thing through the hospital and then we're going to be out on the town all night long. Well, Brooke Army Medical Center has some of the most serious cases from World War II. I mean, men with half of their faces shot off that are going to live there all their lives. We were so shaken up with that that we came back in the evening and toured again through that hospital. One number that we just hum through was called "Come, Sweet Death." But we changed the name, and we announced it as "Come, Sweet Peace." That was quite an experience.

From there, they flew us up to Los Alamos, and we sang in the chapel up there. We sang the Navy hymn and had that congregation in tears. Then the White Sands people sent a bus to pick us up, and the home that I stayed in was where they parked the bus. After the Sunday dinner, the two little boys in that family were just dying to get on that bus. So the driver said, "Well, okay." He went out and opened the door for them, but he said, "Don't touch anything." So, until we left they had a dandy time playing in that bus.

Sergeant Bean was the "housemother" of Barracks Four. They gave that poor guy a hard time. He had the room in the end of the barracks. One night, they attached a bunch of flash bulbs to the light in his room. So when he came back from the club, probably pretty well juiced up, and turned his light on, these flash bulbs all went off. That might be considered a rather dangerous thing. Nobody ever thought of that. His motto was "Always drink them while you buy them." He never wanted anyone bringing beer back from the club. So they would bring empty bottles back and fill them with water and lean them against his door. So when he came in and opened his door they would all fall in. That poor guy put up with a lot.

Cynthia Sommers-Guzevich adds one of the few female perspectives on life in the early days of space research. In this excerpt, she described some of her experiences at White Sands Proving Ground and Holloman Air Force Base.

The original plans were for map making, because there were virtually no accurate maps available. In fact, I remember one time we were going out in a car and Ben [Billup, my boss] said to me, "Let's read the odometer so we know where we start, and let's say we're going north-northwest. Draw a line, and when we stop, we know how far we have gone." This was very crude because at that time, the finer aspects of surveying were not available. It was amazing that when the surveys were finally done, how fairly accurate those maps have been. Alas, at that time, headquarters was the only permanent building, so we were involved a lot in designing of instrumentation sites and the buildings that were required. Many nonpermanent buildings existed on the post at that time.

I got as high in my grade at White Sands as I could, and in a wonderful, big drafting department in Holloman there was an opening. I applied and didn't know if I could get hired since it was all men and many military workers in that department. At that time, Women's Liberation was virtually unknown. Not that I felt I needed it. I thought that if you do your job well, you don't need to be a "liberated" woman. Your job liberates you, and your work liberates you. I applied for the job and I said, "If you are satisfied with what I do in a couple of months, keep me and if not, fire me." They must have been satisfied with me, because I stayed there and enjoyed it very, very much.

To me, working at White Sands and at Holloman was a very exciting time. It was a time of development. I remember when they were doing research on the lunar excursion module. It was a time of working with people that saw into the future. It was working with people who were thinking not of today, but of tomorrow. You never knew from one day to the next and from one month to the next what project you would get assigned to or what development was coming next. You were really on the ground floor of a very wonderful development of the human race.

Len Sugerman was born in 1920 and joined the Army in 1942. He was assigned to the Air Corps and became a technician working on the Norden bombsight and automatic pilots. He was assigned to the Air Force Missile Development Center at Holloman Air Force Base in 1964, where he served as the director of the Central Inertial Guidance Test Facility. Next, he became the Air Force deputy to the commanding general at White Sands Missile Range. Sugerman joined New Mexico State University's Physical Science Laboratory in 1977, and in the 1990s was a founding member of the Southwest Regional Spaceport Task Force.

The Norden bombsight was one of the big secrets that we had. I remember I was just a corporal, and I finally made sergeant. We used to teach, and we would have a .45 strapped on as we taught. The bombsights were kept in a vault, and all of the notes had to be in confidential notebooks the students couldn't take home. They didn't even call it the "Norden Bombsight Class." We called it "advanced armament."

In 1944 we went up to Yale University in New Haven, and we had ten months of technical training. At Yale we had four options: armaments, meteorology, photography, and aircraft engine maintenance. I went through the armaments course. Some interesting things happened up at Yale University. That seemed to be the dumping grounds for all the big movie stars and other noted people that they didn't want to send overseas. The band there at Yale was the Glenn Miller band. This band had enlisted as a group, and every night they would play for retreat on the green in New Haven. All the people in New Haven would come to see the Glenn Miller band playing. Tony Martin was a singing instructor, Broderick Crawford was a gym instructor, and William Holden was our information officer. We had people like that scattered all over. But that was a tough ten months up there. I think I wound up in the best physical shape that I had ever been, because we were trained by the Yale University Athletic Department.

I remember when Sputnik went up, very well. The word came down, we had to put "space" into the title of all our projects. And they got some fathers [priest faculty members] from Georgetown University that knew something about orbits to come down and lecture to the officers, tell them about space. Everything had "aerospace" in the title. The literature showed that the Russians were doing this. If you read some of the reports that Rand and others were putting out, you could see pictures of dogs that the Russians had going up in space. So you knew that they were doing these things. There was a big rush to follow suit.

Born in Arkansas, Thelma Swope moved to New Mexico in 1957 at the age of nine, when her father took a civilian job at White Sands Proving Ground. Beginning in 1967, she worked for three years on the Athena missile project at Holloman Air Force Base. She adds a different female perspective to working in the space research field.

I graduated from high school in 1966. After that, I went to Eastern New Mexico University for a year. A typical kid, I didn't do very well. All of a sudden, all of this great freedom, being away from home, so I just managed to pass. My parents were not in a position to continue to pay for an education with B and C grade level. So they told me, "You will come back and you will pay for your own education, or you will pick a job." So I started interviewing and putting out job applications for an assistant's job. At that time, Holloman was booming with the test business. The test track was the going thing. I put in applications with various contractors out at Holloman, and I got offers from three of them. One of the offers was from a company called Radiation Inc., which was a subcontractor working on the Athena project, and they had other subcontracts with the government. So I took their offer because their offer was a quarter more an hour than any of the other two.

When I went in for my interview, I met with Dr. Soo, who eventually became my boss. During the interview he was asking me all kinds of neat stuff about what my goals in life were, and he was looking at my resume, which was not very long because I was newly out of high school. He asked me, "What is this you have down here under hobbies?" I said, "I have singing, and I have sewing, and I have duplicate bridge." "Well, this bridge thing." Dr. Soo was the neatest guy. He was a refugee from North Korea. He was a doctor of aeronautical physics, and he was head of the department that I was soon to be working in. He said "What is duplicate bridge?" So, I said, "Well, it is a competitive card game. You have to partnership someone in order to win at the game. It is not a solo game. Therefore you have to learn a different language. You have to learn how to communicate with your partner to achieve the highest level of success." And he said, "Well, any woman that can play duplicate bridge certainly can work on Athena missiles." I said, "Oh really?" So, when I got the offer, I jumped at it.

I don't really think that a lot of people understand that document control at that time was, how can I say this without putting it down? It was sort of a theory more than a practice. I had a faith in my office that I was in charge of, which included not only the Athena project, but other projects that were ongoing that were used as part of the research and development of the Athena missile. But people could go in and out of that place, and especially for top-secret documentation, which required a different control mechanism, that was very disconcerting. That is why a lot of leaks happened during that time, a lot of leaks. That is almost a story in itself, as far as how ingenious some people were at compromising their document control.

All I had in my office was a little walk-in safe, and I had a safe within the safe for the top-secret material. But all of the secret stuff and all of the confidential classified were open to the project people. All they had to do was sign a little register, with what document they had taken out. At the end of the day, I had to make sure that that document was back in the safe. Well, what happens to the document in between?

I had to verify people's clearances. We used a badge ID system, and I had a backup file where I could look and make sure. On their badge we had a code; SE was secret, TS was top secret, and CO was confidential. They would have to submit their badge, then I would cross-index it into my backup file, before they could take out a document. When they checked out a document they had to sign two papers, and I had to know where the document was going. It had to stay within their control, and then at the end of the day, or at noon or whenever, the documents for launch had to be back in the safe. It was supposedly a pretty good system.

One day I got in trouble. Dr. Soo, my boss, was the one who compromised the safe. He came in after hours. He was the only other person that had the combination. He came in on a weekend because we were behind, which was typical. He checked out some documents, and he filled in all the forms. But he forgot to put one back. I came in on Monday morning, and I found that top-secret document lying on his desk. You can't cover up that. So the Defense Investigative Service came in and ran an investigation.

They were pretty mean guys. And here I am a nineteen-year old kid. I am going, "Oh, yeah, right! I am going to spend the next twenty years in jail, because my boss compromised me!" But it all worked okay. After the interviews, they had to declassify that document. They took a top-secret document and they staged it down, because it had been compromised. They had no guarantee that another individual had not taken that material. It had been out of control. Dr. Soo was real good. He did take responsibility for it. He did save my job. And he did keep me out of jail. But to a young person who is trying to do everything by the book, it was quite an experience. However it was an excellent experience, because ever since I have been very, very careful about releasing information that is not necessarily for public release.

Dr. Soo didn't suffer any consequences, because he was the primary person in charge of the Athena project. They just gave him a tongue-lashing and told him he would not enter that safe without me. In other words, from that day forward, he and I had to co-sign for everything that went out of that safe. In a way, that was good because it put a backup. And it took some of that responsibility off of my shoulders.

I would assist in some of the drafting, and the changes that were made. I really didn't have that much opportunity to become a physical part of the Athena. I had top-secret clearance. At that time, regardless of what your clearance level was, there were certain projects that you could not discuss with your spouse outside. Nothing. So, many times we would go home and we would eat dinner and watch the news, and we could not converse about one thing that happened on the job. You have one group working on one project and another group working on another project, maybe in the same building. You don't really know what they are doing, and you really don't want to know. You couldn't take the risk of compromising somebody else's job. And yet we would all get together occasionally and have parties. It was real interesting, even during the parties during those years it was all petty talk, because you could not really talk shop at the party.

When we moved from Holloman to [Kirtland] Air Force Base, it was the end of my career because even my husband didn't talk to me. So all of a sudden, I was put in a very new position. I was at where most of the wives at Holloman had been most of their wedded life. I was, all of a sudden, nonplused. I could not have any information. I had been debriefed. I was no longer cleared. I was no longer active. That is a real rough thing, especially when you are spousal with somebody in the industry. I missed that activity. I missed the mental stimulation I was getting.

I directed bridge games. I became a bridge umpire. And I dealt with a lot of upset people. I traveled a lot. I got to go to Las Vegas and Dallas and deal with adjudicating bridge games and running bridge tournaments.

[My husband] Don was so good at PR [public relations], and his foreign technology background made him develop this ability to deal with people and to get the max out of them. They made him liaison officer at Kirtland. Those seven years we were in Albuquerque, I ended up entertaining and setting dinners for all these big muck-a-mucks coming in from California, all of these big contractors. And Don did a lot of the contract negotiations that went on with the laser projects. That is what I did. I became an enabler for him to negotiate contracts.

I moved from Albuquerque back to Alamogordo in '76. Don and I were divorcing.

Chapter 3
Taking over the V-2s

Robert Goddard pioneered liquid-fuel rocketry in Massachusetts, launching his first rocket in 1926. It reached an altitude of 41 feet during its two-and-a-half second flight. Over the next few years, his experimental flights became higher, longer, and noisier. After a 1929 rocket exploded when it hit the ground after a 90-foot-high flight, the Massachusetts fire marshal told Goddard he could not launch any more rockets in the state. Goddard then moved his workshop and launch tower to Roswell, New Mexico, which had a small population, a great deal of flat land, sparse vegetation, and a higher altitude. Goddard left Roswell in 1942, when he went to Maryland to help the military develop jet-assisted takeoff (JATO) for airplanes.

About that same time, a group of German scientists and engineers were working on a missile using very similar technology to Goddard's. The Nazi government supported its development as a terribly destructive weapon. With this large financial and manpower support, the Germans produced a much larger and more powerful rocket than Goddard had. What ultimately became known as the V-2 missile could carry 2,000 pounds of explosives over a distance of 200 miles.

The figurehead of the German rocket developers, both because of his technical skills and his charismatic leadership, was engineer Wernher von Braun. Like Goddard, von Braun was interested in developing rockets primarily for human spaceflight. In 1945, as World War II was winding down, von Braun and more than a hundred of his coworkers surrendered to the United States Army. The American government brought many of them to the United States, along with 300 train car loads of V-2 components, launch equipment, and research documentation. The group became known as the Paperclip Scientists. Many of them, including von Braun, were sent to live at Fort Bliss in El Paso, Texas, where they would work with the Army on further developing missiles and adapting them to scientific use. They launched their missiles nearby at White Sands Proving Ground in southern New Mexico. After most of the Paperclip contingent was reassigned to Huntsville, Alabama, in 1950, the American Broomstick Scientists took over the missile development at White Sands Proving Ground.

Ulysses Hess arrived at Holloman Air Force Base in 1955 as deputy chief of staff for personnel. In his 1986 interview, he summarized the background for the V-2 program at White Sands Proving Ground.

In 1955, we were just getting our foot into space. An awful lot of people thought they were self-appointed experts on space. They could write books about it, but they'd never been there. One of the theories was, as late as the late '40s, if you go into space and you go beyond the envelope of the Earth's atmosphere, there is a possibility that, where the Earth's gravity had dropped to zero, it might pull you right into the sun. That's the old Greek mythology. But in 1952 at White Sands Missile Range, in conjunction with Holloman, they fired a Nike Ajax missile, which is a two-stage rocket. We know that the atmosphere thins out pretty well above 100,000 feet. In fact, you can't breathe above eighteen, nineteen, twenty thousand feet up. So, this rocket went up to over 100,000 feet, and came back. We knew from that, you can go into space and return.

So the US got into a space program. What made it real interesting, very few people stopped to realize in 1938 to 1940, Germany was a way out there in advance on space. We had to do the same thing. We had to go explore space in the '50s. Germany was ready to do it in '38, except Hitler took over all their people, all their hardware and all their ideas, and put them into something for the war effort.

Taking over the V-2s

Since his childhood, Clyde Tombaugh designed and hand built telescopes as tools for his main passion, astronomy. For nine years, he made that sideline his primary occupation when he served as head of optical measurements at White Sands Proving Ground during the V-2 program. He was inducted into the International Space Hall of Fame in 1980.

I worked at Lowell Observatory in Flagstaff, Arizona, for many years. During World War II, I was in the military draft and I was deferred and assigned to teaching navigation in the Navy V-12 school set up at Flagstaff. They were desperately short of people who could teach navigation, to train young men to be officers. So I spent seven semesters teaching navigation to military personnel during the war. Then I went to UCLA one year on a visiting assistant professorship while one of their professors was on a sabbatical leave. Then I was going to go to teaching at Albuquerque. The Lowell Observatory was short of money at that time, and so I thought, for a few years until things get better, I'll do something else and then go back.

I was urged to come to White Sands Missile Range to apply my knowledge of long-focus planetary photography to rockets in high flight by the tracking telescope, which was a new device. The Germans never had anything like it in World War II. Being acquainted with this field of work, they thought that I would be a good person to supervise this particular instrument development. It worked out very well. I worked at White Sands Missile Range nine years with the Army, in the Ordnance Corps.

We had some close shaves. We had rockets fall from a hundred miles, one in particular missed one of our stations by less than a hundred yards. It made a crater 75 feet wide and about 25 or 30 feet deep. But we had pretty good luck. I was the supervisor of the optical section of the ballistics laboratory, responsible for getting the trajectory data. I had about eighty men under my supervision, about half of them were GIs and half of them civilian. I got well acquainted with the problems involved in the ballistics of rockets.

Some research involved increasing the optical power of our cinetheodolites [photographic instruments for collection of trajectory data] and tracking telescopes. And we supervised making similar or larger optics, 6-inch mirrors, to put on the captured German V-2 rockets to increase our range of tracking, which worked out very well. Focal lengths of 7 to 10 feet and so on, which doubled our range at which we could track the rockets.

Changing the paint pattern was another one of the important things that we did then. The German paint pattern did not permit tracking or following these objects far enough in the sky. So what I proposed was to paint most of the missile a flat white so there will be no sheen on it and paint each fin with a different pattern of flat white and flat black so you could identify which fin was which. Since it was white light, they could take out any part of the spectrum they wanted to improve contrast, between the different instruments in the field. This was an enormous advantage, and it doubled our range of tracking rockets. We were able to identify the fins on the V-2 rockets at a distance of a hundred miles on the tracking telescope by the picture at that larger scale. And we found out there were things happening to rockets in high flight that the Germans didn't know about, because we had the detailed pictures to show it.

After that, I came in to New Mexico State University to do a project looking for possible small, natural Earth satellites. Some of us were at Flagstaff for a while; they had a fast Schmidt camera there. Then we decided we wanted to look closer in the plane of the equator. We could not observe that from here because the curvature of the Earth would hide it. So we had to go to the equator and be right under the celestial equator to explore that zone. That's why we went to Ecuador, in South America. The university had the logistics set up for making such a foreign expedition. I sent two men down there with a telescope and fast cameras for a few years.

We didn't find any of those rocks going around, like in Saturn's rings, so we concluded that man could go into space or go to the Moon with reasonable safety from possible collision on the way there. As you know, no one has had a collision. We took over 15,000 films and didn't find a single rock, so we figured the coast was clear. It was not a perfect coverage, but it was enough to sample it, and we felt confident that we knew what we were up against. That was just before the advent of the artificial satellites and multi-stage rockets. The technique that I developed for that was a high-angular tracking rate for cameras. This procedure was later used, even today, by those who are engaged in tracking our artificial satellites. I had forged a new technique here, a new type of instrumentation.

Researchers tried several paint schemes on V-2 missiles to help track their rotations during flight. Courtesy of NASA.

After finishing high school, Joe Gold decided to attend college in New Mexico because the climate was better for his asthma. While studying engineering at New Mexico College of Agriculture and Mechanic Arts (NM A&M), he worked for the Physical Science Laboratory (PSL) on campus. PSL was founded at the college in Las Cruces in 1946 to provide support for the rocket activities at White Sands Proving Ground and, eventually, other space-related programs at various institutions.

I was taking a physics class, and the physics professor, Dr. Gardiner, said to me that he was going to form a laboratory to reduce data for White Sands on missiles they were flying out there. He asked me if I would like a student job. Naturally, I said yes. I worked as a student for PSL, Physical Science Laboratory, for three years. In fact, I am one of their original employees. I went to work for them on May 29, 1946. I went over to White Sands, and I worked on the third V-2 fired in the US, and stayed on. Then, when I got out of college, it was a lean year. Only half of the people that graduated could get jobs. I had actually turned down a job with the Corps of Engineers in Denver because I didn't want to be on a drawing board. That is what they do with mechanical engineers; they make draftsmen out of them. So, people came around from White Sands to interview students. They said, "With your education and the experience you have with PSL, we could use you in our business of collecting data. But we don't have any spaces. If we did, we would hire you right now. But when we get a space we will hire you."

So at the end of school I didn't have a job. I went to Dr. Gardiner, head of the Physics Department and founder of PSL. I told him my story and asked, "Would you have a temporary job in your department?" He said, sure. And so he just continued me on the contract he had with White Sands. He sent me over to White Sands, and said, "Go out there and find Clyde Tombaugh, and tell him I sent you." I already knew Clyde, because I had met him when I was in college. I went out there, and within a month, White Sands had spaces and hired me. I spent the rest of my career at White Sands. I spent the first five years collecting range data. All the V-2s we fired out of White Sands, I either collected data or reduced data on all but two of them.

After about five years in the Range collecting data, I joined the missile test crew and stayed with it the rest of the time. It turned out to be an interesting career. In the missile field, I worked mostly in the optical data reduction, telescopes. The way it works is that you have a very highly surveyed piece of real estate. White Sands is one. You know where everything is, down to the decimal point, so to speak. So having known positions of the data collecting devices and the known position of the missile launch, you can do trigonometry. They have what they call the Range time signal. It puts out a timing signal, which is the standard signal for the whole Range. Every piece of data which is recorded has a timing thing on it. You have azimuth [horizontal angle] and elevation on the upper quarters of the film frame. As the azimuth and elevation tracks, the numbers change. In that reduction you read the numbers, put the adjustments to the difference in the locations on the crosshairs. Now you have angle measurements from that station, and you have time on it. So you take two or more stations. In this way you can get an angle on them. Then you just calculate trigonometric solutions.

When we started doing that, we had these electro-mechanical calculators. They simply add, subtract, multiply and divide. So we ended up with triangles. Well, in the case of right triangles, which we tried to use the best we could, the square of the hypotenuse is equal to the sum of the squares of the two sides. So now we have the squares of the two sides, and we have to take the square root of the sum. That was a problem. But the machines we had would do an approximation, so we could work it out. That is what kids going to school learned to do. Now, the calculators and right quick computers do it, and the things we used to do in a couple of minutes now only take seconds.

I know the first V-2s that came over from Germany were in pieces. The Army worked out a contract with General Electric to assemble the checkout for them, and they fired some for several reasons. One of the warheads was taken out, and all kinds of instrumentations put in instead. That is where they flew the first monkeys, Rhesus I and II. It was very interesting, whether you watched it from a distance and you could see the vapor trail as it went up, or you watched them real close up. I've watched them from almost underneath.

The reliability of those was not good. Plus, the instrumentation was hastily built, and it had a reliability problem. So when you had the missile and the instrumentation both work together, it was unusual.

Many of those missiles, we have had them blow up on the launch or we had to blow them up in flight. One of my early jobs was working at an instrumentation station, the closest one to the launch we had. There was a lot of discussion on what to do in case we have a missile failure up there. I've had pieces falling around me out at White Sands. A couple of them blew up. If you have a missile failure, we always hoped it would fall in one piece. But when it blew up in the area, we would have more than one piece to worry about.

I saw the one that went to Juarez. I was in college, and I was working with PSL. It was near sundown. It was just late in the day, and two other PSL employees and I were standing around, visiting. We saw the missile come up from White Sands. Normally, they would come up almost vertically and bend over to the north. This one came over and bent over to the south. We knew that something was wrong. It just disappeared from us down behind the trees. The next morning we found it had landed in Juarez. We thought, boy, that is going to terminate White Sands.

Gerald Garing began working at PSL in 1952, while he was a graduate student in physics at NM A&M. The equipment they used to analyze data seems primitive today. The Askania cameras he mentioned in this excerpt were cinetheodolites – motion picture cameras that tracked rockets in flight and recorded information about their vertical and horizontal angles from the launch tower on each frame of the film. He also mentioned the WPA (Work Projects Administration), a federal government program that put unemployed people to work during the Depression.

I was born in Gallup, New Mexico in 1931, went on through high school there, then went to New Mexico A&M [now New Mexico State University], starting in 1949. I was doing the Master's degree in '52 when I started working at the PSL.

When I first got there in early '49, there were some Master's degree students there that were going to test X-rays. They had permission to glue cassettes on the fin of the V-2. One of them took me under his wing, and we went out. That was the closest I ever got to a V-2, just right next to it.

He bonded different film thicknesses to the fins. Then they would launch the thing, and we would find out how strong the X-rays were.

At PSL, we got to work on different pieces of data, including V-2 stuff. You would get a big spreadsheet, about half the size of the table. We would write down all the coordinates that we had taken off of an Askania camera that we had taken from the Germans. Then through a formula that they had, they called it a Davis solution, we would run through it and do five solutions, for five different cameras. And you would see where you were in space. It took a good twenty or forty minutes to do one sheet. You got paid by the hour. I think it was seventy-five cents an hour, which was pretty good.

There were hundreds of spreadsheets for each mission, so the student room was full. The things that were unique about that was they used logarithm tables that were like seven-place or eight-place tables. They were written in German, but you could read the numbers. Then later on, somebody was back in Washington and found out that WPA made some logarithm books that were ten places. So we quit using the German books and started using the WPA books. We were like the first people that had ever asked for something that had been done in that period [by the WPA].

When we got further along, about my graduation time, they got the big, IBM 605 computers going. They took up whole rooms. We could run a Davis solution in less than five minutes on punch cards. We worked the night shifts. That was one of my jobs after I graduated. We would go out with the students and take them out to the computers at White Sands and work there. The Army ran them during the day, and students ran the night shift and the midnight shift.

I wasn't aware of the Paperclip people until probably later in the '50s. Right now, I live with some of the Paperclip people in La Luz [New Mexico], four or five people that came over with the Paperclip group. The only thing I can remember real well about that is that if we had a launch, we would see it from the top of the Organs [Organ Mountains between White Sands and Las Cruces]. When it happened, you would drive clear up there, at San Augustin Pass. You would get hundreds of people that were watching it. That was as close as you could get. But then, being part of the physics group, we got sort of special treatment. We would go down and get a little tour with what had happened, and see the launch pad after it had launched.

Lewie "Tiny" Harris was a mechanic who worked on various vehicles and equipment at White Sands Proving Ground during the time when the Paperclip Scientists were working with the US Army on the V-2 missiles they had developed in Germany during World War II.

In 1949, I went to work at White Sands Missile Range. I just about had to camp on their doorstep in order to get a job, because everybody was trying to go to work out there. So I made myself available and then I followed the auto mechanics trade for a while on main base. After about two years, I went to work for the generator division. I worked on generators that powered the stations out on the Range. After that, I went into the fabrication unit, where I worked on all types of machinery and equipment that was used in launching missiles.

I worked on the launcher for the V-2s, the Meillerwagen [a trailer designed to transport, raise, and serve as a launch platform]. The wheels and their bearings would blow full of dust and sand out there, and we'd have to clean that out so they worked. Then when they came in with the Redstone [rocket], I worked on the Redstone launcher. I helped finish installing it. I helped them take it apart, and we loaded the parts on railroad cars to be shipped to Cape Canaveral. It was a tower-type affair, and of course it was much larger than the old V-2 launcher. I would say probably four times as large. They reconstructed it in Cape Canaveral, and it was used in the first suborbital flight.

I was almost too close to one V-2 launch. I was uprange, and I was taking care of generators that morning. I had checked at the office before I left, asked for a Range count, and were there any launches coming into our area that day. So I went to a couple of stations and then I went to the little generator that was running at the target area. I immediately climbed in the shop truck and turned around. I got about four or five hundred feet from the generator when the missile came in. It liked to have turned that shop truck over end-over-end. It caved in the back door. The concussion broke the glasses in the back of the shop truck. It hit within about thirty feet of the little target building.

Another time, they were starting to launch a V-2 and it got up oh, probably 20 feet above the ground and it set back down and turned over on its side and down through the sand dunes it went. It caught fire, and became a real roadrunner. Fortunately, it was pointed away from the blockhouse. It ended up within about quarter of a mile, I guess, of Highway 70 [the highway connecting Alamogordo and Las Cruces].

During his eighteen-year Air Force career, Dr. David Simons made valuable contributions to the field of knowledge in the effects of spaceflight on living organisms, including human beings. He was inducted into the International Space Hall of Fame in 1987.

My father was a physician, and he encouraged me to do likewise. I went to his alma mater, Jefferson Medical College in Philadelphia. Then I came back to Lancaster [Pennsylvania, my home town], where I took a fifteen-month rotating internship. It was then the end of the war [World War II]. They were transitioning to a one-year program. Then I went into the Army Air Corps because I had been put through medical school on an Army student training program and owed the Army two years of time.

I was always quite interested in, not only astronomy and rockets and things like that, but also in electronics and radio. I had become an amateur radio operator, so I was very interested in the possibility of combining my medical training with electronics. I asked the Army personnel people if they could assign me to a research location where they would be building electronic equipment for research purposes, and lo and behold, I was assigned to Wright Field Aero Medical Laboratory. Not only that, but I was assigned to the electronics section that serviced the acceleration unit, which was under the direction of James P. Henry, M.D. They were physiologists concerned with the physiology of both positive and negative human acceleration tolerances and protective devices, ways of increasing man's ability to withstand more than man was built for, that could be dished out by airplanes. So, I was having a great time working with them at the centrifuge, designing and making electronic gadgets for them.

One day in 1946, Jim asked me, "Dave, do you think man will ever go to the Moon?" Well, in college I read everything I could by Chesley Bonestell and [Robert] Goddard and people who'd been thinking in terms of these possibilities. I said, "Why, of course. It's just a matter of engineering design and time to work out the problems. And interest." He said, "Well, what would you think of having an opportunity to help us put a monkey in a captured V-2 rocket that would be exposed to about two minutes of weightlessness and measure the physiological responses to weightlessness?" I said, "Oh! What a wonderful opportunity! When do we start?"

I was made the project officer of what we then called "Project Albert," to put these Rhesus monkeys in the V-2 rocket. I was assigned to Wright Field, in Dayton, Ohio, but these rockets were fired at White Sands Proving Ground in New Mexico. By that time, the Army Air Corps had been split off to become the US Air Force, and I elected to go with the new Air Force. I liked this opportunity to work with research and physiology and to explore new frontiers, so I was in the process of acquiring a regular officer's commission. Since I was identified with the Air Force, I was billeted at Holloman Air Force Base, and we went on over for a few days at a time to White Sands to work on our project there, too.

We had bad luck. The purpose of these rocket flights was to test out the recovery of the nose cones from space with parachutes. We went through at least three flights, and every time a parachute didn't work, they tried something else. I really didn't try to develop any competence or understanding of what they were doing with the parachutes. That was really their business. I just hoped that they were successful, that was all. But they never worked, and so we never recovered one of our animals alive. At least it was a merciful death, because it was a total disaster when they landed, and they never knew what happened.

They were actually asleep during the whole flight. They were anesthetized for a number of reasons. I think the most important one was that if the parachute opened properly, the monkey would get what was, at that point, an unknown opening shock that was calculated to be pretty impressive. And, depending on what altitude and what the position of the nose cone was and its orientation, things can go wrong. If things go wrong a little bit, like if a parachute streams and it comes in either half open or tangles, well, the monkey could land hard but not fatally. We wouldn't want the monkey to be lying there suffering until we could get there. It can take awhile to track down where the nose cone landed and get to it.

We used anesthesia out of consideration for the monkey. It would

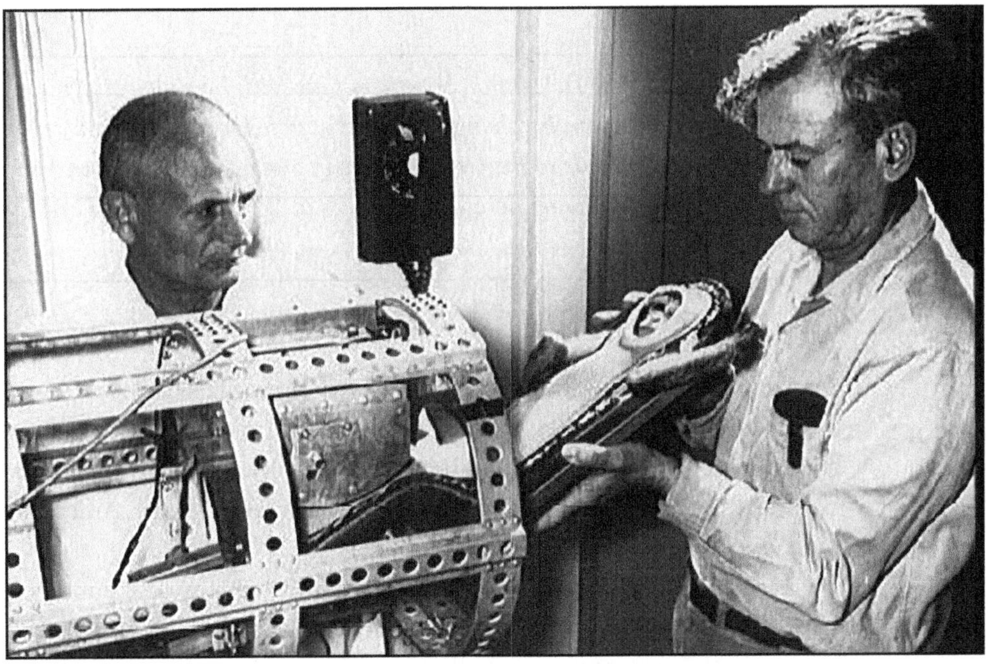

Technicians place Albert in the missile payload chamber before launch. Courtesy of NASA.

have been much easier had we not done that. I think one other thing, too, from the monkey's point of view, the thing we put them in was a seat that they were strapped into, and they didn't have room to wiggle at all. They were totally restrained because our space was as cramped as it conceivably could be, and so was the weight limitation. If they were awake and conscious, this would be a cruel thing to do to the animal if it wasn't necessary.

For those reasons, we did anesthetize the animal. It meant that we had to try to extrapolate what reactions one would get conscious and awake in this situation, compared to asleep and anesthetized. But, if there were any really serious physiologically disturbing factors, they should show up, even under those circumstances. It was a first cut. No animals had ever experienced anything like this before, and at least we learned something about the limits of the problem. Theoretically, we didn't see any, but you never know until you try. The telemetered data indicated that there were no particular problems, that weightlessness was something that was quite tolerable.

During World War II, Cynthia Sommers-Guzevich had unwittingly worked on a program involving the rocket engine for the German V-2 missile. She came to the United States in 1948 and soon found herself working on surrendered V-2 missiles at White Sands Proving Ground.

I started at White Sands in 1950, when White Sands was very much in the developing stages. I worked in the basement of the headquarters building for Ben Billups as a draftsman. Gradually, I worked myself up as chief draftsman. I had no problem, since I had worked with American measurements and American dimensions.

White Sands is old history. People don't know what we did out there. They don't know that we were the kindergarten of the space program. With this V-2 program, you really knew that you were involved in the beginning of the space program. Life at White Sands was very casual. Fun, in a way. I remember one time when I was stopped by an MP because I was speeding, and when I said something to him, he said, "Oh, you're one of the Germans. We're not supposed to give you tickets because they all have their minds in the clouds." So I didn't get a ticket. I was lucky, speaking with an accent.

Lois Godfrey had been a teacher and principal at the Fort Sill, Oklahoma, elementary school. Shortly before World War II started, she decided to try a different line of work.

At the beginning of World War II, I was placement officer in the civilian personnel office at Fort Bliss. My interaction with the Paperclip Scientists began in a very low-tone way. I was placement officer, so people hiring civil service people necessarily got me as the first link in the chain. Major Hamill, who was connected with that group, began to set up the unit, and at that time, it was not well known that the German scientists themselves were there. They had been there, I think, almost a year when he first came to me. There were a lot of whisperings and rumors about what was going on behind those locked gates. As a matter of fact, the story goes that the FBI, sheriff's office, and the police often got calls that said, "Hey, some of your German prisoners of war must be loose because we heard them talking in the grocery store."

In the course of setting up the Research and Development Center, Major Hamill had to visualize the whole structure and organization and then begin to fill it out with warm human bodies who were capable of contributing to that operation. He explained it briefly at the beginning, and then as he came to do more on this part of it, we became better acquainted and I became better acquainted with the project. He would jokingly ask me, "When are you coming to Research and Development Center?" That went on for quite some time. When we got down to the little bit of staffing and hiring, he got very serious. He said, "I'm in trouble. I have seen these people. I know that the housewives can start a fight over a head of cabbage in the grocery market. And you understand how touchy Congress is about funds. The last thing I need is for a big brawl to be publicized widely, and for the Congressmen to say we don't want any more of that, the money is off. I want you to come and help to get these people off on the right track and smoothly integrated, the wives and families, so we can avoid anything like that." Well, that was a challenge. Who could resist? I thought it over and I said, all right, I'll take a chance. We'll do it.

I went there in January '47. That's when the families were really beginning to pour in. I guess there were close to a hundred children of school age. Trying to get some handle on the situation, it occurred to me that school children can do more to civilize families than most anything else, and so I thought I would concentrate on that. But I didn't neglect to do anything else. I tried, through acquaintances and my involvement in various groups in El Paso, to sort of perpetuate the idea of families getting acquainted with families and inviting them into their homes. I knew it was a little early at the beginning. We had to have some preamble there and get them used to the idea. But that was a very informal part of the business.

They told me these children did not speak any English. So, I pondered on what in the world to do with ninety-some school children that didn't speak English. I discussed this with some teachers and other people who would have a good input on that, and we finally got a plan. I asked the

Army to give me a mess hall or a big building like a mess hall. I called the parents together and explained what I thought was the need, and listened to them. To the men. The women never opened their mouths. Very few of them understood what we were talking about. They may have known a few smattering of words, but they did not yet understand English.

I spoke English all the way through, and the men spoke English to me. I didn't run across a one that didn't speak English. I explained what we would have to do. We would have to have that building divided into three classrooms. The Army wasn't going to do it for them. If they wanted to buy the materials and do the work and set it up, okay. They were enthusiastic about it, it appeared to me. I think they were so happy to have a school, and so they consented readily to do it.

I said, "I am going to try to show you how a school operates in a democratic country. I want you to appoint your school board and, together with your board, I will work out a lot of the details of the program. But you understand, when we hire teachers, you have to pay for them. You can assess each parent or each child or however you think it should be done, but the money is going to come from you." They didn't hesitate. They got the partitions in, and I went to the superintendent of schools in El Paso and told him I wanted to borrow about two truckloads of furniture. I explained to him what it was all about, and he was just as cooperative as he could be. So the German men did the work, the Army furnished the trucks, the school furnished the furniture and books and blackboards and globes and reference books. They were most generous in what they provided for us. And we hauled that out.

Then I hunted for teachers. I was lucky, I found three that I thought were just outstanding and could handle something like that. And they had a little time to prepare. So, somewhere around the last of May or first of June, we started school. That was one of the most interesting things I think I've ever seen. These teachers knew how to start with pictures and charts and motions, everything it took to get the children into it. And they got into it and seemed to be very enthusiastic. So the school was a wonderful experiment. I learned, the teachers learned, we all learned.

We followed through with the school into around the last of July. The children eased into the public school, and there was never a ripple that I ever heard of. I went down in advance to the school where I knew they'd be. I knew the principal there, and some of the teachers, and we talked about this. It was no surprise to them at all, and they were very, very enthusiastic about it. The children went to Crockett School mainly, and I have talked to teachers there a great deal about them. They said it was amazing how they fitted right in and became accepted by all the other children. There was no boundary line so far as they could see. Children are children, and they make their own world. And these were such bright children. So that was the crowning experience of the almost three years I spent with the rocket center.

I knew [Ernst] Steinhoff and I knew [Wernher] von Braun. I enjoyed knowing von Braun, and all the rest, for that matter. But von Braun was just really a down-to-earth, personable man, and a short time after he got there, he got his parents to come over. Understand that these people were all living in quarters in some of the old barracks buildings. They were just wooden temporary buildings, but they converted those into apartments for the families.

Von Braun, several times, came by around the end of the day and he would say, "If you don't have anything else you just have to do, wouldn't you please go over and visit with my mother? She's so lonely." And I would go. She was a very aristocratic noblewoman, and his father was a very personable, charming person, too. I so thoroughly enjoyed those talks with them, I spent several visits with them. Then von Braun went back to Germany to bring over his eighteen-year-old bride-to-be and they had the first wedding out there. So that was all very interesting from the human standpoint.

Steinhoff was gregarious, happy, full of energy. He had the world by the nape of the neck, as far as he was concerned. He had so many ideas that he wanted to put into effect. He wanted, for example, to pipe water in from the Pacific Ocean and put it through a process to take out the salt and to irrigate these wonderful deserts. He could just talk so much about his different projects. And he gave wonderful parties, I'll tell you.

Sometimes, the men would talk about their experiences in Germany. What it amounted to in most cases was some parent would come in and maybe he would want to pay the school tuition. We would get to just talking, and one thing would lead to another. I hesitate to say this, but there were times when those men got to telling about their stressful experiences, and they would actually cry. And I would cry with them. It was a terrible thing for them. So many of them said that they never dreamed, they never had any idea, of what was going on in their country.

And they didn't want to be called Nazis; they were not Nazis. That was their attitude. And that is the essence of it all, I guess.

I didn't have anything to do with their setting up houses. They were prepared by the Army, which meant they probably partitioned them just like we partitioned the schoolrooms, and the men and their families furnished them themselves. The wives were shy, so shy at first that it took me several months to get to where one would stand still long enough to try to chat with me. But, gradually, I became acquainted with some of them and they had mustered up enough English to carry on some kind of conversation.

Most of them were not only shy, but you could tell they were very lonely and lost. They really felt that much more than the men did because the men could go to work and forget about it all, and the women were there in the house most of the day. Some of them were aggressive enough to eventually get out and do a little shopping. I explained to some of them how they might go about establishing a bank account with wife and husband checking, and things like that. I told them where a good place to buy clothing would be, and where was a good doctor and things that any housewife would want to know.

They began to be socially active, and many people enjoyed entertaining them. The German people contributed a lot to El Paso. Of course, they received a lot. I think El Paso overcame its antipathy and just fell into the swing of things with these German people. Many of them, I'm sure, still are carrying on friendships.

They still had this war as a part of them. I know that Major Hamill said he knew those people well enough to know that when the planes started over, the children crawled under the bed. And maybe the wives did, too, I don't know. But the children had to overcome things like that. They were right next to Biggs Field, where the planes were coming in and out all the time. They had to really overcome a lot, and I still think those teachers we had there deserve a medal for making them feel comfortable in this kind of a setting. Eventually, the families got very interested in concerts and sporting events and theater, and their natural bent to be involved in cultural activities came to the front. I remember being at some concert where at intermission, I spotted some of them talking to the town people, and that was the nice way to make friends, for them and for the town people, too.

Many of them told me that they learned their English more by going to movies than any other way. They enjoyed that way of learning, and a lot of them got to be regular movie hounds. They would see anything that would come along. I think they got to where they enjoyed exploring the country around here by car, or hiking, or picnics, or things of that kind. And I think once they got started at it, it meant a great deal to them. It was a good part of their life.

Most everywhere they turned, all they had to do was ask for help and they got it. The people, I think, were extraordinarily supportive. What's good about that is that, at the time they came in, it was surprising how much dislike there was. Later, even hearing somebody say, "He's from Germany," well, that was it. The war habit of condemning before being tried was true. People were vehement about it, and that was what Major Hamill was afraid of, I'm sure. And he had good reason to be. But it changed so well and was such a success, I'm glad it happened.

To me, being associated with the scientists and wives and children was one of the most challenging, gratifying experiences that anyone could have. I went into it with cold feet; but I really had a chance to be innovative and meet challenges in my own way, thanks to Major Hamill's loose rein. He just wanted me to do the job and depended upon me to do it. I'm sure someone else might have done it a different way or a little better, but I feel that the accomplishment was a smooth integration into the community, into our country, in most respects. If there were any adverse angles to that, I never heard of them. I learned a great deal from those people, and they learned a great deal from us, and I think we were all better for it. It helped in our little corner of the world. I think it helped at least to tame some of the bitterness that came out of the war.

I think those people could have taught us a lot about the work ethic when they described how feverishly they worked in Germany, how much, how hard, how long hours, everything. Everything they had was put into making bigger, more deadly, dangerous things to win the war. They weren't doing it to win the war, they were doing it because it was a job to be done, and over here, they carried through. We had one, we called him "Professor Wagner," a single man, rather aged and with wispy little gray hair, who never seemed to have a shave or a haircut. They would find him still working in his cubicle around four o'clock in the morning, having been there all night. And they would say, "Haven't you been home, haven't you had your dinner, haven't you slept?" And he would say, "No, what time is it?"

Al Metzger was an important part of the Broomstick Scientists. In this excerpt from his oral history interview, he described a rocket exploding while he was in the blockhouse. He was referring to a reinforced-concrete building that served as the control room for launching V-2 and other missiles. Its walls were 10 feet thick, and its pyramidal roof was up to 27 feet thick. That massive structure protected them from the explosion itself, but the burning liquid oxygen and alcohol mixture that poured out over the ground presented a different threat. A small tunnel carried control wiring from the blockhouse to the launch pad. If the burning fuel had flowed through that tunnel into the control room, it would have been disastrous. However, the Paperclip Scientists had recognized that potential threat when they designed the structure, and they built into the tunnel a series of vertical steps that would block the flow. These were the baffles in the trenches that Metzger mentioned.

I came to Whites Sands Missile Range in January 1951, and I stayed here until January '53. When I arrived at White Sands I reported in to headquarters, and I was sent right away to Colonel Collin, who was activating the First Ordnance Guided Missile Support Battalion. He said, "Metzger, we've got thirty-two scientific and professional enlisted people coming in here, and we are going to set up a Detachment Number One. You're going to be the commanding officer. What we've got to do is to get them into the military regime. Our primary responsibility is going be to assume the assembly and the firing of the V-2 missile with what is left after General Electric has finished their regular program. This will be the first military group who has assembled and fired a V-2 missile." That was a pretty big task. But fortunately, I was smart enough as a lieutenant to know that those thirty-two scientific and professional personnel were smarter than I was. So we worked as a team and we gave them a free rein, so to speak. And they got the job done.

That same day when I got in, Colonel Collin took me down to one building. Mr. White, the General Electric contractor that was there, introduced me around through the V-2 shop and he took me over in the other building. I think there were a few scientific and professional people who came in at that time, and I was introduced to them. We got together and started the work. We went out and looked at all of the piles and piles of parts out in the storage building where the V-2s were. To some of us, it looked like a bunch of junk. But it was a lot of valuable equipment there. Later on, after it was gone through and evaluated by our personnel, they found useable things. Of course, there had to be a lot of interchanging to come up with useable items. As an officer, I was not a direct supervisor of each individual, but I was more like a person they could come to with a problem or they had a coordinating problem or they needed something from logistics. I was kind of like a linchpin or a coordinator. I could tell what to do and when we needed it, but, boy, I sure wasn't about to tell them how to do it.

We had one not-so-amusing incident [in June 1951], when the V-2 we launched lifted a few inches and fell back down and exploded. That is the weekend that we spent thirty-two hours in the blockhouse, I believe, on end. And it ended in that catastrophe. It was a very, very earth-shaking

A V-2 missile is obscured in the metal preparation gantry in the upper right, while the blockhouse is seen in the lower left. Courtesy of NMMSH.

experience. But we knew what the situation was, and we knew what the safety was. We knew the baffles in the trenches protected us. So even though it was an experience in something I had already been through during World War II, to me it was just another explosion. But it was a very costly one, of course, not welcomed by anybody. It was scary, but we survived.

The best experience we had is the TF-1 [Training Flight Number One, the first modified V-2 built and launched by the Broomstick Scientists], which established the high-altitude record of 132.6 miles. We held that until the next day, when the Navy fired another little Aerobee that went up 133 miles. That broke that record, but that was an interesting experience.

I did go out to the TF-1 impact site. I dug for hours, it seemed like, down in the sand until I got the ojive [curved nose of the missile] off of the TF-1. I guess it was 12 or 15, maybe 18 feet down, something like that. We had a pretty good hole until we found that ojive. We were interested in what it would look like.

We worked on five V-2 missiles. That was all we could assemble. There weren't enough parts for any more.

Whenever we ran a test for the Army Ballistic Agency for Dr. von Braun's team, they used to come out to observe. They mounted a V-2 motor on a dolly and engineered it to fasten on to the test stand in order to run a series of static firings which tested the vanes. The graphite vanes which we tested on the static firings were later to become the forerunners for the vanes that were used on the Redstone missile. They also were the precursor on the Jupiter missile. I remember that Dr. von Braun himself brought some of his people out. They wanted to see the erosion of the graphite and observed the actual firing. Of course, our V-2 group was instrumental in conducting those particular firings.

After I left the White Sands Proving Ground and went to Korea, I went back to Huntsville, Alabama, in June of '54. Dr. von Braun was my neighbor, just two houses up from where I lived in Huntsville. I remember when I got up in the middle of the night, as sometimes you have to do, I would look out the window, and I would see Dr. von Braun. And there was a gentleman by the name of Hermann Oberth, who was a rocket scientist over in Germany. I would see Dr. von Braun and Hermann Oberth walking up and down McClung Street alongside my house. They had their arms, of course, like Germans do. They walked with their arms behind them, and they were just talking up a storm. I am sure that many of the decisions that were made for the space program that we see today were made whenever Professor von Braun and Hermann Oberth walked up and down the street. They were such an amazing two people, and me

V-2 missiles left large impact craters at the end of their test flights. Courtesy of NMMSH.

being just a young captain at that time. I watched out the window, and I would watch them, up and down for hours at a time, they would walk and talk and talk. Sometimes, even when I would get up at dawn to get ready to go to the [Redstone] Arsenal to go to work, I would look out the window and they would still be walking up and down the street at dawn. An amazing experience.

John Schoneman was one of the Broomstick Scientists. After graduating from the Illinois Institute of Technology in 1950, he worked for an insurance company that had helped fund his college tuition. During that time, he trained members of volunteer fire departments throughout northern Illinois. Then he was drafted into the Army and, because of his technical education, he was assigned to White Sands Proving Ground.

I ended up with a degree in Fire Protection and Safety Engineering, which is about 50 percent mechanical and about 50 percent chemical, which means you don't know much about either. I graduated in 1950, and I went to work for an insurance company, training and grading volunteer fire departments. I was drafted on November 16 and went to Fort Leonard Wood, Missouri, for my basic training. We were there three weeks, waiting assignment. Finally, three of us were called out one morning. "Report to White Sands Proving Ground, assignment unknown." Best call-out order I ever received in terms of my life.

So the three of us got on the train. [Along the way], I conned the conductor into giving each of us a roomette. So we came out, first class, to Oro Grande, New Mexico. It went downhill from there for a little while. For example, early in the morning, cranking up the telephone and having a vehicle come out and pick us up to someplace. I was born and raised in Chicago, and I had never seen a mountain before in my life. When the sun came up, it was really an experience.

I had arrived in early January 1951, when the Broomstick group was first forming up. It was the V-2 unit. Thirty-two graduate engineers. Lieutenant Metzger, Bob Alley, and I were sent over to work with GE to see what the process was in putting the V-2 together. Then we would come back and talk about this. At this point, this was a strictly an American operation. The German von Braun and others had gone.

In the basic V-2 system, the warhead was at the top of the missile, obviously. Then there was a section with the controls in it. There was a time switch and all of the other relays that controlled the fins and when various things would happen. The main fuselage had the alcohol and liquid oxygen tanks in there. This thrust ring was at the bottom, and the entire propulsion unit was held in that thrust ring. The steam system consisted of a hydrogen peroxide tank, which was quite large, looked like a hen's egg, and a permanganate tank, which was rather small. Those two, mixed together formed steam, which was fed to a turbine.

As the missile was fired, in the countdown you get down to "X minus zero," and you get a preliminary stage where the liquid oxygen and alcohol flowed by gravity. A pinwheel igniter was in there, which hopefully ignited that. If the missile commander thought that the shape of the flame was proper and correct, he called for main stage, at which point the permanganate and peroxide tanks were opened. They were mixed to form steam that was fed into a manifold on the turbine, which spun that. Two large valves were opened by pneumatic valves, and these two large valves brought the full supply of oxygen and alcohol in there. Hopefully, the missile took off at that point.

The first firing we saw was GE's Number 55. I was in the blockhouse along with Bob Alley and the rest of us. I think the missile lifted off a few feet, then it blew up. Apparently, there was some fault in the safety system on board. It crashed to the ground. Everybody was ordered down on the floor. That blockhouse really shook. I was one of the first ones up. You looked out and all you saw was this field of burning debris. What is now a rather archaic control panel that was there, and there were a lot of lights going off and on. But now it was all blank except for one sign in the center that said "Ah shit!" We wired one of those into most of the things. In fact, I have one of those lights on my model train set today. It lights up.

Then we were ordered to build our first V-2 ourselves, which was an experience, because the tanks were out in the field in the back. There were several rooms filled with boxes of stuff. We would go out and pick the tanks that were necessary. They were hauled up into a test stand, and they were air tested. We would bring in the steam propulsion system and a turbine. The turbines had to be torn down completely. I believe there were aluminum spring seals inside which had corroded a little bit. So we had to polish those by hand, put them in, put the turbine back together, air test it. Obviously, it leaked. We had to take it apart again, re-polish the seals or replace them until we got a turbine that was airtight.

The air bottles were hydraulic tested. The propulsion manganate and the other tanks were air tested. The units were brought together out on

GE launch Number 55 exploded powerfully. Courtesy of White Sands Missile Range Museum.

the floor, assembled into the tail section of the V-2 on which the tank stood. And then we went on from there.

Each rocket had a purpose, and each one of them we had to put together by hand. I learned more at White Sands Proving Ground than in four years of engineering school. My ability to work with my hands, to sort of grunt and come up with a solution to something we needed to do. And as Lieutenant Metzger said, he let us alone and we worked things out. Let's make something that will fit this. Let's try to do this. Let's change this. Let's change that. It was good work.

In August, September, and October 1952, the V-2 operation was being dismantled. Boxes and boxes of stuff were now out in the hangar, wherever old rocket parts go as their final resting place. There were a lot of relays that followed me out through the security gate, and a couple of timing switches. We went out to the TF-1 site and dug it up. I got the two pneumatic main stage valves off of those, which I later had chrome plated and made into bookends. I still have them holding up my books. I am very proud of those.

One V-2 was fired at midnight. I think it was the last firing. This was in August of '52. It was the first time I was able to get out of the blockhouse. I went to a deserted camera station, clear sky, and saw this rocket sitting on this stand with the liquid oxygen venting off. It took off into the sky and it became just a pencil of light. That one was scheduled to go about 90 miles, it made it to about 50 miles and telemetry lost [it]. So I don't know if there was any success from that.

One of the firings, there was a group. I wasn't included in the group, but they installed a goldfish in one of the empty antenna pods on the fin of the V-2. Bob Alley was upset, because I think he felt he could get a couple more miles if he didn't have that goldfish weighing down the missile. Each firing was a little different, different time of day, different circumstances, different VIPs visiting.

The air test samples that the University of Michigan ran were one of the most interesting ones, I believe. Two of the bottles came down successfully. One landed and ruptured. We never saw the results of what we did, in terms of the data that was gathered or anything. The work on the static firing had something to do with the Redstone missile. I kept telling my kids about this after the Redstone put up the first satellite, "Hey, I worked on the guidance system."

My wife, Mary, was able to join me. After I was here about two months, I went to the officer judge and said, "Hey, I am married, how about...?" He said, "Fine, go get her." So I did. I had this '39 Buick that my mother said had not been out of the garage. On the way out here, we broke two connecting rods and spent most of our worldly assets to get out here. We lived in Las Cruces. It was a fine community. We had friends with a lot of others in town, visited the restaurants.

There was a Frontier Club here back in those days. I think it was in Mesilla. On the back of the ladies room door, there was a nude superman type thing with a fig leaf, in an appropriate spot, you know, on hinges. And if any of the ladies lifted up that hinge, a bell rang in the bar. In order to get to the ladies room you had to go through the bar. It was a big joke to all the men in the bar. But we had a new couple that had come down here to join the group. We went out there, and she went into the ladies room. She unfortunately lifted the fig leaf, and my wife had to go get her before she would come out.

All of a sudden, the name Broomstick Scientists was there. It was just sort of there, and we used it. I'm surprised we got away with it. It was rather, well it wasn't being military, I guess you might say. But it was morale building. We had excellent morale. The relationship between the regular Army people that were here, I believe, was pretty good. And there were lot more military people here than now. The relationship between General Electric and the V-2 section was excellent. I won a big bet from one guy. He bet me three to one that our TF-1 wouldn't clear the fence. We had all of ten dollars on that particular bet, so I made out pretty well. Those relationships were quite good, and cooperative.

I had an interesting experience during the 2001 Broomsticks reunion. Someone came up to me with a picture of TF-1. He said, "You're fueling?" It was early in the morning and we were fueling. I said yes. "There are two guys on this platform." And I said "Yeah, this where we fuel the liquid oxygen. That is Hashimoto and that is John Schoneman." He said, "You are hooked up to a liquid oxygen truck." And I said, "Yes, we were refueling. You can see the fumes coming out as we are fueling. He said, "Now, if you look up on the gantry, there are three more guys up on different gantry platforms." And I said, "Yeah, they were working on the guidance system and things like that." He leaned over and said, "You guys were crazy to do things like that, you know, fueling with people up on different platforms."

Once, we were doing an early morning shot. There were some very important people here. We reported out about 4 a.m. and fueled the alcohol, which was on the base, fueled the permanganate and the hydrogen peroxide, and then stood around, and stood around. Somebody had forgotten to order the liquid oxygen. It was decided they would have to get it from El Paso. So we were going to have to de-fuel. We left the alcohol in, but we drained the permanganate out and we drained the peroxide out. While Hashi and I were draining the peroxide out, this can of permanganate that was sitting there in the sun overheated and the cap blew off. It blew up in his face, stained his whole face purple and into his eyes. The ambulance came and took him off. He went to the hospital. He was so close to discharge that they let him get discharge early. But he was three days working on his eyes from what had splashed into them. Then we were told to come back and to refuel and to be ready for firing at one o'clock. I was not too pleased with that, as were many. I suggested that maybe I was working under protest, and some very nice officer whispered in my ear, "I wouldn't do that if I were you." So we backed off of that. We refueled and the firing was a success.

I guess we were just like working on the steam automobile compared to some of the things we hear about now.

I was discharged in November of 1952. I had two more years on my work study scholarship as an indentured servant to work off, going back to working in volunteer fire departments. Then I was put behind the desk in the property, casualty, and life insurance business. That industry was very kind to me over the years, to the point that I was able to become the chief executive officer of an insurance company. Our company was one of the first companies to write satellite launch insurance in its primary period. At that time, they were being written in two parts, these commercial satellites. You wrote launch insurance, and then you wrote operational insurance. That earned me a number of trips to Cape Kennedy or Cape Canaveral, or whatever it was at that time, to see these firings.

And with trembling hand, I signed a million dollar life insurance policy for John Young and Bob Crippen, who were on the *Columbia*, the first shuttle launch. I didn't tell my board of directors what I had done until after they had landed. (Laughs). They got back safely, and they sent me a framed patch and signed it. It was hung up in my office. So I stayed in contact. It was just very exciting. This was just at the time they were starting to lease out space on rockets for commercial use. I never insured one that failed, so my record is still good. I've got five V-2 firings, and all the satellites got up in orbit. So, I feel good about that.

In 1950, Wernher von Braun and the other Paperclip Scientists were transferred from Fort Bliss to Redstone Arsenal in Huntsville, Alabama. There, they helped develop the Redstone and Saturn rockets that carried the first Americans into space and the first men to the Moon.

Chapter 4
Balloon Boosters

Helium-filled balloons may not seem high tech enough to be used for spaceflight research, but appearances can be deceiving. In fact, manned and unmanned balloon flights yielded valuable information in the two decades leading up to the first flights of NASA's Mercury program. Some of their most important uses were to evaluate the effects of cosmic radiation on living organisms (ultimately on humans), our ability to build a space capsule with all of the necessary life support and communications technology, and man's capability to endure and function effectively in a cramped vessel at previously unexperienced distances from his home planet.

The excerpts in this chapter illustrate several of the significant high-altitude balloon flights between 1934 and 1960. Unmanned balloons are still being used today to test objects and gather data in sustained excursions to the near-space environment of 120,000 feet (23 miles) above the Earth.

> *Jean and Jeanette Piccard were true pioneers of high-altitude ballooning. They were inducted into the International Space Hall of Fame in 1998. Together, they flew a record-breaking flight in 1934, during which Jeanette became the first woman to fly into the stratosphere. In 1998, their three sons (Don, Paul, and John) gave an oral history interview at the New Mexico Museum of Space History. With the exception of the final paragraph, this excerpt is a blend of the three men's comments.*

My father, I think, remembered practically everything he ever learned in school. So he knew his history that was taught up through the nineteenth century. He knew all his science. He knew crystallography. The kind of things that most of us are exposed to in school to get through the tests, then it drifts away. He was interested in public affairs. He was interested in religion. He attended the symphonies. He had not a musical ear, but that was part of his life. It was a very, very broad, traditional, liberal arts education with emphasis in science and engineering.

Mother started out in the soft sciences but then studied chemistry. She earned her Master's degree in chemistry from the University of Chicago, which is how our parents met. My father had been brought over to be a chemistry professor. They wanted a German, but because of World War I, they settled for a German-trained Swiss. My father was eleven years older than his graduate student. But, as I understand it, it was more my mother's idea of how to resolve this relationship. And they were married in 1919.

My mother had always impressed me with her patriotism, because the reason she shifted from her primary interest was so she could relieve a man for the front [lines in the war]. This was kind of a slogan, that women would go to work so that the men could go to Europe to the front. I always admired her patriotism until one day in 1945, I found myself in Germany at the front as a rifleman. And I thought, "My mother was going to get some guy's cushy job back home so he could get up here?!" I quite changed my view of that.

It is an interesting story on my mother that she would shift to a new area of study and succeed in it. But by the time she got her Master's degree, World War I was over and done with. So married, she went off to Switzerland. It cost her her citizenship to marry a foreigner. Years later when my father was naturalized, the law had changed. Women no longer carried their husband's nationality. So she had to get her own citizenship back. When she was employed by the Space Center in Houston, she had forms to fill out. She delighted in saying that she was a natural born citizen, and they thought the poor old woman didn't quite understand the question.

She did a variety of things. Bob Gilruth, the director of the Manned Spacecraft Center, was one of my father's students at the University of Minnesota. Mr. Gilruth brought my mother to Houston to do some consulting and some public relations. She was often in the offices in

Houston. But she was also often on the road lecturing at different places, high schools and other public places, on the space program. She was very, very enthusiastic about the space program. It was good within itself. And she believed the spin-off arguments that the work done on behalf of the space program spills over for the benefit of other sides of our lives.

I was seven years old, give or take a year, when my father got a letter from his twin brother. Even as identical twins, they were remarkably similar. They thought alike, reacted alike in the same situations. I think about every six months or so, one of them would write the other a letter while they were separated by the ocean, when it took typically a ten day crossing. Anyway, he showed me a drawing in the margin of a partially inflated balloon and explained that for a balloon to go a very high altitude, it would have to be so lightweight that it couldn't be fully inflated at sea level. His twin brother had become a physicist. My dad was a chemist because the employment situation was such that they wouldn't be able to work at the same place if they worked in the same field. It ended up they weren't in the same place anyway.

Uncle Auguste got the personal backing of the king of Belgium. That enabled him to go ahead with his ideas to get measurements of nuclear radiation from outer space and make the measurements from above nine-tenths of the atmosphere. The balloon that he built required a lot of new technology. That balloon was really the bridge between Jules Verne's *Twenty Thousand Leagues Under The Sea* capability of a self-contained capsule for people in a hostile environment, and the modern NASA space capsules – pressurized cabins, self-contained atmosphere, complete atmosphere renewal, carbon dioxide absorption, other body evaporative chemical absorption, oxygen replacement, and all largely automated. Though with people, if they were going to do flying, they didn't really need automation, but controlling the craft from inside the sealed capsule. For a physicist who just wanted some nuclear measurements, it was a bit of a revolution.

Balloonists in the '20s and '30s didn't wear helmets. When Auguste was going to go into the stratosphere, he went into the department store and asked for [wicker] sewing baskets. He found one. Yes, it fits, and it would make a good lunch basket. You are in this spherical capsule. It is a spherical capsule for efficiency and engineering design, but a spherical capsule is not the way to land a balloon, because it is going to roll. So when you come in to land, you take the seat cushion and stuff it in the lunch basket, put it down over your head, and you've got a very serviceable helmet.

From then until the period of the modern airliners, there was incessant demands by my parents and by my uncle that airplanes had no business flying down in the troposphere from city to city, in among the storms and the clouds. They didn't belong there any more than ocean liners belonged in and out on the bays on the seacoast.

The World's Fair in Chicago wanted a real publicity getter for their second summer [1934]. And [who better than] the long-haired, strange, Swiss Professor [Auguste] Piccard in Europe, who had beaten the world's altitude record, not just by a few hundred feet, but from about [the previous record of] 32,000 feet, he took it up to about 50,000. Anyway, they got together with one of the big Chicago newspapers and they got the backing of the United States Navy. A wonderful contract was drawn up. They would take all of the expenses. They would provide one the world's best aircraft pilots, Professor Piccard would fly, and the World's Fair would get a world's altitude record. The great Goodyear Zeppelin Company made the *Century of Progress* balloon. The Dow Chemical Company, which was into metallurgy with the lightest structural material, would make the gondola.

Professor Piccard wasn't available. He was doing physics at the University of Brussels. He had a twin brother in America whose chemistry job had disappeared during the Depression, when a big chemical company abolished the research department that he was the head of. And brother Jean understood the mechanics of balloon flying. He had done sporting balloon flying in Switzerland. He would be able to serve as a technical consultant to make sure that things would go as well as Auguste. He got the contract.

In payment, he was to get the balloon, the complete craft, completely reconditioned for himself after the flight. Things started to get a bit sticky. The balloon wasn't built according to his directions. He wasn't on the craft when it flew. And the flight was a fiasco. Another flight was made which set a record, but under the rules really wasn't entitled to it. They pitched out all of their scientific apparatus, and all of the results of their flight, and crash-landed out of control in the salt marshes in New Jersey.

Finally, it was my parents' turn. Henry Ford came through with a base of operations outside Detroit. [In 1934, Jean and Jeanette Piccard rode a sealed capsule to an altitude of 57,579 feet, setting a new altitude record.]

By that time, I was old enough to actually be a little bit of help. I prepared some of the ballast. One story would show my father's attitude there. In flight, ballast from high altitude had to be dropped through an airlock, and my father's version of an airlock was a bit simpler than his twin brother's had been. But it still involved, after you seal off the upper chamber with the ballast in it, you open the valve and let the ballast trickle down into the vacuum of inner space, up there 10 or 11 miles high. Anyway, they used fine lead shot – ordinary, industrially prepared lead shot for, I suppose, skeet shooting. We just called it birdshot. So fine that falling by gravity, it wouldn't hurt anybody. I had the job of taking it in the rough shipping bags from the lead shot foundry and weighing it out into a more suitable quantity in bags to be loaded in the gondola, the altitude capsule.

I was part way into it, refilling these tailor-made canvas bags, and I said, "Oops." These bags looked nice and neat. The seamstress sewed

Jeanette Piccard and her husband Jean in the capsule they piloted to an altitude of 10.9 miles in 1934. Jeanette was the first licensed female balloon pilot in the United States and the first woman to fly to the stratosphere. Courtesy of NMMSH.

them up with one good seam and turned them inside out from the way they were sewed. So the outside looked neat, and there were all of these raveled edges on the inside. Bits of yarn could come loose and run along with the lead shot when those bags were emptied into the airlock. They might jam the lower valve. I went to my father and he said, "Yep, that's right. Turn the bags back the way they were made. The small number that I had already loaded, more likely we will use those bags last. We will save them for when we are at a low altitude." That was his attitude on technical questions for the entire venture.

On the flight that my mother and father made out of Detroit, one of the problems was to release the balloon. When it had been checked for balance between lift and ballast, my father's idea was, instead of a whole bunch of ground crew standing around close, holding it – it was a heavy balloon for that – have a very heavy rope in each of the four cardinal directions. Then cut those four ropes at the moment the pilot decides to take off. So he took a primer cord. It looked like an ordinary fuse or a piece of clothesline, but it was TNT inside of a black tube. He spliced a length of that into the rope, put an ordinary electric blasting cap on the end of it, tied a bunch of sand bags around it, and tied a bunch of excelsior wood shavings around that. So if somebody was leaning on it they wouldn't get hurt when the rope got cut. It made a sizeable little package out in the middle of the rope. The pilot in the balloon had a piece of wire and a neat little box with four screws sticking up. When the wire from the battery touched one screw, one little bit of primer cord went out of existence. So did a little bit of rope around it, and that rope was neatly cut.

At landing, there was a similar problem. For an ordinary gas-filled free balloon coming down, the pilot judges the air currents and the rate of descent, and, *woops*, there is a church steeple in the way. The parson doesn't like it if you land on – we won't like it if we land on – the church steeple. So, throw out about twenty or thirty pounds of sand. But here is a tremendous balloon going from 10 miles high, a 7-foot sphere, metal gondola with two people in it, a lot of scientific apparatus, a Geiger counter, pilot chamber, air samplers, a lot of ballast left in case this happens. Maybe it's a high-tension tower. You've got to start back up and quit going down, just toss out a few hundred pounds. Several big bags of sand were hung on the outside of the gondola. This was my father's idea. They were made like sleeves. The bottom was pulled up on the inside of the sand, so it was nicely cushioned. The bottom was tied together with a nice little cord with a little electric blasting cap in the cord. All the pilot had to do was to take the wire from the battery and touch the contacts, *bang, bang, bang*, 300 pounds of sand dropped out of the bottoms of three ballast bags.

If he had a problem, he figured out a way to solve it. He was what I call a string and sealing wax mechanic. Later, when I was about eleven or twelve years old, he brought a problem home to dinner when he was working for a chemical company in the explosives industry. They had a chemical that should have been a good product for the company, but the company engineers couldn't figure a way to load it into blasting caps. And Papa made a gadget that he could operate from any length of rope. He made this gadget on a laboratory bench top and demonstrated perfectly safe loading, operated it again and again, never an explosion.

The liquid oxygen converter. Walk in the airport, and here's a man with this breathing apparatus. He's able to go around because he has liquid oxygen, is a result of Jeannette's observations on their stratosphere flight. When they came down, as they were approaching landing or just after landing, mother turned to my father and said "Jean, we are out of oxygen!" The liquid oxygen vessel, the dura flask [vacuum-insulated liquid gas cylinder], is no longer boiling. She was a laboratory technician, trained at the University of Chicago under Professor Piccard. And she thought there is something wrong with the equipment. "We are out of oxygen and we shouldn't be! It is impossible. We had enough. There is no way that we could have used it." And they checked it. There is no oxygen coming out. It was strange. They went about other details. Then several minutes later, Jeanette said, "Oh, we have oxygen." By that observation, they discovered the simple phenomenon that in a dura flask, at a given pressure, the temperature of the liquid oxygen reaches an equilibrium with boiling. If the temperature is high, it boils very rapidly, and the evaporative cooling of liquid oxygen cools it down, and it reaches its equilibrium temperature for boiling at that pressure. When they opened the cabin at 15,000 feet, immediately the pressure in the cabin and the oxygen came to 15,000-foot altitude. The balloon came down rapidly and they landed. It is now at sea level, and the higher pressure means a higher temperature of boiling. But the temperature inertia, since that oxygen in the dura flask is chilled rapidly at 15,000 feet, it stops boiling. There is plenty of liquid oxygen in there, but they can't get it.

That is an interesting phenomenon. What difference does it make?

This is 1934. There is going to be a war. We are going to have airplanes flying at high altitude. There are men who are going to be breathing liquid oxygen. If they are flying at 30,000 feet, then they come down to 20,000 feet, and now the liquid oxygen doesn't boil anymore, they are going to succumb. So as a result of Jeannette's observation, and their discussion of it, Jean then invented the liquid oxygen converter, which is used for the walk-around bottles on the bombers.

The standard balloon, being a rubberized fabric bag held in a net, is the way Professor [Jacques] Charles made the first gas balloon in 1783. We now have additional styles of balloons, but up into the '30s, that was it. It took spheres because spheres were the most economical volume to surface area. But the net hangs like a shopping bag. My father looked at that, and he said, "Why have this redundancy? Why have the skin and the ropes? Why not work them together?" So he made a natural-shaped balloon, and that is how we have balloons today. And then he got the materials for that, to make the plastic. [Jean Piccard flew the first successful plastic balloon, made of cellophane, in 1936.]

So, a little going against the grain [norm] to develop this plastic balloon which enabled the balloons to go much higher, to get into more important areas of the atmosphere for the so-called pre-astronaut work where Joe Kittinger jumps out at 100,000 feet. You couldn't get there in a truck and rubber balloon in a meadow. It would be just too much. The ability to go against the grain is the beautiful part of Jean, and the teamwork of Jeanette controlling and working with him.

[In the late 1940s, Jean Piccard proposed a manned balloon project that would carry a sealed, spherical capsule to a height of 100,000 feet. The Office of Naval Research undertook the project, which was called Helios. Originally conceived as using a cluster of cellophane balloons, the project evolved into one using a single, large polyethylene balloon.] When the time came after World War II, when they started the stratosphere flight for the high altitude, the early pre-astronaut flights, the first plastic balloon flights, this is why it went to General Mills, a breakfast cereal company. The University of Minnesota could not contract for this project Helios because the University of Minnesota, at that time, had a nepotism law. If they had the contract, Jean and Jeannette could not both be on the staff. The University of Minnesota could not contract it. It went to General Mills. General Mills took the contract with the Navy doing that plastic balloon. They had to be together. You had to have Jeannette with it. They were a team, and they did great.

[Don Piccard, one of Jean and Jeanette's sons, also became a renowned balloonist. He recalled that] after Jean died in 1963, Jeannette went on with NASA. People asked Captain Benson, Navy flight surgeon with NASA, "What is space?" He said, "Physiological space is around 45,000 to 50,000 feet." I said, "Captain Benson, would you say my mother was the first woman in space?" He said, "Yeah, yeah, sure, that is great. That is good idea." And when mother went to work for NASA, then they did say for a while that she was the first woman in space. But at a Federation Aeronautique Internationale conference in Mexico City, I had the occasion to meet Valentina Tereshkova [a Russian cosmonaut who became the first woman to fly a spaceship, Vostok 6, in 1963]. Her interpreter was Khrushchev's personal interpreter. She had on loan the best interpreter, there is no question. I was able to say, "Comrade Tereshkova, when I told my mother I was perhaps to meet you, she asked me to greet you." She said, "Thank you, very much." And I said, "Perhaps you don't know who my mother is. My mother is Jeannette Piccard, who piloted a balloon to 57,000 feet in 1934, more than two miles into physiological space. And she wanted me to congratulate you on your marvelous achievement, and on behalf of all the women in America to welcome you to space." This was translated to her. She came back and said, "I know very well who your mother is. I am most appreciative of her good wishes, and please give her all my love." While she was saying in Russian "all my love" before the interpreter could translate it I knew what she was saying, and I leaned forward and gave her a big kiss. So I am one of the few people that kissed both space women. And I did get Sally Ride to give me a little peck one time.

Charles (Charlie) Moore was a physicist and meteorologist who was a pioneer in the field of atmospheric explorations using helium balloons. He participated in Project Mogul (see below), and in 1959, he and Malcolm Ross flew Strato-Lab High IV to an altitude of 81,000 feet in a 28-hour mission. Strato-Lab was a Navy high-altitude balloon program that carried a two-man crew in a sealed capsule to make observations and perform experiments in the stratosphere.

I was a student at Georgia Tech when World War II came on. I left Tech and joined the Army Air Force. I went through one of the meteorological cadet programs. On completing that, I was commissioned and later went to a weather equipment technician school. At the end of the war, I returned to the States, was discharged and returned to Georgia Tech.

The pressure to monitor the Soviet nuclear program became very high in 1946. Previously, the Signal Corps had the requirement to develop a constant-level balloon and had not been successful in doing it. The Army Air Force took over the program in September of '46. I was offered a job when I would complete my degree at the end of the fall semester in '46. I was offered a graduate assistantship at NYU, New York University, in October of '46. On going back to Tech, I began exploring the ways to solve the problem. On talking to one of my chemical engineering professors about the films that were available, I picked polyethylene and reported back to Dr. [Athelstan] Spilhaus that that's the way it ought to be done. So in November of '46, Spilhaus ordered the first polyethylene film to make the balloons.

I then enrolled in New York University's Physics Department. While there, my graduate assistantship was with a program for the Watson Laboratories of the Air Force, having to do with long-range detection of nuclear devices. While I was a weather equipment officer, I'd been involved in balloon work in the Army, and that caused Colonel [Marcellus] Duffy and Spilhaus, the director of research, to offer me the assistantship to develop constant-level balloons for the long-range detection program. I did this while I was a graduate student at NYU.

The pressures for what later became known as Project Mogul were so great that I spent large amounts of time on it when I should have been in school. Balloons were going to be flown over the United States in the hope that we could pick up the sound wave [from an atomic bomb explosion] from Russia. When the Sandstone Operation was carried out in the Pacific in the spring of 1948, I was out from New York University in Alamogordo for several months, flying balloons so that the low-frequency microphones that we were carrying on the balloons could be aloft when the pressure wave from the Sandstone nuclear test passed over New Mexico. As a result, I completed two years of graduate school, but I did not complete my degree.

When we got the job to develop the balloons, we recognized that meteorological balloons that were extensible [capable of being extended] had no stability. There was no way to keep them at constant levels. If you would get a balloon to level off so that it had zero net lift, the sun would heat the balloon up so that it would rise. If it then got rid of some gas so that it leveled off, and cooled just a little bit, it would drop like a stone. So meteorological balloons were not the way to go. We very quickly recognized that one needed a constant-volume balloon. There was a precedent in what we were doing, in that during World War II, the Japanese prepared something like 10,000 balloons. Many of them were made of a mulberry paper, others were made of a silk fabric. They launched balloons from Hokkaido aimed at striking back at the United States. This effort got well under way after the Doolittle raid on Tokyo in '42. By the fall and winter of '44, '45 the Japanese flew something like 9,000 of them, hundreds of which actually made it all the way across the Pacific and landed here. Their payloads primarily were incendiaries and explosive bombs. The problem with their balloons is that they were open and didn't maintain constant level. We recognized very early, in 1946, that we ought to have transparent balloons so that we minimize the absorption of light and sunlight. We quickly selected polyethylene, which was a relatively new film then. The NYU balloon program instigated the use of polyethylene balloons, which is still being used in the large balloons being flown.

At the same time, there was a big program that the Navy was sponsoring for a balloonist named Jean Piccard, who was at the University of Minnesota. Piccard had the idea of making a manned flight with his wife to high altitude. Before the war, he tried to make balloons out of cellophane. Of course, during the war there was no time for such activities, but after the war he was very interested in making balloons out of a film. He picked a film made by, I believe, Goodyear. It was called pliofilm. It was a vinyl chloride film. It was very easy to fabricate. A very nice film, somewhat similar to the modern Saran. The problem with pliofilm is it degraded very rapidly in sunlight. Piccard proposed to get away from that by instead of having a single, he was going to end up with a cluster of balloons. In fact, he proposed to have a hundred balloons, each 72 feet in diameter, flown in the cluster, the equators of the balloons jostling each other. That way, if a single one burst, he still had quite a number there, so he can return to Earth safely. The design of the balloon that Piccard had was very good. The film was not.

In about February of '47, we approached General Mills to make balloons out of polyethylene and were rebuffed for two reasons. One, they were working on the Helios program and they had no time to take outside problems. We, of course, weren't in position to describe the classified, high-priority program we had. So we ended up having a small manufacturer in Merrimack, New York, an entrepreneur named A.J. Smith, make the first polyethylene balloons that were made. The interest from the Air Force was so great to get balloons in the air, that we began flying meteorological balloons just as a test bed to carry low-frequency microphones aloft.

Project Helios, which was the program that Piccard was the scientific advisor and the stimulus for, was cancelled. It was an idea that was far ahead of its time. I guess that there was an interesting association into the war with various people who had agendas and desires that had not been fulfilled during the war. There was a German expatriate named Otto Winzen, and he and his wife Vera apparently came to this country before the war. They were not Jewish, and he was credited with making many pro-Nazi statements in the 1939 period, and later he could not get a security clearance during the war.

Otto was one wonderful promoter, and some way or another he came across Piccard. He got Piccard's idea and talked to the Navy Special Devices Center, who was interested. A lieutenant commander there, named George Washington Hoover, was very interested in physiological problems that Navy pilots would experience if they went beyond the speed of sound, which at that time was a major barrier. Hoover was very interested in instrumenting a missile that would travel at twice the speed of sound. The idea was to drop a missile from high altitude and let it accelerate downward, reaching about Mach 2 until it got below the troposphere, where it will start slowing down again. General Mills, during the war, had used its mechanical division that made cereal packaging machinery, and had really expanded, and believe it or not made Norden bombsights, and made prisms for submarine telescopes, high-quality optical equipment. At the end of the war, there was a desire to diversify, to cash in on the clear technological expansion that would occur, and Winzen talked the vice president of General Mills, a fellow by the name of Arthur Hyde, into setting up an aeronautical research laboratory to build Piccard's balloons to carry George Washington Hoover's missile and drop it from high altitude.

The technology at that time just wasn't adequate. The pliofilm material degraded in the sunlight. It wasn't until the program was just about to cancel that General Mills reluctantly adopted polyethylene.

The first inflation [of a single pliofilm balloon] was made at Ottumwa, Iowa, in the spring of '47. The people just put gas in the balloon, and the balloon had maybe 2 percent of its volume filled, but there was a light wind. It was tied down with a single 500-pound-test parachute shroud line, and that got pulled through peoples' hands and ripped off skin and broke. And the balloon got away. So that was the inflation attempt. Effectively, the Navy recognized that this wasn't something that was readily doable. However, all the cosmic ray scientists, all the upper atmosphere scientists, everyone who had an experiment was attracted by the Navy announcement of opportunity to carry instruments to high altitude [and still wanted to fly them]. So a physicist in the office of Navy research had the happy idea, instead of a hundred experiments under a hundred balloons, why not one experiment under one balloon. So the program got converted from Project Helios to Project Skyhook.

The first Skyhook balloon was flown from St. Cloud, Minnesota, in September of '47. Another flight was made on January 6 of 1948. It's probably the unidentified object that caused the National Guard pilot, Thomas Mantell [to crash]. It was far above him, and apparently he was chasing it in a P-51 that had not been serviced for oxygen. He got above the level of his capability and blacked out and crashed. From my guesses, the balloon was above 55,000 feet, and he had no chance of getting it. Later on the afternoon of January 7, an astronomer in Nashville used a telescope and looked at the object that had passed over Nashville. His report on it is a very good description of the Skyhook balloon. We've had later flights out of Minnesota that would fly right over the Kentucky, Tennessee area, and felt reasonably confident that that Skyhook balloon went that direction. The program at the time was classified by the Navy as confidential, mainly just for control of press releases. The Navy wasn't at all happy about being charged with having caused Mantell's death, but I think the Navy and Otto were really innocent, since this was an accident.

Charles Moore, right, flew the first manned polyethylene balloon in 1949. Courtesy of NMMSH.

Clyde Wilson was the son of a Mississippi cotton sharecropper. After retiring from the Army in 1968, he went to work as a civilian helicopter pilot at White Sands Missile Range and was assigned to Holloman Air Force Base. Both of those facilities played important roles in high-altitude explorations, both manned and unmanned.

One time, they had a balloon go astray, one of the test ones, and it was going to Mexico. They couldn't control it. They couldn't get it down. They thought we could go up there and hook on to it and pull it back. Well, we said, "No. We can't do that." But this guy that went with me in the H-34 [helicopter] that day was crazier than me, and he said, "I'll shoot it down." He lived in a trailer, and he kept all of his weapons in the trunk of his car. He talked the Air Force Test Group into let us go down there and try to shoot it down. They didn't want it to cross the border, and it was close to the border. So he got his big old 12-gauge shotgun and took out a bunch of shells. We flew down there, and we found it. There was C-130 chasing it. A C-130 can't do anything about it. So when we got there, the C-130 had to leave and go to El Paso. We got there and we shot that thing down. I was flying, and he was sitting in the left seat with the window open. I would make an approach on the balloon, come on down, low and slow, and I would get down to it. He would fire it full of shotgun shells. At first, you could see that they were going through. But after it had lost some air, it quit going down. Every time, we would get more air out. We finally got it on the ground. We landed there and let all of our people go out and chase it and hold on to it. The trucks had been following it too. You know, the FAA would not have approved of that.

Incidentally, the way that terminated, we got it on the ground. We got the payload off, but the wind was blowing and we couldn't get the balloon itself off of it. It kept dragging, and it dragged across cactus. Once it got back in the air, we knew we had an airborne problem. Yes, we had the payload recovered, but now we have a free flying balloon, nothing holding on to it. So we got up in the air and we couldn't do anything about it. We called somebody to relay to El Paso to tell that C-130 to get back in the air as quick as he could, because he could at least chase it and warn people ahead of it. It was blowing from west to east, and I gave him an estimate of where I thought it was when it got out of sight from us. By the time he got it, it was a tremendous amount of flying. It landed just east of Dallas, free flying all that time. He chased it until it landed.

Dr. John Paul Stapp was chief of the Aeromedical Field Laboratory at Holloman Air Force Base from 1953 until 1958, when he became chief of the Aero Medical Laboratory at Wright Air Development Center in Dayton, Ohio. While at Holloman, he developed and oversaw the Manhigh high-altitude balloon program.

The Manhigh program was the first field testing of a closed survival and life support system for use in spaceflight. That's where we anticipated it would be used, and we took all systems that it takes to keep a man alive. Some of them were very crude, and most of them were under manual, not automatic, control. So when our first basic explorer of threshold in space went up [on its second flight] from Crosby Iron [open pit iron mine in Crosby, Minnesota] and got up to 101,515 feet, Dave Simons did it all in a 32-hour and 10 minute test. We thought it would be warranted for the systems required for survival in a Mercury project, for instance. So, he did all the precursory experimentation for NASA on that one. They were able to take our findings and improve on them, vastly in many cases, and build a Mercury capsule. They took our impact data and built a Mercury couch inside the capsule for takeoff, landing, and other configurations of spaceflight. So, more or less, we did bootleg research on space before space was authorized as an experimental area.

Dr. David Simons was project officer for the Manhigh project, for which Dr. Stapp was principal investigator. The project's objective was to design and build a capsule that a man could survive in and work in for at least twenty-four hours in a near-space environment. Hovering at an altitude of about 100,000 feet for many hours required using a helium-filled balloon. In June 1957, Air Force test pilot Joe Kittinger piloted the first flight to an altitude of nearly 97,000 feet. During that flight, Simons and Stapp were able to talk to Kittinger, but an equipment failure left Kittinger able to respond only by Morse code. Another equipment malfunction reduced the capsule's oxygen supply to a point that Kittinger had barely enough left to survive as rapid a descent as he could manage. Stapp ordered Kittinger to begin valving helium immediately, not realizing that Kittinger was already doing that. Employing his characteristic sense of humor, Kittinger responded by tapping Morse code for "Come and get me." Simons and Stapp weren't sure whether he was joking or whether he was suffering from a "breakaway phenomenon" that was compelling him to want to stay in space indefinitely.

In August 1957, Simons piloted the second Manhigh flight to 101,500 feet.

After two and a half years in the Far East as a flight surgeon during the Korean War, I was reassigned back to Wright Field, only this time to transition to Holloman as chief of the Space Biology Laboratory, to explore the problems of cosmic radiation exposure to animals at high altitudes, using balloons. Dr. Herman Schaeffer, a physicist at Pensacola in the Navy, had done a number of calculations in terms of exposure to cosmic rays in space that would indicate that there were very significant biological hazards. The concern was heavy atomic nuclei, like carbon and nitrogen, impacting the tissues at speeds approaching that of light. There was enough question about this that somebody had to do some experiments to find out whether it was that bad, not that bad, or worse.

Exposing animals to these particles at high altitude was the only way known to do it. There were no accelerators that could accelerate atoms of that size. The cosmic rays are deflected by the magnetic field that produces the Van Allen Belts at the equator. This meant that experimental flights to really expose the animals to cosmic rays had to be conducted at the northernmost latitudes of the United States, from Minnesota or even farther north into Canada. Fortunately, the balloon flying center of the country then was Minneapolis, [with] General Mills and Winzen Research, and later Raven. The people there were quite accustomed to flying balloons, and so we were able to do our flights with contractors there.

Over a period of four or five years, we gradually learned how to build life support systems that would maintain 200 mouse units' worth of animals in a capsule for thirty-six hours at altitudes up around 130,000 feet. We used mouse units because sometimes we flew mice, sometimes we flew guinea pigs, sometimes we flew rats, and even occasionally monkeys. Life support systems are quite critical in terms of being sure you have provided enough oxygen, and carbon dioxide absorbing capability, and temperature control to cover the load that you're carrying. I remember that a guinea pig was 2 mouse units, and a monkey nearly 7 mouse units. This way, we could add it up and know we had not exceeded our 200 mouse units for the capsule and that it would do well. After about four or five years of this, we'd done enough experiments that it was clear there could be some effects

David Simons, left, and an unidentified man stand beside the gondola of a research balloon made by Winzen Research. Courtesy of NMMSH.

detectable, but they were certainly not as bad as was first prognosticated, and very clearly not worse than was theoretically envisioned.

So, the problem looked like it was one of those things you ought to know about and not forget, but wasn't really a major concern. Except for unusual and violent solar flares, it would not be a problem in near space for things like Moon shots. Now, if you're going to go to Mars and beyond, with transit times of two years or so and round-trip times of four years or more, it involves a much different order of exposure, which may require some reconsideration.

One day, Colonel Stapp, who was my chief there at the [Holloman] Aeromed Lab at that point, said, "Dave, do you think you are ready yet to do a manned flight to check out whether this really is a problem that needs any more study with animals?" I said, "I don't know." I hadn't thought much about it, but it's true, our data at this point looked like it shouldn't be dangerous. He said, "Well, would you be willing to go?" I said, "Basically, with what I know, I surely have no compunctions. I think that it would be very safe from a radiation point of view, and it could very well give us a lot of information we can't get from animals. You can make observations that the animals can't tell you about."

One of the things that we surmised might be observable could be the effect of a heavy-particle cosmic ray penetrating the retina. You should see a flash of light. You would be able to report that, but the animals wouldn't tell you.

So he said, "Well, how much of a job would it be? How much do we have to increase the capsule size?" I said, "Well, let me do a few calculations here and I'll tell you." So, I checked it out and it turned out that a man was 500 mouse units. I said, "We're at 200. That's only two and a half times the size of what we're doing now. That's no big deal. That would be easy enough to do, and we still would know what we're doing." He said, "You now use these 3-foot-diameter, spherical, pressurized containers. How about putting a cylinder between the two hemispheres and making a capsule out of it? You could almost use the same 3-foot size you're using now. It would be about the size of a telephone booth." I said, "You know, that's not far off. That would just about do it." We would have to go to 100,000 feet or so, because we don't get enough exposure at lower altitudes. The air absorbs these heavy primaries and so you have to be not only clear of the Van Allen Belt deflection, but you also have to be clear of the atmosphere to get the exposure. And so, Manhigh was born.

Colonel Stapp was real professional when it came to entering new territory and exploring limits. He knew how to look into what the problems were and prepare yourself to deal with them and not to make mistakes. He said, I forgot the exact number, it was somewhere around eight or ten, these are the things that he would consider adequate preparation and screening to make sure that anyone who might want to do a flight like this would be properly trained and qualified for it. I volunteered myself as the first candidate and Joe Kittinger, who'd been part of our team for many years, also volunteered. Next, we went through the checkout procedure, including some psychological tests at Wright Field and a physiological treadmill test there, a parachute jump, so in case I did have to bail out, I'd have the experience of using a parachute. This way, I would not hesitate to use it if I had to, and would do it right if I did use it. That was the way he approached things.

We did the same sort of shakedown with the capsule. It had to be proven to work before we were going to risk a man's life in it. And so, the project was off to a good start.

We found no health hazards as such, I would say. I was a little disappointed. I really had hoped to see some flashes during the night. I was never really convinced that I saw anything that was impressive enough that I would say it was unequivocally more than background. I did wear track plates on my chest and my left arm. Before the flight, all the hairs under the track plates were black. [During the flight,] we did have two or three tracks that were pretty heavy primaries, one carbon, for sure. And, sure enough, if you extrapolated the path of that carbon on through, I developed two gray hairs in one arm and another gray one on the chest that fit. There were no other gray ones that developed within the six months or a year after the flight. So, there was some evidence that the things that we had been seeing on the mice were extrapolated to a man, to a mild degree. But, gray hairs come anyhow; premature graying is not unknown for other reasons.

The high-energy particles of cosmic radiation are strange beasts. They're so high energy that the lighter ones tend to go right on through and don't even stop. The greatest intensity of radiation is released at slower speeds, as the particle comes to a stop. Actually, you may increase the dose by a little shielding like the aluminum capsule. It slows them down enough that they stop in you, rather than keep on going. So, it's a very tricky business. Shielding in this case is very complex, but essentially, the [capsule didn't provide noticeable shielding].

We were interested in heart rate and that was the main thing that we monitored. With the monkeys, we measured respiration too. Later on, at the School of Aerospace Medicine, we developed this in our manned instrumentation, and we did a lot of things, five channels of measures. But this early Manhigh work was really very, very crude. I don't even think we had body temperature on my flight. On [Clifton] Demi McClure's flight [Manhigh III], we were beginning to develop this concept of instrumentation, so we did get into some of that then.

Whenever we had a manned flight, even with the Sky-Car [two-person, open gondola used for balloon training flights up to 15,000 feet], the lower altitude training flights, we always had a ground crew with a pickup truck that went along. They had radio direction finders that they could home in on where the radio signal was coming from. That was critical when they couldn't see the balloon for some reason. They then tried to coordinate with the person in the balloon as to where the balloon was. We were in touch with them, normally, whenever we wanted to be. On the Manhigh flights, I think we had a half-hour schedule. Every half hour, we called and touched base with each other to see what's going on and what our plans would be.

[Why were you and Joe Kittinger selected for the first two flights?] Nobody else had a chance! (Laughs) I wanted to do it. And Joe was the test pilot type. Colonel Stapp wanted to have him check it out on the Manhigh I flight to make sure everything was working right. And it wasn't, so he did a great job as a test pilot, and we got those things cleaned up and then went on to Manhigh II.

At that time, in the aerospace literature, there were papers by a number of different people concerning breakaway in pilots flying jets at very high altitude in solo flights. They talked about getting into reveries, and identifying with space instead of the Earth, and about having out-of-this-world feelings. At that time, we didn't know quite how to interpret this. Not many people experienced it, considering how many flights were made, but it sounded like something at least some people thought could happen. Knowing individual variability, we didn't know what implications or consequences it might have in terms of pilot competence and responsibility in vulnerable individuals. The very high-altitude balloon experience would be a rather aggravated example of high-altitude isolation. It's a little like being under water in a swimming pool, where you have reduced sensory input. However, in the case of a pilot flying an airplane, you have all the noise of the airplane, and you have the responsibility of monitoring instruments, and keeping the airplane directed, and flying it, depending on how much autopilot you're using. Whereas, up there in the balloon, the pilot is pretty passive, just making whatever observations you plan to do while you're there. The balloon itself requires very little attention. So, we were concerned that a breakoff phenomenon might be intensified by the balloon experience at high altitude. We found no evidence of it among ourselves.

Going up on my flight, I was feeling gratitude for having gotten out of the iron pit [launch site] without whopping the edge of it when the winds got a little high. I got off safely and made it through the first few hundred feet. If a stray radio signal trips the emergency release that separates the parachute and capsule from the balloon, it is too low for the parachute to deploy and too high for a free-fall landing.

Then, on the way up, it was "Boy!" That's a sensation of finally experiencing a long-held dream. Here it finally is. I was just fascinated at the evolving picture. As it went higher and higher, I began to see things like I'd never seen them before, once I got beyond 30,000 feet. Of course, I was used to that altitude from jets. I knew what to expect up to there. From then on, it was all new territory. As I began to get close to ceiling altitude, when the balloon expands to full inflation, I was wondering about what the altitude might be. My high-altitude altimeter had been recalibrated by the National Bureau of Standards to qualify as an official record flight. I guess it was just to make sure nobody did any fudging. I wouldn't know what the altitude would actually be, but nobody told me that. It was telling me I was around 130,000 feet, and I knew that couldn't be right.

So I knew something was wrong with my altimeter, but I also knew that the point of reaching ceiling altitude was very critical, because the balloon must vent the excess gas at that point. It must do so quickly, without obstruction, or it can burst and come right down again. We had it happen a number of times on our animal flights. However, I knew that if I watched the vent tubes, I could tell what was going on, and so, when I saw them inflate, I said, "Well, we reached altitude. I know this is as high as you're going to go." I watched them for about two or three minutes, and they began to collapse again. I said to myself, "Well, we've done it, and we're okay. We're here to stay." So that was a big relief. All I had to worry about now was the landing, but that seemed a long way off.

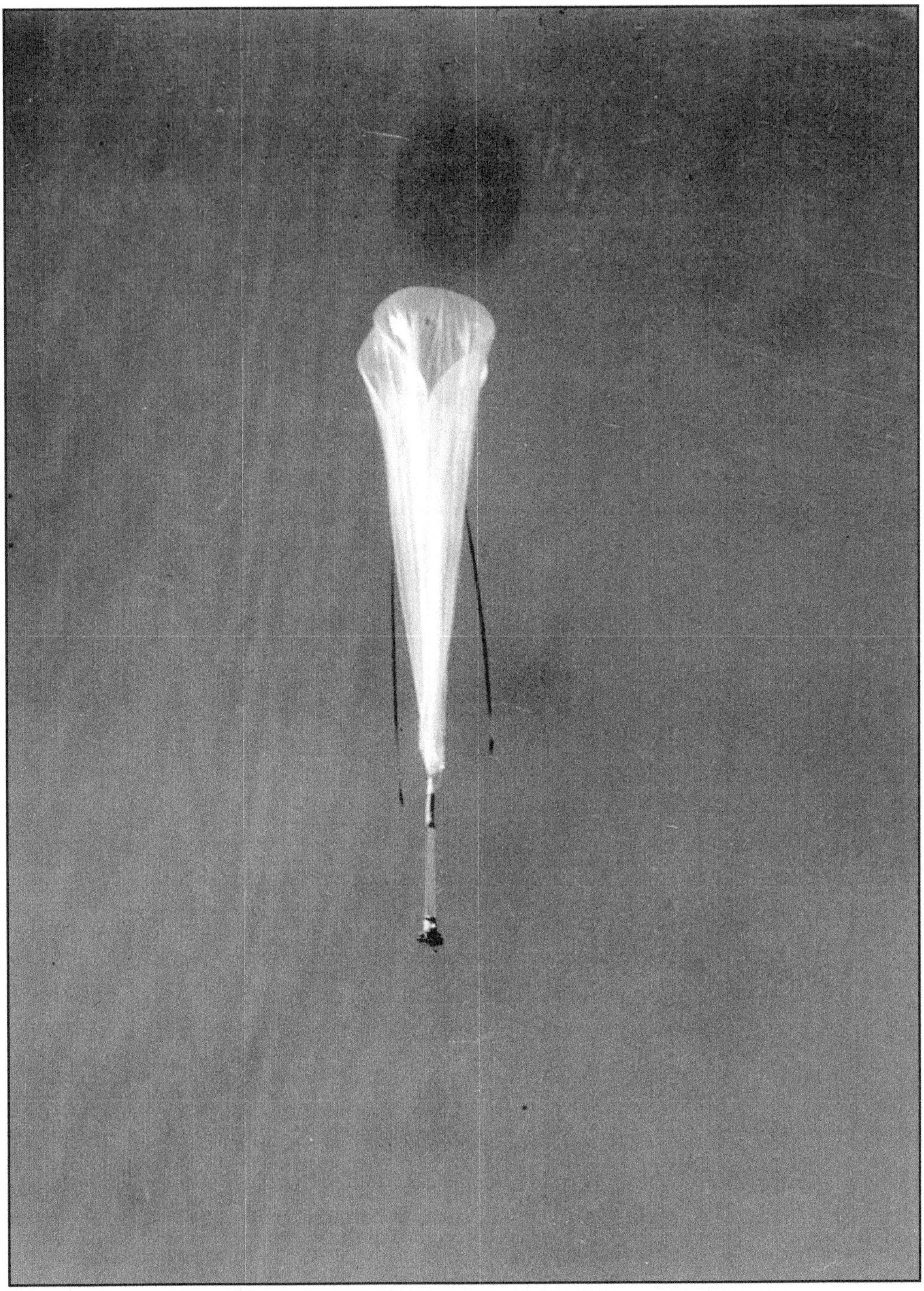

As a Manhigh balloon rose from the ground, its envelope was only partly filled with helium. As it reached higher altitudes and lower air pressure, the helium expanded to fill out the envelope. Courtesy of NMMSH.

As I remember, the six portholes in the Manhigh capsule were something like 4 inches in diameter, equally spaced around the capsule. One of them on my flight was occluded because it was used for a telescope. Another one was filled by an up/down mirror that was split so that through the upper half of the porthole, you looked straight up at the balloon and at high altitude, all you saw was the bottom of the balloon. The lower half looked straight down toward the ground so you could see where you were or what was directly underneath you, including cloud conditions. The rest of the portholes had quite good visibility because you could look from porthole to porthole and, basically, you could see in any direction you wanted to.

Well, there were a lot of observations on my flight! (Laughs) A lot of experiences! (Laughs) The thing that had excited me was the opportunity to make observations that had never been made before. I spent a lot of time preparing what observations might be made. I had gone to MIT at Cambridge and talked with the people there who were doing high altitude research, and with specialists in atmospherics, what phenomena I might expect to see. I learned about the green flash of the sun seen sometimes as it rises or sets.

One of the things that was not anticipated, I remember very clearly, was when you're up above 99 percent of the atmosphere, there is only 1 percent left above you. Looking up, it looks like there's nothing up there. It just looks black. You really don't see any colors. It's very hard to concentrate, to even detect that there's any sense of color at all high in the sky. And the whole blue sky is compressed down to a thin band around the horizon. From there, it quickly transitions into this black, so it feels like you're sitting in a soup bowl with the Earth down below and the sky around you and the open top with nothing above. This makes it rather easy to think of yourself as being as much in space as you are on Earth, and to identify with space.

Looking out toward the horizon, a number of times during the day, I saw thin bands of haze, almost like Los Angeles smog, compressed into thin bands, way off in the distance. These apparently were wind-shear layers or slight temperature inversions in the atmosphere, where dust or ozone or something was accumulating. It reflected and refracted the light differently, so looking edge-wise into the layer, you could see it. Sometimes it seemed further away than others. Of course, that would be dependent on altitude. So this is one phenomenon that was rather intriguing, that nobody had predicted or talked about.

I had always been very interested in astronomy and in astronautics, and one of the things that they warned me about was that the stars might have more color and would not twinkle. Their light would be free of the turbulence of the atmosphere that it experiences as it comes in through the air. And, sure enough, there was no twinkling at all, except right on the horizon, which we'd rather predicted. In addition, the colors of the stars came out much, much more clearly. The blue stars really looked blue, and the red stars really looked red, and the white stars really looked white, and there were many shades of color in between. Every star seemed to have its own specific, individual color. Of course, they do have different temperatures, color temperatures. The spectroscope shows it quite well, but through the air, you don't see this so clearly because of the smearing effects of air turbulence.

I remember clearly when Venus set around ten o'clock that night. I was watching it closely. I was amazed to see the atmosphere of the Earth act like a prism, and it went brilliantly through the colors of the rainbow as it set. This had not been predicted. The ordinary stars didn't do that. It may have been a Venusian version of the green flash, which I saw at sunrise the next morning. I don't know how common these things are at that altitude.

When I first reached altitude, it was clear all the way to the west. I could see the horizon 500 miles away. But as I was going west through the day, there were clouds at 20,000 feet or so. I could see that. Mid-afternoon, I became aware that way off to the west there was something going on. As the afternoon wore on, it became quite apparent that a frontal system and I were going to meet each other. So I wasn't surprised when they said that there was a cold front ahead of me. I got to see it. And it was very, very beautiful.

I was trying to keep up with the schedule of experiments that had been worked out beforehand. We had a large number of experiments, including some with the Air Force on visibility of airplanes from that high altitude. We had quite a busy schedule mapped out, so I had been busy with all of that. And then came the question, "Do you want to stay up all night and go over the thunderstorm, or do you want to come down early?" That was a stupid question to ask me! (Laughs) To come down early?! Besides, what fun to go over a thunderstorm, so we elected to stay up. (Laughs)

At night, over the thunderstorm, looking down at the tops of the clouds, they'd light up like big cauliflowers or brains, to see flashes of lightning light up the clouds from inside while looking down upon them.

I got just a few minutes' rest that night. I might doze for five or ten minutes once in a while. But I had so many observations I wanted to make on the stars, and I wanted to keep track of what was happening to my altitude. As the night wore on, I began to lose much more altitude than any of us had expected. The cloud cover underneath was cutting off the heat from the Earth, and the heat of the balloon was radiating off into space with no replacement. So the helium in the balloon was losing enough temperature to make it contract and thus lose altitude. I knew the clouds were known to reach as high as 60,000 feet. I was between 70,000 and 75,000 feet when I looked out for some stars and realized that they weren't there! I moved my head around a little bit to see what was going on, and I said, "There's only one conclusion. Those are clouds between me and the stars." That meant the thunderheads reached up at least as high as I was, and they were higher than we expected. That also meant that if the balloon got in the area of the thunder, there would be strong updrafts that formed the thunder, with a lot of turbulence. This would likely destroy the balloon. Balloons can't handle turbulent air, particularly at this cold altitude, around 68,000 to 70,000 feet. That's the coldest part of the atmosphere. The polyethylene of the balloon gets brittle. Therefore, I was in a somewhat precarious position.

At this point, I suddenly felt a strong twist of the capsule, and things were twirling around. I said, "Aha! We hit some turbulence, so not only are we in the right altitude range, but we're also in the area where there are gusty winds." I had to choose between dumping some batteries that weren't quite expended yet, which was the only ballast that was left, and get up out of that area, or take a chance on going down through the thunderstorm with a shattered balloon. The latter presented a lot of potential complications that would be undesirable. Unfortunately, releasing the ballast produced a shock wave that also could shatter the top of the balloon, so it wasn't an easy decision.

We had planned to land the next morning, making it essentially a twenty-four-hour flight. But I didn't clear the cold front until around 11:00 or 11:30, at which point we decided to start valving and coming down. Well, the only thing worse than not being able to get the flight off in the first place is to get up and not be able to establish a controlled descent. One of the problems was that our cooling system was a water-evaporating system that had limited capability. You just take so much water along and when it runs out of water, you no longer have cooling. We had learned from the animal experiments that the cooling is critical. We were very worried about how hot it was going to get in there later in the afternoon if I didn't get down.

The trouble is that, when you start down from 80,000 or 90,000 feet – I was around 90,000 then I guess – you pick out a certain descent rate. And whatever it is, when you hit the tropopause at around 65,000 or 70,000 feet, it doubles. I knew I had no more ballast to slow me down when I went to land, so whatever downward velocity I had at that altitude, I was going to land with at least twice that velocity. So I was a little hesitant to overdo the valving. On the other hand, I didn't want to hang around there any longer, either.

I'd valve off for a few minutes, and it looked like it was going to start down and then it would start going back up again. What was happening was that the balloon was picking up more heat from the sun and rewarming, expanding. I didn't really know how much correction to make for it. After about half an hour of this, we weren't making any progress. I began to do bigger bites. Each time I'd do a little bit longer, then each time it'd do the same thing and finally, at one point, it really started down. Sure enough, by the time I could establish what this descent rate was, and really see what had happened, it was about twice what I'd planned for. As expected, when the balloon went through the tropopause, the descent rate doubled. I knew I was in for a hard landing, and that was correct, it was! (Laughs)

[Did you reach a point in your own mind where you were almost ready to say let's just cut the balloon loose and come back on the parachute?] I wasn't very enthusiastic about releasing that parachute at those altitudes because there was one thing that we did know. When the chute opens at such high altitudes, you get a very, very heavy opening shock. I didn't want to expose the capsule, its instrumentation, and the life support systems to that if I didn't have to. The reason I didn't panic and do a parachute descent was because it didn't get hot. I was okay. Oh, well, I was hot and sweaty. I was uncomfortable, but I didn't feel that I was getting hot enough that I was seriously impaired in effectiveness, which was probably wrong. In retrospect, I think I was impaired in judgment and effectiveness. But at the time, I thought I was all right. It was a combination of fatigue and heat. That was one of the reasons, actually, that we were so careful to include the temperature monitoring on the next flight, because we did know temperature was critical. I became concerned that the pilot is one of the last people to ask about how capable he is to make decisions. You need to have some additional, objective evidence that tells you to what extent you should trust the pilot's judgment. Joe Kittinger's comment earlier on Manhigh I to "come and get me" didn't help matters any on that score! (Laughs)

Actually, the only time in the whole flight that I panicked was when I landed. I don't know that this has been told before, but I'd forgotten one detail in landing procedure. This is an example of the sort of thing that, in retrospect, made me question my judgment and how well I was functioning after being in the capsule forty hours, or something like that, by the time we finally landed. I forgot to turn off the lithium hydroxide air scrubber (the carbon dioxide absorber). That lithium hydroxide is a beastly irritating chemical. And I also had on board a bromine fire extinguisher. We knew that, if I had a fire, it was the most effective thing we could find, but the fumes were lethally toxic. I knew that if I used the bromine fire extinguisher, I could only do so with my faceplate on and breathe the oxygen from the supply of the capsule, and not breathe the air in the capsule.

As I said, we had a hard landing. It was hard enough that everything in the capsule that could possibly fly loose had collected into a mixed-up heap in the bottom of the capsule. All of a sudden, I had this terrible, stifling sensation that every time I took a breath it would burn. I thought, "Oh, my God! That damn fire extinguisher broke loose and is releasing bromine, and here I am, in this closed capsule! I've got to do something about that quickly!" So first I scrambled around looking for that small capsule of a fire extinguisher and couldn't find it, of course, in all the mess. Then I tried to get the face opening of my pressure suit over the porthole that I had opened. However, my pressure suit helmet configuration wouldn't let me get close enough to it. I still kept smelling this same damn stuff. I couldn't get fresh, clean air. I had to undress to get the helmet off. That meant I would have to take my pressure suit off, and it required somebody's help to do that. So, I had no way of getting the helmet off.

I thought, "This is a stupid thing! Here I'm likely to die in this darn thing because of this crazy fire extinguisher business, after having made a successful flight!" There would be a sequel to this that became all too real for Commander [Victor] Prather of the Navy. [On the 1961 Strato-Lab High V mission, Prather survived a ten-hour flight to 113,740 feet wearing an experimental pressure suit in an unsealed gondola. The capsule landed in the ocean, and Prather slipped out of the recovery helicopter's rescue line and drowned.] These things do happen. I remembered that there was one other emergency procedure that I could turn to. The top of the capsule had squibs that cut all the wires, and I could push a manual release and push the top hemisphere of the capsule out. Making the decision wasn't easy. I had been thoroughly warned ahead of time that if I did that, it meant more expense. We'd have to rewire the thing and that's a lot of trouble, so don't do that unless you really have to. I said, "Well, I think I'd better do it because I don't think they'd like to save the wires and pull my remains out of here, so I'd better release it." And I did. With much effort, I forced the top off and crawled out. There was never a breath of fresh air for which I was more grateful than that one! (Laughs)

Later, I realized it was only dust stirred up by the hard landing. It was temporarily very irritating, but not all that hazardous. I simply was very tired and not thinking clearly, and I didn't even think about the lithium chloride. Had I thought about it and turned the switch off, I'd have been okay. For Manhigh III, they went to potassium hydroxide for the $CO2$ scrubbers because that's not as toxic.

We were really worried about the cabin atmosphere and the possibility of fire. We did a lot of tests, different combinations of nitrogen and oxygen, and helium and oxygen. Our tests showed that we dare not have a 100 percent oxygen atmosphere at ground level. The hazard was frightful. We were caught between a rock and a hard place because, on the one hand, we did not want to change atmospheres when we went to altitude. We wanted to ground-test at the same atmosphere we'd have at altitude. On the other hand, we wanted to have a reduced pressure atmosphere at altitude, because it made fewer pressure demands on the capsule. It could safely be made lighter. So we finally decided to compromise on a combination of helium, nitrogen, and oxygen. The addition of helium improved the fire characteristics at ground level because of its higher heat conductivity than nitrogen. It also increased the rate of heat loss from the capsule at night and raised the pitch of my voice a little bit. Otherwise, it didn't have any remarkable effects. It was perfectly inert, so you were in fact, if anything, in less danger of decompression sickness in case of sudden loss of cabin pressure. This worked very well for us. I was disappointed that the NASA program didn't pay attention to this issue. We write this up very clearly in our tech reports, that 100 percent oxygen atmosphere at ground level was seriously dangerous, a real *no-no*. We decided early on that that wasn't the answer, and we had to find some alternative.

[The pilot selection for Manhigh III pretty much set down parameters of what Colonel Stapp and you thought would be needed for astronaut selection, right?] Well, Colonel Stapp, to my knowledge, was not thinking in those specific terms, but it turns out that his requirements for pilots of Manhigh flights were very similar to the evaluation of astronauts. So it worked that way. I should point out in this regard, all the credit goes to Colonel Stapp. We really were a team, in the sense that he took full responsibility for all the financial cares, and all I had to do was worry about the technical design and operational characteristics. He left that up to me and Winzen Research. The pilot responsibilities were mine, of course. The overall management and supervision of the project was his responsibility.

After Manhigh III, I was an advisor to the Mercury program. I was on a panel that gave recommendations to them with regard to radiation hazards in space. I wasn't involved directly in any of their flights. I'd been transferred to the School of Aerospace Medicine and was developing another kind of program at that point.

[Looking back, thirty years after your flight, where do you think Manhigh fits within the history of space exploration?] One thing it certainly deserves credit for is that we made that flight about a month before Sputnik flew, and in that sense, we were truly space pioneers. Before the flight, usually, the official attitude within the Air Force and generally within the country, at that time, was that space research was a misuse of scarce dollars, that it had no military or civilian use, and that we were a bunch of visionaries who were squandering money. (Laughs) Also, when Sputnik did fly, our flight did help Americans feel like we weren't totally behind in foresight and interest in the future. There was some offset to the impact of Sputnik. We had something to point to, also. These things fire the imagination of the young people.

We were not the only space pioneers. The X-15 rocket flights were also penetrating space. We got a lot of publicity at the time. We contributed significantly to space consciousness. How much else did it contribute? You never know. The lessons and the experiences and what we did gets melded in with everything else people know, and they use it. Sometimes they don't realize where it came from, and sometimes they ignored our experience and reinvented the wheel. But you have no way of measuring this.

There's one little story that really tickled me, talking about emotional reactions. I think that one of the most comical things that happened to me on the flight occurred after landing. There I was, in this partial pressure suit, which was unknown in those days. At this time, no one had even seen a partial pressure suit, except a few people in the Air Force. Here I was, standing in this strange partial pressure suit beside the 8-foot long, 3-foot diameter gondola with all these telescopes and boxes hanging on it, with a parachute lying on the ground off the end of it. This about ten-year-old boy and his father were the first ones to reach the scene. I was standing there, and really glad to see somebody and to have someone to talk to. At that point, an Army helicopter was circling in preparation for landing. So, as the boy approached with his father, the boy said, "Oh! Oh, look! LOOK, Daddy! There's a helicopter!" There I was, just having arrived in a space suit and it didn't faze him a bit! (Laughs)

As non-commissioned officer in charge of the Space Biology Branch at Holloman in the late 1950s, Ed Dittmer was actively involved in the Manhigh flights. He worked with the pilots to ensure their safety during the balloon flights.

I reported to Holloman on the first of February 1957. By that time, I was a certified aeromedical technician. When I first arrived, I reported to Dr. Stapp. He was then a captain, I believe. He was the Aeromedical Field Lab commander. Then I reported in to Dave Simons. Colonel Simons was a captain back then, too. He had the Space Biology Branch, and he was doing quite a bit on upper atmospheric research with high-altitude balloons made by Winzen Research.

I reported in to Simons, and he said, "We're having a flight in the morning so be out here at three o'clock in the morning." I came out to the lab at three in the morning, and of course, not familiar with all the technical aspects of it or anything, I thought, "What did I get into here?" At that time, we were flying some guinea pigs and fruit flies in a small sphere capsule, and Master Sergeant Pierce reported in there with me.

He had this potato and he said, "Here, peel this and quarter it." It was food for the animals. That was about all we got out of that, just observing everything and watching everybody do their job.

Simons wasn't the easiest guy to work for, but he was good and you respected him. He worked hard himself. He pushed himself to the limit all the time, and I had a lot of respect for him. A lot of guys didn't like him, they couldn't stand him because he was very quick. If he didn't like what you was doing, he would tell you right in front of everybody. He wouldn't call you aside or anything, he'd just hit you with it. But I respected him and I got along, I think, real good with Colonel Simons. We went on after that to develop more flights of the sphere type, and I got acquainted with the package and how it was assembled and what to do on it.

What they were wondering about at that time was how you'd react in zero gravity in space orientation and how you'd orient yourself. So they thought by going under water, we could simulate that. We fabricated a seat on a long pole that sat down in the swimming pool, with a gauge up over the top. We'd strap you in the seat and spin you around and see if you could orient yourself. The project officer on that was Dr. Grover D. Schock. He had the first Ph.D. in space biology from the University of Illinois, a real sharp individual. We fabricated the instruments to do the underwater research with. This was done in the pool at the Blind School [School for the Visually Handicapped, in Alamogordo], and we had an old pool at Holloman that used to be the officers' pool that was not in use and had been condemned for swimming purposes. We used that for some time. Then we took it to the YMCA at El Paso and used their pool.

They didn't know anything about zero gravity at this time and what effects it would have on the human body. We went into the airplane aspects of it, too. [Aerospace medicine pioneer Dr. Harald] von Beckh had it for some time, then Dr. Schock took it over. We'd fly an F-94 on a full-hour mission. We'd pull maybe fourteen to twenty trajectories, what they call "parabolic trajectories." You'd go up to about 30,000 feet, then you'd drop over and during the arc, you'd maintain up to about 45 seconds of zero gravity. Then you'd go straight down and then you'd hold in a spiral at about 6 g's and pull out at 9 and back out. During those parabolic trajectories, we would perform different programs. One time, I was eating peanuts, and peanuts would float up over my nose. And we tried micturating [urinating] in a jar and, of course, it would float up. One time, I flew about fourteen trajectories in the morning, and we were going up in the afternoon again.

I was flying goldfish in this bowl and you could see the fish come up out of the bowl and swim in the water and, of course, catch them as we came down into the g's. That was really quite an experience.

Captain Kittinger was the test pilot for the lab at that time. He would fly the missions for us. We'd go up and go through all these trajectories, and we'd burn out the trim tabs and lose all the hydraulic fluid so, of course, we'd have to come in on an emergency landing. We were doing this for some time, and finally the safety people said they're going to have to modify them airplanes.

Joe Kittinger was the pilot, and he'd say, "Now, if this thing goes *kaplooey*, I'm going to punch out and if you don't leave when I do, after that canopy leaves, you're going to be sitting here alone with a big airplane." That was quite an experience, that zero gravity. They'd take you up and really wring you out. There were several guys who couldn't fly them because of their stomachs, and fright, too.

Prior to Manhigh, we'd done a lot of research with mice and guinea pigs and so forth in balloons. We put track plates on their head. They were concerned with cosmic radiation at that time and what it would do, how it would affect the brain in a human being. We'd put track plates on the head, and we had a little monitor we'd made. We'd cut a little square track plate and we'd monitor the head of the guinea pig and keep it in a stable position. We'd X-ray it before and then put the track plate on and, of course, X-ray it afterward to see if we had any movement. But these track plates, we sent them up in high-altitude balloons in a small sphere capsule and check and see if there were any cosmic radiation hits on them. And they determined that there was really no problem in sending them on up. All these prior flights, of which there were many, I chased them all over the country, recovering their packages and data and so forth.

Then we went into the Manhigh program with training the man. Dr. Schock was training for a Manhigh pilot flight, and he and Mr. Winzen, the head of Winzen Research at that time, were on the Sky-Car flight, which was a training flight for balloon pilots at that time. Myself and Sergeant Hensley were chasing him on the ground with a pickup, and they run into weather. They had a storm coming behind them and big Lake Superior in front of them. So they wanted to set it down in a hurry, which was logical because they didn't want to get out over that lake in a storm. Winzen hit the release button several hundred feet up. The cargo chute that was supposed to bring them down streamered, and they just went straight down. Of course, we were on our way, and we saw what happened. We had radio contact with them, but we lost that. Fortunately, the state police got in behind us for speeding, and they followed and we went to the scene. They landed out in the middle of a pasture. I got over to the Sky-Car, and they were both pretty conscious, disoriented but conscious. Dr. Schock had cut his throat from ear to ear. So I slapped some gauze on him and held him tight. The police called an ambulance right away with their radio. The ambulance was there within fifteen or twenty minutes, and they got him to the hospital.

The sudden jar caused an aneurysm, and they had to do some operating on him. I stayed up with him that night and most of the next day at the hospital. They let me because I was a medic. Colonel Hessberg and the crew from Holloman flew down, and he did recover. Winzen just had a few broken bones out of it. Dr. Schock recovered, and he went on to be a professor at the Air Force Academy. That took him out of the pilot staff, but I think he would have made one heck of an astronaut. He was one fine man.

We had several that we trained in this area. There was Captain Collins from Wright-Patterson, who was a parachutist; Lieutenant Clifton McClure, who was an F-104 pilot; and one or two others I've forgotten. Schock was eliminated because of the accident. Colonel Simons was the project officer, and he wrote all of the specifications and so forth. We all helped in designing the Manhigh capsule. It was a sealed cabin environment so that they could live at those altitudes. We made several trips to Winzen Research [in Minneapolis] and worked there on the capsule and the intricate parts of it. And we had the pressure suit on the pilot, and the helmet, which was the old antiquated type.

I was the project NCO [noncommissioned officer] in charge. My job was to take care of all the pilots, as far as getting them prepared to take off and checking the physiological data on them and so forth. We also had a medical officer, of course, and I helped fabricate different items and design items. At that time, money was really tight. We weren't on an open budget like NASA and the others. We were on a shoestring budget, so to speak. We had to be real careful with funds. In fact, we'd run out at times, and Dr. Stapp would run up to Washington and get us a little loot to work with. If it wouldn't have been for Dr. Stapp, I don't think any of those projects would get off the ground. He really hustled and worked, even though he wasn't directly involved in Manhigh, he was involved from the top. He was really involved and concerned in the early days of it. And, of course, lab commanders changed through the years. I had several as we went on into Mercury.

Dr. William Douglas, another physician, was directly involved with NASA's first manned spaceflight program, Mercury. He offered insight into the personality of David Simons and the value of the Manhigh project.

Dave Simons was in my class at the School of Aviation Medicine back in 1948 to '49, and I got to know Dave quite well there. When Dave was the commander of the lab at Holloman, I came to visit him. Late in the afternoon, he was showing me around and finishing his paperwork. We were going out to his house for dinner and he said, "Excuse me, I'd like to take a nap." So he took half a dozen folding steel chairs, opened them, put them along the side of the wall and laid down on them, put his arms over his eyes and instantly was asleep. I was reading a magazine there and I cleared my throat, and he suddenly awoke, looked at his watch and said, "Darn, I meant to sleep thirty minutes and I've only slept twenty-five." He could just set his clock to wake up. He got up and said he needed some coffee and he took a cup and put two or three spoonsful of instant coffee in it and some dry milk and some sugar and went to the tap and let the hot water run into that cup, stirred it and gulped it down – not as something to enjoy, but something to stimulate himself and get some energy. And away we went. Dave has been an unusual man, one whom I admire greatly.

It's hard to say how Manhigh contributed to the Mercury program. Mercury would have gone ahead anyway, even if there had never been a Manhigh, but this should not detract from Dave's accomplishments. We put a man in a terribly cramped cylinder. Put him in there, put him on the balloon, sent it up to 100,000 feet in the air, where he could see the curvature of the Earth, where he could look down on thunderstorms, where he was in a very dangerous and hazardous position in case the balloon ruptured. Whether he would have gotten out of that capsule was debatable. But when the detractors of Mercury said that you are going to put a man up there in a capsule and orbit the Earth three times at ninety minutes per orbit, he is going to suffer from sensory deprivation and isolation, confinement, all these terrible psychological problems. He is going to be exposed to cosmic radiation. Well, we could always point back to Dave Simons and say, look, here's a man who was in a much more cramped container that was much more hazardous than that. At the time, people couldn't imagine putting a man on the end of a rocket and blasting him off the Earth. It seemed much more hazardous than to put him up in a balloon, but not many people had seen how thin that balloon was and what could happen to that balloon.

We could point back to Dave's experience and we could say that here's a man that was in a tiny, tiny compartment, up longer than the first few Mercury flights would be. He didn't suffer from sensory deprivation, isolation, confinement, any of these things. He's a normal human being, functioning well, making scientific observations, communicating back to the Earth, sending this valuable information back. So we could point to the role of man in space.

We could point to Dave's experience and say that is the kind of person we want. We already know that people can survive these things, and not only survive, but function and return completely normal. That was a great contribution that Dave made. John Paul Stapp was sort of Dave's mentor, I think, and Dr. Stapp's great heroic work on the [high speed test track, see Chapter 6] sled was one of the other things that we could point to and say man can withstand these accelerative forces. Stapp has done it much higher than the Mercury people will ever experience and survived quite well. Each of these people, and the other people that participated in Manhigh, we were able to point to them and say, "Look, they did it." So each of these real pioneers contributed greatly to the program.

Dave had cosmic ray track plates taped to his body. I recall, long after Manhigh, he showed me a white hair on his arm. Long before that, they had sent black mice up and as time went by, hairs naturally fell. And when they grew back, they grew back white. The implication was that a cosmic ray had destroyed the little cells that made the black pigment. Dave showed me a white hair on his arm around what was normally pigmented, and he said that this was an example of it. He said that a cosmic ray went through the track plate right at that spot where that white hair came out. We could point to these and say that Dave had clearly been exposed to cosmic rays and they didn't do any great damage to him. So, we know that men are exposed to them and we know now that people who have been up in the shuttle close their eyes and they see these little flashes of light, which are presumably caused by cosmic rays stimulating the retina or maybe even just making a flash somehow in the vitreous humor or the aqueous humor. I don't know the true reason for these flashes of light, but these things were sort of predictable. They were predictable on the basis of Dave's work and the other pilots that went into the Manhigh program.

Dr. Rufus Hessberg, chief of the Biophysics Branch of the Aero Medical Laboratory at Wright-Patterson AFB, had participated in Manhigh II as the paramedic to attend the pilot, Dave Simons. He was also in charge of monitoring the pilot of the Manhigh III flight in 1958.

I arrived for the third and last flight of Manhigh, which was McClure's flight. That came about six or eight months after I got to the lab. One of the major differences between Manhigh I and II, and Manhigh III was the fact that we moved the launch from Minnesota to Holloman, where we felt we had better launch crews, better weather prognostication, more of the support personnel and equipment needed for that, and we used the resources of Holloman to support the ground operations for that. My feeling is that we had better tracking vehicles and aircraft by working out of Holloman than a, literally, field operation in Minnesota.

The scientists on the ground were asking McClure to make observations. That was a science team that was made up of individuals selected for the whole Manhigh program. They were adding questions. As a function of what they learned from Manhigh I, they would add on or substitute new questions for II. The same thing for III. The science team stuck together the whole three missions and they operated in that mode, sort of like your science group on any of your space missions. They had their own organizational element. They had their own leader. The leader interfaced with the operational people, and it was through that interchange that certain things got scientifically introduced to the program. But the ultimate concern always was whether anything was being added or tried that would in any way jeopardize the mission.

The medical evaluation of the Manhigh pilots on the ground was done by telemetry of physiological parameters that could be measured at that time. It was basically EKG, which gave the pulse. We also had a stretch band that gave us respiratory rate, but it was very rudimentary in those days, and it was analog. The event of digitizing physiological information for more reliable transmission had not been developed. I'd say that about 30 percent of the medical information was obtained by air-to-ground voice communication. You could tell an awful lot just by conversing with whoever was up there. We had the backup physiological information, sporadically. It wasn't always reliable, depending on the terrain features at the receiving end, where the capsule was at the time, and which way the capsule was facing at the time.

The evaluation was done mainly by Simons and Stapp. I had also been involved in the Manhigh II ground physiologic parameters, so I was part of the team. By the time McClure flew, I had the ultimate responsibility, so I was heavily involved.

I first realized McClure was in trouble when the temperature began going up. I forgot to mention, we did have a temperature probe, so we did have body temperature. And that was our basic clue. I had to make the decision to bring McClure down. His reaction was disappointment. Of course, all these guys are so highly motivated and they think they can tough out things. But when you're responsible for somebody's safety and well-being, you have to override their motivation and their willingness to push on when they shouldn't.

Clifton "Demi" McClure piloted the Manhigh III flight. He applied for the Mercury astronaut program, but was rejected because he was too tall. However, he was selected to pilot the third flight of the Manhigh balloon project. During that 1958 flight, he endured near-fatal heat prostration. Three factors contributed to his rising body heat. One was that someone either forgot or decided not to place a dry-ice cap on the gondola prior to liftoff. Dry-ice packs had been used in the first two Manhigh flights to keep the interior of the gondola cool until it reached a high enough altitude for the onboard cooling system to function. The other factor, as he described in his 1990 oral history interview, was that shortly after he was sealed in the gondola, his emergency parachute accidentally opened in his lap. Rather than report the incident, which would have caused the flight to be aborted, he worked for three and a half hours to repack the parachute, twice, in the tight confines of the capsule. During that effort, he began to perspire. The partial pressure suit he was wearing would not allow the perspiration to evaporate and cool his body. Making things even worse, the drinking water dispenser and the capsule cooling system malfunctioned, and the cabin temperature reached at least 97 degrees Fahrenheit during the flight. McClure was awarded the Distinguished Flying Cross for his Manhigh flight, and he was inducted into the International Space Hall of Fame in 2001.

I was born in Anderson, South Carolina, in November 1932 [when Franklin Roosevelt had just been elected President]. I got my nickname right there that day. Everybody in the South had a nickname. They give you a nickname or they shoot you, one of the two. So you have to have a nickname. They called me "the first Democrat" and they shortened it to "Demi."

My father taught me to fly when I was in high school. He came back from World War II and was on the GI Bill. He got him a plane, and I flew the plane before I drove the car. As a matter of fact, the first day I flew, he decided to give me my check ride a year before I was old enough. See, there wasn't a big problem about age back then as much. People drove cars, they might have had a license limited, but when a kid was responsible enough to drive, the parents usually would let them drive. I never did get to drive because we didn't have no money to buy me a car. But, some of my friends did in high school. When I got ready to solo, my dad knew that I was a year younger than the license said I had to be, so he couldn't give me a license. But he soloed me.

That day, I had a sheared shaft on the magneto. I never will forget it because I took off, and my dad and the mechanic were standing in the middle of the field in Anderson. I took off and turned crosswind. I looked back at them and the motor started running rough, sputtering and going off. I kept pushing the power up, and the plane wouldn't fly any more. It was just a little Piper Cub, anyway. I was told a number of times not to land downwind. That's one thing you don't do. Anyway, I turned downwind and the engine wouldn't run and I pushed the power up and it wouldn't hold it in the air. I knew I had to lower my nose. So I lowered my nose and I kept looking. Can I make it all the way around and come in like I should? You know, you're doing a solo and you're kind of new at it. I decided that this thing wasn't going to make it, so I turned downwind and landed going downwind. Made a good landing on the grass, which is where we were taking off and landing. Boy, was my daddy mad! He said, "What are you doing, landing downwind?" I had full power on it, trying to taxi it, and I kept trying to tell him that I had full power. The mechanic had run off to the plane, and this door flopped open on the little Cub. The mechanic said, "Cliff, he's got full power on this plane right now. It was running rough." So that got me out of it.

I went to Clemson College in South Carolina, and got a degree in materials engineering. At the time, it was a military school, and I took Air Force ROTC there. I was commissioned in 1954. Then I had a delay of one year to get my Master's degree in materials engineering in 1955. Then I went into the service. I had already decided I was going to be an astronaut, but there was no astronauts. I had figured all that out, and I started taking astronomy courses and things at Texas A&M.

I took a training command job because you could fly in the training command if you wanted to, and I wanted to. I went to instructor's school at Craig Air Force Base in Alabama. So I got at a base where I could fly a lot. Now, at a base like a training command base, half of the pilots work a full day, starting real early in the morning until noon or something like that. And then the other half comes in about noon and work all the way into the night. So the planes fly some fifteen or eighteen hours a day. I could fly both shifts. Nobody else really wanted to, but I did. I'd fly my early shifts, take a couple three hours of nap, and fly the night flight, and volunteer for some extra training on students and on the weekends. A lot of people didn't want to take their students on cross-country because they were tired of it. You know, it was hard to sit in the back seat of a T-33. It was a terrible seat back there, it just wasn't comfortable at all. So I would always be able to figure out somebody who wanted me to take one of his students. I built up a lot of time in a hurry. I built up something like 800 hours in one year, or a year and a few months. So even though I was just out of flight school, I was approaching this magic 1,000-hour mark, which you have to have before you can be *anybody* in the Air Force.

While I was at Bryan Air Force Base in Texas, I was taking astronomy at Texas A&M. When I took this astronomy course, I could only take half because it was scheduled at ten o'clock in the morning. Every other week, I'd be on an early morning shift and I'd be flying then. So my wife became the first female to attend Texas A&M that I know of. She went to class for me on my off weeks. Before I signed up, I told the professor that I would not expect to be let out of any of the tests or anything and I would take my grades as they came, if she could just take notes in that astronomy class, which she did. And in that astronomy class, I found out he thought that we were going to fly in space, too. I mean, it was very few people thinking that then.

That was in 1957. Nobody thought that we were going to do this. Buck Rogers was in the comic books, that was it. My call sign in the training command was "Astro," and I had this little rocket on it that I had designed myself. One morning, this professor came in and he was just pacing back and forth. He couldn't stand still. He said, "I told you, I told you, I told you." He couldn't even get it out. Most of us didn't know what he was talking about. I hadn't heard it on the news. He said, "Yesterday, the Russians put an artificial moon around the Earth." The President was talking about the artificial moon, and nobody could get a new word for it. It was just an artificial moon, which is a long word, particularly for a journalist. So *satellite* became the later terminology, but everybody was talking about a moon so human beings could understand it on the corner of the street.

The satellite excited the professor, and he knew that people were going to be next. Anyway, he said, "You're going to be transferred where?" And I said I had no idea, but good friends of mine with an engineering degree got assigned to Turkey as sewage control officers. He said, "I called somebody this morning to see if you could get a job working with him." I said, "Who'd you call?" He said, "Colonel Stapp. He said he'd think about it and let me know." I went a week or two, I don't know how long it was. I was at home. It was Saturday morning and I had tried to get a flight, but I had been cancelled. This call came from the duty officer on base ops on the weekend. He said, "There's this colonel down here that wants to see you and said you better get down here." I said, "What in the world would a colonel want?" I thought I had done something. So I went down there, and there was Colonel Stapp. That was really neat.

He didn't really make me an offer. He talked to me a little while, and he asked if I would like to work out there at Holloman Air Force Base, and I said yeah. He said, "Suppose I wasn't there and couldn't guarantee you anything even if I was?" I said, "That don't matter. If I get close enough to it, I'll do the rest." So the next thing I knew, in came a dang set of orders for me to go to Holloman. After I got out there was when they assigned me something to do. They assigned me to the solar furnace project office. It wasn't doing what I wanted to do, so I thought, well, I better do this the best that can be done and get this other deal going anyway. I went out there and saw what all they did. Around the base, the reputation was really two types of people. One was people that understood what was coming and smart enough to see the future a little bit. There were two or three of those for every 200 or 300 or 500 of the other kind of people that said that's a bunch of crazies. So, you had to have the stamina to withstand current opinion, which was adverse to what you intended to do. Dr. Stapp left about the same time that I got there.

I always thought that I should try harder because he sent me there. In reality, I was never even entered into the Manhigh program. You see, the program by that time [preparing for the third flight] was defined among the people that knew it well enough to know that it was very valuable personal experience. They were thinking different than the first people like Joe Kittinger. He was doing it because he loved to fly. There is a big difference in these kind of folk. There was a long line of colonels and light colonels and majors, and when you step down a bracket, you multiply by about four. They had a fairly good selection procedure or test, but they also had another thing. They were looking for a baseline. They always talked about baselines, and that is, they wanted to get more data on each set of tests because they didn't know what tests were valid and what tests were unnecessary for spaceflight. Physical things a person should be, or you should use as a criteria to pick somebody out.

And that was what they needed baseline [measurements for].

They had a lot of candidates. So I just kind of entered myself by entering as a test subject. And then I entered myself by flying a C-47 on the weekend up to Minnesota, up to where they launch the balloons with monkeys and things. So everybody on the project got to know me. I kept saying, "You know, I'm the best guy. I can do this job a whole lot better than anyone you got here." But that didn't help a lot. I should have kept my mouth shut, being reasonable. But anyway, I'd do these tests and I'd try to do better than anyone had done. I didn't know how well any of them had done, but I would do these tests and I would work so hard to do better because I wasn't on the project. I didn't have any guarantees of any kind either. And slowly, they began to eliminate candidates by one method or another. Grover Schock hit the ground and it like to killed him. But anyway, we did all these tests, and in the meantime, I would get my flying the same way. I wasn't assigned to fly, I was assigned as an engineer, so I'd have to get my flying in. But I found interesting things to do like being a target for some of these missiles.

I also went to Lovelace Clinic, in 1957, I believe. The U-2 pilots were also going through Lovelace Clinic. After a while, they recognized I was a candidate [for Manhigh III], even though I had not officially entered my name on the list. And I started getting some treatment as a candidate. Then they called out to Holloman and said, "Could you transfer him permanently here to the Aeromedical Laboratory?" I was transferred on this project as a subject and potential pilot. So then I had the official duties to do this all day long, every day, all day, Saturday and Sunday included.

They have a lot of these different criteria that you did. Like, up at Wright-Patt we did heat stress. They weighed you sitting in there with no clothes on, in a balance at 160 degrees Fahrenheit. Nobody could have told me that the flight was going to be more impressive of a heat load than that. You get a very high heart rate and a lot of sweat, and the surface temperature starts to rise and your body loses control. He didn't run you into heat prostration, but you were supposed to get to 90 percent of it.

Why did I pursue this? That's just what I wanted to do. I don't even remember when I started wanting to do it. I'm telling you, that's all I ever wanted to do. If you want to know the truth about it, this is my own place to simulate space flight.

There was a period of time, it must have been a couple of months, which was a fairly long time, that I knew that unless something happened to me, I was going to make this flight. In which time, I switched to what they called primary candidate and started doing a large number of low-altitude helium balloon flights throughout northern Minnesota and South Dakota and North Dakota and Iowa.

My Manhigh flight was cancelled twice after we had gone through the process of getting into the capsule. The third time, they had left off the CO2 cap. It was not on it when we launched. What I didn't realize was that we were at 26,000 feet pressure altitude. At 26,000 feet, it isn't going to take the heat away from your body with convection like it does if you are at sea level, and that's a big difference. I was in a pressure suit, which restricts the main part of your body from heat transfer by sweating.

The parachute that I packed was added by Colonel Hessberg after the fact. It was not designed in there. It would not work at the altitude I was going at and was totally useless in my opinion. I didn't want it in there. It was in my way. It was terribly in my way. The cabin was, remember, a minimum volume, by design. It was what you could barely put a human in, otherwise they would have made it smaller. They knew they had a limitation on weight and volume to launch people. You see, nobody gave Colonel Stapp the data like they do today – "We've got so much size of a rocket coming and you can count on this much weight." He didn't have all that. So he said, "What we're going to do is test the minimum that a guy can stand. And then we'll say, hey, you can't give them less than that." Then Colonel Hessberg put this parachute in because he was a parachutist. I couldn't talk him out of it. He was afraid he would be criticized if something had happened and I had no way to get out, because he had my interest at heart. But he was wrong. I talked to him about that, but it didn't do any good.

The parachute came open while I was still in the sealed capsule inside the assembly building. This was the worst burden anybody had laid on me all my life, this excess thing which was intended for my benefit. It was just an incredible burden. I had no idea.

Now, you pack a parachute with two little metal cones that had holes in them. Then they folded the final flap that closes the envelope of the parachute and makes it into a little neat package. That's closed by the little metal rings of the outside fabric, going over these cones. And then the two pins went through. And this thing was hanging upside down. Nobody ever realized it. It was put in there at the last minute, so we didn't have any time to test it. That was the problem. Had we had time to test it, somebody would have said let's put it in with the pins down, which is the only logical way to do it. But it was an afterthought all the way around.

These pins dropped out. I didn't do anything to hit them, I was just

Clifton "Demi" McClure inside the Manhigh capsule. John Paul Stapp's instructions for designing the capsule included making it as small as possible for the mission. This left little room for the pilot to maneuver. Courtesy of NMMSH.

sitting back there resting, and the pins had been working their way out, but they didn't have that far to go from the beginning. And the flaps hid them a little bit, you couldn't see them, and suddenly they fell out and, *poof!*, there it was, right in my lap. This was the last day in the project, and we didn't have enough money to fill the balloon up again. Everybody was talking about cancelling it, anyway. I said, hell, I owe this country more than to just act like a good soldier, because all of this was unnecessary. If they hadn't put this damn hindrance in here, I wouldn't have had it to deal with it.

So I said, by god, I had practiced parachute packing, I had bailed out three times at El Centro as part of my test. I had gone in on my own, and I said I wanted to learn how to pack parachutes. They said, "It said on your orders just to jump out." I said, "Either you can teach me or the Navy can teach me, because I am learning how to pack parachutes. And I got the whole weekend anyway." They said, "Yeah, but we want to go home." I said, "Yeah, but you'll help me if I ask you." They liked me, so they stayed there and helped. So I officially learned how to pack parachutes from the best parachute-packing people in the world, the Naval Parachute Test Center and the US Air Force Parachute Test Center at El Centro. I had already done that. So I wasn't a neophyte at this, and the parachutes that I had packed were a lot harder than this one.

Everybody criticized me after the fact to some extent. They all thought it was great on one hand, you know, because I didn't tell them. The way I looked at it was that it was unnecessary to stop this project for such a damn nuisance. And while they said, "You weren't supposed to take the authority to remove one of your safety systems," I said, "Nobody ever asked me. I didn't initiate the question, and you should have asked me if my parachute was still good." They said they wouldn't have had any reason to ask that. And I said, "Well, I did it and that's all there is to it." So there was some criticism after this, but the whole thing was I couldn't open the capsule to throw it away. I couldn't tell anybody, because they would have opened the capsule and then stopped the project. And they were already inflating the balloon. They had lost a balloon from weather the day before this, ripped it to shreds.

So I was just sitting there figuring it out, and I said, "I am qualified to pack this chute. I will pack it and see how well it is packed." Considering that I returned it to normal, and I would put it back on in the proper way. I wasn't the one that put that parachute in there, but I was the person that turned it around to make it fit the proper way so the pins wouldn't fall out anymore. So I repacked it right there in my lap, and I didn't have to let it all out. It was already in a good package and I proceeded to keep it in as good a package as I could, you know, perfectly packed partially by the guy that packed it before. I reinstalled the outer cover on it with the pins and the cones. I checked it all out, and it was a perfectly good chute like the people at the Parachute Test Center out in El Centro had showed me.

Oh, I did have to repack it twice. I put something in wrong. I had to take it back apart to do it right. See, there was not enough room to do this, and when I had done it, I just said, hell, to be honest to Hessberg who put this damn thing in here as my commander, I must do this right or tell him. So I took it back apart and did it the second time. I put it in the correct way so it would never come loose again. And I didn't have to worry about it anymore, except for it being in my way.

When I finally took off, it was great! It was incredible. They let the thing go, and it just went so smooth. The ground started dropping away, and you couldn't see the balloon from there. You had an up-mirror. Anyway, it was just a really great feeling, because when you got to 50,000 feet, nothing changed, nothing slowed down, you know. You just kept on going to 75,000, kept on going. By then, I could see the expansion in the envelope really big above me in that mirror, and it was just incredible. I was seeing this deep blue sky, dark black sky for the first time. I mean, I wasn't getting to fly in space, but....

I reached 99,900 feet. My altimeter said 105,000 feet. You have to understand, Colonel Simons had a registered altimeter, and he would not allow a registered altimeter on this flight. I believe that I went higher than 100,000 feet, but the honest truth is subservient to Colonel Simons. [Eli] Lack Beeding said it was 99,900 and I didn't argue with him, but I believe it was 104,000 plus 500 or 600 feet, and that's my honest belief. But I wouldn't have gotten the record anyway, because it was an uncertified flight.

I realized I was in trouble right after I got to altitude. See, at that time in your life, you had just gotten your training as a pilot, bullets bounced off, and you could jump to the moon without a pressure suit. You were ready for combat. But you really weren't. You were ill-equipped for combat, in reality. But this is an attitude they get in training and it infuses you, whether you realize it or not. I thought I was very considerate and understood flying a lot better than a lot of people that I was learning with. I think I realized I was in trouble later than the ground people did.

The medical people that were on the ground knew the effectiveness compared to normal. See, I thought I had to get up to about 160 degrees in order to feel it. But I stored 35 percent more heat than I did on that test, so I was somewhere like 33 percent into heat prostration by anybody's standards. I stored enough body heat to just have been wiped out, and that's what they knew on the ground. I didn't recognize the temperature as it impressed the human body under the conditions. So I felt like it would be all right, but they recognized it.

They had a bunch of scientific observations for me to do. See, people couldn't understand why Joe [Kittinger] had never seen any stars in the sky [during the first Manhigh flight]. So I was to look for the stars in the sky and look for the green flash on the west at sunset. We had some astronomy-type things, sort of basic observations. I was to describe what you could see, and what I was looking at was like a fluorescent bulb all the way around on the horizon. It obscured the horizon, visually, because that's the atmosphere. So for the first time, I could see how thick the environment is on Earth. I could see that it was very shallow. It wasn't very deep at all, just 100,000-plus feet. And they had a spot photometer for me to take actual luminosity readings on these different spots around. We also had a missile firing, and I was supposed to see how well you could see that, but they cancelled that.

Being at 26,000 feet pressure altitude was not factored into the ability to draw water from this storage compartment that was based on a slight vacuum. The straw was a clear plastic tube, and I saw vapor lock in there every time I would pull water through. Nobody realized that water was going to be so essential. So to get it to flow at the rate I needed it, it would vapor lock the water. I was saying, what are those bubbles doing? Every time I would pull on it, needing more water, these bubbles would come and the water flow would stop. That's only after the cabin heated up. And the more it heated up, the more the problem got worse. Of course, it doesn't take much temperature difference to cause a big vapor difference in a 26,000 foot cabin altitude. You can get vapor lock real easy at altitude. Anyway, that caused a reduction in how much water I could draw in, and it wasn't coming in at a rate I needed to keep my temperature under some kind of control from that kind of effort.

I got about an hour or two into the observations. The flight was supposed to be thirty-six hours, and it was only a total of thirteen hours. Most of it was descent.

I didn't want to come back. I was absolutely certain I could still fly. Of course, I couldn't have. I would have died if I had stayed there. I was over the San Andres mountains, very steep mountains, and if I had hit in the gondola at 8,500 feet with a 4,000 foot drop after that, rolling down a mountainside, there wasn't any good way. See, that cabin had two things in there. It had a liquid oxygen container that was fragile. It had lithium hydroxide in a lot of water. It was caustic, saturated with lithium hydroxide. I was going to get severe chemical burns all over me if I hit and rolled down. I was going to have the possibility of loose liquid oxygen coming in that cabin.

There was some discussion, at one point, of cutting me loose from the balloon. I had a cut-down override switch which they had ordered taken out after Colonel Hessberg got there. I don't want to be complaining about Colonel Hessberg. He was a fine person, as far as I was concerned. But on the other hand, he didn't have much background information on the flight. I knew the condition I was in, but the radio quit working and I couldn't mash the switch any longer anyhow. I dropped the spot photometer, and it messed up the switch.

I could see my heart rate rising above 180 every time I would raise my head. I had a heart rate running 180 from the heat overload. I was in good physical shape, really good. But I could see that I could kill myself by trying to do a pushup, if I had wanted to, just by my arms. I could have done three pushups and probably died right there. Because my heart rate would go above 180 in a minute. I mean, it would go immediately when I just lifted my head. My body temperature went to 108 degrees Fahrenheit by rectal thermometer, and it was accurately calibrated before and after the flight. That's the highest recorded rectal thermometer body temperature without any permanent damage. Of course, there's some argument about that!

This is an open-bottom balloon. Its design altitude is when the envelope is full and the excess that's carrying you up vents out the bottom of the balloon. When you open the valve, which may be a foot or two in diameter at the apex of the balloon, particularly in the daytime when the sun's heat is coming on it, it's going to make up the difference in lift. There's almost no gases drifting out because there's only lift pressure. It's just way too slow. You just had to sit and wait. And frequently, there were things that were happening that countered this, such as solar input that caused the rest of the gas to take up the thing, and you weren't going down. You might even go up while you were valving. It takes a couple hours of venting one of these valves to start down.

It didn't seem to be any problem to me during that time, because it was early in the flight. I thought, I hate to do this, but, you know, the boss is the boss. And I'll follow his commands as they give them to me, regardless. And I did, every time they told me how much to valve. Of course, I knew that they wanted me to come down, from calculating this.

When it got to be late in the day and I had so much heat stress, reaching up to valve off would send my heart rate up. I would get a very bright color in my field of vision, and, I mean, it was a real overload. Just the simple thing of lifting your hand and putting it on top of a switch, was a heat overload and a physical overload. The only thing I could do was just sit there and try to relax as much as I could. I could make one or two clicks on the microphone after this, and that was about it.

I knew exactly the control I had, and I knew what velocity I needed and at what altitude. I had it all figured out ahead of time. I was just controlling my descent. It was dark, so there was no chance I got solar heating on the balloon. It was the opposite of solar heating, so I was just clicking off ballast. I knew exactly how much vertical speed would be dropped by ballast discharge if I waited long enough. So I would drop it just exactly that much and I knew I had that coming. I was willing to wait, even though I knew it was going down too fast. I would see it level off, reduce the velocity. Then in a few minutes, it would pick up again a little bit because the bag was cooling, and I would kill [drop] another battery or two. The battery bolts were not exactly right and they were not dropping, but I could tell what was happening. I knew the bolts were failing because of the sound and I could feel the cabin wasn't shaking. So whenever it wouldn't shake, I knew I had to drop another one right away to make up because none had dropped. In other words, there were some faulty explosive bolts. About half of them, exactly. Which was just another little complication, but it didn't matter.

They had a time-frame camera for the subject camera, and every so often they had a red strobe. I never knew exactly where I landed, except I knew I was on the side of a mountain and I stayed upright in spite of being on the side. Anyway, it was a reasonably accessible place, but not immediately. The chopper had to land some distance away. So I had unfastened the top ring when I saw Lee Lewis running across the field. I pushed the cabin top off, and it really felt good getting out in the fresh air.

Whew, I can even feel it today! He started climbing up, and I had to cut some of the wires that were holding that top so I wouldn't get entangled. But the last wires I hadn't cut were those holding the capacitors for the flash unit. Well, I knew it would shock you like crazy to cut those wires. I had my wire clippers in my hand, and Lee climbed up on the capsule on the up-side to balance it a little. And he said let me cut this for you, Cliff. And I said, "Lee, you're going to shock yourself. Don't you realize you could get killed?" He said, "No, let me have that." I said, "Let's figure another way to disconnect it." And he just reached up and cut it and it just knocked him off the capsule! *BLAP!!* He was laying on the ground and he said, "I guess you told me, didn't you?" I said, "Yeah, I told you! You shouldn't have cut that. We could have gotten them loose down here somewhere without getting shocked." You can get killed with one of those capacitors, even though it was battery powered. It didn't kill him, but it knocked him right off the capsule!

There were a lot of jokes made about [possible physiological effects of the flight]. Some people would say, "That's how come he's the way he is," kidding me, or something. But other people would say, "I knew him before, and he was the same then."

[Sometime later,] Simons called me over and said, "Would you like to fly in space?" And I said, "Yeah." He said, "You don't have to think about it any more than that?" And I said nope. He said, "Wernher von Braun can do it. He says he can do it in thirty days." But he said, "The Air Force and the Army and NASA are into a big hassle about it. Don't tell anybody I even went to the place." I said I wasn't going to tell anybody anything in the first place. He said, "We're going to put this capsule on a rocket and provide all the controls and the pilot. And we can let you do it since you have already done this Manhigh thing for us." But it never came. They never did it. It would have surely changed my life a lot had that happened. Whew! I knew exactly what all this meant. Nothing has surprised me about how far we have come. It was really frustrating that I could not continue, and that really bothered me.

I've been working hard to go into spaceflight ever since then, and I ain't there. I won't admit I won't ever get there, but maybe I won't.

Meteorologist Bernard "Duke" Gildenberg was involved with balloon programs at Holloman beginning in 1948. He worked on unmanned programs including Mogul as well as the manned flights of Manhigh and Excelsior. Excelsior was a high-altitude balloon project at Holloman headed by Joe Kittinger, who had piloted the first Manhigh flight. The main purpose of Excelsior was to develop and test a multi-stage parachute system that could be used at very high altitudes by either airplane pilots or astronauts during an aborted launch. Kittinger piloted three Excelsior flights in an open gondola, at increasingly high altitudes, and parachuted out each time. On the third flight, he stepped out of the gondola at an altitude of 102,800 feet. It took him 13 minutes and 45 seconds to reach the ground. The "covered wagons" Gildenberg mentioned were trailers with tall sides and a canvas top that protected a balloon during the early phase of inflation. When the balloon reached a certain size, the canvas top was removed and the balloon floated in the open until inflation was complete.

After World War II, I went to New York University Research Department School of Engineering – it's called the Guggenheim School of Aeronautics, actually – with Charlie Moore [see his oral history excerpt earlier in this chapter]. Charlie was going there for his graduate degree. I was still an undergraduate when I started working on Project Mogul, unknown to me because we didn't call it Project Mogul. Charlie and I thought the name was classified, incidentally, up until recently. But we don't even remember the name. We didn't call it anything. We had a bunch of others along with it.

In the service, I started out like Charlie, wanting to be a pilot. So did everybody else. I got part way through aviation cadets, and then got washed out on a back physical. How strict they were varied from month to month, depending on how many pilots they needed. I happened to hit it at a time when they had an oversupply of pilots, so I got washed out. You pick a selection on the air crew and a selection on the ground crew, in case you wash out. I had a nasal problem, supposedly, so I couldn't get on an air crew either. I was on the ground crew, and I just picked weather, since I had a course in high school. It determined my whole career.

My first security clearance was in 1944 at the Pentagon. I worked at the Pentagon for about six months as a meteorological cryptographer. I had one detail for a while where I was decoding messages from the resistance fighters in Europe behind the lines. I'd keep up with them for a while in a certain position, and all of a sudden they would stop. We didn't hear from them again. We didn't know whether they got wiped out or if they were just changing positions. We also worked on decoding messages to and from bombers in route to bombing Japan in real time. We were actually trading weather reports with the Germans via Sweden, which was neutral. During the war, there were no weather reports in the paper. They were all classified. But you needed a certain amount for, at least, commercial transportation. So we made a deal with the Germans that we would trade so many weather reports for an equal amount of their weather reports. I was involved in that a little bit at the Pentagon, selecting reports they were given and so on. That was kind of interesting.

When I got to the university after the war, on the GI Bill, the research department advertised for people who had some experience with weather balloons. So myself and Stan Greenfield went in together, because we both had radiosonde experience. Charlie Moore hired us.

When I was still a sophomore, the first summer, I came down to Holloman in 1948, working on Project Mogul, among other things, with Charlie Moore. Actually, we started at the research department in late '47, at least myself and the future Dr. Greenfield. We were building what turned out to be equipment for Project Mogul. We came to Holloman in '48, and we also flew Mogul flights in '48 and '49. I don't even remember the damned payload. I was part of the launch crew, and I was also doing the weather forecasting. I was too damned busy to pay attention. The only thing I remember was that it was Columbia University handling the Mogul payload. It was just sounding equipment.

By '51, I joined the Holloman Balloon Branch. The first project I worked on was more Project Mogul, except we didn't call it Mogul. We called it the Atmospheric Sounding Channel. I guess it was classified. We were on twenty-four-hour call to report at any time in order to get a mission up to coordinate it with a test out in the Pacific.

In '47 we flew the first polyethylene balloon from Holloman. And in '48 we were mostly poly balloons, but we were experimenting with a few compounds of rubber and so on. But, like, 90 percent were poly balloons by that time. They were mostly 20 feet in diameter, which was the standard size in '48 and '49. Finally in '49, we got our first 72-footer, which was the next size up, just a midget nowadays. But we were so wary of it, we got extra people from the base in order to help us launch it – something that a two-man crew could take care of now, but our techniques were so primitive.

We were part of what was known as E and A, electronics and atmospherics, which included the Aerobee [missile] and the high speed track [see Chapter 6]. We were all together. Then we moved to Building 850 in 1952, when it was completed. All of the German scientists that were at Holloman were in there with us. They were working on the track. We had this stratosphere chamber in there also, and the Aerobee rocket. And they gave us supplementary help on our balloon program.

The Aerobee did a program where they dumped some sodium compound way up in the high stratosphere that lit up the sky. We also did that with a balloon in 1957. I can remember people in Alamogordo on the rooftops, all set up to watch it, because it covered the whole sky. Even the people in Phoenix could see it. We did a duplicate of [the Aerobee dumps] with balloons, but it was a much smaller cloud. We had a bunch of programs as we duplicated it. We flew the same payload aboard the Aerobee as in the balloons to get data points at different altitudes. Not necessarily at the same time, but we basically were doing everything that the Aerobee was doing.

The meteorologists in our balloon branch worked for the Aerobee every time there was a mission, making wind forecasts and getting wind data. We had some balloon programs that were parallel to the Aerobee. We had at least two programs where they were flying the same payload that we were flying, and we had a launch at the same time. So we would inflate and have the balloon ready, waiting until they took off, and then we would take off a few minutes later. We had one where we went to a lot of preparation and got everything inflated, ready to go, and the Aerobee blew up. It was a mess! It blew up on the pad. We were getting ready, you know, "Three, two, *boom*! Hey, where is the Aerobee?" They said, "Well, go ahead [with the balloon] anyway, and we will get a data point off of it."

Skyhook was just a name for the poly balloons. It was a Navy program from the start. The way this whole thing worked was that the Air Force and the Navy both decided that they wanted Skyhooks developed for research and development as a result of the Japanese balloon bombing program. In fact, I remember we had the reports from Japan on our desk in Japanese with English translation next to it. That was the reason they started this whole thing, figuring it might be a good vehicle for research and development.

The 119-L project [also known as Genetrix, a secret program in which unmanned, camera-carrying balloons were sent over target areas in the Soviet Union and other hostile countries] was only declassified this decade [the 1990s]. It was so secret, we are not even 100 percent sure what year it was declassified, but it was this decade. We did the first test flights at Holloman in 1952. Then we did crew training. Moby Dick was unclassified. It was a predecessor of 119-L, besides being a cover-up.

In '51, first I did Mogul. Then, when we more or less completed Mogul that same year, I was sent as an Air Force coordinator to Wright Field, Ohio, where we were launching the first test flights on Moby Dick. There were competing universities, and one of them was New York University, my old crew. I knew everybody there. I was supposed to be there as an observer, and I wound up working with the crew, "Hey, give us a hand." I was supposed to be there just taking notes, and I was working twenty hours a day. Tufts University, under Dr. Howell, was competing for the instrumentation. Plus, we were learning how to make the balloons function, launching techniques, and whatever. Then I came back to Holloman, and we did more Moby Dick test flights for two years with 119-L.

Charlie Moore came down for some 119-L tests with us, and other 119-L tests were done up in Minnesota. But they preferred here, because of the classification. We had an armed guard twenty-four hours a day. The whole thing was covered with a shroud. You would take it to the launch site and ten seconds before the launch, they took the shroud off. But you couldn't tell anything by looking at it, anyway. The instrumentation was all inside. It was kind of ridiculous, but one way or the other, it was pretty heavily classified. I think some of the original plans were to go all the way to the East Coast with it. We have one or two that we cut down short, because we were having problems on board, and even there we sent out helicopters with armed guards to the recovery. So that became part of the UFO story too [see Chapter 13].

We did test flights, though not all of the test flights. And we did [launch and tracking] crew training for Moby Dick, which turned out to be crew training for 119-L also. After it was all over, we got some of the crews back. All of a sudden, we had 120 extra people we didn't know what to do with, back from 119-L. I did some of the forecasting for evaluation of 119-L. For Moby Dick, we had three sites on the West Coast. One was at Vernalis, which is near Frisco, and another one was at Edwards Air Force Base, and another one was at Tillamook, Oregon. We would release three at a time, and they were set for three and a half days to traverse the whole country. We were using that to get information on wind patterns for traversing the Soviet Block eventually with 119-L. That wasn't stated, but that was what it was about. So in fact, on the 119-L payloads we test flew, we had a Moby Dick insignia. We had all three crews at Holloman, and we were using covered wagons to launch the balloons with. So we took out for the West Coast in covered wagons. (Laughs)

The balloons were a big part of it all along. In fact, we were in trouble with [President Dwight] Eisenhower. He hated us, because of the lines of communications more than anything else. We had problems with the first balloon, which was supposed to fly the thing at 90,000 [feet], and at the last minute they switched to the smaller balloon with a flight at 60,000. We told Eisenhower, "There is no sweat, they can't touch us." But that was at 90,000. We start flying at 60,000, and we started getting shot down. Not all of them, but a fair percentage got shot down. So Eisenhower said, "Stop!" Then they released a few after he said stop. One way or the other, it was a big mess. But they found a few [Soviet] targets that they had not known about before. The people who did the camera work did use the same cameras on the satellites, and the crews that did the recovery, did the recovery on the satellites also.

We did [a test] at White Sands that was a riot, out at the [White Sands National] Monument. We got there about midnight, and there were a bunch of trailers there, visitors, and they were all asleep. So we got everything prepared. We tried to stay quiet, so we wouldn't wake the people up. We got everything prepared, and then we had the dummy on this little porch on the outside of the gondola, standing there in a spacesuit. And we had a search light on it, a spotlight. Then right around dawn, when it was still a little dark, some of the people stirred and got up and looked out and saw this thing in a spacesuit and it looked like a spaceship. And God knows what came out of that!

Anyway, we had a parachute failure, and we had a freefall from 80,000 [feet], with about 1,000 pounds, [and it] landed right in back of the drive-in theater, *boom*! It was about seven or eight o'clock in the morning. We quick got there and dug everything out and got the heck out of there. We didn't damage anything, luckily.

I think at least twice we got cut down by the radio station playing

"Tiger Rag." It happened to synch with our telemetry tone.

I was involved with all of the Manhigh flights. Doc [Dr. William Douglas, see Chapter 9] checked out all of the NASA astronauts, as a result of his experience with us on Manhigh. He said he wanted me since

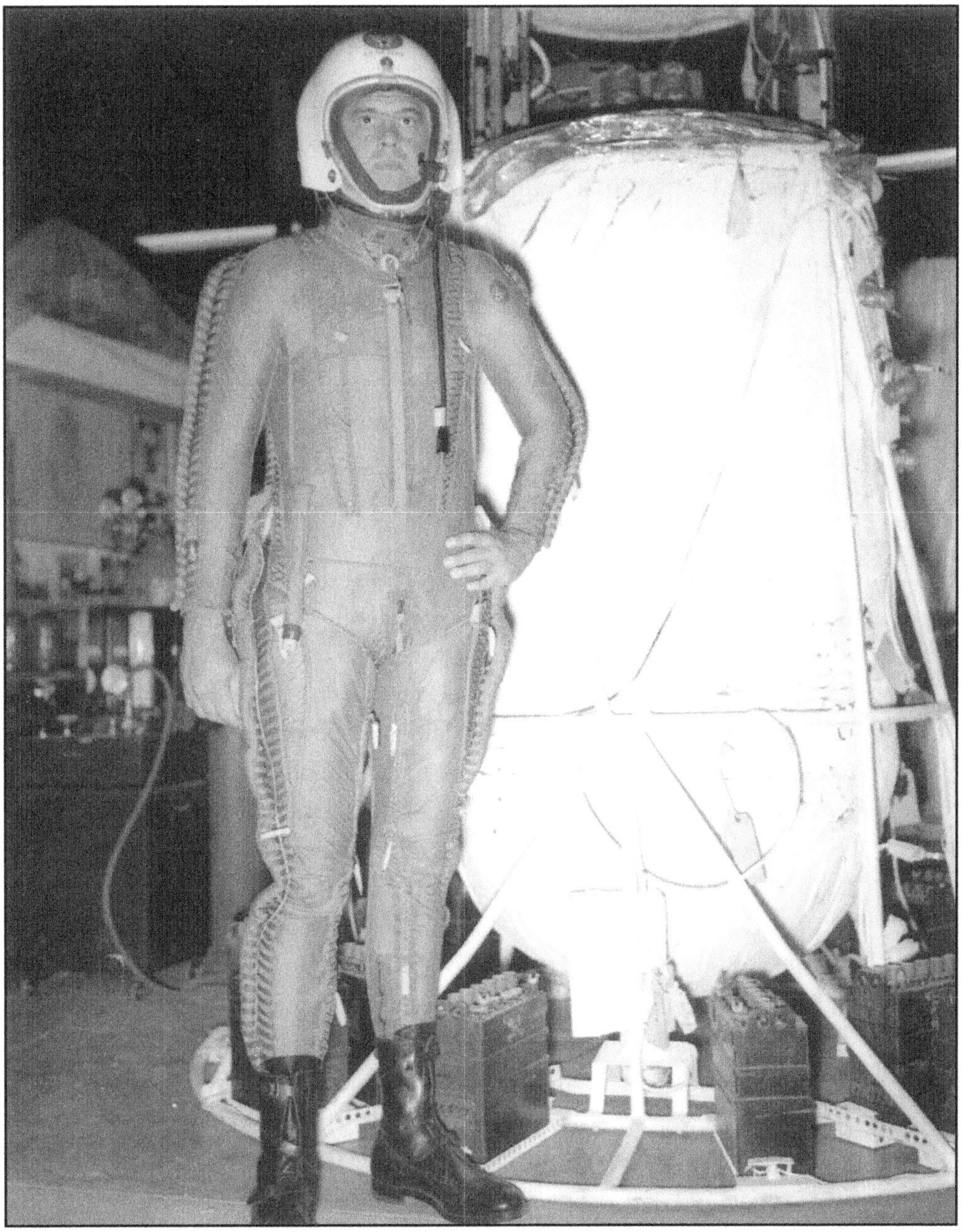

Clifton McClure stands beside the Manhigh capsule, wearing the confining pressure suit he would wear on Manhigh III. Courtesy of NMMSH.

I was a balloonist, an amateur astronomer, an amateur photographer, and a whole bunch of other things. I can remember one of the dangerous parts of this is where there is an unknown factor in the stratosphere with cosmic rays. You might wind up being sterile. I said, "Well, since I am free and a bachelor, that might be a good deal." Although, we had been doing test flights with animals about the same time to analyze the effect of cosmic rays on mice and hamsters and the whole works.

As part of the training on Manhigh, we had to get our balloon licenses. I am not sure we *had* to. We did it on general principles. It was before [modern] hot air ballooning, and there was very little sport ballooning, if any, done in the States in those times. We got in our flight times, and then the three of us, me, Joe [Kittinger], and Major Simons went to an FAA office down in El Paso to take the written test. The written tests were for the fabric balloons in 1930s. Luckily, I had the textbook as part of our research for developing the poly balloons, so I could study what the other lunatics did before our time. We went and took the test. They left the three of us alone, so we would say, "What is Number 4?" We all passed the test. But it was ridiculous anyway. It was for controlling the old fabric balloons instead of the new poly balloons.

We all soloed out at the White Sands, me, Joe, and Major Simons. That was a ball, in jump balloons. We were all pilots, but neither of us had ever flown a balloon before. So to save time, we'd do it in a small jump balloon. They weigh you off until you weigh zero, and then toss you up. You go up, *wooooh*! It was a ball. Then it would start down, and we had a bag with ballast on board, and you would just take a handful and watch the ground coming up, *chuh, chuh, chuh*. It was one of the best rides I ever had.

We were all pilots, but we had never done any vertical landings. The first couple of times, we weren't sure how fast we were coming down, because we weren't used to that kind of approach. So we kept thinking, well, discretion is the better part of valor. We kept on ballasting more than we really needed to, to start with. But it was a damn ball. It was around sunset. It was beautiful lighting out there. It looked like we were landing on Mars. We had a light wind, and every time you'd go down toward a sand dune, you would go *woosh*, like that. That was a ball.

I didn't want to come down. But they tied a long rope onto my belt. Doc got hold of my rope and started working me down like a big fish. He says, "Hey, we can't stay here all night. You're down, buddy." Anyway, the three of us soloed that way, just to get a feel of balloons. Then we started accumulating time for experience, plus we had to get our licenses.

We were doing a flight that was supposed to be an all-day flight, launched from Truth or Consequences [a town north of Las Cruces]. We didn't have enough experience with manned balloons to start with, and particularly in New Mexico, where with the heat, turbulence can be just like landing an aircraft. When you come in, it is like this *harrumph.* We were going to get out and let Joe solo. We started out in the morning; it was perfectly stable. It was beautiful. And then around noontime, we were down around San Antonio [New Mexico]. We started down, and then we hit a thermal, a downdraft. And we really hit hard. Then we started up again, and I could tell this one was going to be even worse. So Joe, like a monkey, sprang up into the rigging. He had terrific reactions, and I just reached over and caught the edge of the gondola, because I could see it coming. Major Simons was in the middle, and he had been thrown to his knees by that first impact. He didn't have time to get a hold.

When we hit the second time, I was fine, except he got thrown right into me. At the time, I was hurting, but I wasn't that bad. Dave said, "Oh it's probably cracked ribs. If one is broken, we can't stand another impact like that. It could puncture a lung. So we have to stay up until it quiets down. When does it quiet down?" I said, "Sunset." So we stayed up for another four hours using ballast over the mountains back of Socorro. Finally we had a landing. I wound up with three cracked ribs and a vertebrae that is still a little bent. It took me out of the program. It turned out it was a good thing, because I was supposed to do that first flight, so Joe took it. We learned right away, you don't do science to start out.

We had valves and ballast, so we were able to more or less determine where we were going. In fact, there is more control than in a hot air balloon. That is another interesting part of my history. Raven Industries, with their hot air balloon development, had to get FAA approval of the vehicle. So they had a test flight set. It required burner off at 10,000 feet and then a landing that was no faster than a parachute. They were going to do three test flights with it. Their crew chief was a great pilot, but he decided to go back to Alaska to fly. So J.R. Smith [Smitty], a Raven executive from the old NYU bunch, called me up and he says, "Hey, I know you are a pilot and have flown balloons. Do you want to take this?

So I says, "Well, yeah, if I can get away. Get me the specs." I started working on the specs, figuring on what I'll do. I am going to bail out if the descent velocity got too much and I couldn't control it with the burner, et cetera, et cetera. Then I got delayed because we had a hot program come in. Finally, he called me up and he says, "Our crew chief has come back in." That was the first [modern] hot air balloon which, incidentally, was classified. The Army wanted to see if they could take tanks to the front lines and drop them off. We know there is no way we could do it today even. But they were evaluating it with a hot air balloon.

Then the Army demonstrated it was impossible. Raven Industries said, "Well, what are we going to do with it next? Let's make a sports balloon out of it." That is when the business on the test flights came up. I remember Smitty saying to the vice president of Raven, one of the old NYU bunch, "I went up in a hot air balloon, and there is not as much control as in a helium balloon. In my first flight I landed in a swamp up to my neck. You can have it! I got kids." I know you don't have as much control as a helium balloon, because you have that thermal lag with the burner. And they don't carry ballast, per se. With ballast, Joe used to come over telephones lines, and wait until the last minute to jump over them. He said, "I am going to have some fun with this balloon. It is so slow after flying jets, I want to jazz it up a little bit." You could do that with ballast. You can't do it thermodynamically. It just takes too long.

Then we went on to Manhigh I with Joe. That was up in Minnesota. I was on the Winzen launch crew, plus I was doing the meteorology. I can still remember all the cheering when we broke altitude records with Joe. I can remember everybody cheering when we read off the altitude, "We went up to 75,000, *rah*!" Then we finally got to the top, which was about 90,000 feet. Some more cheering, then he ran out of oxygen. That was the classic event where we said, "Joe, open up the valve and start coming down." And then he calls back, "Come and get me." (Laughs). Everybody stared at one another. Some were wondering, "I wonder if he has run out of so much oxygen that he is goofy already?" But the rest of us said, "Well, knowing Joe, he is just kidding." We called Colonel Stapp. He got on the mike and said, "Captain Kittinger, you will start down now." "Yes, sir." That was Joe all over. He was a ball. So we got him down.

Everything worked out okay. We found out what the problem was, and then we did Manhigh II. Actually, it was launched by Winzen's crew, but we supported it from Holloman with aircraft and with telemetry and tracking vans and a whole bunch of other facilities.

Manhigh II was interesting because in those days, we *were* the space program. We had all the important newspapers in the country following us around, including *Time* and *Life* writers. It was a real interesting experience. We got to talk to those guys a lot. I watched them do their stories when it was all over. They were racing to see who got the front cover. They couldn't both get the front cover. That was the agreement. So they both got a whole bunch of coffee and cigarettes stocked up in their hotel rooms and worked for ten hours straight, getting this story out. Then they telephoned the whole thing in. *Life* got it.

The Manhigh gondola looked like a garbage can. You are inside a garbage can for thirty hours. We had a special diet set up for us. We were able to urinate, as I remember. But obviously, there was nothing else you could do. That was no problem. We were all pilots and used to flying solo for long periods. When you were younger you were able to keep it in. Ostensibly, among other things, they used the capsule to check for claustrophobia. None of us had a problem with that. The main reason was because we didn't have a specially designed spacesuit for Manhigh. We were using the jet pilots' pressure suit, which was okay for a short jet flight of four or five hours. But for thirty hours it was murder. We were so uncomfortable in that pressure suit that nothing else would bother you.

The pressure suit was tough enough, but then we had all these telemetry sensors. It took them about half an hour to wrestle you into this thing. You had long underwear underneath and then a rubber bladder. When you were all sweated up from getting inside this pressure suit, there was no place for all that sweat to go. So for thirty hours you were all sweated up. You didn't realize it until you'd take your suit off thirty hours later, and you're still wringing wet from the day before. By the time you got through, oh man! Anyway, also we were checking out your response over time to the various instruments on board. We had to go through emergency procedures on the cockpit. I was the first because I was going to do Manhigh I. So I was the first to find that the tireder you got, the quicker your response was to punching the stuff on the board, because that was all you were concentrating on. Earlier on, when you were fairly alert, you would think about other things like writing your report, et cetera. "Oh my God, there is a light. I need to go into it." But the more tired you got, the more you concentrated on the dashboard, so your response time actually picked up. As I remember, I lost about seven and a half pounds in that thirty hours, just from sweating it out. But beyond that, there was no particular problem. The experiences we got on checking out me and the other pilots went on to NASA. Doc checked out all the initial NASA astronauts, and they used some of the experiments from those.

But we were it. We were the space program. Everybody was asking for autographs. We were in all the papers. We had headlines in the papers. People forgot about that. They [the Manhigh pilots] were the original astronauts.

Manhigh III started out in Minnesota. But when I got up there, I did some analyses. I decided that the only *go* situation we had that met the specs for that October in Minnesota was a hurricane off the East Coast. (Laughs) So I persuaded them that the best month for launching at Holloman was October, and we are just not going to make this in Minnesota. We got 500 people here wasting TDY. I persuaded Doc immediately, and he talked Winzen into it. So we got down to Holloman, and we had one problem with the first try, but the second shot was perfect. It was dead calm all morning. Got off straight off. And Manhigh III was right out of the movies, too. In fact, I was talking to this screenwriter, trying to get him to make that part of his movie. You know Hollywood. What he was going to do was to take all three and make them into one mission.

Manhigh III was the one that was the closest to the movies, in a sense. We started getting into trouble toward sunset with a temperature problem on board. We were on a cargo aircraft, circling below it. It must have been bigger than a C-47, because we had a long table with a bunch of scientists and doctors on board. We were relaying up to McClure to get our technical information. I was doing the meteorology, asking him to look in different directions to see what he saw, et cetera. Then he ran into trouble, and that was the end of the scientific part of it.

But then there was a real Hollywood kind of drama, where we had to get him down alive. The doctors said, "First thing, cut him down in the parachute." We told them, "Oh no, the opening parachute shock at 100,000 feet would probably kill him. We are going to have to valve him down." They said, "Okay, but start in a hurry." We were all around the table, looking at one another. They kept on saying, "Faster, faster, this guy is not going to make it!" We had to get down to the tropopause, where the minimum temperature was, that would cool it off. But they kept on saying, "Go faster!"

Otto Winzen, the manufacturer, and myself were doing the valving, and we said, "You can't valve too fast because the balloon can break." The balloon design is not for descent. It was made for an ascent. All of the aerodynamic characteristics are wrong. They are more likely to fail during a descent than during an ascent. So we had to limit how fast we could get him down. We had remote and manual valving capabilities, so in case the guy passed out, we would take over. And we were giving him the valving amounts, anyway. I had a table for him in case we lost contact. But basically he had enough things to do, and we were doing the valving.

The doctors were finally saying things like, "We will give you five more minutes, and you are going to have to go on the parachute because the guy is going to be dead anyway." Looking around the table, everybody was real strained, and I thought, "God, this is like a Hollywood movie. You've got five minutes or this guy is dead." Finally, he barely made it, and the temperature started to drop degree by degree. But it was interesting.

For protection against the extreme cold at high altitudes, Joe Kittinger wore multiple layers of clothing over his pressure suit on his Excelsior flights. The extra clothing and safety equipment doubled his weight. Courtesy of US Air Force.

McClure was an absolute monster. I mean, he was built like a boxer and he's got a Master's degree. We were buddies. We used to fly together all the time in his glider. He told me afterward, "You know what I was doing up there? I was trying to think what kind of condition each one of you guys were in so when you gave me ideas I can evaluate how good they were." He says, "I know how 'so and so' functions under stress, and I am not sure I am going to buy anything he would say." You know, he would have been the perfect astronaut. And the only reason [he didn't become one was] because he couldn't get F-100 time, because he wasn't base ops. He was such an outstanding young man.

The other thing in that Manhigh III drama was that he was right over the San Andres [mountains]. They were feeding me data, and I did all kind of analyses and decided it looks like a pretty sure shot, the balloon is going to drift off and land on a pole line road, right alongside the San Andres. So I told them that, and then afterwards he said, "You know what? I thought there was a possibility you were telling me that just to make me feel good. So I didn't buy it a hundred percent. I was thinking of an emergency procedure just in case I really did land in the mountains." That was the kind of a guy he was. Well, he was test pilot quality. So that was really a drama, and we did get him to 100,000 feet and we did get a certain amount of data out of it.

We had great plans for Manhigh. We wanted to do a gondola with an outside porch, and then have the guy do kind of a spacewalk, go out on the porch and check the pressure suits. That was before Joe did it on the Excelsior. We actually had run some plans along with Winzen, but we just ran out of money. NASA was getting most of the funds. But it was a prelude to space, and experience that Doc had in checking us out was used for checking out the original NASA astronauts. All of the experience we got with the Manhigh gondolas was applied to the first NASA manned rocket shots. Dave took all of that data and went to NASA with it, and they designed some of the early stuff. I think even part of Gemini was based on our data from Manhigh. So, we really contributed with the first space capsules.

Anyway, that finally got straightened out, and then we got into the flights of Excelsior. Joe was a test pilot, so he wasn't that excited about it; same thing with McClure. They considered it a little hairy, but doing acro [acrobatics] on high-speed jets was just as hairy. I don't know if Joe says it in his book, but particularly the last flight, we knew that was going to be the real dangerous one. Everybody was uptight. He arrived a little later than he was supposed to, and he goes around to all the various groups calming them down, singing a song, patting them on the back, "You guys relax, you might make it." He was the one that was going up. The guys were just crazy about Joe. He did that to a lesser extent on the other flights too. He was as happy as a jaybird, as they used to say. And the crew was just crazy about him. That's why they were so uptight. He was just one of the guys.

An equipment malfunction during the Excelsior I flight nearly cost pilot Joe Kittinger his life. His experimental multi-stage parachute system deployed too soon, and the shroud lines wrapped around his neck. Kittinger lost consciousness. Floyd Fannon, who was working in the field at Holloman, saw Kittinger plummeting toward the ground before his emergency parachute deployed automatically.

I remember once, we were out setting up for an Army mission near Denver Site, which is out in the middle of White Sands Missile Range. We hear them counting down to a guy coming down in his parachute from a balloon. Some of the other fellows were at main radar on it. They'd keep track of where he was. I looked up just as Kittinger came out of the clouds, and this is the day that he got tangled up in his chute.

He's spinning and he's falling like a rock. You wouldn't believe it, you know. All I could think was, I left my shovel home. How are we going to dig him out of there? I can't believe I was going to dig him out, and get him back on his feet so he could go on the next mission. I kept thinking, what am I going to do? I haven't got a shovel, and I don't think anybody else brought one. How are we going to get him out? The ground's pretty soft out there. I had convinced myself that he was going to hit in a puddle of mud and we'd just dig him out. You know, it would definitely kill him. We just thought, well, he'll hit out there in a mud hole, and we'll dig him out, and we'll go again tomorrow. We thought that he was indestructible, and he almost proved it.

No chapter on using balloons for space research would be complete without including comments from Joe Kittinger. During his years at Holloman and Wright-Patterson, he piloted the first Manhigh flight, developed the Excelsior program, and piloted its three flights. His experiences in those programs are well documented in several books, including his 2010 autobiography, Come Up and Get Me. *When this author contacted retired Colonel Kittinger, he said he preferred that his contribution to this book be in the form of a new interview that would include his recent involvement in the Red Bull Stratos program. In the final flight of that program in October 2012, skydiver Felix Baumgartner broke three of the records Kittinger set during his Excelsior III flight in 1960 – highest parachute jump, and longest time and longest distance of freefall. Kittinger, however, still holds the record for the highest balloon flight in an open gondola. Kittinger was inducted into the International Space Hall of Fame in 1989.*

Working in the early days of the space program was exciting, because there was a lot of people that didn't believe that we were ever going to go into space. When we first started our research back in the late '40s and the '50s, people thought that space was something that we would never, ever go into. It was Buck Rogers. It was something that was not going to happen. But there were visionaries like Dr. Stapp. My part of the space program was because of Dr. Stapp. He gave me the tools and the opportunity to do the work that I did. He was a visionary, and thank goodness we had visionaries like him. But there wasn't very many of them, I'll tell you that. It was a very difficult task to get funding, to get any support, because people had other priorities for their money and they never thought we were going to go into space. It was a real challenge to get support for the programs back in the early days, because people thought we would never, ever do such a thing as go into space. But Dr. Stapp pressed on and knew that there was a need for this information, and gave people like me the opportunity to do research.

There has been an awful lot of space history written in New Mexico, and it's still being written. We just made the Red Bull jump out there in October [2012]. That was a real contribution to space. I personally picked that site over others because I love New Mexico, and I wanted to see it done there. Of course, that was the best place to do it, also.

The purposes of Excelsior and Red Bull were different. I was doing it for air crews. I was doing it to come up with a means of escape from high altitude, for aviators and astronauts. Felix's project was not associated with rescue for astronauts and pilots, but more of testing a new generation full-pressure suit. When I made my jump, I had a total of thirty-three jumps. Felix had over 2,500. So there was a big difference in the experience level and in the objective of the two programs.

But we both used balloons. We used the Holloman balloon crew, who had been launching balloons since the '40s. And we were still using New Mexico. There was continuity there, too.

My jump was made in 1960, and this was made in 2012. There was a big difference in the technology in those fifty-two years. He was in a pressurized capsule, and I was in an open gondola, so that was a big difference, also. He used the next generation full-pressure suit, and I used what air crews were using at the same time. So there were a lot of differences, and there were a lot of similarities.

I volunteered for the Stratos program. I knew how to do it, and I knew the problems. Going back in history and being a part of a very interesting space-type program, I knew what Felix was going through, and I knew what he needed to concentrate on. I designed the life support system in the [Stratos] capsule. I was active in the engineering aspect of it, from the very beginning of the project. It was almost a full-time job, and it was a very interesting four years of my life.

Chapter 5
Bopping the Daisies

The early 1950s were an exciting but challenging time in the history of spaceflight. The United States was developing rockets that would soon be capable of carrying men into Earth orbit, but many questions remained about the human ability to withstand such a trip. As described in Chapter 4, the threat posed by cosmic radiation and the potential psychological trauma of living and working in the isolation and confinement of a small capsule separated from the familiar gravity and orientation of the Earth's surface were examined using high-altitude balloons to simulate the near-space environment.

Another serious question was whether the human body would withstand the acceleration and deceleration forces involved in rocket launch and reentry. Dr. John Paul Stapp investigated that question between 1947 and 1958, first at Muroc Army Air Field (now Edwards Air Force Base) near Lancaster, California, and then at Holloman Air Force Base in Alamogordo, New Mexico. Some of his experiments used a high speed test track (see Chapter 6), and others used equipment designed by Stapp himself. This chapter deals with two of those devices, the Daisy accelerator (or Daisy track) and the Bopper sled.

The Daisy track was a set of two rails, set 5 feet apart, that were 120 feet long. Various designs of sleds could be accelerated down the track and then brought to an abrupt stop. Because compressed air powered the acceleration, the track was named after the Daisy air rifle that was popular in the 1950s. A subject (dummy, animal, or human) could be strapped to a sled at various angles, facing either frontward or rearward. The sled would be accelerated for 42 feet, after which it coasted until it impacted a 4-foot-long, water-filled piston that brought it to a sudden stop. In most tests, the subject sustained a deceleration force between 10 and 36 g's. However, in one anomalous test in 1958, Eli "Lack" Beeding rode the sled facing rearward and survived an astounding 83 g's. Lengthened to 240 feet long in 1961, the Daisy track remained in service until 1985.

Another device designed by Stapp was the Bopper sled. It used a 20-foot-long track. A stretched elastic shock cord (commonly known as a bungee cord) pulled the sled down the track until it hit a pair of spring-tensioned brake pads. It could produce a deceleration up to 30 g's. The Bopper was used primarily to test restraint strap systems.

Ed Dittmer was working as a medic in the hospital at Great Falls Army Air Base (now Malmstrom AFB) in Montana around 1950, when his superior officer offered to send him to an air evacuation and aeromedical training course in Alabama. By the time he arrived at Holloman in 1957, he was a certified aeromedical technician. In this excerpt, Dittmer described not only the Daisy accelerator and the Bopper, but also another device used in deceleration testing at Holloman during the Stapp era: the swing seat.

I reported to the Test Track Division in February 1957. Being a medical technician at the time, there was a clause in my AFSC [Air Force specialty code] to assist medical officers in research, and that's how this assignment came along. This assignment was open at Holloman, so I applied for it and got it.

At the time I came, there were several phases of research going on in the Biodynamics Division, and they also had the Space Biology Branch. I was assigned directly to Dr. Simons in the Space Biology Branch. Being a crew member on flying status, I was able to participate as a subject on anything that Dr. Stapp deemed necessary for human subjects. However, on any of these tests, Dr. Stapp was "number one" to ride them. He'd take the test first; he'd ride the sled first. Any experiment that came up, Dr. Stapp was number one on the list. He never let anybody ride it unless he tried it out himself and made sure that it was comparatively safe for everybody. Of course, there was always a risk in volunteering for any of these things, but Dr. Stapp took the first one, so everybody else just fell in easy.

The Daisy track was in the Biodynamics Branch. It was for testing seat belts and different hardware. Later on, we went into a lot of the space hardware and space suits and everything, to automobile seats and seat belts, on up into airbags. In the early days, when I first came there, I was assigned to the Space Biology Branch. Occasionally, they'd need a subject on the Daisy track, the Bopper, or in the swing test. Being on flying status, that was my hazard pay, also. So, I did volunteer a few times to ride it. However, I wasn't a steady rider on the sled. My ride on the Daisy was a very low g and it was just a smooth, short ride. I really didn't have any of the harrowing experiences and hard rides that a lot of them took. In fact, I rode that thing just a few times.

The Bopper was a short sled. That was for restraint testing and it was run by bungee cords. You'd wind it up and shoot it like rubber bands, during the slack periods on the Daisy, when you were running animals, and so forth. See, we had a crew of people set up to ride the Daisy, and in order to keep them qualified for their hazardous duty pay, we'd run them on the Bopper and give them their g's on the Bopper with seat belt restraints and sometimes shoulder harnesses. We took it to a school fair and demonstrated the seat belt fastening. I think Colonel Stapp was the forerunner of the seat belt restraint system. In fact, before they ever thought about putting them in cars, he told us he'd get the belts and we could put them in our cars and wear seat belts. He was that enthusiastic about them. He knew that they were good restraints and that they were going to save a lot of lives. So he encouraged everyone to put seat belts in their cars. We had them in long before they came out with them in automobiles. And a lot of this developed from the Bopper. You ride it a few times and see how those seat belts do hold you in place. I don't know of anyone who was hurt on it.

I was personally involved in the airbag studies [on the Bopper]. Out of the several runs we had on there, the airbag seemed to be the best system going, because we could run them down there at 35 miles an hour and (*CLAP*) like that, they'd go into that airbag. Had that airbag not been there, they'd have been hurt. We never had a failure. The airbag worked every time, and they'd go into the airbag on these high-velocity runs, and they'd get a little redness, which is called erythema, on their arms or wherever they hit into the airbag. And that was the extent of the injuries.

The swing drop was set up with an old B-17 seat and a lap belt. They had a crane sitting there, and the swing was on two cables. The crane

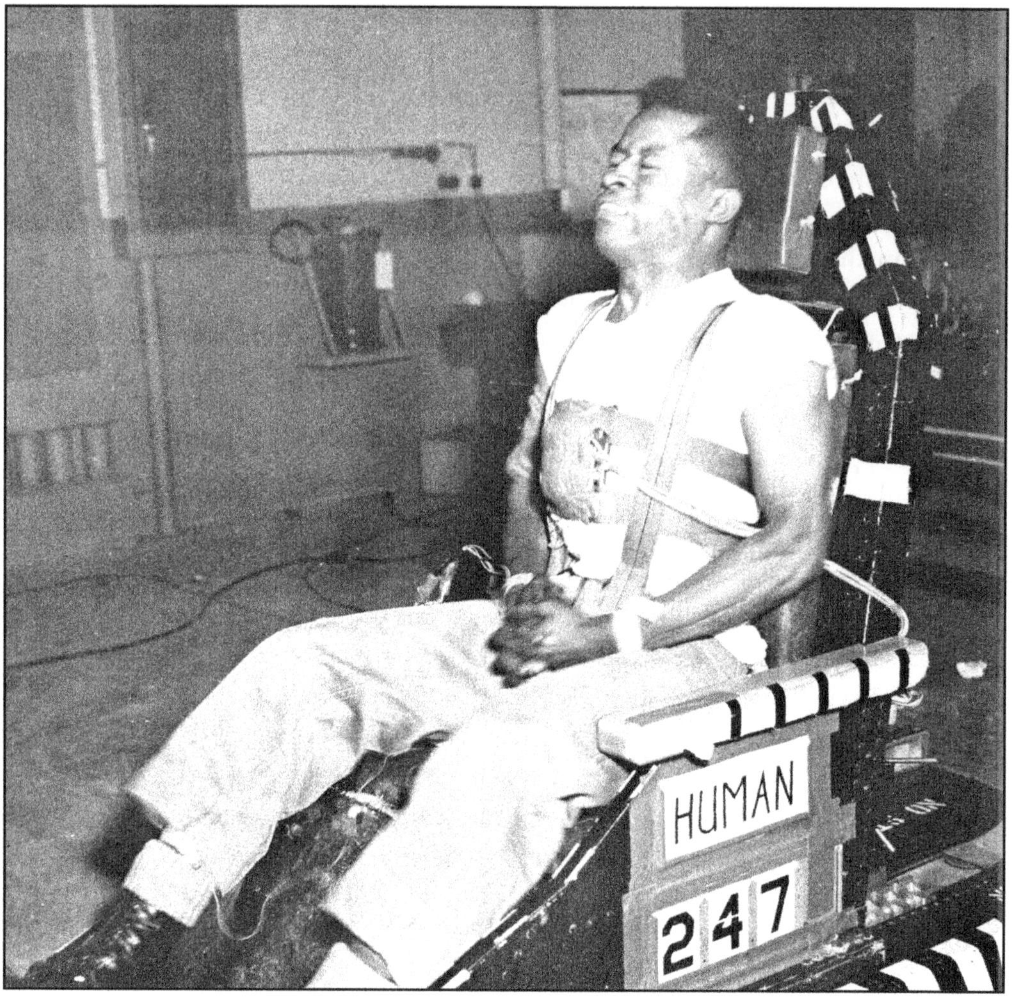

Although less powerful than the Daisy decelerator, the Bopper could produce substantial deceleration forces, as lab technician Alton Yates experienced. Courtesy of US Air Force.

would pull it up and they'd measure the distance in the air with a rod. The seat was hooked to what we called a "dead man's cable," which, when it got to the end of the cable, it stopped. So they'd drop the swing, and it would hit the end of the cable with the instrumentation, and in that way they got the g-forces. They'd start you out at 10 feet, 20 feet, 30 feet, 40 feet, and on up. This was the early days for seat belt testing, and of course, when you'd get to the end, you'd jackknife. They had these accelerometers and different instrumentation. At the bottom of the drop, they had these different electrodes set up so that when you hit, it measured the g-forces on the instrumentation recorders. Again, Colonel Stapp rode it first at the different levels before he'd let anybody else. I was second to ride the swing. We just ran them for a few weeks. It was quite exciting, a pretty good thrill to ride. It was different.

After graduating from high school in 1961, Earl Cline began working as a small arms repairman on rifles and machine guns. When he joined the Air Force, he spent four years at Holloman Air Force Base, where he participated in Daisy track tests.

I got my draft notice in November of '63. I enlisted and went into the Air Force in January of '64. I did basic training in San Antonio. Then I got sent up to Lowry Air Force Base in Denver to be a guided missile control system technician. The only thing I learned in tech school was that a capacitor retained a charge. I never made it in electronics. In August of '64, they sent me down to Holloman in Personnel. While I was down there, I thought, I have to do something more than just shuffle papers around. A sergeant, I think his name was Rodriguez, was picking the people that rode the [Daisy track] sled. He was talking about it one day, and I asked him if I could get a shot at riding it. He said he would put my name on it. He called me up about a month later. Told me to come out and I could ride.

I think he had to know you for a while before he would give you an opportunity to do it. And it *was* an opportunity to go out and do something like that. It was totally different than anything I had ever been exposed to. I didn't get to ride it enough. I only really got to do six rides. Then they sent me to Andrews Air Force Base, which shortened my career on the sled. But it was fun.

The first day out [at the Daisy track], we went in and they talked to you about what it could do to you and that type of thing. And then they showed the movie of Captain Beeding with his famous ride and the aftereffects of it. We went down to Fort Bliss and went through an incredible physical. I have never in my life been examined like they did down there. Basically, they called you one day and told you to come out. The Bopper was what we were indoctrinated on. That was like an 8-g ride. You just got on it, and it was the end of it. They did that to keep you qualified every month.

There was extra pay for riding the sled. I think it was either fifty-nine dollars per month, or sixty-nine dollars per month.

I did 25 g's on the Daisy track. We were facing on a 45-degree angle, and that did a few things [to me physically]. None of the people that were doing it when I was, dropped out. We had a lieutenant who went for his first ride on the swing seat, and they had a crane there. He asked what the crane was for. We said, "Well, when they put you on there and it shoots you down, it knocks you out. And they don't have time to mess with you. You can see all the people here that was riding. So they catch you with that crane, lift you up and pitch you in the corner until you come to." After he rode it the first time, it wasn't a big deal. The 8 g's, that was nothing at all. It was when they did some of the other testing, it got a little more serious, I guess.

One day, they had a tough schedule. I wasn't out there that day, but they always tested the sled before they put a human on it. And when they shot it down, the seat kept on going when the sled stopped because they didn't put a pin in it, or something sheared. I'm not sure what happened. But the guy that was the first after that was white as a ghost, from what they tell me.

You were just led out there, and they put you down in the thing, and you really didn't have time to think about much of anything.

They would call you, and they would say "Tuesday morning at nine o'clock you are going to be doing some rides." So you would just tell your NCO [noncommissioned officer], I was enlisted, that you were to ride on Tuesday morning. You would be out at the sled and get prepared. You wore, like hospital scrubs, those green things. I have a hairy chest, and I had to be shaved, or I had to shave myself before I went out, because of all the EKG connections, and stuff like that, so that they wouldn't have any problems with them falling off. And they put a wiring harness on you. I mean, they tested every vital sign, I think, in your body. They plugged you in to monitors before you went out, took the data. Then they took you out and put you on the sled, and hooked it up out there, so they monitor everything while you were going down. And then you just sat in there. They taught you how to brace yourself to absorb the impact, so your body wasn't flying all over the place. You had to lock your heels in. They had little cleats stuck up on the floor of the thing, and you took your heels and you locked them in there. You clamped your hands together and held them between your legs, and just shoved yourself back in the seat as tight as you could. And that's the way you sat. They did a countdown from ten to zero. *Poom*! You were gone. And the next thing you knew, you could hear voices off in the distance, and there were people unhooking you, trying to get you out of it. You were unconscious. You were only out for a little bit, but it would knock you out.

When you went out, everybody went into a big room, and you changed in that room into the scrubs. When they were doing the Bopper, everybody was just all kind of hanging around, and talking, and whoever was first in line got on and they sent him down. They set up some tests to do, to check the space helmet that was going to be used in the Apollo project, for its ability to withstand impact. When I did the helmet test, there weren't very many out there. I think I was out there by myself when I went into the room to get ready for that.

[The Daisy track ride] was over so fast it was unbelievable. I mean, the countdown would hit zero and you could feel yourself accelerating. And, like I said, in no time you heard voices in the distance, and they were there getting you out of it. They would take you back inside to the same big monitor thing and plug you in there, and test your vitals. They had to pull your shirt off. When they took my shirt off, my back was just a mass of, you know what a blood blister is like, my blood was all drawn to the surface. I think that was on a Tuesday, and then they asked me if I would come back on Thursday and ride again. I guess they wanted to see what would happen, I guess, to the body after two of them that close together. Or, whether my body healed itself in time. I went back on Thursday and ran again. I did 15 g's the first time, on October the 16th of '65. On October the 18th I did 25 g's, and it kind of messed a few things up.

[The Daisy accelerator had an omnidirectional sled.] Some of the people, they turned them with their butts facing in the air. They were facing down on the track, and shot them straight forward that way. One of the guys got a ruptured eardrum. One got a rupture on his stomach. I wound up, when I came up for discharge I had a torn heart valve and one eye just slightly skewed. You can't see it, to look at it. They wanted to send me back to, I think, San Antonio to the hospital and run me through a bunch of tests. I just told them it wasn't worth it. So I went ahead and got out. You definitely gave them a blank check.

I really enjoyed it. I enjoyed being a part of a bigger picture of the space program. Even if it was a minute, little part, it was still a part of it. It gave me something to be proud of.

Dee Gragg was born on a cattle ranch in southern Oklahoma. After finishing high school, he worked for two years with his father, building Aero Commander airplanes before he was drafted into the Army. "I had been interested in journalism and went to college one year, thinking that I would be a journalism major, but ran out of money," Gragg said in his 1999 oral history interview. "In the meantime, before I got back to college I discovered engineering."

After the Army, I went to the University of Oklahoma, where I got my Bachelor's degree in mechanical engineering. Then I went to Sandia Corporation in Albuquerque. Due to the nature of their work, I won't discuss what that was. But it was one of those things that you didn't talk about when you went home. Not only didn't you talk about it, but you hoped that nobody would ever know what you had done, and what you had done would never be used. That wasn't very satisfying to the soul, if you will. So after a few years, I moved on and went to work for the Aeromed Lab at Holloman Air Force Base in 1968.

One of the first programs that I worked on there was with Colonel Stapp. Colonel Stapp had instigated the idea of what was called the golden shells. The idea was, as Colonel Stapp once told us, "People don't fly airplanes and drive cars standing up." So he wanted these built in a sitting position. And we did this with another gentleman named Ed Herzberg in Wright-Patterson Air Force Base. He was the anthropologist. Stapp provided the money. He was in Washington, DC, and I was the contract monitor at the Aeromed Lab at Holloman. Neither one of them is an engineer, so I also ended up doing the engineering on the program.

What it consisted of was two sets of three shells in the sitting position. [One shell for] a 95th percentile male. This meant this male is larger and heavier than 95 percent of the population. [Another shell for] a 50th percentile male, who would be right in the middle, and a shell for a 5th percentile female, which means this female would be larger than only five percent of the population. So we kind of covered each end and the middle with those. We made two sets of shells; the shells themselves were made by Alderson Research Company out on the West Coast. They were just thin, plastic shells in a seated position. They were sent to Sierra Engineering Company. They filled in one set of those with meat, and they froze them, sawed them apart, and did the weight, CG [center of gravity], and MI [mass index] of the human segments.

The second set, which was not sawed up, was given to the National Bureau of Standards to be used as a standard by whoever wanted to make dummies for sale to the government. That was assigned to me right after I came to the Aeromed Lab, and I worked on that the whole four years. I think that the last thing I did before I left the Aeromed Lab was I approved the drawings to build those. At that time, there were just two [crash test] dummy manufacturers, Alderson Research Company and Sierra.

We used dummies mostly with the Daisy [accelerator track]. In fact, every morning we were going to run humans, we put a dummy on the sled and ran the sled down the track with a dummy first to show that everything was functioning correctly that day, before we put a human onto the sled. My first test that I saw at the Aeromed Lab when I was brand new, turned out, coincidentally, to be the first time that the lap belt and the shoulder harness that you see in all the cars now was ever tested on a human. That was the first test that I saw.

While I was there, one other interesting program I worked on, on the Daisy track, was the Apollo program. This is the one with the three astronauts. They worried about what if the capsule they were in came down [and] impacted on earth rather than water. If it was under the chutes and it comes down into water, there is not much of an impact, not much of a problem. So they worried about what if, on an abort or an unplanned landing, they came down on earth? I was the principal investigator there. They wanted us to use three humans for the test, but we had second thoughts about that. We were afraid to put one in the middle, afraid we couldn't get to him fast enough if he had a cardiac arrest, or that sort of thing. So we ran it with two humans. We had a doctor on each side of the track, who would run up when the test was over to attend to the test subjects to see if they had any distress.

The program I spent the most of my time on was airbags. I began researching airbags right after I got there in '68. We did the programs in '69 and '70. At that time, we were really worried about what might happen to a human subjected to a rapidly deploying airbag. We had the

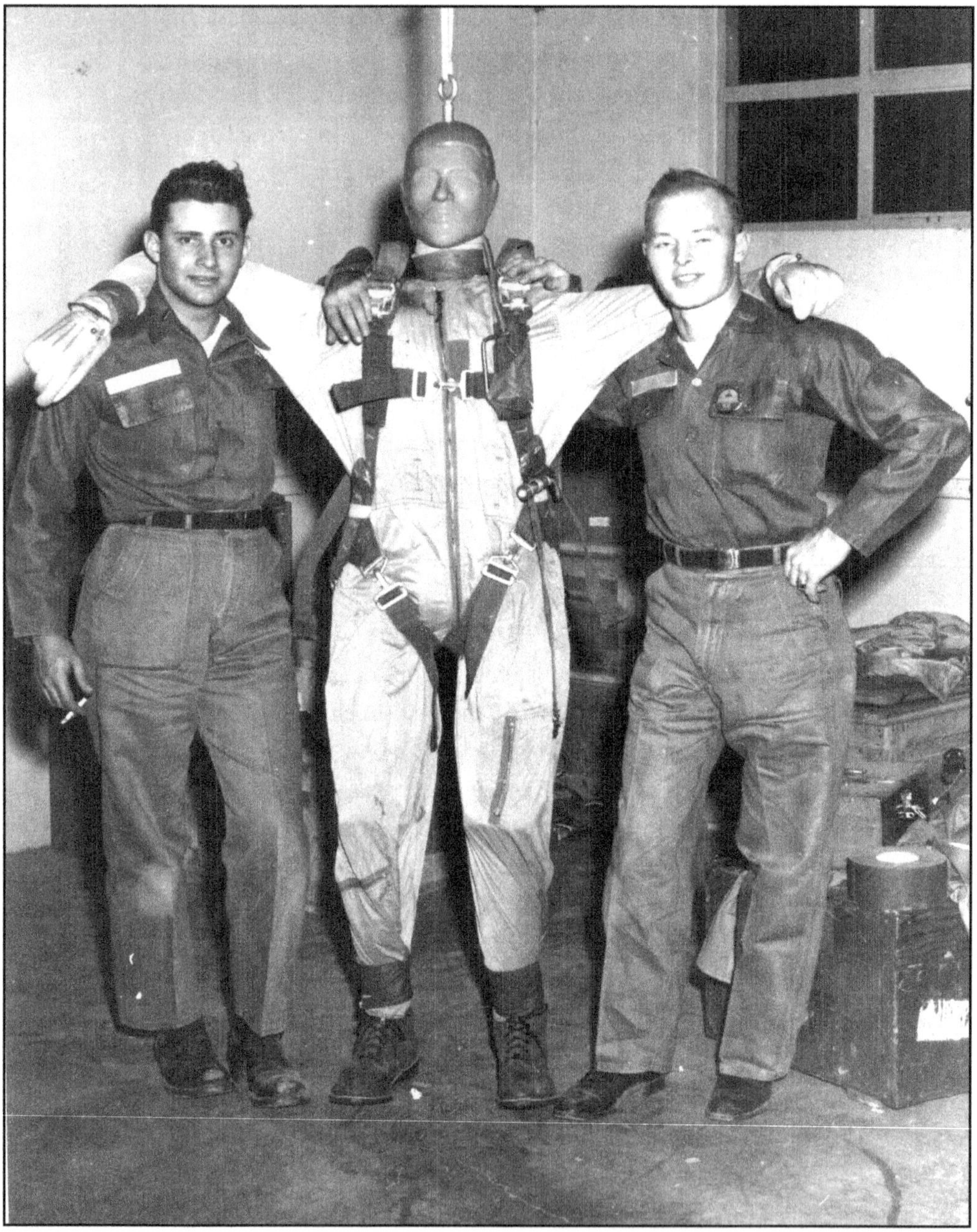

Crash test dummies, like Sierra Sam (middle), were used extensively for high-altitude balloon experiments as well as deceleration sled studies. Courtesy of US Air Force.

sled subject all padded up, we had a baseball chest protector, and padding all around, and safety glasses and so forth. The [US] Department of Transportation was paying a lot of the cost of the program. There was

Three human volunteers ride the Daisy decelerator, facing rearward. These tests were used to evaluate designs for Apollo space suits and the couches that would be placed in the Apollo spacecraft. Courtesy of US Air Force and NMMSH.

no problem at all. The first test was pretty much a non-event. As we went further, we had him sitting straight up. As we went further, we had the subjects lean closer and closer into the airbags. The most they got out of it was a little bruising, a little reddening, one bloody nose, those kinds of things. The deployment of the airbag is sort of a violent reaction, but you are trying to save the person from a much more violent reaction, namely a car crash, so all in all it worked really well. In the last test, we had the subject just a few inches off the airbag. He got a pretty good whap out of that. To me, he seemed liked he was stunned. He was never unconscious, I don't believe. But he seemed like he was stunned and he had a rather severe headache for the rest of the day. I am not a doctor, but I suspect he had a mild concussion. He was back to work the next day, so no ill effects that we know about.

The airbag program consisted of three parts. One was with living humans, one was with baboons, and one was with anthropomorphic dummies. The humans, there were forty of them. They were Air Force officers and enlisted people from Holloman who volunteered to do this.

And they got extra duty pay for doing it. We ran each of the forty humans three different ways. We ran them with an airbag, we ran them with a lap belt only, and we ran them with the lap belt and the shoulder harness. So that was 120 tests that we did. Those tests are all reported in a number of different places, but mainly the fourteenth, fifteenth and sixteenth Stapp Car Crash Conferences. Through that, we were able to determine that the airbag was very effective in reducing the load on humans, the load we put on the lap belt and so forth. And we always recommended, and it's what you see in cars today, in addition to the lap belt they have the airbag also. They do different things, and they complement each other. We came to this conclusion thirty years ago.

The baboon program was to establish the LD 50 level, that is, at what point 50 percent would live and at what point 50 percent would die. We could do that with baboons. They were all anesthetized, so they had no pain through it. After the tests, the animals were sacrificed.

The dummy program was somewhat similar to the human program, but we could only take humans up to certain g levels we kind of called the "outs level," if you will. Then we would have to stop. We didn't want to injure the humans beyond a bloody nose, a scrape, or a little burn, or something like that. But the dummies, you could go ahead and take them up to higher and higher levels. So that was the dummy program, to go to the levels we couldn't with the humans.

Al Zaborowski served three duty tours at Holloman Air Force Base, the first two with the Aeromedical Field Laboratory, and the last one at the high speed test track [see Chapter 6]. In this excerpt, he described how he first came to Holloman. In his description of the Daisy track, he explained that, although the device was designed to be accelerated by compressed air, administrative and funding complications delayed full implementation of that system. Initial tests were conducted using a catapult system powered by explosive, ejection seat cartridges.

[Zaborowski graduated with a Bachelor's degree in mechanical engineering from the University of Buffalo in 1958. While he was there, he was involved in the Air Force ROTC program.] I was the commander of the Arnold Air Society, which is kind of like a fraternity.

We had speakers come in. One time, we had Dr. Stapp come in. He was in the area, and he came in and gave us a nice talk. I introduced myself and that was it. After graduation, I interviewed throughout the Northeast. I had an interview with the Ford Motor Company, Curtis Wright, United Aircraft, Cornell Labs, and the photographic activity at George Eastman. I selected the Ford Motor Company because I knew they were doing things in crash research, and I had met Stapp.

I took the position in the Research Center at Ford which was in Dearborn, Michigan. I started out a little bit in automotive crash research, then they moved over to human factors. We were doing such things as how do you drive the car? Where do you look when you drive the car?

I was with Ford for not quite a year, eight or nine months. Here I am, an engineering background and a photographic background, human factors research background, automotive crash, so what does the all-knowing Air Force do? They assign me to cryptographic school to eventually go to Greenland at one of the radar sites. I said, "No way!" Of course, I didn't know that you could do that as a second lieutenant. I was dumb. So in my dumbness I wrote them a letter, and of course they said, "Oh no, the needs of the service come first. You're going to Greenland." So I said, "Oh no!" That's the first brush, and I expected that kind of answer, so I wrote them another letter in more detail. I put in regulations and all this, and about the Air Force's interest in having graduate engineers and that's when that activity started off.

So they withdrew that assignment and assigned me to a motion picture and photographic school in Orlando, Florida. I was going to be the instructor. I said, "Oh that's great!" In the meantime, I kept running across Dr. Stapp's papers. I was using them as some background research. So as a lark I wrote him a letter. I said, "If you ever have an opening at your facility at Holloman, please consider me, and this is my background." Well, he wrote me a letter back and said, "Oh, we would be delighted to have you. I don't have the position yet, but I'm going to Congress." They were talking about what the crash problem was costing the Air Force. You know, train a pilot, and today I'd say it's about two million dollars, but even in those days it was still a significant sum. You get a pilot trainee and he's ready to fly, and he doesn't die in the airplane, he dies on the weekend in his car. So Stapp says, "You know, we have a direct correlation here."

He was doing some of that work at Holloman in addition to the human tolerance studies and his original chimpanzee colony.

Then I got a telegram from him. He said, "Stop, stop everything. Don't go anywhere before telling me. I will have a slot for you at Holloman when I come back after this Congressional investigation." He did. So my time was extended a little bit, and I came to Holloman on the 18th of February 1958, in a snowstorm.

It was my first time in this part of the country, first time west of Detroit. I knew I was going to Alamogordo, of course, and we had a wonderful research library at the Ford Motor Company. They had a brand new McNally Business Guide, and one of the criteria was the number of gas stations per county. Otero County had twenty-three gas stations. I passed twenty-five gas stations just on my way from home to work. So I knew it was sparse, but that almost is too sparse.

My first tour at Holloman was very, very busy. That was from '58 to '61. I was going to be the chief of the Test Section, Biodynamics Branch. Captain [Eli] Lack Beeding was my immediate supervisor.

Now, the Daisy track was not up to snuff. It was only [about] a hundred feet long. That was not enough room, so we extended the track another hundred feet [or so]. We used ejection seat cartridges to propel the sled. The instrumentation was terrible, archaic – old instruments, operating out of a trailer, of all things. Didn't have a blockhouse, didn't have a place to prepare vehicles. There was no way to take a blood pressure except before the test, and then the flight surgeon would run off, and you would take the blood pressure after the test.

Electrocardiograms were kind of sensitive in those days, and to have one really done well you had to have a screened room. Obviously, you couldn't do that. You did an EKG in the animal house, and then after the run you did another EKG. Well, that's just too long. So Lack was also a visionary there, and he had established a contract with some instrumentation manufacturer to build a one-lead EKG system. This was a box about the size of a Kraft cheese box. To protect it to 30 g's, that thing weighed about thirty-five pounds. It was extremely dense. We didn't mount it on the man, but on the sled.

By the time we were finished, we had a building for the air gun. We had built a blockhouse, an instrumentation room, and the countdown house, and it was protected from the dust. There was an interesting thing. We put this all together and then we were getting extraneous signals on our data. We could not figure out what that was. This was an irregular beat, not like a 60-cycle sound or some harmonic of the 60-cycle. So we said, "What is that?" We started tracing the lines, and a guy had a radio turned on. He was listening to the radio and he saw changes in this extraneous stuff, and he says, "We're picking up the radio station in town."

What was the solution? We dug a hole, and this is very, very difficult. This is all caliche. Once you get into a foot of this, it is hard as rock. So we got the Army to come in with their posthole driller, and they put in a shaft about 50 or 60 feet deep. We put copper sulfate in there, put in a grounding rod, and it went away. That was our natural ground, then, for the Daisy.

I was a guinea pig, riding on the sled. I went through the physicals to do this. Did an indoctrination ride on the Bopper, which was this rubber powered, 10-foot track with mechanical brakes that you could set.

We added a good photogrammetric system where, by using markers on the individuals, we could actually get a photographic record, and that would be hand reduced, and we would then have a space-time relationship. We could calculate a g [force] and a velocity, and all that kind of stuff.

We had developed a seat that was rotatable up and down. That was patterned after the configuration of a typical ejection seat. Those were the seats that Lack and I and other people rode on. I remember Lack's ride. It was number 335. My run was the week before, and it was 332 because there were two other people. It rode three. There were 332, 333, 334, and then Lack was 335. Those associated with the program were on first. That was Stapp's [mandate]. He'd say, "If you have confidence in your study, then you have confidence to expose other people to it, but you have to prove that confidence to yourself." You know, we're operating in the nebulous area. We're doing human tolerance work. I had reached 54 g's and Lack had reached 83 or 84. These were backward-facing runs.

I was in the hospital [when Lack did his run], straightening out my back. I was surprised that Lack was so injured. We ran an animal after that, and the animal's autopsy [showed] the lesions in the bear were similar to the pains that Lack had reported. Lack complained about tenderness in his esophagus. The bear's esophagus was bruised. He had tenderness in his chest. The lungs on the bear were bruised. He had tenderness in his liver. The bear had a small cut. So it correlated very well. That was the last run in that series.

We ran chimpanzees and bears. Bears because of the spinal column, and pigs because the internal organs were similar to humans'. The chimpanzee is a pretty good surrogate for man. A difference in weight and various other things, but nevertheless, a very good surrogate.

Fortunately, Eli Beeding was facing rearward when he experienced 83 g's of deceleration. If he had been facing forward, he probably would not have survived. Courtesy of US Air Force.

Dr. John Paul Stapp earned a Ph.D. degree in biophysics in 1940 and an M.D. degree in 1944. As a flight surgeon in the Air Force, he developed a special interest in studying the effects of acceleration and deceleration forces on the human body.

I first came to Holloman Air Force Base the twentieth of April 1953. At that time, we had already planned a program to work out at Holloman. This centered on two activities. One, a 3,500 foot launching track for Snark missiles that Northrop had built at Holloman, which would be used for deceleration studies. And the other, a balloon project already in progress with Dr. James P. Henry (at that time, Lieutenant Colonel James P. Henry) and Captain David G. Simons. They were located in two temporary buildings. This was a tenant project on the base.

One of the active participants was Lieutenant James Telfer, a fascinating individual. Once, in the summer of 1952, the cooling system failed. He was growing fruit flies for genetic experiments in which they would be exposed in balloon flights to cosmic radiation to see if it caused any genetic changes. He called the Base air installations for emergency help. They started giving him the brush-off, so he wrote a famous memorandum to them stating that unless they wished to be responsible for the deaths of 30,000 Drosophila melanogaster, they would have to come and repair the system. They didn't want to be blamed for the death of anything, so they immediately showed up and repaired the air conditioning and cooling system, and the insects lived. Second Lieutenant Telfer enjoyed one more triumph of his way of doing business.

I came to Holloman as projects officer and director of research and became chief of the Aeromedical Field Laboratory. Eventually, it became a separate laboratory in the Air Search and Development Command, until I was transferred to chief of Aeromedical Laboratory, Wright Field, April 20, 1958.

I drew plans for the Daisy track in June of '53. It took about a year to get it built, installed, and running. We had a small staff. The biggest it ever went to was thirty-nine people, yet I marvel at the amount and kind of work they did. We had the cosmic radiation balloon flights. We had Dr. von Beckh and his zero-gravity exposure of animals during Keplerian trajectory flights in a T-33 jet trainer plane. Dr. von Beckh sat in the back seat holding his animals in his lap and recorded their reactions. The Daisy track and the long track served for deceleration and impact experiments.

Later, about 1955, we began a motor vehicle crash program. The first two projects were to test a demountable rollover structure for open-top, rigid-framed military motor vehicles, such as a weapons carrier or Jeep, accomplished on a weapons carrier. An energy-absorbing bumper was developed and tested under contract with Professor J.J. Ryan of the University of Minnesota Engineering Department. He accomplished his contracts, and we were able to take this weapons carrier with his 21-inch-stroke hydraulic-mounted bumper, raise it with a crane to 58 feet and drop it nose-down on a wooden pallet. The bumper would take impacts up to 40 g's. The vehicle would rebound and come down on its wheels and could be towed away. Of course, the engine wasn't working because we drew it from salvage. On his own, Professor Ryan invented the telescoping steering wheel column to absorb crash energy. Material was crushed inside the column as the body pushed the steering wheel three or four inches into it. That preceded airbags. We had lots of other little projects. I can't even recall all of them that we carried out.

Chapter 6
HSTT!

One of the most valuable experimental devices for testing both aircraft and missile components, from the 1940s until the present day, is a high speed test track (HSTT). The device to be tested is attached to a rocket-powered sled that glides along railroad-type rails and can be brought to either a gradual or a sudden stop. Among the benefits of this device is that it is less expensive to test a missile's flight characteristics than an actual launch would be. In addition, the missile is undamaged during the test (assuming it worked properly), whereas it would be damaged when it impacted the ground after an actual test launch. Components or instruments can also be tested individually before they are flown on a missile.

Dr. John Paul Stapp developed a novel adaptation for high speed test track experiments to investigate the question of whether the human body would be able to withstand extreme forces of acceleration and deceleration. He used dummies, animals, and humans (including himself) in tests that provided valuable information for occupant ejection from aircraft as well as the effects of manned rocket launch and reentry.

Stapp began his experiments on the test track at Muroc Army Air Field (now named Edwards Air Force Base) near Lancaster, California. He took the first human-carrying ride on the track in December 1947, beginning with a cautious 90-mph run. During the next seven months, he rode the Gee Whiz sled fifteen more times on the test track at speeds up to 150 mph, stopping in as little as 19 feet, and enduring up to 35 g's in backward-facing runs. In 1953, he transferred to Holloman AFB in Alamogordo, New Mexico, which had a longer and better-equipped test track. His forward-facing runs on the Sonic Wind 1 sled at Holloman were even more dramatic, culminating in a 632-mph ride that came to a stop in 1.4 seconds, during which he sustained a peak deceleration force of 46 g's for a fraction of a second.

This chapter includes recollections of people involved in various aspects of work on high speed test tracks. Because of his leadership and personal involvement, Dr. Stapp comes first. [See more portions of Dr. Stapp's interview in Chapters 1, 4, and 5.] The following excerpt begins with his work at Holloman on aircraft ejection seats:

The test track project was outlined before I arrived, and we intended to do it as it related to the F-111 and other supersonic aircraft. The issue was, should we continue to have deployment of canopy and open ejection seats, or should we have capsules to enclose occupants and protect them from windblast and impacts of air and tumbling during escape? I was, in particular, interested in extending open ejection as far as possible.

The escape problem has two aspects. In flight training, you go all out in any way that you can to save inexperienced trainees. That is the first priority. For combat aircraft and combat pilots, the consideration was to attain the compromise with combat hazards. In other words, dead weight of a capsule versus much lighter weight of an ejectable seat – which he had to have in there anyway – might allow more bullets, more fuel for longer range, being able to defend the craft longer, and in most instances, improve the chance of reaching friendly landing sites for rescue. From my evaluation of accident and combat data, it was not as essential to provide escape at maximum speed and altitude, because the aircraft could slow down to below sonic speed before losing too much altitude. If they didn't have that much time, the plane was going to burn up or blow up, with little or no chance to survive. If they could hold out for less than a minute, they could eject and survive.

We had to know, though, where was the absolute limit of exposure to windblast and prolonged deceleration at very high altitude and high speed, prolonged by low drag. This could be attained by a ground simulation on the 3,500-foot track using a rocket sled. We did three human experiments and over eighty animal experiments. With anesthetized chimpanzees, we exposed the subjects to the same conditions that would occur at

36,000 feet at 1,800 miles per hour. With the animal subject, we got 1,355 miles an hour actual speed on a run 2,300 feet long at ground level, and exposing the animals seated in a steel seat with no windshield, but with a fully enclosing helmet over their heads locked to the headrest, and using a nine-ounce duck overall suit. This was on a 22,000 foot track [built in 1953] at China Lake Naval Air Station. That experiment resulted in a ram pressure of 4,000 g's, or 4,000 pounds per square foot stagnation pressure and 440 degrees Fahrenheit stagnation temperature. The garment was totally shattered. The only unburned areas on the subject were under the straps, which remained in place. Of course, we euthanized the subject without allowing him to wake up from that one.

We immediately invented a suit that would be protective. In the next experiment, we used thirteen-and-a-half ounce Dacron sailcloth for the outer garment and borrowed ideas from the Air Force fire suit used by firemen in putting out aircraft fires. They had about sixteen layers. We had an insulating layer under the Dacron sailcloth, then a reflecting aluminum layer, another insulating layer, and an inner layer of long john, cotton underwear. We tried that out on dummies and chimpanzees and found that it held together, never did fail, and that it protected from heat and windblast at Mach 1.5. That was as high as we continued to certify. I don't think it was ever used by the Air Force, but at least they had it available. We lost the argument on capsules versus open ejection. Later on, we won because they discovered what we had objected to in capsules was really so.

At what we called Muroc Army Air Base, there was a 2,000-foot track on which we did 73 human experiments between 10 December 1947 and 5 June 1951. We did 3 on the long track at Holloman with human subjects, and we did 5,012 human experiments between 20 April 1953 and 31 December 1970 on the Daisy track. A total of 5,088 human subject exposures without one single fatality, not one irreversible or disabling injury. This does not include thousands of tests with the Bopper. We explored the hazards of crash impact in both flying and ground vehicles with a perfect safety record. That is the achievement I am proudest of.

Germans did ejection seat tower drop tests and ejection seat firings with non-volunteering subjects during World War II and found that they could produce fractures at 24 peak g's at rates of onset higher than 500 g's per second. We went as far as we could with human volunteers before resorting to anesthetized animal experiments using higher impacts.

Animals were chosen for closest resemblance in mass, size of organs, and configuration with the human. We used three species of animals to work with. Chimpanzees, because they are two-thirds of human size on the average and are primates corresponding very much to man. Swine, although their configuration is not altogether like human, their organs are the same size as humans'. The pathology of trauma from impact can be very accurately related to measured force and related to human accidental trauma. After surveying all available large mammals, our veterinary pathologist chose the black bear because it has a recurved pelvis like that of man. Other primates have flat pelvises and, therefore, a belt won't hook on the pelvis of a chimpanzee as it does on a bear. Chimpanzees have just two curves in the vertebral column and neck. An 85-pound chimpanzee will have a size 22 collar, and he can't move his head more than 45 degrees to either side of his midline. He is built to fall out of trees. In vertical drop tests with anesthetized chimpanzees, a vertical jar of 60 g's or more produced no injuries. Bears have a limber, flexible vertebral column like that of man, and can swivel their heads almost all the way around. They sustain spinal fractures at 35 g's in vertical drop tests, the same as human impact failure tests.

The first run we did at Holloman in which I took a ride was on March 21, 1954. We used seven rockets to reach a speed of 421 miles per hour in five seconds, followed by deceleration of 0.79 seconds, a plateau of 14 or 15 g's with a peak of 21 g's deceleration. I did not have a helmet for that one because a fixed windshield protected me from windblast. This first test determined that the restraint system we were using was adequate. Then on the 7th of August 1954, we did the second run. This time, we wanted to simulate supersonic jettison of the canopy. To do this, the windshield had hinged double doors that would pop open inwardly when the sled contacted a cam on the track at the at the point of burnout. The cam would trip the trigger, releasing these doors, and they would pop open at maximum windspeed, abruptly blasting the occupant protected by a helmet secured to the headrest of the seat. This occurred at 505 mph. I got my 2.7 seconds at 10-12 g's. This duration did overlap the time it takes to get a centrifuge up to 10 g's, demonstrating no difference between the two devices in relating the effect of linear deceleration to that of circumferential rotating deceleration on the centrifuge. The ram pressure encountered when the doors popped open was 707 pounds per square foot.

On the 10th of December 1954, staying just subsonic and using nine rockets at 45,000 pounds of thrust, the sled accelerated to 632 miles per hour in five seconds and decelerated in air drag for about seven-tenths of a second before entering the water brakes, coming to a stop 1.4 seconds after burnout. Going 632 mph to a dead stop in 1.4 seconds, surely, the world's record for drag racing. This was with no windshield at all, the helmet locked in place to the headrest, with Plexiglas visor to protect the face from windblast. At 632 miles per hour, the windblast was 1140 pounds per square foot.

John Paul Stapp was the only human to ride the high speed test track at Holloman Air Force Base. Courtesy of NMMSH.

[What sort of things ran through your mind when you were riding Sonic Wind 1?] Well, the same things that ran through my mind the first time I rode the Gee Whiz back at Muroc. That sled could get up to 242 miles per hour on four rockets. We usually used three for human tests, because that would get up to 180 miles per hour, then a brief coasting period, and we entered mechanical friction brakes. We could generate up to 55 g's with these friction brakes, simulating an airplane crash. We did progressive increments of decelerated force by 5 g's, then we did change of brake pressure and brake sequence that would build up the peak faster. We called that the "jolt," which was calculated by dividing the peak by the amount of time it took to get up there. It took a thousandth of a second to get to 10 g's, that would be a thousand g's per second rate of onset. If you got that up to 1,500 g's per second, it was considerably rougher for the same peak than at a lower and slower rate of buildup. We found that to be a very significant factor.

The main concern of someone taking a ride was to remember and record mentally as much as he could while it was going on and put that in his subjective report. At the same time, we had some instrumentation, not nearly as much as we could have today, particularly when we had acorn miniaturized radio tubes and radio transmitter equipment mounted on the sled. We had to protect those acorn tubes with felt of rubberized horsehair so that they could take up to 100 g's without beginning to shatter. Selecting and schooling subjects to give us as much information as they could was very important. Of course, we couldn't get much out of a chimpanzee, and there were some *people* we couldn't get much out of. At any rate, that was the discipline we tried to have, getting the subject to remember and record accurately as much as he could.

What was going on in my mind when I was ready to be launched on the Sonic Wind sled? I held a lanyard in my hand and turned the camera on the prow of the sled on so it took a face-on view of me during the run. On a certain count of the countdown, I had to pull the lanyard and always did. Of course, you're in a depressed state because you know it's going to hurt and you don't know how much. You just hope you come out alive. It makes you very alert, and when the countdown is going on, your pulse rate, blood pressure, and heart rate go up. The takeoff is a tremendous jolt, because when 45,000 pounds of rockets come on in 67 milliseconds and it really rams you in the back, you feel like the sled is going off the rails and taking off in a steep climb. Then, of course, is the phenomenon of the semicircular canals [in the ears] being stimulated and sensing 45 degree vector. In the first two-and-a-half seconds of the last run, I lost vision very rapidly, a visual blackout with no loss of consciousness due to the 19 g's of acceleration draining the blood back from the vessels to the retinas. Then, at five seconds, when we reached burnout, there was an immediate total reversal and we had about 19 or 20 g's of retardation by wind drag. That brought back the blood to the retinas and vision to vivid yellow and brilliant flash. I could see the water brakes briefly before we hit them, and I was able to see the first splash of water. Then the buildup of pressure into the retinas caused a redout and, in fact, I couldn't see anything for eight minutes after the sled stopped.

They took me off the sled. I couldn't see, but all those hands got hold of me and I felt myself being lifted out and put on my back on a stretcher. While I was looking at the sky and trying to blink, the blood that drained out of the retinas drained back and I could see. But then I discovered I was having double vision because the muscles around the eyeballs were swelled up, and the eyeballs didn't track together. They got me back to the hospital, and I had two lunches because I had missed breakfast. From then on, I was busy on the phone.

I can tell you that the sensation of relief and elation after that run was the biggest hazard euphoria I've ever experienced. I really felt like I was ragged but right. After this, they'd all have to support my project. Well, I didn't know it, but they weren't going to support my project, they were going to put me out of business. [Stapp had wanted to continue his project by riding the sled at a speed of 1,000 miles per hour, but his superiors denied him permission because of the danger.]

Test pilot Joe Kittinger worked with Stapp on several programs, including the Holloman HSTT. In fact, for the December 1954 test, Stapp assigned Kittinger the task of flying a jet airplane at low altitude with a photographer on board to record the sled run. It was a tricky job. Kittinger described the preparations in his autobiography, Come Up and Get Me*:*

> *[Stapp said,] "I need you right there [at the start of the sled run] at 350 miles per hour when the rockets fire." . . . "Just let me practice it, sir. I'll figure it out." I got a T-33, put the photographer in back, and flew a dry run. . . . I missed it a little bit early. We tried it again, and this time I was a little bit late. We made twenty practice runs that day. I never missed it by much, but I never hit it exactly right, either. I made five more series of practice flights over the next few weeks, trying to get the timing down so I would hit the launch end of the track at just the right speed at the right moment. . . .*
>
> *On the appointed morning, with Stapp's support crew ensconced in a concrete blockhouse 100 feet from the track, I came in hot in the T-bird, right on the deck, hoping like hell we'd hit the launch close enough to get the pictures. I was going to get only one shot at this. . . . I heard the countdown as we approached the track, and things felt good to me. This time I hit the launch site perfectly at precisely 350 miles per hour just as the white smoke boiled out of the rockets and the sled Stapp had named Sonic Wind shot past us in a blur. It reached a top speed well in excess of 600 miles per hour. The photographer got some tremendous shots. It was the only time our timing was dead on – but when it mattered, we nailed it.*

Kittinger got to know Stapp well as they worked together at Holloman Aeromedical Field Laboratory and later at Wright-Patterson's Aero Medical Laboratory.

The Aeromed Lab was kind of off on their own, over at the side of the [Holloman] base, and they weren't really involved in anything other than research that was directly related to space and crash protection. An awful lot of emphasis Dr. Stapp had was on crash protection. He was instrumental in the first research on lap belts and shoulder harnesses, restraint devices. Really, when you think about it, the contribution that Dr. Stapp made to crash protection was probably equal to what he did for the space program. And it was right there at Holloman.

He was one of the most busy men you've ever seen in your life. He was doing all this type of research on automobile protection, and aircraft ejection, and restraint, and high-speed deceleration. And he had the Manhigh program and other things, and he had all of them going at the same time. He was an extremely busy man working on a variety of different projects for our nation.

Jack Glasco was forty-two years old when he went to work at Holloman for the contractor Land-Air. As a photographer, he documented many kinds of experiments, including high speed test track runs. Holloman's test track was lengthened from its original length of 3,550 feet to 5,071 feet in 1956 and lengthened again in 1958 to 35,071 feet. Later additions brought it to its current length of 50,788 feet. It consists of a pair of rails spaced 84 inches apart. For the final 20,200 feet, there is a third rail between them, spaced 26 inches from one of the outer rails. This allows for a variety of configurations for two-rail and monorail sled tests.

I went to work out at the test track in 1960. I'd go there two or three times a week, at least. I photographed some double-track sleds and some single track. At that time, contractors for the government were trying to design any kind of a vehicle that would fly in space and would soft land on either the Moon or a planet. I worked on various programs, but one in particular that I worked on several launches for was called the DynaSoar [an experimental spaceplane developed by Boeing for the Air Force beginning in 1957; the program was cancelled in 1963 before it went into production].

During that time, Land-Air hired a man by the name of Tex Helm. Tex was a very good cinematographer. One time, he and I were assigned to the high speed track on the same run. I'm up in the bed of a pickup with a still camera. Tex is on the ground with a Bolex camera on a tripod right by the tailgate of the pickup. They pushed the button on the DynaSoar mounted on the sled. Here it comes at, what, 400 miles an hour, and there is a loudspeaker system that tells everybody what's happening. As the sled got about half a mile from where we were stationed with our cameras, the sled just disintegrated, flew apart. The chassis was made mostly of quarter-inch-thick boiler plate steel. One piece of that was going through the air and heading right where Tex and I were standing. Of course, he is down behind the camera, and he can only see what it sees, and all of a sudden the loudspeaker system started telling everybody to look and see where the pieces are going. Tex came out from behind his camera. Here is this 300-pound piece of steel heading right towards him. I mean, he dropped everything and took off running. The last I saw of him, he was still running over the hill. It landed a hundred yards short of where we were. I was in the bed of the pickup, and I had a little bit better perspective on it than he did.

We never knew what was going to happen when you pushed the button, so we were all advised to keep our head and eyes moving when the sled was running. Many times, pieces did fly out to where we were on camera row. It was a parallel road to the track for servicing camera stations. We weren't supposed to be closer to the track than that, but some jobs did require that we move in a little closer. We were vulnerable there. We could have been hit.

Terry Van Wormer was twenty years old when he enlisted in the Air Force in 1955. After six months of technical training, he was assigned to Holloman as an airborne radio repairman. He left the service in 1959.

In 1961, I hired in with Autonetics, testing on the Holloman high speed test track for a year and a half. It was 35,000 feet long then. They were sled testing the Minuteman guidance system. They trained me to operate a Wild theodolite. It was a surveying instrument. At that time, they had some photo autocollimators [optical instruments for measuring angles] which were situated alongside the track. My station was at 17,350 feet. It was after the engines were firing with acid. After the engine shut, you would have a coast period. Then they would run it into a water brake and slow the sled down. When it came by my station, it was going about 200 feet per second. Theoretically, when the sled came by, we would take a picture and be able to determine by the crosshairs the orientation, the pitch, the roll, the yaw, of the platform.

Sometimes, the water brake would be a little too deep, and the sled would come out of the water brake and cut my breakwire to open my camera lens and then stop before it got to the breakwire to close the camera lens. So you didn't get a picture. Sometimes it would come by too fast. You had it set up to figure the speed and where to set the lens, and it would go by so fast you couldn't get a picture. Once in a while, it would work.

Sometimes when you went out to the track, you were in the middle of a dust storm, or in a rain, or hotter than blazes, or colder than blazes. At two o'clock in the morning, it is dark out there. And you hear all kinds of creatures running alongside the track, little pack rats, but you didn't know what they were. You would go down there, and you would take your flashlight and shine it on your target. But you had to walk back in the dark. And, boy, you would hear these little things. They liked to crawl up on that cement at night, lay there where it was nice and warm, right next to the track.

They had a test track at Edwards that they ended up bringing the rails here. That is how we got the extension to 50,000 feet. The personnel from track went out there, and they cut the rails in lengths that they could fit in a train or trucks, I don't know which. But then, when they got it here, they had to weld those sections back together, and grind them down.

That track was under tension. We used to have the rails break. If you get a real sudden temperature drop, the rail would break and it would separate about six inches. You would have to go out there and cut it off, at least two or three feet each side of the break. Cut it, go get another section and bring it here, and then weld it. They would put these big banks of heat lamps to heat up the rail, and it would kind of creep together. Then they would weld it, and then they would let it cool, and it would stretch back out. Grind it smooth, realign it with theodolites. That would be a three-day, round-the-clock operation to get the rail fixed. They would break one rail, and next week they would break another rail. It was a job, but not a fun job.

Clyde Wilson served as a helicopter pilot in Vietnam, ultimately retiring from the Army in 1968. He came to White Sands Missile Range in 1968 as a civilian employee and was assigned to Holloman as a helicopter pilot. The oryx he mentioned in this excerpt were the descendants of African antelope brought to White Sands Missile Range between 1969 and 1977. The New Mexico Department of Game and Fish sponsored the relocation of the animals to provide big game hunting to a region of the state that did not naturally offer that opportunity. The original herd of 93 oryx has developed to about 3,000 now. Since 1974, annual hunts have helped stabilize the population.

We flew safety for the test track missions. We would go out there and fly around it ahead of time to see that nobody got on it. That would have been the worst thing, if somebody walked up to it – sabotage, or somebody coming up to bother it, or somebody just lost. And then you had to worry about the oryx being there. They would fly bird chasers, because they worried about birds. So they would try to scare the birds away before the missile firing. A cow would wander up. So we would check that ahead of time.

And then when it would fire, we would go down the track, and we would orbit it again until all the people got there and returned to safety. In the meantime, we wouldn't let anybody get to it, just for security of the item.

Richard Finch worked for civilian contractors for government missile and rocket programs. "Most of my career has been in troubleshooting – see a problem, solve it. I think any time I ever worked at a desk for three months, I was ready to bail out. Just too full of energy, and I've got to get on and do something else. I just couldn't sit still."

In ’69, PSL [Physical Science Laboratory at New Mexico State University in Las Cruces] made me an offer. I lived on Stanford Avenue in Alamogordo, and they paid me per diem to live in my house. They gave me a PSL car to drive, and they put me to work on the sled track, collecting data. We did quite a few things there. We did shell firings, bomb drops, rain erosion tests.

April 6,1969, I was doing the consulting at Holloman on the Chaparral [a surface-to-air missile capable of reaching a speed of Mach 1.5], the fastest thing ever launched up to that time. The major who was in charge of the project called me in and said “What’s the secret to going fast?” I said, “Light weight and lots of power, and in case of a missile like that, no aerodynamic drag.” So we strapped a Nike missile, without the fins on it, to the sled. And we bolted a one-inch-diameter tungsten rod to it. The tungsten rod made the shock wave. We were, at that time, running in a vacuum. This was not the fastest Nike, but it was the fastest land-based Nike. We did 7,772 feet per second, or 5,299 miles per hour. It was going to go faster, except the slippers burned off. The only purpose was to see how fast it could possibly go on the sled track. I was also sitting at the viewing area, which was a thousand feet from the sled track, when they lit it. You could see it go for a millisecond, and then all of a sudden here goes the shock wave. And then you see this wave come out around it, and it’s gone!

Wally Adam came to Holloman in 1976 as operations manager for the high speed test track. After three months in that position, he was named director of the track. Adam held that position for five years.

At that time, we had the biggest variety of testing, I think. Maybe it was that way before I got there, but we really had a wide variety of testing. There were some very exciting and interesting programs that went on. The test track was being extended when I got there, from 35,000 feet to 50,000-plus feet. The last end of the track had three rails on it, and the east rail was gauged with the center rail on that portion of track to be the same as the rails at Sandia Laboratory. They thought there would be more interplay between Sandia Laboratory in Albuquerque and the test

A mishap during a rocket sled run on the high speed test track could bend the rails. Straightening them again was crucial for keeping the next sled run from jumping off the track. Courtesy of NMMSH.

track than actually ever developed. So they had that narrow gauge track, and then they extended the regular, I think it's 84-inch gauge track, out to 50,800 feet.

That's interesting, because that's a continuously welded rail that's aligned with a reference line that's probably the most surveyed and accurate reference line in the world. That thing gets surveyed every year, and on top of that, because they ran a lot of very accurate missile guidance systems down it, you could tell from those systems where every little wiggle in that track was. I mean, if there was an eighth of an inch out of alignment in that track someplace, those guidance systems would pick that up. So that thing is really accurate. And being continuously welded and in tension at any temperature below 120 degrees, that thing really is straight and true. Except in the first little bit of it, where there's a curve in the guarder that you can't correct.

There was also, I think a mile and a half of rain field where they could run rain erosion tests. There's a side rain field, where they ran ballistic rain tests. That came about directly to support the Army. They were trying to find out what happened to artillery shells when they fired them in the rain. They'd go down to Panama and wait and wait for the rain to come – really costly in terms of lost time. So they developed that track-side rain field so that they could have rain on call, up to the kind of rain they get in Panama in the rain forest.

We had a sled designed with a nose that was V-shaped but had little places to put in pieces of nose cone material. You'd have probably two or three dozen different little two-inch squares or so of nose cone material, and you'd run that through this rain field at about Mach 2 and see which of them withstood the erosion of the rain. The photography out there was good enough that when you had that rain field going and that sled going down, you could stop the raindrops in flight, so you could really catch the effects on camera.

We were doing some of that testing and helped develop the nose cone materials for the reentry vehicles. We had a high speed test program at the time that I was out there, where we were trying to push the envelope in the speed of the tests we could run and recover, because there was a lot of interest in what happened at the higher speeds. Those were monorail sleds. The way the sleds worked at the track, there were wrap-around slippers that wrapped around the rail. They slid down the rail, and you'd get a little film between it and the rail. We had done a test with the high speed rocket motors where I think we recovered a sled that had gone, like, 7,500 feet per second, that's Mach 7, 7.5. They recovered that sled and they were trying for Mach 8, or 8,300 feet per second.

Both the tests that I was involved with where we did them at 8,300 feet per second, the sled got up to 8,300 feet per second and it came down to about 7,000 feet per second and then departed the track. What has to be happening, the nose of that slipper is about 5,000 degrees Fahrenheit, and there aren't any materials we know that's going to take that kind of temperature for very long. Those things were just going at that high speed until the slipper absolutely got plastic and melted. They hold up through the peak speed, but then as they came down, the slipper just melted off and the sled left the track.

Since then, they've gone [close to] Mach 9 on the track, but that's for impact testing off the end of the track. What they do to get the high speed on that is they cover a mile of the track with a plastic bag they fill with helium. When they fire that sled, it hits the helium at something like 4,000 feet per second or so, about Mach 4 or 5. And when it hits that, it goes like crazy because the drag of the helium is so much less. It gets up to Mach 9 in that last mile before it hits the end of the track and goes off and does its impact test.

We had some impact tests that were very interesting. We did some impact tests for the Navy on hulls of ships. They build up big pieces of a hull of a ship and they impact it. When I was out there, we weren't using any helium bags. We were strictly running in the air. We were only getting about 8,000 or 8,500 feet per second at impact. The biggest thing we probably tested out there, when I was there, was the B-1 [supersonic, strategic bomber]. It's a requirement that all Air Force recovery systems be tested on a high speed test track before they can be put in service, all ejection systems and everything. Prior to my coming, we tested the capsules for the B-1 and F-111, and during my stint we did the testing on the ejection seats for the B-1. The capsule proved out not to be a very good system for recovery, so they did ejection seats. Those tests went very well.

A lot of people don't realize that there was ejection seats on one space shuttle. We did testing on the space shuttle ejection seats. That was on the space shuttle that was drop tested from the 747. They wanted to drop that with people in it, and they had to give them a way to escape. So we did one test that was interesting on that one, where John Young and the other astronaut that was going to fly were out watching this test. It had a unique characteristic in that the ejection was to take place at 2.5 negative g's. They were decelerating two and a half g's, and that's probably what led to part of the problems. That space shuttle is like a thermos bottle; it has two shells. They had primer cord around the shells, and you had the two hatches. The primer cord burned off, the two hatches ejected, and the seat then came out.

These astronauts were watching dummies come out of this thing. The first seat ejected, the dummy's parachute's not working right, so he's kind of flailing. And the hatch came out like a Frisbee, spinning along, and hits him right in the leg, chopped his leg off. Not good! The second dummy comes out, and also there's Frisbeeing hatches from that guy. His parachute opens, but he has a Mae West [a parachute malfunction]. He has a cord over his chute and also that hatch comes along and then chops his parachute out from under him. Both of the dummies in the tank. The guys are watching this, and we just reminded them that's why we test, so we get these out of the road.

One incident was with the SM-2 Navy missile. It kind of showed that track testing, relative to other types of testing, was relatively inexpensive. These SM-2 missiles need to be tested at a speed of about 5,700 feet per second [3,900 miles per hour]. It was like 12, 14 inches in diameter, on a big monorail sled. The least little thing, at that speed, will go unstable. They ran the sled down the track, it gets up to 4,000 feet per second and departed the track. Something happened to destabilize it. The Navy looked at that and said, "Okay, we got the data we wanted, that's great. Costs about $50,000 to do a sled, run another one." But we don't know why it went off the track. "We don't care, run another one." Ran another one down the track, the same thing happened. Got a little call from headquarters saying, "Hey you had an accident down there." This last one, I think, took a chunk of the rail out, which wasn't a big problem, that happened all the time. We had guild people who were good at doing that, welding that rail back together. So they did a third one at the Navy's request, and then we heard from the systems commander. He says, "What are you guys doing, tearing up your track and losing sleds, having all these accidents?" They called it an accident. We said, "Wait a minute. The Navy knows their chance of losing it is high. For the amount of money they're spending, they can't get the data that cheaply anyplace else. They don't care if they lose a sled as long as we get through max velocity." But they did kind of hamper the program a little bit, because there were a lot of people at the headquarters that didn't understand the operation.

I guess the Minuteman missile is a good example of why you can do that and lose that sled. The Minuteman missile guidance system test was conducted by the Air Force. They had thoroughly tested all the components on this system, put this thing together, got it ready, and flew it on a missile. And the combined guidance system didn't work. It blew the missile. That cost $14 million to do that launch. So they came to us and said, "Hey maybe we ought to have tested this integrated system before we put it on. How about you guys doing it?" We had about a million, million and a half dollar program. We did more tests for them than they could ever imagine, and really did a lot of development for a million and half dollars. Now, that sounds like a lot of money, and it was for the track, but compared to a $14 million missile launch, where you got a few seconds of test, it was really cheap. That showed the advantage of that test system.

One of the things we developed out there which was kind of fun was, there was a truck that rode on the track. You know, like the railroads had. That was a very interesting thing to develop because we said, "We'll go up and talk to people at the railroad development center in Pueblo [Colorado], they must know how to do that." They'd never done it. They had no idea. On our way back, we came through Alamosa, Colorado. The guy that was with me was Hans Rasmussen, who was one of the Paperclip bunch. And Hans says, "I know this guy, I think, over here. Let's go talk to him." And so we went over there. We saw a guy a wondering around and he had one of these pickup trucks that they run on the rail. We said. "How do you do that?" So they gave us a real good briefing on how you run a rubber-tired vehicle on the rail.

What he told us was things like, you have to have the right percentage of the load carried by the rails for the pickup, otherwise it will lope, and things like that. So we went from there and got some guy to develop a rail system for one of our tractor trailers and put it on the track. We were doing these Minuteman guidance tests, and we'd run that thing down the track. Then they'd hook on with that vehicle and tow it back while they're still taking data, so they got the going and coming data. That gave them a lot more accurate data for their error analysis.

Fresh out of college in 1951, Marvin Weber went to work in the chemistry lab at Holloman Air Force Base, analyzing rocket propellants. Ten years later, he transferred to the high speed test track operation. By that time, the track had been lengthened to 35,000 feet (6.6 miles).

The last few years when I was associated with the chem lab people that stored and transferred liquid nitrogen and oxygen, I became associated with the test track. The track was into the liquid rocket business, and we developed and operated to support them and the propellants for the engines as well as set up and operate the pressurization systems for the engines. They were cold gas pressure systems. I started at the track in 1961.

One of the things that I did a lot was to go around the country and raid all the Army storage areas for surplus rocket propulsion. We'd bring them to the track and build a sled to hold the thing then take them up to the horizontal test stand and fire the thing to figure out how to fire it, and stage it and so forth. Then we'd measure the performance so we'd have a current performance information. We did an awful lot of rocket motors like that. The main problem we had with those was exterior rusting of the case. We had a bunch of motors that came from Korea, and they had been stored in launch position, fins and everything on them. Well, the fins were a dissimilar metal, and galvanic action set up there. There was pitting in the rear end of the motor. We stewed over those pits. We finally took a batch of bad motors and hooked them up to the test stand and fired them. They all worked perfectly, but we measured those pits and they were about half the thickness of the case. We had a good high-pressure hydrostatic test facility. We sealed those motors up and hydrotested those motors, and they still went about two times the operating pressure. So we didn't worry too much about the pits.

We did a lot of funny little testing programs during the Vietnam War. Our weapon systems weren't made for that much rain. A lot of our ammunitions were exploding in the rain rather than from impact. The rain drops, rather than the target, activated the fuse. That really prompted a lot of work.

I was there when we used the helium-filled tunnel for the first time. It was a labor-intensive operation to get that plastic bag on the track and sealed enough that it would hold helium. Of course, the helium is lighter than air so it would sort of raise and support the bag. But, oh my, the weathermen were against us. (Laughs) It was a difficult, difficult thing to do, but it did help a lot when the sleds got in the bag. It knocked that drag down.

The alignment process [on the test track] was a continuous thing. Sometimes there were special alignments, when you did an impact test or something like that, that we wanted to be real sure for. Or we had indication that there was a bend in the track due to the sled failures, so we'd go out and do special surveys.

The main problem with the rail slipper was the impact, because the sleds tend to fly, particularly those high-speed runs above 5,000 feet. The sled would bounce up and down, and bang into the rail at that real high velocity and leave what we would call rail gouges – wide stripes where it looked like it had just been shot at with a high velocity bullet or something. They filled them by welding them and grinding them. We've had some accidents out there where the rail was physically broken.

We worked a lot of impact tests after the Vietnam War. We were developing warheads and shooting through 3, 4, 5 feet of concrete, inch-thick or more steel, and all that sort of thing.

I liked the tests that were the fast ones. We were anxious to get to Mach 5, Mach 6. To get those motors to survive and reach max velocity was always a challenge. Even at the lower velocities, we had to get some thin-wall rocket motors to get to that velocity, to get the efficiency. And we knew that there was molten metal flying all over the place. In fact, some of those cases that we did recover had gouges in them from the track debris and molten metal. Those slippers literally burn up in that heat.

Boy, those guys worked hard out there. We'd skim the ice off of the track. Did a lot of things to get the water bags [for braking]. We were using plastic-filled tubes on top of the rail for the monorails, and slurry mix that we applied to the rail. It's a silicone compound, and it comes out a nice white, kind of like toothpaste. We had that tank, and we'd mix the stuff up and extrude it on top of the rail just before the firing. It was much better and more reliable than the bags because you could control the depths very well.

Dee Gragg worked at Holloman as a civilian engineer from 1968 until 1997. He started with the Daisy track, and then moved to the high speed test track. During that period, Stapp was not at Holloman, but the Daisy and high speed test tracks were still being used for research on ejection seats, safety belts, and airbags.

In 1972, I moved out to the high speed test track. Ejection seat testing is mostly what I did at the high speed test track. The track length is 50,788 feet, which is about 9.6 miles. It's aligned to within about 15 thousandths of an inch. We have to realign it and grind it to keep within align. At slower speed tests below Mach 1, that is not particularly important. You go a higher speed test, Mach 4 and above, then you've got to have the precise alignment, or the slightest out of alignment will probably cause the sled to break the rail and come off the rail.

The first ejection seat test I started working on was the B-1. We would take the forebody, either a real forebody or, usually, it was a forebody made for us because it needed to be a little stronger and beefed up, put slippers on it. We had rocket motors and pusher sleds and this would push it down the track to the correct velocity. We would typically run tests from static zero, just sitting on the track, up to 600 KEAS, Knots Equivalent Air Speed, that is knots that have converted for temperature and pressure (the standard temperature is 59 degrees Fahrenheit and the standard reading for [air] pressure is 29.92 inches of mercury). Typically, we ran about 25.9 inches of mercury there at the track, and at various temperatures. So we would make that reduction to an equivalent KEAS, and that way you could make a comparison with other aircraft speed or other sled tests. We typically ran up to 600 KEAS and would run maybe from six to ten or twelve tests.

It is economically feasible to operate the track using over-age rocket

Workers clear snow along the Holloman high speed test track in preparation for a sled run. Courtesy of NMMSH.

motors, so we'd get those from the three different military services. They were all solid propellants, improved Honest Johns and Sergeants were used. We had an engineer whose other duty was to keep us supplied in rocket motors and to figure out where they were about to go past their age limit that you would use for their regular, initial, mission. We would procure those to use to propel sleds on the track, saving ourselves, the government, the contractor, and everybody a lot of money by doing that. On occasion, for some special application, we would have to buy a rocket motor. That turns out to be very expensive, maybe $10,000 to $15,000 more per rocket to add to the cost of the test, whereas the other rocket motors were just the cost of transportation. That makes it economically feasible to operate the track, the fact that we can use those over-age rocket motors.

Some of those were, typically, over as much as thirty years. We never had that much trouble with them, but there is a little bit of difference in firing. Sometimes the motor will fire hot, and sometimes it will fire cold. But there is not that great a difference. If you are firing eight or ten of them off, then randomly they kind of cancel each other out. So they were pretty reliable. We would run the sled down the track, and then at a predetermined place on the track we would fire [the eject mechanism]. Some shots we ran with hatches or canopies, and others we just ran with those off, depending on what we were looking for.

Stapp said that God had humans in two different models and that we had extensively tested one, namely, the male, but have done virtually nothing with the female. We were going to be the first to test the females. There were not female enlisted and officers running around thirty years ago like they are today. So we had convinced about eight or nine nurses to do this. They were pretty gung ho about it, but then, I don't know if it was the politics or what got to it. We got shut down by the headquarters in Brooks [Air Force Base in San Antonio, Texas], and never got to do it. It was finally done for the first time at the Aero Med lab up at Wright-Patterson Air Force Base some years later.

Now, in the really early days, we didn't have any problems with females flying in ejection seats because none of the services would let them. The few female pilots that we had flew in fixed seats in cargo planes and that sort of things. About 1980 or so, we started permitting females to fly in all the combat aircraft, so they were flying in ejection seats. And no tests had ever been done for females in ejection seats. Numbers of us worried and wondered about this thing. To be a pilot, you have to meet a sitting height so that you will be able to see out of the aircraft, and you can't be so tall that you bump the top of the canopy, and you have a weight limit. We didn't know what we had out there in terms of females, what sizes they are. Nobody asked me to do this. I just had the idea that somebody should look and see what we were training, what sizes these females were. When they did their physicals, these dimensions had all been taken. It's just that nobody had ever put them together. The females were scattered around at, I believe it was six Air Forces bases. I had made friends with one of the medical technicians at one of these six bases. He greased the wheels for me to call the other medical places. I didn't get the names of the females, but I got the numbers of who they were.

I found out there were 109 females out there who were training and were riding in ejection seats or anticipated flying ejection seats. I did a survey of their heights, their sitting heights, and their weights, and published this paper. What I found is that most of them met the 34-inch sitting height, but they were grouped down toward the low end of the sitting height, so they could see out over. And most of them could just barely meet the 64-inch stature. They were kind of clustered at the low end. But the thing I found that was disturbing was their weight was low. We were testing, at that time, for a 132-pound, 5th percentile male, thinking that we would cover everything except the lower 5 percent. But it turns out that 55 percent of the females had a weight lower than that. Now, that's important because you are putting the same rocket [ejection] motor under them no matter who they are. So the lighter the female, the more they were going to get their spine compressed. So we felt that there was going to be a greater probability of injury to the spine in the females.

In fact, one of the training bases worried about it enough that when they sent the female out with her gear to the aircraft, they had a lead weight they installed in the survival kit. But they discontinued that after a while. I think it was kind of a discrimination thing. They were treating the females different from the males. So they just sent them out at the lighter weight.

There was a more surprising development that came out of this that I didn't discover myself, but heard about later. It turned out that the females weren't the only light ones flying. We had a number of males who were very light, and some of them were flying under waivers. It turned out that we had more males that were of a lighter weight than we had females, because we had so many more male pilots. So then it was a question we had to address for the males as well as for the females.

One way of addressing it is with the new CREST [crew escape technology] seat, which is a seat that had the capability of sensing the weight that it was accelerating. It would adjust its rocket motor thrust based upon the weight it saw as it accelerated up the rails. But that seat has never gone into production, and I'm not sure if it ever will. The program was about to wind up at the test track about the time I retired.

Once, a [female ROTC] group was touring one of the training bases, and they were giving them incentive rides. This particular ride involved a rather small Oriental girl. The pilot who was going to take her up was sitting beside her and he had told her, "If anything happens and we have get out, pull on these handles and I will tell you when to get out." "Okay." They went up, and they flew around and landed, taxied back in, and they parked. The pilot said, "It is time to get out," and she pulled the handle. Out she went! But a strange thing happened. Instead of going a hundred feet and coming down and being killed, she only went about 30 feet in the air, stayed in the seat, came down and hit the tail of the aircraft, bounced off and sprained her ankle. Instead of being dead, she had a sprained ankle. Well, it sent the Air Force into a panic, because at a very light weight of about 90 pounds, she should have gone very high in the air and come down and been dead. What really worried them was they were afraid they had a catapult out there that wouldn't have enough force to accelerate them and perhaps miss the tail or not get them high enough to get a good chute [deployment].

From that, they ran a test program at Holloman. We had to come up with a lightweight female dummy of 90 pounds. Our lightest weight dummies were 5 percentile males, and they were 124 pounds. So we had to make a lighter dummy. We made some of the parts out of aluminum and took out some of the instrumentation that was weight. So we got down to about a weight where she was. We conducted the test there at Holloman, and we could not repeat the 30-foot failure. They all went up about 80 to 100 feet like they were supposed to. If it went to a hundred feet she is dead. You can't survive that. In the work I've done and the research that I have studied there at Holloman, about 50 feet is where you expect to survive. Fall from about 50 feet, and you expect to survive but to be injured rather severely. Past 50 feet, you really start expecting fatalities. That is even coming down in dirt. She would have come down on an asphalt tarmac. That problem was never resolved. We hope that there is no problem. We hope that was just a one-time fluke.

We own the world's land speed record, at Mach 8. That was done in 1983. I think that at that time, nobody thought that record would stand, but it is still standing today. [Note: Gragg made that statement in his 1999 interview. In 2008, a new land speed record of 6,453 mph, or Mach 8.6, was set on the Holloman HSTT.] I didn't have a lot to do with it personally, but I did edit that technical report. I had to go back and study what had to be done, in order to do a good editing of it. I computed the Mach number, and it did come out exactly 8.0.

That was in the helium tunnel, which is a bag about 4 feet in diameter. It goes through about two miles of helium. Of course, the speed of sound is much lower in helium. So when it burst out the end of the helium bag into the air, at that instant it's at Mach 8. It's pressurized only to the extent that it will keep the bag filled out. If the wind is blowing you can't do it, because the wind will blow the bag over and spill the helium out. You can't run a high-velocity shot with very much of a crosswind anyway. We look for under 3 knots [3.5 mph], and that means some of the high velocity stuff we have run after sundown, like eight o'clock or so in the summer. Or we will run them right before sunup. These are the times when there's the least amount of wind.

Unprotected, even the best steel just catches on fire. It gets so hot and burns right up. So we have, in addition to the most heat-resistant steel we can come up with, we also have coated asbestos to give it as much protection as we can for as long as we can. And, of course, going through the helium helps some too.

It was in the *Guinness Book of World Records*. They put it under railed vehicles, like we were a train. If you look through the *Guinness Book of World Records*, there is nothing else that ever moves across the surface of the Earth anywhere near as fast those sleds did. So when I did the track video, I started to refer to it as the "world land speed record." And nobody has ever contradicted that yet. Nobody ever thought the Mach 8 would stand. We thought we had technology to about Mach 9.

We had a track director who was really an enthusiastic guy. But he was a pilot, and he wasn't really conversant with the technology that it would take to go Mach 10. All of us that had some knowledge, we really felt probably about Mach 9 was as fast we could go with the current technology, with the materials we've got and the ways we know to coat them and protect them, and these sorts of things. But he wasn't having any of that. He thought we could do Mach 10. It was just his enthusiasm more than anything else. So, among us it was kind of a joke. Not really at his expense, because he was really a nice guy. He was putting out there what he believed in, and so he had those signs put up. "Think Mach 10." I just followed along with that and my car tag says "Mach 10."

One of the reasons for the length of the track is to save sleds intact. We have occasionally saved a sled at Mach 7, but nothing above there. As you get heavier and heavier vehicles, like the ejection seat stuff, you really like for them to coast out. It's easier on them. They will coast out to maybe 30,000 feet or something like that, where you retrieve them. When you are stopping high velocity sleds, you really can't do that much above Mach 1 because you tear the sled up with braking. So you got to let it coast down, mostly by air friction, to about Mach 1. Then you can start braking it. There are different methods of doing it, cold cream on the rails, and water between the troughs, and various things such as that.

The cold cream is called super goop. It has a consistency of cold cream, and it's biodegradable. It evaporates right away. You have to use it within about an hour of the time when you applied it. So one of the last things we do is put that material on the top of the rail to slow the vehicle down. We have a machine that mixes it and applies it. It applies it in ten different shapes, going down from very small [thickness], like the end of your finger, up to large amounts about 3 inches in diameter. That way, if your sled is going really high velocity, you will start with the very lightest first, and then graduate on up. If you think there is still some doubt in stopping it after that, then you will put the water sausages on top of the rails, to go ahead and bring it down to a stop.

[In 1984, ten people from Mainland China arranged to visit Holloman to see the high speed test track. They were interested in learning how to use a test track for evaluating ejection seats.] We gave them a day's tour and let them ask questions, and we showed them things. The rules from the State Department were that we couldn't give them anything to take back, any documents or anything. But they could take whatever notes they wanted to. So they had about half of these people sitting at tables writing furiously all the time, taking as many notes as they could. The thrust of that was to build their own test track.

They had a test track. They built it somewhat like you would build a train track with ties and rails. They said they were getting up to about

Mach 1 with that, which we thought was admirable, that you could get up to that speed with something like that. But they wanted to go faster, and they wanted to do the kinds of tracks that we had. [The group leader] was asking a number of questions, but one of them was, "How do you keep the people off of the track while you test?" I said, "Not much of a problem. We are 15 miles out in the desert. There are no people there." He said, "Well, in China there's no place that is 15 miles from anyplace else, where there are no people there." So, a big problem they had was getting the peasants off the track and back out of the way so they could test. It was an interesting visit with them. I enjoyed meeting with them a lot, and I tried to help them as much as could.

Joe Mapes, a mechanical engineer, was hired to work as a civilian at Holloman Air Force Base after working for ten years at the Naval Ordnance Test Station at China Lake, California. For the next twenty-four years, he worked on the high speed test track at Holloman.

I came to Holloman in December of 1956. They offered me a promotion to come here from China Lake. It helped quite a bit that I had some concept of a test track. I hadn't worked on the track at China Lake, but I knew what it looked like. I was probably the first one that they came across that didn't say, "What's a test track?"

When I got here, the new extension of the test track was under construction. They were building the whole extension part, going from 3,500 feet to 35,000 feet. They didn't have the rails on, or any of that. We went on from there. The original rail was a standard railroad rail. I think it's a 90-pound rail. Now they're using heavy-duty crane rail.

They'd have problems crop up relating to the building of the track every now and then. I got involved in that. I found out that the Army Corps of Engineers was building the track for the Air Force. And the way the work on these things went, the Corps of Engineers were making all the decisions that were needed along the way. They made the drawings and had designed things the way they thought they should be, but none of them had seen a track either, so that didn't help.

They had survey work going on all the time to check the alignment of the track. And the tensioning part that was originally installed, they had big hydraulic jacks that were the tension of the rail. Another thing that was new was the welded rail. The Edwards track was in existence before the Holloman track, but they had 10,000 feet [increased to 20,000 in 1958]. The Air Force did most of their testing on that track before Holloman came along. This welded rail idea really made a big difference. Another difference between the Holloman track and the Edwards track is difference in gauge. The Edwards track is standard railroad gauge [four feet, eight and a half inches], and the Holloman track is seven feet. That was another advantage for big payloads.

[Several kinds of stopping mechanisms were used on the test track.] We had some three stage, or more, of the different kinds of water brakes. We had some Styrofoam troughs at the side for some, and some nylon straps to catch a hook on top of it at one stage, and two or three different methods to stop at the lower stages. One of the final stopping schemes that we used was that we had a Styrofoam trough that fit on top of the rail, and we'd put two or three water sausages in the Styrofoam holder. At first, they used to put a plastic tube on top of the rail and tape it on with duct tape, but that was sort of primitive. And the tape that held the sausage caused lots of trouble. It would sometimes damage the test team. It wasn't a good arrangement, but it would work for certain things.

They used sand braking at China Lake [sometimes]. I don't think anyone ever tried that at Holloman. It was a pretty crude scheme, but it worked. They had a box of sand on wheels, and they'd hit an opening in the back of it. They could meter it out, and they put down this pile of sand. They could taper the depth of it, and they could vary the width. The sand braking scheme had one serious drawback, and that was the mess that it made on the sled. You'd find a black box [instrumentation] that was tight and solid, but it was full of sand when you got through with that sand braking.

We had several tests [using an old railroad siding at Holloman]. The test item was a simulated nuclear reactor. The reason for the test was that they were concerned about what would happen if this thing was in orbit, then came back in through some condition, and hit in the water somewhere. What it would do to this reactor, whether it would make it radioactive, or break it all into little pieces, or what it would do to it. They came to us with that problem, but they wanted to impact in the water. I've forgotten now exactly why we went over there to that siding, except that we had to make a big water tank. We had one about 10 feet cross section and maybe 30 or 40 feet long. It had a lot of water in it.

We had to have photography of what the trajectory of this thing was in the water, so we had to have a transparent side in the water tank. We had to get rid of all the sled hardware so that the only thing going into the water was this simulated reactor. It was not too big. We wound up suspending that with some nylon cords in the middle of a special sled that we built that used a HVAR [high-velocity aircraft rocket] motor. We had to get rid of all the motors and the sled and everything, so it had to be an expendable sled. Then we made sort of a highway culvert at the tip of the sled, allowing the test device to go free.

I think what they originally planned, they were going to have the impact point close to a fence. Late in the game in the program, we were looking at that, and said, "Well, we'll keep that rancher over there from coming out to watch this test, and sit on his fence right there where it was going to impact." We were considering his land as part of Range, really. It was going to be aiming right for his land, and he could have a cow or horse standing there, and we might have wiped him out. So we reversed everything. It was still fairly close to some of those buildings.

I think we ran three or four of those. But before we even fired one of them, we had a big problem. That was the water. I think we brought tank trucks out to fill this tank that was partly canvas, and a plastic side for photography. Anyway, we got the thing filled with water, and the photo people set up. They were focusing and checking things, and they said, "We can't work with that water." You couldn't see. It was too murky. They put an object in there that would be similar to the test item and took a couple of pictures. They were pretty sorry. They finally solved that one by getting water out of the officers' club swimming pool. They had a filter, so they had clear water for the pool. They went over there in a couple of big tank trucks and got the water. It worked all right.

I remember we had a real good picture of one of them where the test item went into the water. They had some cameras up close, but they had one back to show the whole layout. And right in the foreground of this picture was a big jack rabbit sitting there with its ears like a Bugs Bunny movie.

There were a couple of tests where we had some expendable length of track on the end. I think we learned that from Edwards. They had a test that may have been some kind of an impact test, but anyway they blew off the whole end of their track as part of the test. It hadn't been planned, of course. But then we decided if we ever do one like that, we need to have an expendable section of track at the end. So their experience saved us a lot of agony.

Going over microfilm records from the early days at Holloman, everything was classified. They got to the point to where practically none of our projects were classified. ARDC [Air Research and Development Command] discovered that it was pretty hard to classify that much space and that many people. The things that were classified were kept quiet, but it's hard. A couple of officers came out once to set up a meeting about a closed test. They started by saying, "What's the fastest sled you've got, a big sled to carry a payload of about 200 pounds?" They were talking about a quick test of some kind that they needed to do right away. We brought out all our data and pictures and things. They said, "Well, that would be great, that looks fine, that sled you've got. How soon can we schedule a test?" Ordinarily it takes several months, negotiating over money, and things like that. They were interested in doing it in a couple of weeks. So we went into a huddle and then said, "We can do it, but these rocket motors are the most expensive ones we've got, and the thing will cost around $250,000." They said, "Let's go! What's the big holdup?"

We thought maybe these people were some kind of con artists that were just sent out from Wright-Patt to con us in some way and then go back and write a nasty report of some kind. It turned out they were authentic. They gave some code names. I think there were two majors, or two captains. One of them's name was Blossom. That was the name they introduced. He even had a name tag, "Blossom," and the other guy was named "Flowers." They were supposedly either from the Pentagon or Wright-Patt. They couldn't give us any address or phone number where we could reach them. They said, "Here's a number at Westinghouse. If you want to reach us, call this number and ask for Blossom or Flowers." That was one of the reasons we thought it was sort of a phony. We weren't used to working much with highly classified stuff. We actually did about three of those tests. We furnished the sled and the velocity, and they did the rest. They seemed to be pleased with the results. We were pleased too, because they did pay their bills.

The way Holloman got into the rain testing business was through a Sandia test. Sandia had a test involving rain erosion, and their track was not long enough to handle it. So they came down here, and they built a test section. They had to put in a pumping system and spray nozzles, and a lot of water mains and pipes. When they left, rather than remove it, Holloman made arrangements to keep the system. Then later, it was improved quite a bit and lengthened, and they used it on a lot of other tests.

One time, they had an explosion out there at the pump house. They had a big transformer and a couple of big electric motors that ran the water pumps. For some reason, this transformer blew up, blew oil that cools the transformer. It was a big one. It blew up like a bomb, and it happened when there wasn't anyone in the building. There was one man that was close to the building, but he didn't get hit. It really scared the daylights out of him. He was leaving the place, and he got about twenty yards away from the building, and the thing blew up like a bomb. I think it knocked him down, but he jumped up and ran. He found out that he was barefooted. His shoes were still back there. He jumped out of his shoes! I remember being out there within a couple of hours of that, and we were looking at all the damage. It took a long time to get that back in business again, because it damaged just about everything in sight.

[At night, the track rails remain warm and attract small animals. The water in the braking section of the track also attracts animals, especially birds. At the extremely high speeds the sled runs, an impact with even a small creature can cause a disaster.] Bird attacks are a favorite topic at the track. I don't know whether they have any newer methods, but at various times we tried a lot of different things out there. It turned out to be a real serious problem. Everybody ribbed me, it was sort of a big joke, and humorous, worrying about these little doves or sparrows or house finches.

There was one guidance test that was really a problem. It hit a bird, and the bird penetrated the side of the sled. It was one of the big missiles, and it penetrated the guidance equipment. It's the part that they put together in the clean room. This bird penetrated the inner sanctum of the whole system. You've got a lot of real delicate parts, you know. Here was the remains of the bird right in the middle of all of that. That stopped the program for a week or two. We had to go back to the factory and redo the whole thing. That one wasn't a joke at all.

They have a little monorail sled out there at the track that is sort of a trophy. You can see the imprint of a bird in the steel, about a quarter-inch steel. You can see the whole outline of the bird, and it looks like a fossil, sort of. That was a high-speed monorail that hit him. For some reason, the birds like to sit right on top of the rail, right in the flat part of the rail. If you send the bird-warning monorail sled down there, if there's any birds there, it will get rid of them. They don't come back for a few minutes, anyway. That has some drawbacks, because sometimes there's instrumentation set up for measuring velocity or triggering cameras or even some of these blast tests, and that monorail going by would trigger some test event.

So that's been a pretty effective way, but they tried a lot of other things. They tried playing tapes. They had speakers along the track, and they play different sounds like machine gun fire, hawk calls, and owl calls, and all kinds of things that were supposed to scare birds. They used to try noise makers, and at one time they even had the people from the track with shotguns go down with the truck just a few minutes before the test and shoot in the birds that were on the rail or close to the rail. They sometimes would get twenty or so. That's not a satisfactory scheme, either. It was fairly effective, but by the time a couple of truckloads of people with shotguns would get to the north end, some of the birds would be back at the south end. So, that didn't work.

Then we once had a contractor from New York. They have a big problem in the cities with the pigeons roosting on the buildings. They had a chemical preparation called "Roost No More." They sprayed some of this preparation along the track, and it wasn't very successful, either. In this climate, the stuff doesn't last too long, and it's sort of like a flypaper. It's supposed to irritate the birds' feet so they won't stay in one spot, but some of this was so thick the bird would get stuck in it, just like flypaper. It was fairly effective, but it just wasn't practical for over any period of time, and it was pretty expensive too. It was a different situation than a building in New York, quite a bit.

One of the things that I was involved in, the Air Force has some people that study bird impact problems full time. They had some up at Kirtland [AFB in Albuquerque], and we got some of them to come down and look at the problem. They did some good work. One of their schemes was to set up sort of a bird sanctuary off to the side, a bird feeding station here and there. They did a study to catalog what type of birds were causing the problem. We thought it was mostly doves, but it turned out the house finch was the most common bird out there. They made a good report about analyzing what types of birds, and how many, and which of them are migratory that come to here in certain months. They gave us a lot of good information, but it didn't really solve the problem. The feeding station idea didn't work, or very much, because the birds like the water that was in the track. [The braking system consisted of a trough of water between the rails. A scoop on the bottom of the sled would drag through the water and slow the sled.]

In 1984, I got a chance to work as a consultant for a couple of weeks to the Taiwan military army. They were interested in track testing. I went over there and stayed two weeks. They had a small track that they'd built and used some. The gauge was 80 centimeters [31.5 inches]. They wouldn't let me see the track, even. They showed me some pictures of some of their tests. It turned out that they were planning on building a track 10,000 feet long. I got the impression that they might have been working like our defense department does. In order to get money for this track they had to have some studies made, sort of like impact statements.

It was very interesting for me. I think they got their money's worth from me, because they had some real basic things wrong with their idea. They took me to a site where they were planning this track. They're pretty crowded on that island, so they don't have many suitable locations. One of the things I mentioned to them was the earthquake problem. It turns out they had already thought of that quite a bit, because when I mentioned earthquakes they showed me the results. They had done a computer study where they had sensors all over Taiwan, and they had picked out an area for the track which was better than most. It had been a sugar plantation. They said the reason they could get the space was that the price of sugar at the world market was so low that they could afford to use this sugar cane field for a test track.

Right across the middle of it was power lines, big transmission lines. I mentioned that, and they'd already made plans to re-route this power line. One they hadn't thought about was they were right on the edge of the town. It was the southern tip of Taiwan. It's near one of their main seaports. They flew me down there on a commercial flight onto the south end, showed me the site, wanting me to comment on it. They had a safety problem there. The place they'd picked for a launch point was right alongside part of the town. You could ride out on a bus and be within a hundred yards of their launch point. All kinds of buildings and people were nearby. I suggested to them that they needed to really do something different there. The thing was, they hadn't even thought of that one. They had thought of just about everything that I had mentioned, the earthquakes, and power lines, and the drainage, the slope of the land, and different things, but this safety thing hadn't even entered their mind, apparently. They were going to have this launch point right in somebody's front yard, more or less.

They did agree that this could be a big problem, they just hadn't thought of it before. Well, they hadn't been around a test track. It's hard to visualize the blasting you get, or what happens if things go wrong and a rocket motor takes off in free flight or something. I don't know how they solved that. I'm not sure they built the track in that location, but they may have.

I recommended to them that they come over and see some of these tracks if they possibly could. Starting from scratch it's pretty hard to visualize these tracks and some of the problems. So they did come over here. They went to China Lake and they came to Holloman. They gave me a call, and I went with them out there at Holloman. We talked about the rails, and I told them we got some from Bethlehem Steel. They said they had some steel metal, and they could make their own rails. They could, but there was some real simple things that they hadn't thought of, like the safety problem and different materials and cameras and things.

The other thing that they had mentioned to me that I helped them with was a little out of my field, but I was familiar with the problem. That was getting a good electrical ground around the track for the data collection business. You need a real good ground. They had problems at China Lake with that. They asked me if I knew anything about that, and I helped them. They said they got a lot of noise in their data. I asked them what kind of a ground they had. They said they had a 6-foot copper bar pounded into the ground, which ordinarily might be a pretty good ground for a lot of things, but with this kind of business they needed something better. I think at China Lake, maybe at Holloman, they drilled a well for a ground and got down to the water table. That idea seemed to really excite them.

Chapter 7

Animal Pioneers

Research using animals was a bittersweet chapter in the preparations leading to manned spaceflight. On one hand, the information gained from animal experiments was crucial to the success of manned missions. On the other hand, some of the animals gave their lives during or after the experiments.The oral history interview excerpts in this chapter give a behind-the-scenes glimpse into the behaviors and attitudes of people who were involved with the animal experiments.

The United States was not alone in its use of animals to prepare the way for human spaceflight. The Soviet Union sent dogs on suborbital flights as early as 1951. Laika, the first dog sent on an orbital flight, rode in Sputnik 2 in 1957. The flight lasted 162 days, and then the satellite burned up on reentry into the Earth's atmosphere. Laika had died about six hours after launch because of stress and severe overheating of the vessel. Other than that one flight, the Soviets' other suborbital and orbital flights carrying dogs were planned to recover the animals alive, and most did survive.

In more recent years, the United States and several other countries have sent a variety of animals into space. Temporary passengers on the International Space Station have included worms, frogs, mollusks, butterflies, spiders, and transparent fish.

After basic training, Albert Poitra was sent to Holloman Air Force Base in New Mexico in 1949. He worked with several missile programs, including the Aerobee rockets that carried monkeys and mice on suborbital flights.

I remember the animals that they had, these little rhesus monkeys. They had them in the nose cone [of the Aerobee], and I guess they were wired up for sound. I'm not too sure exactly how it was recorded, whether it was from the nose cone to the ground or recorded on board. I think they measured their heartbeats and their physical aspects. I still remember the cages where these little guys were at. One was named Pete. I think that's the one that went up. I used to bring him slices of bread from the mess hall. We weren't supposed to do it, but I took a liking to him. You couldn't get too personal with him because he'll bite you. They were kind of nasty little animals. They wanted to be running free.

Jim Diehl grew up on a farm in Indiana. A year after graduating from high school, he joined the Air Force in 1950. He was trained as a security policeman and spent several years on duty in Germany as well as in the United States. In 1963, he was assigned to Holloman Air Force Base, where he was the operations sergeant for 200 security personnel and sixty civilian guards. Some of his guards were assigned to the Bopper sled area and the high speed test track. As he mentioned in his 1998 oral history interview, animals were involved at Holloman in both official and unofficial roles.

At the Bopper sled area, my guards wanted me to come out to see the bear out there. I said, "Oh there is no bear out there." And they said, "Oh yes, there is. We are watching that bear." So I went out and saw it. Sure enough, they had a bear out there in a cage. It was interesting.

Our squadron commander wanted to go out and shoot coyotes all the time. He said, "There are too many coyotes around here, and we have to shoot them." He never did tell me to go out and shoot them, but they used to come up around the trash cans at the housing area and get in the trash cans. That was always a problem. Skunks came around to the commanders' houses at night.

In 1956 and 1957, David Vigil was working for the Denver Fire Clay company, which was supplying instrumentation for high speed test track sleds at Holloman.

I happened to be here when they were putting chimps on it. They would always tranquilize the chimp. We would watch the track. When it would stop, the chimp would sit there a minute, then it would shake its head. The crowd would go crazy. It was kind of funny. One time, something happened and there was a delay before they opened it up. They got him out, and he actually came back to life and started running across the desert. They had to go chase him. He didn't want any more part of that.

A chimpanzee rode Sonic Wind 1 on the Holloman high speed test track four months before John Paul Stapp took his first ride on the sled in March 1954. After Stapp's third run on Sonic Wind 1 (see Chapter 6), chimpanzees, other animals, and dummies were used almost exclusively as subjects. Courtesy of NMMSH.

Don Larsen was twenty years old when he joined the Air Force in 1961. After basic training, he was asked to choose one of three fields: air police, cryptographic communications systems, or aircraft weapon systems. He picked the third option because "it was the only one that said anything in the description about airplanes." After armament school, he was assigned to Edwards Air Force Base in California, where he worked with the rocket sled test track.

We ran a lot of chimps down the sled track. Of course, this is related to the space program, and we had the chimp in the sled doing different things under the g forces. They had the animals trained. That was a lot of fun. One of the funniest stories I remember about this, and they had it on film. They had a chimp one day in the sled, and he was a little upset, and somebody gave him a banana. It turned out to be a mistake, because they forgot it, and closed this thing, and they fired it. They had this film, this chimp, he had this banana. He's eating it, and the g forces hit the banana and it goes all over his face. Needless to say, when you tried to give this chimp a banana after that, he would get real upset, to put it mildly.

Have you heard of the famous chicken gun? This was a pneumatic-powered cannon. We had an air compressor hooked up to the thing that would put a pre-charge on an actuator. You had a solenoid release, and it released this cylinder real fast. We would use live chickens, and here I am on the front of the *Los Angeles Times* holding two cluckers. We'd get in big trouble with the animal people today, because these were live birds. We'd get them from the local chicken farms, which were all around Edwards. What they were using this for, they'd fire these chickens at different velocities into aircraft canopies for bird strike. Bird strikes were every bit the problem then as they are today, particularly if they hit the wind screens. Usually, it would incapacitate the aircraft pilot when that happened.

G. T. Lamberth worked for Land-Air, a contractor serving both White Sands Missile Range and Holloman Air Force Base. From 1960 to 1962, he worked with testing on the Daisy accelerator. Among the chimpanzees who rode the Daisy track (see Chapter 5) was Ham, who took the first occupied suborbital flight of NASA's Mercury Program in January 1961.

Ham was one of our stars [at the Aeromedical Field Laboratory]. He was something else. One of the veterinarians had a little old scooter. He bought it and had it shipped in here. He let me ride it sometimes. Many times, I'd put Ham on the back of it, and he and I would run around the compound.

The chimps were social. In fact, we had one that would get loose out in the compound once in a while. We used to have an old trailer at the shop. It would come into the trailer. It had the wrenches all haywire. That little wart, he would take them down one at time, look at them, and hang them back up.

I was always chewing on a toothpick. I would walk up to the cage, and this one little chimp would come up. I'd step close to the cage, and he would reach through and get that toothpick out of my mouth and start chewing on it. Well, my boys soaked some in cinnamon oil. So I walked up and got close to that cage, and he got that thing out. He would suck on it and he would run get a drink of water, suck on it and run get a drink of water.

In 1958, Dr. Rufus Hessberg and Dr. John Paul Stapp exchanged locations, with Hessberg taking command of Holloman's Aeromedical Field Laboratory and Stapp assuming leadership of Wright-Patterson's Aero Medical Laboratory. With Hessberg's arrival at Holloman, a new program began to take form, as he described in his interview:

Most of the laboratory effort at that time was strictly the investigation of man's tolerance to impact, the very rapid-onset, short-duration type of acceleration. However, when I went out there, I was charged by my boss at Air Force Systems Command to develop, if possible, a program which would allow the lab at Holloman to train and study chimps, with the idea that they could be a surrogate for man to test out the Mercury capsule. The idea was to show that the design of the Mercury capsule and the whole flight program for Mercury was going to be tolerable and within human limits, and it was strictly a "safe for human consumption" program.

That was an interesting little phase because, at that point, Project Mercury was under the Space Task Group at Langley Air Force Base in Virginia. Literally, NASA was in its infancy. NASA was born in the summer of '58, and my transfer was in the summer of '58, so I was faced with going into Langley and convincing the old NACA engineers. [The National Advisory Committee for Aeronautics (NACA) was the predecessor of the National Aeronautics and Space Administration (NASA).] First of all, educating them a little bit about the physiology of flight, and then convincing them that it would make good sense, before they exposed a man to the Mercury flight program, to let us fly a chimp first in the suborbital mode and then in the orbital mode. Once they accepted that as a viable option, then we had to go to work and start training the chimps for that program.

The idea for the chimp tests originated in the minds of Don Flickinger, who was at Air Force Systems Command Headquarters; Jack Bollerud, who was commander of the lab at Wright-Patt before Stapp came in; and I suppose, partly my belief that it made good sense to test that capsule with something other than man as the test subject.

Most of what NASA knew about space medicine at that point in time was with the military individuals assigned to NASA. That was where all of the space medicine expertise existed. NACA had nothing to bring with it to NASA on that. It didn't take too long for them to begin to realize that we were talking about real concerns – not problems, just concerns – and that we were confident it could be done. But like everything else, we always like to expose a dog or a chimp or a bear or a pig on the track to see what would happen before a man did it. And it makes good sense to us, in spite of the animal rights people these days. We didn't have to contend with much of that, thank goodness, in those days.

Nor did we have to worry as much about what we did with humans as research subjects in those days. Never hurt anybody or killed anybody. I shouldn't say we never hurt anybody, but we didn't do any severe damage to anybody. That was because we approached everything cautiously. We worked our way up into higher and higher impacts or higher and higher g forces, and we always preceded it with seeing what would happen to the properly selected animal prior to exposing man.

The chimp program began in late '58, after NASA made the decision they would go with the flying of a chimp in Mercury prior to the suborbital flights. That was when we really went into full speed ahead. I can't tell you the census of our chimp colony at the time, but we had just begun bringing in the real young chimps into the colony. In parallel with the commitment to support the NASA program, we had to modify our colony arrangement. We had to dedicate one of the buildings to be strictly a quarantine facility, and that required some minor modifications. I don't think that the chimp colony, when I arrived there, was more than about sixteen or eighteen chimps, maybe six or eight of them being adults, which were not usable for what we were about to embark on. I think we had eight young chimps at the time. Then we began a very concerted program to acquire more young chimps. We developed the quarantine thing, and we began immediately to start developing physiological and medical files on the chimps. I'm talking about their blood type, their blood chemistries, their growth, their development. We recorded anything that we could on those chimps and had exhaustive and marvelous records. We proved through this kind of work how similar a lot of the physiological and biochemical factors are between *Homo sapiens* and chimps. I don't know how many people realized how similar a lot of those factors are until recent years, long after we got through with that program, and what a marvelous surrogate the chimp is for a lot of human problems that need to be investigated.

All of this was new research. Most of the body of knowledge about chimps, at the time we began, was growth and development and performance. When we went to Yerkes [National Primate Research Center], which was *the* primate colony and the source of all knowledge in this country on primates, we found out they had done none of the physiological parameters on them, did not know what a normal blood biochemistry profile should be, did not know what certain X-ray growth and development closure of the epiphyses, et cetera, should be. Had no

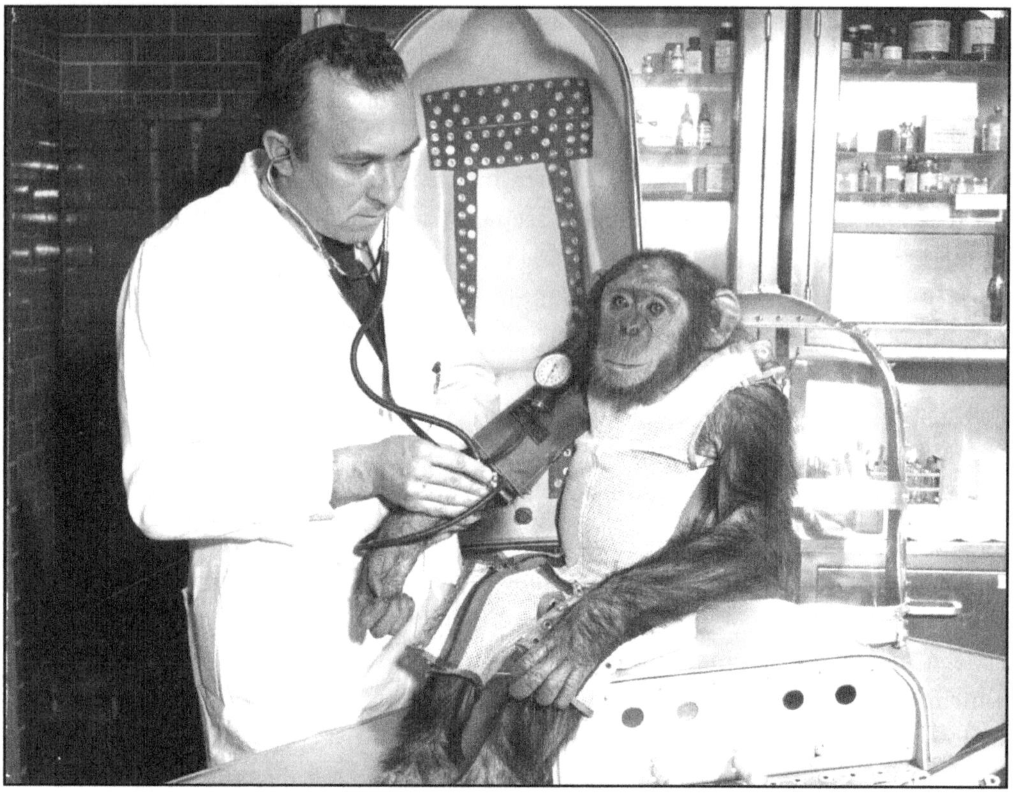

Besides using chimpanzees in deceleration experiments, Holloman researchers collected and analyzed important physiological information about the species. Courtesy of NMMSH.

data on tooth eruption. I just can't begin to tell you. We began doing EKGs on them, and would take their temperature and know if there was a diurnal variation. And blood pressures were not known. I mean, they knew a lot about growth and development, behavior, but none of the factors that go along with that and none of the physiological background. We literally had to start from scratch on that.

The group we brought out from Wright-Patt were marvelous at training. They brought the best equipment there was available for automatically recording performance parameters. They developed the methods and techniques, and the little training chairs that the chimps sat in, the consoles that were presented to them. All of this, by the way, with the idea that it had to be in the same size, shape, and weight parameters that were going to be for the flight program. In other words, the chimps, when they once started training, were training in an environment that was going to be totally familiar and recognizable and translatable from the laboratory to the actual Mercury capsule. We trained them in everything you can think of. We had an airplane dedicated to us from Holloman, where we could fly them and get them familiar with the vibrations and the disturbances, if you wish to call them that, which you can have in flight but don't get sitting on the ground on a solid foundation. We flew the chimps into Wright-Patt and exposed them to accelerations on the centrifuges so they would know what the acceleration of launch and reentry would be like. All of this was done in training capsules the same size, shape, and configuration that was going to be used in the Mercury capsule. We flew the chimps into McDonnell Douglas in St. Louis, and the capsules for the chimps that were the flight candidates were molded just the way they did for the astronauts who flew into St. Louis and got molded for their couches.

The fidelity of the training and the environment of the training was the best that was humanly possible all the way through the program, with the idea that if anything strange, or abnormal, or different, or variances in the flight performance occurred, we were trying to eliminate every possible factor beforehand, other than the true aspect of weightlessness, which is the only thing that we couldn't do on the ground. Parabolic flights, to us, just weren't practical because of the complexity of having to do that for only a matter of seconds. And that really wasn't going to be the answer that we needed. So we did not do that, because we wanted that to be the first and only different aspect of their flight program.

We had what we called a Bopper [see Chapter 5], which was an elastic-driven 20-foot sled thing. We used that to bop them, to give them a simulation of landing in the capsule. But we did not run the chimps on the test track.

We were going to get EKGs and blood pressure and temperature, but the main thing was to see if their performance was degraded in any way. This was done by presenting them with a panel that they were used to from Day 1, so that it would not be strange. We had their performance records. It was automatically recorded and timed to reaction and the number of mistakes they would make, the number of pellets that were delivered for reward. We had squares and circles and triangles, and they would have to hit the lever under the object that was different. When we presented them with two squares and a triangle, they hit the lever under the triangle. Or two circles and a square, and they hit the lever under the square. These were all randomly programmed and presented so that they couldn't learn the pattern. They are so darned smart. We learned early that if we didn't have a random presentation, they learned it and they wouldn't even have to look. They learned the timing, they learned the sequence, and they would just sit there and hit the lever without looking because they knew what was coming. So we had to make it a totally random, mixed performance.

We first got it down to eight that were candidates, then down to four. We had a prime and a backup for the suborbital, and a prime and a backup for the orbital. We worked with those four and honed them for the actual flights as those flights were coming. The selections were made, basically, on the consistency of their performance and their general well-being. All four chimps were in good physical condition, no problems there. But if one had an erratic tendency, we would pick the one that was going to have the most reliable data for us.

The veterinary personnel were strictly at Holloman in the lab under my command. Jim Cook was in charge of the animal colony and the health and well-being of the chimps, and totally in charge of whether the chimp would or would not have a certain procedure on a certain day. When something was needed, he was the final authority. If he told me that so-and-so chimp should not go out for training today, that chimp didn't leave the colony. If he said this chimp needs a procedure done, that procedure was done. He was just absolutely outstanding. He had two vets working for him that were also superb and dedicated to what we were doing, and instrumental in making sure that our entire program went as planned. We were relying 100 percent on the selection of the proper chimps for the proper flight.

Those little chimps are the most affectionate, loveable creatures. They are mischievous, they love to play tricks, and they are just like any adorable child, almost, in that you can become overly attached to them. But as long as you understand what you are doing and what the purpose of the program is, you can manage to divorce yourself from the emotional aspects of it and not only get the job done, but be very gratified when it's all over and you get the big smile and the big hug.

[Where were you during the chimps' Mercury flights?] At Holloman, biting my nails and sweating it out.

During the time Ed Dittmer served at the Aeromedical Field Laboratory at Holloman Air Force Base, he became involved with training the chimpanzees for NASA's Mercury program.

That came under the Space Biology Branch, and I was the noncommissioned officer in charge at that time. In the early days, we got these small chimps from Africa, and we started a training project. At that time, I really didn't know; I figured they were going up, likely using missiles, because they were training them to sit and so forth. I knew there was some ulterior motive. I never questioned anything because a lot of things at that time were classified, and the less you knew of classified, the easier it is to keep it classified.

We'd get these little chimps about a year old from Africa, and we'd start training them by putting them in chairs, just a little metal chair. We'd set them about four or five feet apart so they couldn't reach each other and play. We had a little web jacket of nylon that we put over their chests just to tie them in the chair. Then, we'd give them fruit and stuff. We started out only keeping them in there, maybe five minutes a day, and then we'd increase it the next week. Finally, we could set them there all day and they'd just sit there and play.

[One of the chimpanzees in the training program flew in a Mercury program suborbital flight about two months before the first American astronaut took a suborbital flight. That chimp's name was changed from Chang to Ham just before his flight. Dittmer talked about how that name was chosen.] The **H**olloman **A**ero**M**ed is what they came up with, but also at that time our lab commander's name was Colonel Hamilton Blackshear, and they like to think it was named after him.

There were several people in on the training of Ham, as far as the actual training of the psychomotor and so forth. The Comparative Psychology Branch had a big part in it. The Veterinary Department, of course, were always there and available if they were wanted. It was quite an elaborate deal, because Colonel Stapp had chimpanzees for years prior to this, and he was very, very particular with the colony. Anyone that mistreated an animal or anything else, they were out the door. He didn't put up with any nonsense with his animals, and I think it was through him that all this care developed. The Veterinary Department's primary purpose was

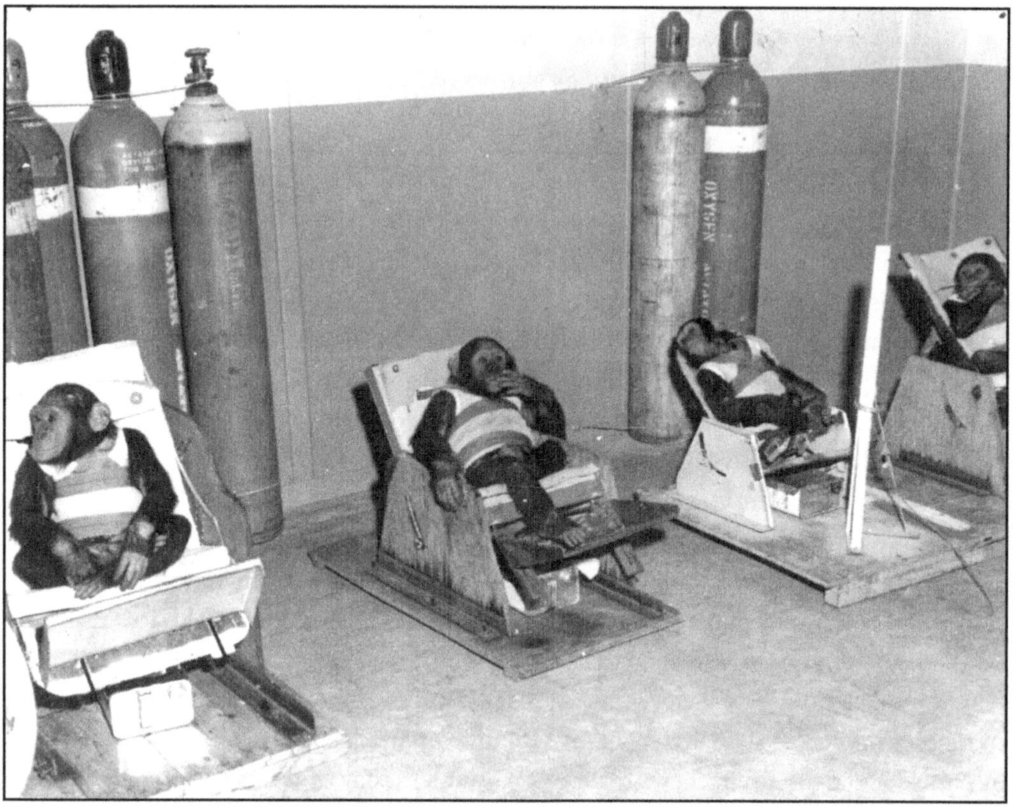

Out of an original group of forty chimps, eighteen continued with training. Ultimately, six finalists were selected for the first suborbital flight of the Mercury program. Courtesy of NMMSH.

to take care of the animals, the health, the housing, and feeding and so forth. Of course, they were concerned about the research aspects, to make sure the animal wasn't being mistreated in any environment. If he was in a deprivation study, so to speak, they wanted to make sure that his physical condition was okay to go through with this.

Before you checked an animal out of the Veterinary Department, they would take him in and give him a complete physical. Then they'd turn him over to you, and occasionally they'd come by and see how they're doing on the program. We had a wonderful vet staff. They were all very professional people, and they didn't want no monkey business as far as the animals were concerned.

In the training aspect, after the chairs, we started out with aluminum couches, with the Mercury couch, and psychomotor [personnel] would take over with the hitting levers and so forth – so many times for a drink of water and so many times for a banana pellet. Then, our lab would do the metabolic studies on them. We just didn't take them down to the Cape and throw them into a missile. We'd go down there early and get them used to the environment of the climate and all this, and work them down there in the couches. We had a little compound there, a little trailer with all our comparative psychology instrumentation and data collection, and we'd work them for weeks on end and collect the data just like we would on a regular flight. Then we'd prepare them. My particular job was the medical aspect of it. I'd put the electrodes on for the EKG in flight and the respiration sensor, hook the plates to the feet for the shock [a mild electric shock let the chimp know it had made an incorrect response], suit them up, and secure them in the capsule, then put the lid on and check them out and accompany them to the missile, put them inside, then go back and watch the data.

I had a very good relationship with Ham. I know he liked me. I'd hold him and he was just like a little kid. He'd put his arm around me and he'd play, you know. He was a well-tempered chimp, a little different from Enos. No one ever held Enos. If you had him, he was on a little strap. Enos was a good chimp. He was smart, but he didn't take to people. A lot of people had the wrong impression of Enos. They said he was a mean chimp and so forth, but he wasn't really mean. He just didn't take to cuddling. That's why in any pictures you ever see of Enos, you don't see anybody holding him.

In fact, we had to develop a system to take blood pressure during flight. We were out at the University of Southern California on their centrifuge, and we were doing a cut-down to put a catheter into the artery for the blood pressure sensor. We were studying under gravity. Take it up to 6 g's or so to see how it would react. Dr. [Jerry] Fineg was doing the cut-down, and I was pacifying Enos with banana pellets. I just happened to be looking over at Fineg and I got my hand down too low, and Enos got my finger between his teeth. Boy, I tell you! He finally let go, but.... Anyway, on that centrifuge, I'd ride the center of the centrifuge and watch the instrumentation, around and around a few times at high speeds. The cord came loose from the capsule and flew around and just missed my noggin. *Phew*! Then we did centrifuge studies with them out at the centrifuge at Wright-Patt, and we did some, I believe, at Brooks [AFB in San Antonio, Texas], too.

I was with Ham all the way up until the actual suborbital flight.

Lieutenant Ward came up to the gantry with me and we put him in the Mercury capsule. I had a twinge of anticipation, because I had been down there when some of their missiles had exploded out over the ocean. You know, they were off track and they just hit the button on them. I didn't want this to happen to him. There was a lot of anticipation, and I would sweat. And happy, jubilant when he came down and we got the word he was okay. When he came back to the compound, boy, he run right to me. He was quite a chimp.

I felt it was a good move to send a chimp up first. With all the work we'd put into it and all the training and so forth, we felt they should give it a shot, which they did do. The orbital flight, I think, did give them quite a bit of data for the manned flight. The suborbital one did too, but the orbital flight gave them quite a bit of data that they prepared before John Glenn took his flight. So, they really did pioneer it and do some good. I think that without it, they might have had some problems. And fortunately, they were all brought back safely. Enos died about six months after the [two-orbit] flight, but it wasn't due to any aspect of the flight. He had developed shigellosis, which is a dysentery which they get from Africa. Once it gets hold on them, it's just impossible to treat, almost. That's what he died of.

Photographer Jack Glasco had many interesting assignments documenting the rocket and missile development work at Holloman and White Sands. Once in a while, he had a chance to take pictures of living, breathing participants such as the chimpanzees that were being trained for the early Mercury spaceflights.

Ham was actually in the Aeromed section, which was in the north area of Holloman. That wasn't my primary job. We had two, three, or four photographers, still and motion picture, that were assigned permanently at Aeromed. They did a majority of the photography there. But occasionally a job came up that either they didn't have the time or the equipment for, or several things were happening that needed two or three photographers at one time. So then the people would pull us out of the Land-Air group at Holloman, and we would come over and help to complete the photographic job, whatever it was. We worked with seven, or eight, or nine of the chimps that we had names for. They recognized us, and we were friendly with most of them. We could handle them, and they would sit on our arm and put their arms around us and hug our necks.

Ham was a friendly and loving animal, because he would go to you. Of course, it helped if you had something for him to nibble on, a banana pellet you could feed him. Two or three of those chimps were very docile animals, and I got so well acquainted with them that when I walked in, they would climb in my arms and put their arms around me. I enjoyed that. Billy Joe liked to be caressed and held and be rubbed on his arms and back. He posed for pictures, like he enjoyed it.

As a physiologist at Holloman's Aeromedical Field Laboratory, Freda Herrera was closely involved with the chimpanzee colony. In this excerpt, she mentioned prominent people who were involved with preparations for the manned Mercury flights. General Bernard Schriever was commander of the US Air Force's Air Research and Development Command in 1959. Dr. Harald von Beckh was a prominent aeromedical researcher, particularly in low-gravity conditions, who worked at Holloman in the 1950s and '60s.

I was around the chimps every day. The majority of the chimpanzees at that time were young. Being in the administration, we had to go back daily into the research areas to answer questions from headquarters or to prepare reports and data, so we were constantly working with the project officers.

The main credit has to go to veterinarians. We had the cream of the crop of young men that were just graduating from veterinarian school. They came out to a whole new species that they probably hadn't even mentioned in most of their education in veterinary medicine, and they had to learn what to do with the animal, [how to tell] who was a healthy animal, what diseases affect them, how to feed them, how to care for them. Without them the program wouldn't have gone. It really was one of the main contributions to the program. We had some psychologists that came out and said, "I don't want to do any work with a primate."

They felt it demeaning, and they were fighting the program. Whereas, the vets came out, and they were young and wonderful fellows, but they got into it. There has been many a biological research project [elsewhere] that the whole thing has fallen through because the animals had this or that strain of a bacteria, and they had never been tested. They weren't healthy animals, some of them in the radiation. So how did you know what the radiation did to them if they weren't even healthy specimens? We had chimpanzees that came in with tuberculosis.

Sergeant Dittmer and all the other airmen that trained the chimpanzees deserve the most credit. Yes, the psychologists set up the program, but it was the airmen that went day after day after day, and said, "Come on, Ham, learn how to push this button, red means this, and that." This went on for months. You know how hard it is to train a dog, or even toilet train a child. You have to just stay with it and stay with it. There must have been at least sixty or more dedicated airmen. I don't even know the total that were there throughout the years that I was there. They constantly, constantly cared for those animals.

Probably the most important element of the program was General Schriever. A lot of people may not have realized how much he knew about what we were doing, why we were doing it, how we were doing, what were the holdups, and so forth. He came every chance he had, to make a tour of the lab. When they would call and say, "General Schriever has five minutes on Wednesday between so and so," my job was to go around and find out which division wanted to give a short briefing to him, and where. Then one of the things was constantly walk the routes so we knew how much time they had in each place. I would say that in the seven years or so that I was at Holloman, he must have come twenty-five to thirty-five times. For a general officer, that is a lot of times. He had the missiles on the West Coast, so anytime that he was going through, he would stop and he would talk to us. He knew what we were doing, and a lot of people since then have sort of degraded the work with the chimpanzees, but that man never will. He knows how vital it was, because we proved man could live in space. We didn't know that until Ham went up.

We had a standard signal that if any of the chimps got out, we just went on the PA [public address system], because frequently we had visitors throughout the lab. We went on the PA and said that freshman class was over. That meant grab an airplane strap, run out, and catch the first chimpanzee you can, and tie him to the fence or an automobile or something. We had a few of these occasions. The most dramatic, though, was the time when Samson got out. Samson was this huge, magnificent specimen. He was probably the best, the biggest, the oldest, because usually we didn't have adults. His thumb was as big around as my wrist. When we got him, he put his hand right through the concrete block wall. When he got out, they didn't go and get an airplane strap to tie him to the next pole. They let him walk across the desert. They got the Jeep, and they got the dart for tranquilizers, and after a while they gave him a shot, and he just slowed down. They couldn't put him out completely because nobody could pick him up. They had to get him on the Jeep, and take him back. He was a beautiful animal.

We would use the big, wide airplane straps that they tie down cargo in airplanes. We just put that around the wrist of an animal. Now, it was Fred that I saw which dramatically displays the strength of the chimpanzee. He had four straps, and two around his middle, and six airmen, they're trying to hold Fred. He's jumping up in the air five and ten feet, and they're holding on to him. I mean this is one chimp, a little one. He wasn't quite as big as Enos. Enos, they said on occasions, was even stronger than Fred.

Ham was beautiful. A couple of years out of my seven years there at Holloman, we did have another WAF officer, that's what they called us at the time. We were Women in the Air Force. Her name was Jane McElroy. In the front office, we all wore white lab coats because we were in and out of the working area. Somebody called and said, "What's the value of the chimpanzee?" Jane needed to get some information, so she went back there. As she was walking along and heading toward the project officer in this one room, she felt this tap on her derriere. She turned around, and there was a beet-red airman ready to fall through the floor, and he had the strap on Ham, who was giggling. Ham had come along and patted her. She told the airman to relax and laughed, and went over and shook hands with Ham. It's a little startling. He was a love. He never had a mean streak in him.

It may have been a Senator or something who asked, "How much is a trained chimpanzee worth?" We said, "$25,000." They said, "Okay." Whoever it was who had asked the question came back then and said, "Well, justify that $25,000 figure." So we sat down, and we said, to buy them, to train them, to feed them, to house them, the veterinary, the medical. And we came up with the figure of $25,000, which was really a very moderate estimate. We could have given a much higher estimate than that.

The chimps didn't get credit for being smart. Some of the chimpanzees would recognize the particular way I would walk. Every tour we had, I had to time it, and as I would walk around they would think, "Ah, the next couple of days we're going to have a tour," so they developed a very nasty habit. I swear, a half hour to an hour before the tour, here would be the chimp just sitting in his corner. Here would be the guest and everything coming by, and just as they got in front of that particular animal's cage he'd take one leap to the front, and spew water all over. They could hold a full mouth of water for an hour. They were pretty smart, because they didn't spit me or any of the other airmen that happened to be around. They spit the most brass, the most gold on the hat, the most decorations, always the highest ranking person on the tour.

The one time Dr. von Beckh got sprayed by the chimpanzee, he was in the washroom washing diluted Clorox all over his hands, his face, his back. I think that one had planned on doing von Beckh for at least three or four years, and he got him. See, as the rest of the tour would go by, and look at the chimps, von Beckh would always step back. He knew they loved to spray. Oh, they would sit there, and laugh and laugh and laugh. You know, just like any kid: "I got you, I got you!"

I knew all of the chimps. One of things that I did in the Plans and Programming Office was I had to have a complete layout chart showing this chimp goes to the psychologist this month, this chimp goes to the vet this week, this chimp goes to so and so. So I knew every one of them by name, and where they were being used, and why. Probably the time I screamed the most, because I like to scream, was when we lost Enos. He died of a contagious disease.

Besides the fact that man had to live in space and perform in space, there was one series of experiments that they used with some of the older chimpanzees in which they did do a decompression [reduction in air pressure]. They wanted to know if the physiological body could withstand decompression. So the chimpanzee's capsule was decompressed. He threw up, passed out. They recompressed it. A few minutes later he woke up, cleaned up all the mess that he had thrown up, and went right on. So we did prove that a very minor decompression was still possible.

As I said, General Schriever made these wonderful tours through. This one time when he came, I was at the tail of the party poking the commander, saying, "It's the next exhibit; keep moving," so we'd stay within the time limit. This one time, we went into this big room with locks, and there's eight or ten airmen all standing around perfectly at attention There's only supposed to be two or three in each one of the different project offices. We had the presentation, went on to the next one. As we started to the next program office, I noticed airmen running all around the hall. The whole tour was very fine, and General Schriever went on his way. So I went back to the lab, and I said, "What was the matter? Why were there ten airman in that one project office?" They said one of the monkeys had gotten loose just before the presentation. A monkey, not a chimp. Monkeys were fierce. So they gathered all the airmen, and they said, "If the monkey comes down from the ceiling, jump on the general and protect him." Well, fortunately we didn't have to jump on the general and protect him. This building wasn't built for a research center, and it had these very high ceilings. It took about two or three hours later that day before they finally got this monkey out of the ceiling in the room.

Bill Holsey was born in Fort Cobb, Oklahoma, in December 1927. His parents were sharecroppers who lived on a succession of farms, raising cotton and corn. Holsey worked as a civilian machinist at Holloman beginning in 1956.

In October '57, the Russians put up a Sputnik. It wasn't probably a week after that until a staff sergeant and a captain came in on a shot. They were from Wright-Patt, and they wanted to know if we could build a box for them to train chimps. It has got to be strong. No engineers were involved. They told us about the size they wanted. They wanted a box with one light and a lever on it. So we built it. It just kind of started from there and lasted nine years, building things to train chimps.

Ham was there. We had another chimp there that was fully as smart as Ham, I thought. He was probably a little younger. I can't remember his name, but he was a pet too. I said, "Why don't you train him?" He had a heart murmur, and they wouldn't send him up. But we built lots of different things. We built two cubicles, overall they were about 6 feet high, and they had Plexiglas in two sides. And then we built one just opposite the Plexiglas. So you could stand on this and see through. They put a chimp in each one, and we had nine buttons just like tic-tac-toe. They could play each other. Whichever one won got the banana pellet. Well, one chimp was the best. I think Enos was one of them. He would win every time, and he would look at the other one and just clap his hands. The other one would just sit over there. They knew what was going on.

I was sent up to Pennsylvania to a Naval air station where the big centrifuge is up there. Astronauts did some training on that. They sent me up there to build something for a chimp to ride on that centrifuge. So I took some measurements and whatnot, and I built a deal for that. I also built one for a centrifuge at White Sands. They put him on that and pulled him up to 10 g's, I believe. You know, even as strong as a chimp is, if their arm gets away from them at 10 g's they can't get it back. It was a fairly small centrifuge, maybe 15 feet diameter. We made a seat for him to sit on there, and strapped his arms. They had a video camera going in his face, and he worked the lever, going 10 g's. You know, a lot of things we were doing were just to find out what would happen. I also ran some rapid decompression tests in case the spacesuit busted or something. I think we made about three different setups for this to go in a chamber over at Guidance. He decompressed them rapidly.

There were some Ph.D.s who came down from Wright-Patterson that worked with the Holloman staff on what kind of test they would decide. They would decide on a test, and then they would come tell us what to build. Whatever they decided they wanted, we tried to make. Like the mechanism on the lever, they had to be spring loaded to come back. We ordered a bunch of springs, and we designed brackets to hold them. All the time, they were saying these have to be strong. So we made them pretty strong. I think they did break a few of them, even at that. It was all real interesting. When we would start to design something for them, first we would go out and take an inventory of what material we had. This was working without going through an engineering department. They were in a hurry. They were pressed. The pressure was on to get a chimp up.

I used to go home at night and lay in bed and figure how I was going to do something the next day. Sometimes it would work and sometimes it wouldn't. I was really in to it. I really enjoyed it. The shop boss there didn't know what we were doing, because the guys would come over and tell us what they wanted. There was nothing on paper. I mean, we didn't have a drawing to say here is what we are going to do.

As personal physician to the Mercury Seven astronauts, Dr. William Douglas accepted the value of sending a human surrogate on a suborbital flight to ensure that it would be safe for a man.

[What did the doctors learn from flying Ham?] They were important flights, and we proved what we already knew, I think. That was that exposure of the human being or the human surrogate (namely, the chimpanzees) proved that man could survive the space environment with impunity and that he could function there and could do all the things that would be required of him. So, we just simply proved what we already knew. I wouldn't, for a minute, have suggested then, or even now, that we should have eliminated those flights. They were essential. It was the smart and intelligent thing to do, to prove what we already knew intuitively: that all would be well.

Cynthia Sommers-Guzevich was not only an intelligent woman who achieved leadership positions at White Sands Proving Ground and Holloman Air Force Base, but she was highly skilled in several hobbies, including painting, making dolls, and training and judging show dogs. Because of her experience with dogs, she and her husband were offered a chance to lead a unique program that used dogs to find missiles and payloads after test flights. After the missile dogs program at White Sands Missile Range ended in 1965, she looked for a different type of work. She took a job in a drafting department at Holloman, where she had a chance to interact with a different animal species.

The problem that had been on White Sands for a number of years was that the payload of a rocket was lost due to the fact that a small rocket would come straight down and bury itself in the ground, or the payload was ejected and was lost out in the vastness of the White Sands Proving Ground. The military couldn't find it because a rocket that's in a hole is no different than a rabbit hole, and the objects of the different projects were many times buried in the sand. So a man going out there, or maybe hundreds of people going out there, looking for that object could not find it.

One of the engineers heard about a man in Hyde, Maryland, by the name of Wilson Davis, that was very involved with dogs doing various aspects of detection. He contacted the man, and he was brought out and gave a demonstration of the capability of dogs with good noses. Dogs that were not tracking dogs, but air sniffers. What good is a tracking dog like a bloodhound? A missile doesn't leave a track [on the ground]. He came to White Sands and gave a demonstration and sold them on the capability and the possibility of using the dogs in that type of work.

He was contracted to train twelve dogs for this. Two of the dogs, Dingo and Count, the original missile dogs, were shipped to White Sands and turned over to the military. Because the military was in charge of that and the trainers had no idea of what the whole situation was about, the dogs failed miserably. They ran out into the desert and chased rabbits and never recovered anything. The problem was not the dogs, but the lack of training by the handlers.

At that time, my husband and I were very active in kennels and breeding dogs and showing dogs, and Wilson Davis contacted us. How he found out about us, I don't know. I guess it was because I worked at White Sands at the time. They called me in at headquarters and asked me if I would be interested in this. It sounded very, very exciting and so I said yes. I stopped my job and they shipped both my husband and myself to Maryland. We stayed there for two weeks, training with the dogs. We were trained with the whole program and what would be the thinking behind it.

What we had to do was bring dogs that were absolutely superior in their hunting ability to White Sands and retrain the dogs from a live object, rabbits for instance, to a dead object, a missile part. This was very difficult at first because you have to realize that the desert is filled with livestock – pheasants, lots of quail, and lots of rabbits. This was fascinating to the dogs, but we had to retrain them on this one deal, which was a scent. We couldn't even train them to an object since the scent was applied to various objects, many of them explosives and dangerous to the dogs. The scent is called squalene and it is made from the liver of the basking shark. Engineering could design a measuring instrument into a rocket or a missile, sometimes the size of a cigarette pack, put the scent on it, and shoot it up into the air. It would come down, and we would look for it. We did lots of research on this scent itself, the amount of scent we could use, the time span involved, how long the scent would last. It was very interesting that there were objects that were shot up into the air a year after the scent was put on them, and the dogs would still find it.

Sometimes an object would bury itself into the ground, and the dogs were trained so that they could go out and look for an object that was buried. Another day, an object would be shot up into the air that retained an explosive portion of it, and they were told to only point to it. The next day, it might be out in terrain that was impossible for humanity to get into, and they could retrieve that object. So, the aspect of training was varied and great.

Of the twelve dogs, some of them were virtually untrainable in the terrain because they could not put up with the heat and the sand burrs and the desert. It was strictly a matter of terrain. One of them, the biggest Weimaraner I had ever seen, a big horsey thing, we think he was half blind because he would go out and he would see an unusually formed bush and would stalk it. You would have to remind him that he was not hunting bushes, but was after a scent. We had reasonable success with eight of the twelve dogs. One of them was a fighter; he was a Shorthair and if he would see a Weimaraner, he wanted to kill him rather than go after scent. So we had to give him up right away. But eight of the twelve dogs were successful. Not as successful as Dingo and Count. They were successful 100 percent. You would put them on a scent, especially Dingo. He was the ugliest Weimaraner I have ever seen. He had everything wrong in the breed, but boy, did he have a nose and a brain! He could figure out things. In fact, he was my policeman. If one of the dogs would make a mistake, I would send Dingo out and tell him to correct it, and he would shake him up.

The program began in 1961, and it lasted until 1968. The dogs were purebred dogs, Weimaraners and German Shorthair Pointers. The government purchased those dogs, and they were registered in the government's name. But the dogs lived with us, and we had the personal rapport with the dogs. They lived in our kennel and our two stars, Dingo and Count, actually lived with us in the house because if all other dogs failed, you could always count on Dingo and Count. They were really the stars of the program.

One time, we went up to Gallup [in western New Mexico] in the deep

Cynthia Sommers-Guzevich with Count (left) and Dingo (right). Courtesy of Cynthia Sommers-Guzevich.

winter because a missile part was lost that was of dire importance. We hunted and hunted in deep snow to where the dogs left bloody tracks because the icy snow cut their pads. They kept on hunting and hunting, but they never found it. Then in the spring, a farmer found that part way off from where they had sent us. Sometimes the measurements just weren't what they were cracked up to be.

The dedication of the dogs to hunting was something that really gained our respect. We hunted at part of the [White Sands] National Monument in 120 degrees Fahrenheit temperatures. One lieutenant collapsed and we didn't know how we could get him back in, and the dogs kept on hunting. So, because of lack of understanding, or lack of feeling for the animals, the dogs had to hunt by order. You have to realize they were closer to the ground where it is hot, much hotter than what you as a human experiences. We had to devise something, so I came up with a little jacket with pockets on the side. We would carry ice with us and stick ice cubes in the pockets. They had little hoods so that their brain would be on ice, so to speak. But we could not give them water. The minute you give them even one drink of water, they would be finished for the day. They just wouldn't hunt any more. We may let them lick an ice cube and send them out again.

We would hunt them an hour on and an hour off. Many times, we were out there for twenty-four hours because we would be looking for one object, and then we were called in to hunt for something else. Not only did our way of thinking on the hunt have to change, but the dogs' way of thinking. On occasions like Armed Forces Day, we had to give demonstrations on TV. We would have to have generals coming in from Washington, and we had to give them demonstrations. I remember one time, it was very comical, a general said, "I am a golfer. I am going to hit out a hundred scented golf balls into the desert and see how many they can find. But you must blindfold the dogs so they can't see." And I said, "No. I just put them in the car and you can beat out your golf balls." He beat out all those golf balls. They were freshly scented, and when I released the dogs, that was like *candy*! They tried to pick up two and three at the same time! After they found about half of them, he was convinced that they could do the job.

What people could not understand was that a dog would not go to the object to be found in a straight line. The scent travels with the air, and many times the dog would seem to go out of their way. We did one project for the Atomic Energy Commission in Tonopah, Nevada, and the dog went way, way out of his way and one of the engineers said, "Well, this dog is stupid. This is not where it is at all." I said, "Give him a chance. He is just circling with the way the scent travels." And he did find the object.

One of the most intricate things we found was a measuring device that was supposed to be used in the Moon program. A jet was supposed to drop it at the small missile range and when the jet landed on Holloman, the device wasn't where it was expected to be. The engineering department decided that it was somewhere between the small missile range and Holloman. Somewhere out there in the desert in a 40-mile stretch, that object was lost. It was a very important object, so they said no matter how long it would take or at what cost, they needed that object. We went out there and in less than an hour, we had it. It must have sent out a wonderful scent because when I sent out the Weimaraner, he immediately took off like a shot. I had to climb on a vehicle to finally spot him. I thought he was out of control, which amazed me. He was just standing out in the desert. Finally I went out there to him and scolded him, and he put a foot in the sand and scratched the sand, and there was the object. How do you apologize to a dog? They were all amazed.

Occasionally, some of the rockets would go off range or would go into an area where there was no way we could get into with a Jeep or any other of the vehicles. We had to teach the dogs how to ride a helicopter. When Count, the German Shorthair, would see a helicopter, he was always a bit on the frightened side. He was so bold, but helicopters were not his cup of tea. But Dingo became a real fan of helicopters. You have to realize that those helicopters were the bubbles with two people in them and no doors. I was strapped in and I had to hang on to Dingo. He would sit there between my legs and look down and just loved it! They would drop us off in areas where there was nobody there, and it was a wonderful experience because it was in areas where we found old Indian camping sites over on the other side of the mountain towards Engle [west of White Sands Missle Range], or at the Jornada Range. The animals were really exposed to many aspects that normal dogs never see.

There was also some experimental work done for White Sands up in Albuquerque, and this was explosive work out of a cave. It was done in the winter time. I just let the dogs out to exercise, and one of my dogs immediately went under a big pine tree. I said, "He acts like he is on scent." The project engineer said, "Impossible. The object impacted on the other side of the mountain. Tomorrow we hunt on the other side of the mountain." The next day, we couldn't drive there and so we had to park at the same place that we parked the day before. Immediately the dog went to that same bush again, and I said, "Look, my dog is on scent." [The project engineer] said, no, it was impossible. We went over to the other side of the mountain and the dogs never reacted to scent. We hunted and hunted all day. When we went back to the car that night, there he was again [at the bush]. I said, "Is it an object that the dog could pick up?" He said, "Absolutely." So, I gave my dog the signal that if he is on scent, whatever he found he could pick up. He had it within minutes and he brought it back. It was under that tree because the whole shot was a failure.

Many times, we were sent in one area and the dog wanted to go in another area. We, as human beings, don't realize the capability of their noses. The project engineers would say, no, it can't be there, it can't be there. But it was there because maybe it was a misfire or the impact was predicted in the wrong area.

White Sands, at the time we had the program, was very, very proud of the dogs and really pushed their capabilities. So every time you turned around, we were contacted. We were published in *Time* magazine and *Life* magazine and *New Mexico* magazine and the Sunday supplement of the *New York Times*. The dogs were greatly publicized and rightfully so, because people had no idea what was being done with them. They were just dogs, but in finding some of the instrumentations that were decisive in the failure or the success of a missile shoot, they added to the advancement of the space program.

I remember one time, we were waiting to go on a shoot and a group of the astronauts came in. Everybody was there with a new dollar bill to get their signature on it. They asked me, "Don't you have a new dollar

Dingo and Count were trained to point to objects that were dangerous or fragile, or to retrieve objects when it was safe to do so. Courtesy of Cynthia Sommers-Guzevich.

bill?" I said, "No, maybe you would like to have an autograph of a paw print of a dog." So, the dogs got to meet the astronauts. The astronauts thought the dogs were wonderful, and they said if we ever land on the land [instead of the ocean], I hope you send the dogs out to find us if we get lost.

They had a new colonel coming in to White Sands in charge of recovery. We and our helpers were the only civilians in an all-military department. He did not want any non-military, and the man extremely disliked dogs. This was not due to a failure of the dogs, but to the nonrecognition of the man in charge of the capability of the dogs. He said he could send out hundreds of men for each dog, and this is why the program collapsed.

When the dogs were no longer involved, they said the recovery rate went down from 96 percent to 23 percent. That should have told them something. Years later, we got contacted to reissue the program again, but there was just too much involved. The dogs were eventually sent back to the training agency and I understand that all but the two dogs, Dingo and Count, were sold. Dingo and Count remained at the training agency. Dingo was used as a demonstration dog by the training agency and Count was sent to Vietnam. They used the scent in areas where they thought the Viet Cong would come through, and he could point out what was the Viet Cong and what wasn't. I have no idea what happened to them, and I really don't want to know. We were very specially attached to the dogs, and we had to make an emotional judgment on what to do with them. The government was willing to give us those dogs, after all the fortune they had invested in them. But the dogs, in our estimate, were so valuable in their capabilities that it would have been a tremendous waste. We had to think of the use of their capability and so we, under tears, sent them back to Baltimore.

It was a very interesting program because we were exposed to the various points on Holloman, the work that was done on the [high speed test] track and the work that was done with the chimps, designing cages for the chimps and working on additions to the track, which was all a very unusual and very challenging job to be exposed to.

The problem with the chimps was that they were really smarter than the engineering department. The problems they had with the chimps was that they could open any gate, and they could figure it out just faster than the engineers could design it. I remember being out in the chimp colony many times and seeing [people] chasing them because one had just gotten loose.

I also had a friend that worked with the chimps and made the little space suits for them. She said that once a chimp had a space suit on and was taken for a ride, he knew what it was all about. So when they developed a new space suit, she had to make them some very colorful little jackets. They would be fascinated with the colors and the new design and, unbeknownst to them, they were being measured for the space suit. It is amazing when you work with animals, like we did with the dogs and then later on when I saw the chimps.

The physician of the chimps took me out there one time when he had the babies. The minute they saw him, they stuck out their lower lip and pointed to the lip. I asked what are they doing? He said they wanted their vitamins. They loved those vitamins. He said, "I'm the vitamin man." It was fun.

To me, working at White Sands and at Holloman was a very exciting time. It was a time of development. I remember when they were doing research on the lunar excursion module. It was a time of working with people that saw into the future. It was working with people who were thinking not of today, but of tomorrow. And it was just a greatly exciting time because it was a time of research and development. You never knew from one day to the next, and from one month to the next, what project you would get assigned to or what development was coming next. You were really on the ground floor of a very wonderful development of the human race.

Chapter 8
Pilotless Rockets

Missiles and rockets are close cousins. Missiles are powered by rocket engines, as are rockets. The difference is in their purpose. Missiles are used as weapons, while rockets are used for peaceful missions such as research and spaceflight.

Although this book is about *space* pioneers, it is appropriate to include a chapter about missile programs. Technological advances made with missiles would be applicable to rockets, and the experiences of people working on missile design, testing, and tracking are similar to those of people developing rockets. Furthermore, some of missiles were used for experiments related to manned spaceflight. For instance, the Aerobee missile was developed for high-altitude atmospheric and cosmic radiation experiments, and it sometimes carried mice or small monkeys to test the effects of cosmic radiation on living creatures (see Chapter 7).

Like their counterparts who worked on rockets, people working on missiles were dedicated, hardworking, creative, and adventurous. Willis Brown, an employee of Army contractor Douglas Aircraft, was assigned to missile testing at White Sands Proving Ground in 1952. He worked on several missiles, including the surface-to-air Nike Ajax and Nike Hercules. Brown captured the spirit of missile pioneers when he said, "Once, we fired twelve Ajaxes and three Nike-Hercs in one day, just to see what we could do."

Floyd Fannon was assigned to run a logistics supply center at Alamogordo Army Air Field about six months before it was renamed Holloman Air Force Base in January 1948. He worked for North American Aviation, a civilian company, on the NATIV (North American Test Instrument Vehicle) project.

The blockhouse for the NATIV had just been completed when I first got here in August of '47. The tower was lying out there in all its individual pieces. One of the things that I had to do was inventory it, see if I had everything. Then they decided to make it twenty feet longer or something. We had enough steel to build two of them. That didn't seem to be any big problem, except the tower bent, and they couldn't get the missile in it. So we had to figure out some way to line up the tower so something could go down the guide rails. Some guy went out one day when we were having trouble. We couldn't get the jig, or mock-up, to work. We figured if the jig wouldn't go up and down those rails that we had to guide it with, well, the missile wouldn't either. The guy went out there and stayed all day. He come back in and he said, "Go get me a whole bunch of 1-inch turnbuckles." Whoever saw a 1-inch [diameter] turnbuckle like that?

So I looked around, I don't know from turnbuckles anyway. I went over to my old buddy that runs Boeing. He said, "Oh yeah, I've got a whole bunch of them over here." When Boeing had first got out there, they kind of scrounged the area pretty good, and they picked up good stuff that they knew somebody would need someday. Anyway, we got these turnbuckles. They put in about eight of them, and cranked them around, and got the tower straight again.

A crew prepares a NATIV missile beneath its elaborate launch tower. Courtesy of NMMSH.

Robert Montgomery was an electrical engineer working at the North American Aero-Physics Laboratory at Inglewood, California. In 1952, North American sent him to Holloman Air Force Base to work on the NATIV missile, a modified and smaller version of the V-2. The Askania theodolites he mentioned in his oral history interview were a brand of optical instruments used to track missile and rocket flights. A theodolite consists of a small, mounted telescope that can be rotated in horizontal and vertical directions to measure angles that described the flight path. The Askanias were actually cinetheodolites, which filmed the flight as well as recording the trajectory data.

They had set up Askania theodolites out there during the NATIV program. They built a little outhouse-type thing to house the instrument when it wasn't in use. There was a little railroad track built from that out

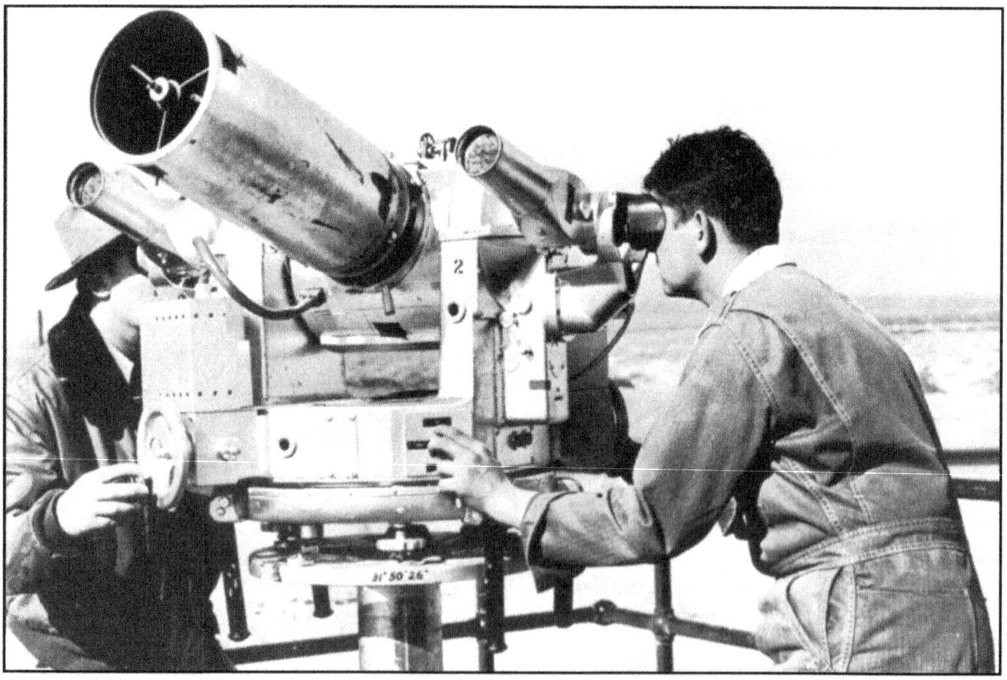

One of the Askania operators tracked the vertical movement of a missile through one telescope while the other tracked its horizontal movement through another telescope. Working together, they were able to follow the flight path and film the missile's behavior. Courtesy of US Army.

to the pedestal where the instruments set up. The four Askania operators stood out in the sun for hours and hours while they were waiting for missiles to be fired, and delays in firing. By the time they are ready to track the missile firing, the Askanias had gotten so hot from the sun you couldn't touch it. So they submitted a requisition to the home office for beach umbrellas to shade the Askanias, but an explanation wasn't provided with the requisition. You can imagine, when the California office got a requisition for fifteen beach umbrellas for people out on the desert, what they thought!

Frank Conway was born in Worcester, Massachusetts, Robert Goddard's original home. Conway finished vocational high school about eight years after Goddard had moved to New Mexico. They never met, but Conway eventually worked with a variety of rockets at White Sands Proving Ground (WSPG), about 170 miles southwest of Roswell, where Goddard conducted his liquid fuel rocket tests between 1930 and 1942. The Viking missile he mentioned was a high-altitude atmospheric test vehicle. Several series of Aerobee missiles were also used to carry scientific payloads. Hermes missiles were American modifications of Germany's V-2 missiles. The Loki missile was designed as an anti-aircraft device, but it became popular for launching atmospheric testing devices. Honest John was the United States' first nuclear-capable surface-to-surface missile.

My first job was a V-2 double shot on the little static test stand. We had two hulls set up, less the guidance and warhead, and they had them manifolded together. They made a 120-second run on the V-2 motor, which was absolutely perfect. It didn't hurt that motor at all. It was ready to be fired again, which was enlightening. I was the only American up in this little static test stand. All the rest were German scientists. I was just kind of an NCO janitor, you might say. I had nothing to do with the test except I was physically there.

I went through the Viking program. In the time I was here, they kept switching me all the time, so I would get a little experience on each one of these things. I was on the Aerobee, Aerobee High, Loki, and Hermes A-1 [projects]. I think Hermes A-1 was what we called the V-2 that had fins on the front so you could take it up and fly it horizontally. They called it Hermes A-1 as a kind of a cover-up for what the missile really was. It was a security measure.

Then Honest John was getting into the game. WSPG had a cold box in a wooden building, a test chamber for temperature, and they put an Honest John in there and dropped it down to 40 below zero. They took it real quick across and launched it. It left a whole bunch of unburned propellant. Solid propellant was lying on the ground all over the place. So I picked up a bunch of it, took it back out to C Station. I cut it up into pieces about like a business card with a hacksaw, took a 35-millimeter film can and made a flying saucer out of it. I put eight bolts to hold it in place with big washers so it wouldn't spread out, then punched holes in the bottom and on the side, and tipped them a little so they would be a venturi that was tipped. That would make it rotate. So we had a countdown, twenty, nineteen, eighteen, so on. We had it sitting on some concrete block with some of the original propellant underneath. We lit that, and it went through the holes and ignited the internal ones. Well, that thing took off, just beautiful! Spinning as it went up and everything. Back in those days, the radars were solid metal, like reflectors. Didn't have the mesh type stuff we have today. And that thing went up and whammed a big dent right in the side of one of the radars. We had some tall explaining to do on how that radar got messed up. But that was the last missile, the flying saucer that we launched at C Station, the only one. But it worked.

About late '53, we were working on the prototype for the mass production model of the Nike Ajax [the US Army's first operational surface-to-air missile]. We had some very successful shoots. One of the problems they had was, they changed contractors in the guidance packages. Gyros and gimbals broke off due to the thrust of the booster, so you didn't have anything spinning in there anymore. Well, this thing was trying to correct itself. So what did it do? It went around in a circle, and it landed about 150 yards from where it was launched. And it had a warhead in it. We were scared to death of that thing. We thought it was going to land right in our area again. They got that corrected. That was a pretty interesting part of history.

Wayne Priest was born in Louisiana and reared on a farm in Mississippi. He joined the Navy amphibian service during World War II when he was seventeen. He participated in the Normandy invasion and brought General Patton's army from North Africa to Italy. Then in the South Pacific, he participated in the invasions of Saipan, Iwo Jima, and Okinawa. He was heading to Japan when the war ended. After four and a half years, he left the Navy and returned to civilian life. He soon found himself involved with the space industry.

I never graduated with a college degree. I got with a young company, Stratoflex, and through experience and hard knocks, I made my way in the industry. We were dealing with high temperature and flexible plumbing. Started off with rubber with wire braid reinforcing, aluminum tubing, stainless steel tubing, and graduated to Teflon. Teflon probably was the most aerospace-type product, because we worked with it, perfecting braiding over it. It fit right in to the aerospace industry, particularly aircraft. I was in engineering and sales. That led me into the marketplace, and into aerospace development, design, testing of aircraft.

My first office start was in Birmingham, Alabama. This was in 1950. Redstone Arsenal was one of the first places I went to. I was called there by the man that had purchased some of our product. So I drove up. I was almost twenty-six years old, a pretty young fellow. He invited me for a meeting because he had some engineers who had questions about Stratoflex products.

After much fingerprinting, picture taking, X-raying of my briefcase and my sample cases, I was escorted into a long hall, and there was a long conference room. Dr. Wernher von Braun was sitting at the end of the table. I had read the history of World War II, and I knew that he had been in charge of the space program in Germany. The guy that was with me was with a group of engineers, both German and American. They began questioning me about my product. I was eager to know where they were going to use it and how they were going to use it, to make sure that it was used properly. Flex hose with a rubber tube has limits on it. I wanted them to know the limits. We were just getting in, in '51, with the development of Teflon products in America. Many, many questions. Here I am, sitting with my briefcase taking my notes principally of the questions they were asking me. I was trying to answer their questions. Dr. von Braun was talking, himself. He asked direct questions. And a lot of his engineers were questioning me about everything, the properties of the rubber, the wire, the fittings, every detail about the product that you can imagine.

I have long since recognized that was my first entree that something really big was going to happen, with no knowledge of what it was going to be. My experience up to that point with the company, which would have been about two and a half years, had been Delta Airlines, Cessna Aircraft, Beech Aircraft, American Airlines. Here was my first quiz about something much, much bigger than that. I had no idea of what the future was going to hold, but I sat there and answered all the questions.

In fact, the guy with me said later he was surprised that I didn't get nervous. He began to question me. "Did you know where those guys came from?" I said, "No, I really don't, except I know they were from Germany, a lot of the doctors." He says, "That was the German missile development program." I began realize, and I was intrigued by it. About four or five years later, I was up in the Detroit area, and I got called to Chrysler, who was manufacturing the Redstone missile. I began to piece together that what we had done at Redstone was now in missile form, and the missile was now being manufactured. At that time, I was living in Chicago. They would call me, and I would go over about the alignment of flex hoses, the curvature, how it should be coupled, and how the fittings should work. Many, many questions. I spent many days at the Chrysler Redstone facility.

I began to piece together that, boy, this is the real thing! That thing is going to blast off someday! Then when Alan Shepard took his first trip in '61, it all began to dovetail. I spent a few years in the Chicago area, in marketing and sales and working with customers, in aviation, principally. Then the company sent me to the West Coast and made me in charge of sales for seventeen states. I was right in the middle of Douglas Aircraft, North American Aviation, Northrop, and General Dynamics down in San Diego, which evolved into astronautics division.

I was learning new things about our own products and how they could be applied to the aerospace industry, more so than just Boeing up in Seattle. Boeing was really expanding and pushing the envelope as far as components for an aircraft. But when you talk about a component for an aircraft, it's a different thing than talking about a component for a missile.

Yet we were having to use the same technology that we had used in aircraft development. So we began to get much more difficult questions about higher temperatures and higher pressures. In an airplane, even today, the highest pressure is 3,000 psi [pounds per square inch]. But in the missile program, we were being asked if that same product would handle 5,000 or 5,500 psi. And the highest temperature that was needed in an aircraft, at that time, was approximately 250 to 350 degrees. Teflon will adequately handle 500 degree temperature.

We could see that we had to do a lot more work in developing plumbing for Convair Astronautics. That was one of the first places on the West Coast. They asked us to help work out some fuel problems with hose. A portion of the line would be flex hose, and a portion would be hard tubing. We had to join it together with the least weight possible. There were so many unknowns. No one knows the unknowns. Even their own engineers couldn't answer our questions about what the product would have to do. First, they weren't going to tell us. Second, some of it had to go through test cycles, a lot of testing in our own lab and in their test facilities.

We got involved in the Atlas program there at General Dynamics Astronautics. Then, there was Rocketdyne, which was part of North American Aviation. They were having a lot of line failures and running their blast tests from the desert out there, just an ordinary B-nut, a nut that you connect a fluid-carrying line. They came to our company, as well as perhaps to others, and asked us to work out a nut that would be self-locking. I am talking about stainless steel, which is difficult to machine and to work with. Our company, along with SPS [Technologies] up in Pennsylvania, we worked out, secured, and completed the test firing with just an ordinary B-nut that you have on a flex hose connected on a rocket engine. When the harmonics and the vibration of the blast would start, you could watch that B-nut and it would start moving, loosening. Then all of a sudden, it would just come loose. We had fires and everything else. So that's why it was necessary to tighten and to perfect the technology on just common B-nuts. We had taken these things for granted in airplanes. As it turned out, we got it perfected, and that nut became standard on hard lines and also on flex hoses.

I was getting to know the Air Force people in research programs. I had access to these places, being secret clearance. But with the Titan [two-stage liquid fuel intercontinental ballistic missle] program, I suppose we were more deeply involved, because AMF (American Machine and Foundry) had the silo construction responsibility and also the door that would have to open to let the missile come out. We were pushing technology hard there, because we were talking about great big Teflon hose, three-quarter-inch or one-inch diameter with 5,500 psi operating [pressure]. Those doors would open quietly but rapidly, at high pressure [above the vertically stored missle]. Then at the proper time, it was fired. We had to do a lot of research and development. Over about an eight or nine month period, we were constantly testing in our own laboratory in Fort Worth.

I have got to tell you about one instance, on the Titan program in Denver. One day, we had one of our technicians from Fort Worth out. We were sitting down in the silo on a little shelf, because about halfway down the missile, there was a workstation. It was no more than two feet diameter. We were sitting there conducting these tests on the missile's hydraulic system. We were getting extreme surges, and they were popping hoses and tubing everywhere. So we had the system all set in and were running these tests.

All of a sudden, we got the word that there had been a lockdown. They closed the top of the silo. There was a nine-missile complex in the Titan program, where you could go underground and live with our crew and everything. We were told that our site was under surveillance. I suppose they suspected sabotage. We were locked on that little ledge for thirty hours, the technician and I. They would send food up to us or down to us, going to the bathroom facilities. But after thirty hours they secured it. We never knew what caused it, but there was something that went wrong and they were suspect. They finally secured the area. That is a little experience I will never forget.

Bill Hermann (see also Chapter 2) was drafted into the Army in 1950. After basic training, he was assigned to the Scientific and Professional Personnel program, which later adopted the name "Broomstick Scientists."

I signed up for an after-hours course taught by one of the professors at New Mexico A&M, in missiles and rockets. I enjoyed it. Then out of the clear blue sky, I was given orders to report to JPL [Jet Propulsion Laboratory] in Pasadena, for the Corporal missile project. That was my basic assignment throughout my whole career [at White Sands], going back and forth to JPL. I spent over half of my Army time at the Jet Propulsion Labs at Pasadena. The Corporal was a liquid propellant, ground-to-ground missile. It had fin-stabilized guidance. We came back [to White Sands] for the launches. The countdown was pretty exciting. Sometimes we would have a hold, at "T minus one" or whatever it was, because there was some malfunction, but I don't remember any long delays.

My assignment was to work on the guidance system, which at that time was still vacuum tube technology. I can still remember those vacuum tubes. They were small, but they were special, hardened vacuum tubes. One time, I wanted to do a drop test on them, to see whether they really were that hard. I dropped the thing on the glass end of it and it malfunctioned, so I knew they weren't to be dropped in that way. I got into a bit of trouble for it. I think they were about nine bucks apiece in 1950 dollars, which was pretty significant.

Jed Durrenberger was a civilian photographer and mechanical engineer who began working at White Sands Proving Ground in 1951. Clyde Tombaugh (see Chapters 1 and 3) was his first supervisor. Along with Tombaugh and Walter Haas (another prominent astronomer), Durrenberger was an early member of the Astronomical Society of Las Cruces. During the 1950s, that group participated in Operation Moonwatch, a project of the Smithsonian Institution Observatory that enlisted amateur astronomers to try to spot and track the first artificial satellites. In this excerpt from his 2003 oral history interview, Durrenberger referred to meteor tracking studies that took place at WSPG before and after World War II. That was a good location for meteor watching because of the dark skies and the excellent camera equipment available at the facility.

The man who had been the originator for the meteor-tracking bit was also at White Sands. He and I got to be good friends, primarily through Clyde Tombaugh. One night we were talking, [and saying] wouldn't it be nice if we could make an artificial meteor out of a Loki missile. The thing was so small and so fast that you couldn't see to track it photographically for gathering information on missiles. The General Electric people liked the idea, and the Cal Tech and Jet Propulsion Lab people liked the idea. They fixed up some Lokis with a quarter-round window on the side. The General Electric people boost up a small projection lamp, which would fit inside and shine out that window. We ended up aluminizing the part of the glass around the bulb like it was a mirror to focus all of the light outside. At that time, I didn't realize that the Loki missile spun at a high rate of speed when it first separated from the dart at launch. But that gave us "*bip bip bip bip bip*" as the missile rotated and the small window would be visible.

We made some cameras over Christmas vacation at White Sands that we used to photograph a thing up against a star background, to try to end up with some trajectory information on the Loki. The Jet Propulsion Lab in California made five Lokis, and three out of the five worked. The bulbs didn't break with the shock of launching. That is what began a period of over three to three and a half years where I spent almost every moonless night at White Sands firing Loki Fireflies. The Loki would deviate so far, some of them would go over the mountain. So we had to restrict our firing times. In those days, Santa Fe had a steam train that came down from Albuquerque, a passenger train, connection between Albuquerque and El Paso. It would arrive at the Las Cruces station a few minutes after ten at night. So we had to quit firing about 9:30 and wait until the train went by before we could fire more Loki missiles over the other side of the Rio Grande.

We needed more cameras. So at Christmas vacation we gave up having a vacation, and in the machine shop in the basement of a building at White Sands, we made some ballistic cameras. With the absence of much developing money and money to buy things that were available commercially, we improvised an awful lot of things that were basically sound and were used to gather some of the early White Sands data. The British had a bunch of reconnaissance cameras, which we used in World War II that I happened to get the lenses out of for the cameras. Much of the stuff that we used was improvised, made of stuff that we could scrounge. We didn't have multi-million budgets, but we got a lot of things done.

Bill Hansen grew up in ranch country not far from Alamogordo. As a college student, he began working on missile tracking at WSPG. After finishing college, he worked for two years in the Physical Science Laboratory (PSL) at what is now New Mexico State University. PSL did data analysis for the missile tests at WSPG. Then Hansen worked at White Sands, first as an employee of contractor Land-Air and then as a government employee.

In 1950, I went over to Las Cruces, and started at the New Mexico College of Agriculture and Mechanic Arts, as it was known then. I needed a job. They were starting what they called a co-op program, where the government would hire students for the summer. I got in that in '51 or '52, in optics out at the Range [White Sands Proving Ground].

I had no prior experience with that. I was just trying to make a living. We used Askanias [cinetheodolites], which did five frames per second. One guy cranked azimuth and one elevation. You took orientation before you tracked. It would get real accurate position for the trajectory of the missile.

At the time, we were working with V-2s, Corporals, and I think Honest Johns, Little John, and some of those type of things. Before I finished school, I still needed a job. I started to work for PSL, Physical Science Laboratory. I started out in data reduction in the basement of old Kent Hall. We were reading telemetry records from Corporals. You did it with a ruler, and you read this squiggly line and recorded them on paper. I remember the old calculators were Friedens and Marchants, electric. You punch in the things, and it would reduce the data and get the information.

I got a job with Land-Air in '56. We had Aerobees, sounding rockets. They would go real high, 250 miles up. They would take different chemicals and release them in the atmosphere. We had this big shot with analine. After so many of them, you get kind of lackadaisical. They briefed us about the danger. The analine was highly toxic. They fired that thing, and it only got up about 200 feet and it blew. I was sitting there controlling. I had given a briefing on the danger of it. When that happened, right away, I said, "Get under your trucks, get under anything. If you see any of this brown chemical, get it off as soon as you can." One old boy called out from under the truck and said, "Now I want to tell you something, Queen 6, I've got brown stuff all over me." He was talking about what happened when he got scared!

About '58, I left Land-Air, and I became a scheduling representative to the Air Force. I could make a lot better money with the government. I went to work for them in mission control. You would clear airplanes on the Range, off the Range. You actually controlled the mission on the floor. I was new, and I knew very little about airplanes. I was sitting there and I got this airplane call wanting a clearance across the Range. Fine, we cleared him across. I said, "What is your altitude?" He said, "Above 60,000 [feet]." I thought, "Ho ho ho. We don't have anything that will go that high." So I asked him to stand by. Through the intercom, I talked to my boss, and I said, "So and so called in, gave me his call sign, and he says he is above 60,000." My boss said, "Clear him across." After it was over, I asked, "What in the world, who is pulling whose leg on what?" He said, "Oh, you don't know about that?" I said, "No, you didn't tell me. What am I supposed to know?" Then he told me we had airplanes that could fly that high. It was a U-2.

The Air Force wanted all the airspace they could get. FAA said you can have down to 24,000 feet, but below that we are keeping it for light aircraft, and airlines and that kind of things. We had some situations where an airplane wandered in front of a missile that was radar guided, or a heat-seeking missile that could have knocked them out. But once you got in touch with them and told them what the situation was, they would plead ignorance. What do you do? You get them all out of there. But we have had to hold missions for them.

One of the better missiles I thought we had out there was the Hound Dog. We had a corridor from Del Rio, Texas, up the Rio Grande, coming in over Fort Bliss, going all the way through the Range and up through Mockingbird Gap, with impact at the Stallion Area. One fine day, one was going through the Gap and it didn't quite make it. You are not going to believe this, but it hit a rabbit! So the Hound Dog thing.... We had a big laugh.

On some tests, you had to turn on the cameras at least minus fifteen seconds to get the trajectory. When you heard the ninety-second tone, you would look at your clock and keep up with it. You would give the countdown to all of your troops, all of the instrumentation people. Then when you heard "Baker, Baker, the tone, fifteen seconds, cameras on," you would turn them on and they would go. Sometimes, they wouldn't fire. [The countdown would continue], "plus ten seconds, plus twenty seconds, no fire, no fire." Okay, you are going to run out of film. So you say, "Cameras off." About that time, one of the optical guys would say, "[The missile] just fired." Then you would get a nasty letter. Okay, the next time you think, "I am not going to take a chance on that. I will let those cameras run." They don't fire. They come around again right quick, and you're telling your guys, "Reload, reload." And so they are reloading, and you say, "We are not ready." Well, everything is right. *Boom*! They would fire. We wouldn't have optics. We would get a nasty letter. "Why didn't you...?" The boss would never understand. "You were out there to get it, so...?"

Albert Poitra was in the Army Air Forces when it transitioned into a separate branch of the US military, the Air Force, in 1947. In 1949, he was assigned to Holloman Air Force Base, where he worked on missile recovery.

I worked on guided missiles, surface-to-air, air-to-surface, surface-to-surface. There were so many different programs that were going on. The [main] one I actually worked on was helping with the Matador. I remember doing some wiring. I'm not a skilled electrician, but the guy told me just hook this blue wire on this blue wire and this, that, and the other. They simulated a launch with the Matador by using a flashbulb. So if everything worked and the flashbulb went off, that means the JATO [jet-assisted takeoff] unit underneath the Matador was going to work. The Matador was launched at zero speed, and they revved the engine up as high as they could go, and the JATO would take off and away they went. Of course, they used it on a track, too, a couple times, to test it.

When I came, the test track was in the process of being built. I remember the first time I went out there, I could see these guys sitting in behind what looked like a dugout from a ballpark. I said what are these guys sitting down here for? "We're just about ready. The warning signals are up, 'X minus so-and-so,' and we had to clear the area or get in the blockhouse." These guys were a good 30 to 40 feet underground. They had a series of mirrors that you could look right there, and it looked like they were sitting right in front of you, but they were underground. It was a series of mirrors going up like a huge telescope, and they were observing the countdown. That was the blockhouse.

I don't even think the cement was dry when they put the first Matador up on the sled to fire it. Our job was to go and retrieve it. We went way out in the Alkali Flats once it was fired, to pick up the parts and see if they could land it.

[Did you get a chance to actually see the launch?] Well, if you want to call a cloud of dust as actually seeing it, that's what we saw, because we were at Alkali Flats. But it never left the sled. The report I got was somebody had forgot to pull a couple of vital pins on it. It was kind of tilted about a 30 degree angle on the sled. What was supposed to happen once the engines were revved up, the sled was to fire, and the ball and socket setup was on the Matador. The socket was on the Matador and the ball was on the sled, and after it got up to a certain speed, the socket was supposed to fall down and it was supposed to leave the sled. But it rode it clear to the end. So we didn't retrieve that one.

I also worked on the Aerobee missile. After I got indoctrinated by the ground safety people showing the hazards of the (red fuming) nitric acid and the aniline, then we used to have to wear protective clothing while refueling and de-fueling. It consisted of a rubber suit. I never could find one that would fit me because I was always the smaller guy, so it draped on me. A red rubber suit, zippered type, boots and all, and you could melt in there in the desert sun. But they also had a mesh that went on the outside of this suit, and you'd stand under a shower and the webbing of this outer suit would soak the water up. So, actually, the only antidote to red fuming was the water, in case a hose busted or we overfilled or whatever. The breathing apparatus was a safety appliance that they used to use in the mines. That was before they had the pressurized Scott air packs that they have now. We used to have to wear those suits, and there were showers close by. In case of an accident, you could hit the showers in a minute. You put your feet onto this mesh floor that the shower would turn on, so it was pretty well guarded.

I remember this one shot that I was involved in went up 21 miles. That was headlines: Aerobee rocket from Holloman Air Force Base goes up 21 miles. I remember sending that clipping home to my ma. Later on, we had the V-2 with the WAC Corporal as the second stage, and it went up quite a ways, then you're talking outer space. But 21 miles, that was really top of the notch stuff. Who'd ever thought that later on they'd be orbiting the Earth and going to the Moon?

Marvin Weber worked in the chemistry lab at Holloman in the early 1950s. Some of the chemistry work was well planned, but as this excerpt shows, sometimes accidental discoveries led to better performance of propellants.

Another fellow and I were roommates, and we spent the last six months of our college career trying to find a central government type job or direct commission or anything, because it was during the Korean War. We thought maybe we could find a good deal somewhere, and the other fellow found a job at Holloman. He came down and worked as a mathematician with Dr. Steinhoff. I didn't have a job yet, but I came down, and two weeks later I went to work at the chem lab in Holloman. That was in the summer of '51.

We concentrated strictly on analysis of the propellants that were on the Base at that time. We got a lot of extra jobs brought in, but our real mission was to take care of the propellants that were in storage at Holloman. They were the rocket propellants that were brought there originally for the old Aerobee program. The fuming nitric acid was not easy to handle, but handling the aniline and furfuryl alcohol weren't bad. They weren't volatile, but they were toxic. Not furfuryl alcohol, but the aniline. So we had to handle those pretty carefully. Some of those propellants had been shipped in there and had sat around for several years, particularly the fuming nitric acid. They didn't know much about storing that fuming nitric acid, so they shipped it in stainless steel containers, which is the best thing they thought they could use. It turned out that the acid attacked the chromium in the stainless steel. It turned from a nice orange liquid to a black, syrupy mess. There was some that we call leakers. They'd transfer them from one barrel to another.

That old black fuming nitric acid leaked when we were firing the Aerobees. We'd use it, because that was all we had. And we'd get this fantastic performance. Now, we didn't realize we were getting that until we received fuming nitric acid in nice solid, pure aluminum drums. They store it very well. It didn't attack the aluminum. But the performance of the rockets dropped off terrible! It turned out that the stainless steel that was absorbed into the acid acted as a catalyst and gave us a much better burn. So it wasn't very long until they were coming around wanting for us to put the iron back in. (Laughs)

I don't know exactly how that fit in the time frame, but I can remember in that old lab, we used to do a lot of testing for the Aeromed folks with the survival systems for the monkeys and the canisters that flew on the balloons. We used to sit up all night with one of those things with the blowers going, and we had some subjects inside, measuring the CO2 and oxygen levels all night long to determine if the system's working right. We did do a lot of work with Dr. Stapp and his crew on the balloon launches, for the subjects that they flew.

Jack Strehlow worked for North American Aviation in rocket development. In this anecdote, he mentioned the Snark missile. It was an early version of an intercontinental guided missile developed for the Air Force in the late 1950s by Northrop Corporation. Then, he described an episode that happened when he was working at Santa Susana Field Laboratory in Simi Valley, California.

The Snark missile had a real funny navigator, it had thirteen telescopes, very complex. They had had thirty failures out of thirty tries, when they got the production contract. And of course, the guys who were involved and who knew about what was going on in the program kidded about that. They said, "We now called the waters around Canaveral 'the Snark infested waters.'"

In 1952, I got tasked with the job of developing part of the equipment for testing rocket engines at Santa Susana. I remember one of the things I did in particular, because it came back to bite me many times. They wanted to have a documented record of when some of this switching was done. And it was in milliseconds. So I investigated all of the tape recorders that were available, paper tape, and came up with one that moved at, I don't know, it was either 10 or 20 feet a second. And we had twenty of them in this small room there at the test site at Santa Susana. I only saw this once. They would run a test, I forget how many seconds, you know, it was 20 seconds or something that they were recording, and when they opened the door the room just exploded with paper. The whole room was filled with paper coming out of these tape recorders. Each of them had twenty pins, as I remember. Years after that I got a phone call, "What made you do this?" They wanted a recording that gave it to them in milliseconds increments, and that's the way you got it.

Leonard Scoby was hired as a civil service employee at White Sands Proving Ground in 1954. He was a surveyor, part of a small group assigned to work at Holloman Air Force Base.

The surveying we did normally was, we would put positions on cameras all over the Range. When the cameras would photograph missiles, they could run a triangulation on to the missile and tell you exactly where it was at any given time. They needed to know the exact location of each camera and the orientation of each camera. So from where the camera was located, we would survey the eight targets that were around it, every 45 degrees. We would tell them exactly the azimuth and distance out to those targets, then they could get their orientation from that. The cameras were what they called cinetheodolites. Each frame includes the picture, the exact time, exact azimuth, and exact elevation angle. So if you know how to read them, you can get all of this information off every frame of film.

This missile range is the most highly surveyed piece of land in the world. Sometimes, we would shoot [i.e., survey] wooden targets that were out about a quarter of a mile. We had to survey these because they were on telephone poles, and, you know, with the horses rubbing against them they moved. So, we would have to survey these about once a year. They would go out and tighten the guy wires on the telephone poles, and then we would survey the targets.

I think the strangest job that I ever got in on was when we surveyed the feathers. I am not sure how these feathers were dispersed, but they wanted to know how far the feathers would fly. I think it was probably to find out in case of a gas attack, what sort of a pattern the gas would make, by looking and seeing what type of pattern that the feathers made. Man, we worked on that thing for days, trying to get to the extent, the outside boundaries, of where these feathers were flying. They were blue, and we'd be looking for feathers so we could survey them.

We used to align the Honest Johns and the Pershing missiles. Then when they would fire, being on the north end of the Range, we would go up and survey the impacts, wherever they were. We knew about where they were supposed to go because we would stake the target beforehand. So we would go out into that area with recovery in the helicopters. We would find the impact and survey the impact. Then they would know where they ended up and where they started, and they would be able to compute the trajectory and the speed and all of this good stuff.

One of the strangest things that we ever got involved with was when what we called the AFSWEP area, Air Force Special Weapons out of Kirtland [Air Force Base in Albuquerque], would drop a bunch of bombs. It was kind of hard to know exactly what aerodynamics and stabilization were, so we put in a geophone field around this target. As the bombs were released from these airplanes, they would have exact release time. They would come down and hit, and they would have a smoke charge. We would spot this smoke charge with the cinetheodolites. Then we would go in after all of the bombs were dropped and try to find them. When we found the exact hole, we would survey that point, so they would know exactly how far they were from each one of their geophones. Then they could tell exactly from the time it would be dropped to the time it would hit, all this good stuff. It was real interesting.

When they found out that we were only about 300 feet from where the impacts were going to be, they said, "Oh, you can't do that!" I remember one. We were getting ready to survey it. And here it came in, *swoosh*! It was about 30 feet [away], so it was kind of scary, but not too bad.

One time, I went to Utah Test and Training [Range in western Utah], where they had a cruise missile come in. They had an awful time trying to figure out where it was, so we went up in helicopters. We finally found the impact. There were no survey control points anywhere around. So I took a topographic map, and tried to identify exactly the mountain it was on, and where it was on that mountain. I gave them approximate coordinates of it, at least they found out about where it was. That's all I could do on that one. If we'd had good distance measuring devices like we do today, then it would have been a lot easier.

At one time, they had what they called a co-use area [at White Sands Missile Range]. The ranchers could live there all the time, in their places. But if they had a mission that had to go into that area, they would notify them a week ahead of time and say, "We need you to be out of here." Then they would pay them for that time that they would be out. One time, we had one missile that we went up looking for. It had just missed the triangulation station on top of a peak, went over and almost hit this old boy's house. He was a little bit perturbed. He would have been paid for it if it had done a lot of damage. I don't think we have ever killed a cow up there, even. Things have come a little close, but nothing has really done any damage. We have knocked a few fences down up in that area. They got fixed back up.

John Cox went to work at Bell Aircraft in 1941, after finishing high school. He served in the Army during World War II, and then earned a degree in aeronautical engineering. After working on several experimental airplanes, he moved into rocketry. As he described in this excerpt, that could be quite exciting work. The RASCAL missile he talked about was a supersonic, air-to-surface missile developed by Bell Aircraft in the late 1940s. The missile's name comes from RAdar SCAnning Link, its guidance system. The Shrike missile he mentioned was a preliminary version in the development of the RASCAL.

I was working for Bell Aircraft on a guided air missile commonly called the RASCAL. In 1953, I was assigned to a job that they called a missile engineer. They made a Shrike missile ahead of that, which was a small one, and they cheated and used 21-inch torpedo pieces in the fuselage. The RASCAL was 48 inches in diameter. Anyway, they found out that when they were making the Shrike, they had to assign somebody to each missile as traveling historian. Classifications being what they were back at that time, there were no notes. It all had to be in your head. So they continued that same thing starting with, I think it was either the fifth or the sixth RASCAL. They didn't need it for the first few because they were structural test vehicles, drop vehicles with nothing in them, just a shell. But then, starting with number five or six, that's when they started to put the electronics in, and that's when we started to have problems.

I came to Holloman originally in January of '54 with RASCAL missile number twelve. I had to basically shepherd the thing through testing down here, get it hung properly on the B-50 and hopefully launched and going where it was supposed to. In the case of number twelve, it didn't work, but the engine blew up. It was dropped from the B-50, and the secondary guidance system took over and homed in on the ground. Sometimes we did two a week, and sometimes we didn't do any for a month or two.

We used to take off in the B-50 and circle. We'd go down south, just north of Juarez. Meanwhile, the electronics of the missile are working, and the guidance operator of the missile is trying to use the radar that was in the front of the RASCAL missile to make sure he can pick up the proper aim points on the ground. If he did, he switched over the radar that was in the B-50 and tracked everything, and we got to the launch point. Theoretically, if everything was working right, we dropped the missile and away it went. Of course, I said theoretically. We had some of them that actually worked right.

After it hit the ground, we always sent a ground crew up to pick up the pieces. If it was really apparent from the instrumentation that was on the missile and on the ground what happened, it was no problem. But if it wasn't readily apparent, then we went through the autopsy business.

We knew what happened to number eleven. It exploded underneath a B-50 while it was sitting on the ground. Two bundles contained the high-pressure nitrogen so we could expel the fuel and the acid from the tanks. They ruptured, cut the missile in half, dropped it into the pit. It was loaded with acid and fuel.

The time between when number eleven blew up and number twelve was flown was about four months, because they had to redesign the complete two-bundle system that was used in all the missiles. But they wanted to fire because we had number twelve down here and number thirteen and fourteen. We had three of them left down here at that time. They sent number twelve back to the main plant. They took the radar out and they put Taylor-Wharton bottles in there. Taylor-Wharton bottles were the large storage bottles that you get any of your gases in, and they used that for the high pressure system.

They went operationally dormant in November of 1957, right after we fired the last one at Holloman, which was number ninety. That one blew up also. Air Force chase pilots used to be down here, and there was a gentleman by the name of Joe Kittinger [see Chapters 4 and 6]. People know him. Crazy fool, he bailed out of a parachute a hundred and some thousand feet [actually, 102,800 feet] and did other things. But he's a damn good pilot, and he was flying. He was a pilot in the F-94 that I was in, and we were right up there where we belong, getting ready to take pictures and everything else of the RASCAL as it lit off. We started while it was hanging from a B-47, and then it was dropped. After it dropped so far, the thrust chambers would light off. In this case, the thrust chambers lit off and blew the tail end right off the back of the missile. Joe Kittinger and myself were right there. We got hit by some of the pieces of the missile. Joe says, "What do we do?" I said, "Lower it down. The son of a bitch blew its ass into us." That went out all over the air because people on the ground wanted to know what happened. We caught hell for that because of the language in the air. They complained to the FCC. The FCC tracked it down, and I caught hell for saying the damn thing blew its ass into us.

Anyway, that's beside the point. When Joe asked what to do, I said, "Let's follow the s.o.b. down," and he did. We had a little vibration on the way down, and when we got down and finally landed, the wings on the F-94 just drooped. The missile was painted fire orange, and we had orange marks all over that doggone F-94 where things had hit. Tip tanks, leading edge of the wings, the fuselage, you name it. They scraped the F-94. We had closer than that, lots of times.

One case, over here at Holloman, we were flying in the B-50s back in 1954, '55. One day the field test director, Frank, was flying in call sign Bell 1, which was the guy who actually flipped the switch to launch the missile. I was flying in Bell 2 on the left-hand side of the B-50. We were up there and we had a photographer, Oscar, going uprange. We always had a photographer uprange so in case the missile did get up there, he could get pictures. Anyway, Oscar's going up there, and he'd had a hell of a night the night before. He wasn't even supposed to be uprange that day. His boss was supposed to be up there, but his boss got sick the night before. So he had to cut short his celebrating the night before and go in. He was a little bit on the woozy side, shall we say. He got up there, and the boys up at another area were practicing with some of their dummy nukes. The missiles still went off, but they didn't explode. And Oscar would drive along in this GI three-quarter-ton [vehicle] all by himself. He came up on the radio and says, "Bell 1, Bell 1, I've got a problem." And Frank asked him what his problem was. He says, "The bastards are shooting nukes at me!" I had just taken a mouth of crackers out of my inflight meal because we'd been up there for about three hours, and I couldn't say anything. Frank had just swallowed a couple, but he still had a few crumbs in his mouth. He gulped and he says, "Oscar, repeat after me, 'Our Father, which art in heaven....'" They finally got ahold of the Army and told them to shut down because they were shooting missiles and they were impacting where we had people.

The RASCAL was launched originally from a B-50. Then we went to B-36s, and then B-47s. We had what we called a profile flight, which was a flight of almost 3,000 miles with a missile hanging on the doggone carrier, which is the way it was supposed to be when it went operational. We used to fly chase on those too. On one of them, we were flying chase on a B-47 with an A-26. An A-26 is an old twin-engine attack bomber from World War II. Believe it or not, we used two of those for chase when the B-47s were on their profile missions, which includes aerial refueling and everything else. We were up around Albuquerque one day and taking pictures while the thing was being refueled with a SAC [Strategic Air Command] tanker. We didn't have any oxygen in the doggone A-26, and the B-47 tanker kept going up higher and higher and higher. We finally came to when we were about 18,000 feet and there's not much oxygen up there. But we made it back.

In early 1959, I went to work on the Agena rocket engine. The Agena rocket engine was a multiple start engine. When we first started out, we had two charges of powder so you could get two starts out of the engine. We finally wound up with five charges to be sure it started. That's the only missile engine that was ever delivered into the Air Force with a money back guarantee. Bill Smith, who worked on the power plant of the X-1 and wound up to be vice president of the rocket division of Bell Aircraft, guaranteed every one of those things the Air Force would fire. "If they don't work," he says, "bring them back and I'll give you another one free." Now that's not as bad as it sounds. If the darn thing is sitting up there at about 150,000 feet, or a couple hundred feet or maybe a hundred miles up, how are they going to bring it back? They fell for it and they were happy.

Pilotless Rockets

Raymond Trigg's mother died when he was ten, and his teenage years were "pretty tough," so he did not finish high school. He enlisted in the Air Force when he was sixteen years and seven months old. He said he accomplished that with "a change of a birth certificate, and a photocopy, and a recruiter who was willing to accept that paperwork." His first assignment was working on the development of the launch facilities at Cape Canaveral, Florida.

I went into the Air Force in 1956. I was in communication construction. The first project that I really recall was putting the cables in to the blockhouses at Cape Canaveral, to Launch Site 17. That was one of the very early launch pads, or, complexes, they called them back then. That was, probably, 1957. There wasn't a whole lot there. The bulldozers had cleared areas, but there were areas that we had to go to that needed clearing.

[On a typical day], we'd get up in the morning, go to the mess hall, get on the bus, and go out to the Cape. One good thing about duty there at the Cape was we ate with Pan-American World Airways people, and we had to sign on the dotted line for the lunch. We were eating for the civilians, and we would sign, and they charged the government ninety cents for our meal. That was the good side of it.

The side that wasn't so good was that we were actually doing all of the dirty work for the civilians. We would dig the ditches. Down at the Cape, the water level was just like a foot or eighteen inches, so, you dig a little bit and you were standing in water. Then you got to put cable in it, pull it into the transmitter side, receiver side, blockhouse, whatever. And then, there was the climbing of those towers, putting up those rhombic antennas. All of that was hard work. We had [wooden] poles that weren't stepped, and they were ninety footers. So, you would have to put your hooks on and hook them [i.e., climb them using boots with long spikes]. One thing they'll teach you is, don't look up. Looking up, if the clouds are moving by, you will think you are falling, and you will panic. Keep looking at that pole and don't look down.

Another interesting thing about crawling on some of those towers down at the Cape, if there was a thunderstorm off in the distance, you could feel the static in those towers. And when you got to the bottom, you didn't step off and leave one foot on the tower. You just kind of jumped off, so if there was any static in there, you wouldn't be the direct ground.

Michael Hadfield had a Bachelor's degree in communication electronics, and had worked in that field in the Marine Corps. After he left the service, he worked for General Motors' aerospace division. The Thor missile he worked on was the first operational ballistic missile deployed by the Air Force. Ballistic missiles follow a natural flight path, as a bullet does, rather than a directed flight as a guided missiles does. Thor missiles evolved into Delta expendable launch vehicles that propelled satellites into orbit in the early 1960s. Two versions of Delta rockets are still in use. Hadfield recalled interesting stories about test launches at Cape Canaveral.

I interviewed and accepted a job with General Motors to work on their skyphone project for commercial airliners. When I reported in, however, that position had been filled and they said it would be three more months before another one opened up. How about working over on the aerospace side of it for a while? Well, in my total lack of knowledge at that point, I thought, what could be interesting about shooting a rocket up in the air and guiding it? I soon found out. I went to work on the Thor missile program. I started in the system analysis area and immediately got into some extremely interesting work.

Our systems analysis group was actually a systems analysis and advance systems design activity. We got into all facets of the operation. What was being designed there was being built and launched at Cape Canaveral. What ideas were coming along for the next generation were our responsibility to come up with. As a consequence, I had some very interesting opportunities to watch launches at Cape Canaveral.

One day, I was up on the roof of a Cape Canaveral hangar where our laboratory was located. A place where the security folks kind of didn't want us to be, but a great place from which to watch missile launches. Our competing program, the Army Jupiter missile, a 1,500-mile ballistic missile weapon system, was launching that day and we saw them light off the missile. It moved up and, gosh, got all of maybe fifty or a hundred feet before it tipped over and crashed and burned on the pad. That was my first experience with something that was less than successful.

Then a couple of weeks later, I was back up on the rooftop again. This time they were launching a Polaris missile. The key difference between the Polaris missile and the Jupiter was that the Polaris had solid rocket fuel engines, whereas liquid fuel engines were in the Jupiter. Well, the Polaris took off and it looked like it was just going great. It climbed up to altitude and started to pitch over to start its downrange program. All of a sudden, it changed its mind and started to do a figure eight over Cape Canaveral. At that point, we thought it was probably prudent to get off that roof and find a safer place. So we watched the rest of that launch over our shoulders as we were clamoring down the ladder of two stories. That missile went on to do some strange things. After circling all over Cape Canaveral, it turned back onto Cocoa Beach, the community just to the south of the Cape. It aimed at a mobile home park where a friend of mine happened to live and, unfortunately, he was there that day. He looked up only to see this Polaris missile coming straight at him. He felt the best place to be was at least under the trailer, so he burrowed like a rabbit down there. The missile sailed over the top of the trailer and, fortunately, landed in the Banana River, hurting nothing but a bunch of fish. This was back in the 1958, '59 period.

Then I had a chance to be very directly involved with the next one because it was our bird, the Thor intermediate-range ballistic missile, which was the Air Force's answer to the 1,500-mile Army missile. We were having a night launch. Night launches are always very dramatic because you have this big white or silver vehicle standing up there on the launch pad. They train all kinds of spotlights on it, and it's just impressive, either up close or from miles away. We were very up close, because we were on camera station number one, which is the first one downrange of the launch pad. Exciting things happened there. We got all set up. I didn't have any official responsibility. I was an observer, so I had my binoculars trained on the base of the missile. It lit off and started lifting up. It looked really beautiful as it went straight up, straight up.

That was a problem. It wasn't supposed to keep going straight up. It was supposed to pitch over. Instead, it kept going straight up and the range safety officer blew it. When you blow a liquid-fuel missile like that one, it's very spectacular. It doesn't run around up there; it just plain fills the sky with fire and smoke and everything else. Well, in this case, that sky was directly above our heads, which motivated us to quickly exit camera station number one. I got in the vehicle I was assigned to and had the engine turned on before the driver got in. Fourteen of us cleared that camera station in probably something like fifteen seconds flat and didn't stop for three miles down the road! Of course, curiosity being what it is, we all went back there at three or four in the morning, after we had calmed our nerves down a little bit. We used that as an opportunity to look for some shrapnel that might be an acceptable souvenir, and there was some there. But that's as close as I ever came and, frankly, as close as I would ever want to come.

After that, we did have a series of very good launches. They were thrilling, because we were challenged by the upper level managers in the Air Force and in DOD to see just how good a missile could be made to perform. We had our specification, which I can't talk about specifically in terms of performance numbers. It was a fairly tight specification for those days, just as the Army's was. They took two of the USAF systems and called them their gold-plated ones. My boss got the first one, and I got the second one. Our jobs were to go into the factory and select every piece and assembly that went into that particular system, and install it in one of those missiles. We had a little bet going as to who was going to get a more accurate launch. We also were two of the very few individuals that knew all the latest calibration procedures, software techniques, and everything else for optimizing the system performance. My boss, Glen Quasing, got his system all fixed up, and I supported him on that back in the factory. Then he went to Santa Monica, California, and participated in the integrating of this system into the missile. This was an exercise that we normally didn't get involved in, but certainly it was very interesting.

We then went from Santa Monica over to Cape Canaveral, where it was put up on the launch pad. They went through all the pre-launch operations and calibrations and got ready to shoot. The day of the shot came, and everything was going very well during the countdown. I know Glen must have been praying hard and had his fingers crossed, too. It lifted off and moved on 1,500 miles downrange. It beat the specification by a factor of three to one, which was a significant achievement. Of course, that was a challenge then. I wanted to beat Glen. We had a beer bet on this. So I took my system through and out to Santa Monica, then down to Cape Canaveral. When our time came, again, we had a smooth countdown, a good launch, and a 1,500-mile downrange flight. We beat the specification by four to one. So I went back to Milwaukee and collected my beer. We all felt pretty good, including the Air Force, who knew what the full design capabilities of the weapon system were. That was one of the most exciting experiences on that particular program.

There was another story that comes out of those years at Cape Canaveral that's worth telling. It shows what dedicated people can really do to help make their program a success. One night, we had planned a twin launch of the Atlas ICBM [intercontinental ballistic missile] and the Thor IRBM [intermediate range ballistic missile]. The Atlas had some bad luck for almost six months, and they were very anxious to have an excellent shot. We hadn't had quite that much difficulty, but we also wanted a good shot. Atlas went first. They got the countdown to something like T minus eight minutes, and they called a hold. They had a problem with a valve that got stuck up in the plumbing connections that they have up inside the bird. They didn't know how they could handle it. They could not do it remotely, so this one technician from General Dynamics, quite a heroic chap when you think about it, volunteered to help. He was very tall and very skinny, so he shed most of his clothes. They greased down his body, and he crawled up inside that missile to get to the valve and free it up. He did that successfully. I don't know, he might be Chairman of the Board of General Dynamics now. (Laughs)

Later, I went to work for Honeywell in the advanced systems analysis group. I didn't understand how many other things those engineers did. The first assignment I had was working on the electrically suspended gyro. Without getting into the details of it, suffice it to say that this is a hollow rotor. When you spin it up to speed, you want it to be perfectly spherical. Otherwise, you're going to suffer the introduction of error torqueing, which degrades the performance. The problem was that it wasn't spinning up to be perfectly spherical, and they weren't quite sure how to machine it to what they call prolate at rest. In other words, tuck in the equator of the gyro so that when you spun it up it would grow into a sphere.

This is when my appreciation of history really got another nudge. I went to the classical text in mechanical engineering and analysis to see about how to form a rotor body like that. One of the foremost scientists at the time, Dr. Timoshenko, had an excellent book, but it was all on thin wall designs. I didn't have a thin wall, I had thick walls. So I went to the other extreme and found another text that had a good discussion of solid theory. Well, I didn't have a solid either. I had hollow, thick walls.

So I got on the phone to one of our research scientists in Minneapolis, Dr. Wally Ito. He could usually be counted on for some very good advice. He came through again, because he said, "Back in the late 1800s, there was a chap over in England who did some work in this area. It might be helpful to you." It turned out that this man's work was published in 1879 in the Cambridge PHi Losophical Transactions. His Master's thesis at that time would almost be the equivalent of a doctoral thesis today. I knew where it was; the question was how to get it.That's when I experienced the fact that the women in an aerospace organization are often among the most important and valuable people you've got. Don't overlook what they can do. In this case, I went to our librarian and I said, "Virginia, I need a copy of the Cambridge PHi Losophical Transactions of 1879, and it's really critical to me. I have no idea where to suggest you look." She said, "Oh, that's my department." A week later, Virginia called me back and said, "Mike, your book is here." I went into that book and reviewed everything this gentleman had done, which was excellent work. Then I proceeded to adapt it to my specific problem, and about three days later I was able to turn the solution in to my boss. He had given me three months in which to work on this. Not claiming any credit for myself, without that reference work, I'm not sure I'd still be working at it. I was able to deliver the whole thing in three weeks. That effort formed the basis for the prolation of all rotors associated with the hollow rotor, electrically suspended gyro production. It is used to this day and has been proven out at gyro and system levels. The result has been a major navigation upgrade for a B-52 D, G, and H, plus an F-117A stealth aircraft.

[Later, Hadfield worked on the Centaur missile guidance system.] The Centaur was an upper-stage liquid hydrogen type vehicle set on top of the Atlas. The object was to provide enough additional boost to be able to put things not only in orbit but to put them out to the Moon and so forth. We were having troubles, as people often do in transitioning from the developmental phase to the production phase. Part of it was communication, part of it was lack of knowledge on the part of one group as to what the other group was doing and why. One day, the program manager called me in and said, "I'd like for you to take a special assignment. I want to put together a, we'll call it a system test and evaluation group. You'll have all of the system production test engineers, production test engineers, and their supervisor. You'll have quality engineering, reliability engineering, systems engineering, electronics engineering, and platform, or inertial measurement unit engineering." I'd have some of the top people from each of those areas, and they all would report to me as part of this forty-man test group.

My ground rules to my people were very simple. I said, "There is no such thing in this group as one person succeeding or failing. We all succeed or we all fail on the strength of what we all do collectively and individually. There are no organizational lines in this group. We all belong to the same organization. We rub elbows every day. We work together to give whatever it takes." The group I had, which was a very good group of people, responded very positively to this, and at the end of eighteen months we had accomplished the following. We were shipping all systems at least a week or two ahead of schedule. The systems were staying in the field. They were not coming back because of malfunctions. And we were supporting a spacecraft flight program that was second only to Apollo. It was called the Surveyor. Surveyor was basically an unmanned Apollo. We had to put seven or eight of them on the surface of the Moon, collecting data, sending it back to us, before we could be comfortable with putting man on the Moon.

What I'm talking about started in 1959. Our heavy pressure came about in 1963 to '64. Today, in 1998, that same space booster system is still being used, and there are advance orders on the books going on years from now. So any time you can take one system like that and have it do the job for thirty-nine years plus, that's something to be proud of.

After a stint at Mountain Home Air Force Base in Idaho, Wally Adam returned to Wright-Patterson in Ohio, where his first assignment had been. In 1968, he earned a Master's degree for which he wrote a thesis on performance enhancement for liquid fuel rocket engines. One of the rocket propulsion systems he worked on was electric propulsion, or ion propulsion. Robert Goddard, inventor of liquid fuel rocketry, had experimented with ion thrusters in 1916 and 1917. He saw it as a potential propulsion system for interplanetary travel. The first ion thrusters actually used in space, in the early 1970s, kept geosynchronous satellites in their proper orbits.

When I graduated from college in 1958, I worked for three months as an engineer at Wright-Patterson Air Force Base. It was fun. I worked with the missile program there. My project was to fire a rocket with heated red fuming nitric acid as the oxidizer. Nobody had tried heating that up to 400 degrees. That was an interesting program, even though I didn't ever get to fire the rocket when I was there because my time was too short.

They told me that they wouldn't do this pilot project with red fuming nitric acid that was heated unless I had a solid emergency shut-off valve that would instantly shut off the flow in case there was a problem. So I went back to my engineering studies and designed a valve with a Teflon tip that would expand a little bit and close that off instantly. That was kind of fun. I got to do an actual design of something that would take that kind of abuse. That red fuming nitric acid is very corrosive stuff, and heated to 400 degrees it was really corrosive.

From Vietnam, I went to Colorado Springs to the Air Force Academy and spent four years as a professor of aeronautics at the Air Academy. I'm not an electrical engineer, but the course that really became my specialty was a course in electric propulsion. Electric propulsion was a new field, and the academy was giving a graduate level course in electric propulsion. One summer after I taught the course, I think the first time, I had an opportunity to become a visiting faculty fellow with Case Western [Reserve] University in Cleveland and work with NASA Lewis Laboratory [in Cleveland] in the electric propulsion branch. They were the renowned experts on electric propulsion.

People in that laboratory developed what's called the Koffman thruster, which was named after Harold Koffman, who was head of that lab. The Koffman thruster was a mercury bombardment ion engine that could be used for space station keeping or as primary thrusters. They had this plan for an engine that would go to Mars as a continuous thruster. So that was fun!

There's one interesting point that I learned from that electric propulsion program and NASA Lewis. When they started the program, like in the '60s, they had about twelve different pHi Losophies of how we can build an ion engine. They pursued all twelve technologies for about two years. Then they looked at them and said, "We made some progress in these six. Those six aren't yielding to solution, so we throw those six away." You couldn't prove scientifically that those wouldn't work, but they just weren't yielding. So they took the ones they had left, and they pursued them another couple of years. Then they said, "Okay, which of those are yielding solution?" Again, we can't prove which is the best or the worst scientifically, but we know what's yielding to solution, and about four of those weren't yielding. So they ended up with the mercury bombardment engine and the cesium thruster as the two that seemed to be yielding to solution, and that's the two they developed. I thought it was a very clever approach for researchers to take, because they didn't have to disprove the ones they didn't use. That would have taken a lot of time and money, and you wouldn't develop anything.

Art Reiners volunteered for the Navy flying program after he graduated from high school in 1942. He left the service after World War II, but was recalled in 1951 during the Korean War. He was in charge of the ground control approach program at El Toro Marine Air Base near Irvine, California, for three years. He left the Marine Corps, and he was recommissioned in the Air Force in 1956. After accumulating enough service years to retire, he entered the civilian work force.

I had gotten out of the service and I got very interested in the space program through the Air Force. So I went down to Rockwell, which was at that time North American Aviation. I applied for a position in 1962, and they called me the same day and said, "You got a job." They sent me to Marshall Space Flight Center in Huntsville [Alabama] to work with the NASA people, to learn all the techniques of how to put cables together, and different wiring and soldering, and molding, and the various types of techniques that were being used on several different programs, Saturn and Jupiter and all that. We had everything there at Huntsville at the Redstone Arsenal. For about two and a half to three months, we had pretty intensive training in various types of things. We were able to design some new techniques on how to protect the cables when they were fired so they didn't burn in half. We called them umbilical protective coatings. They were very helpful in getting me a really good start. I came back to North American Aviation, and we built up a training department to send some people out to do the certification work on the birds.

I was in and out of Marshall for about four years. At Marshall, we had about eleven different sites we were responsible for, including the Cape, where we went down and showed them techniques. We were developing all kinds of course material to teach people how to build the Saturn I and the Saturn V. We designed the first ovens that were used to roll the whole engine in and bring it up to temperature to melt the nickel that was used to weld the tubing together for the cryogenic materials. I helped North American Aviation develop a patented ignition system. They were having trouble with leaks on the Saturn I, and we designed a method of being able to put the system together without leaks.

[Wernher] von Braun was not a real easy person to meet, although I sat at the same table with him in the executive dining room at Huntsville occasionally. He was not a person who would just introduce himself and meet you as a social activity. He was strictly a business type person. Very typical German scientist, you know, "Don't bother me with these little things. I've got big things on my mind." And he did, there was no doubt.

I don't think I ever heard too much pessimism at Marshall. Most everybody felt the program was going to work. You just didn't talk about the things that might go wrong. You talked about the things you wanted to get done, and I think that probably had more to do with the fact that everybody was working in pretty good harmony. You'd get an occasional pessimist who would sit around and bad-mouth the program because of some little thing, but generally speaking, I think they thought the thing would work from the very beginning.

We worked on the J-2 [engines for the second stage of the Saturn V rocket] before we worked on the F-1 [engines for the first stage]. The reason for that was that there was no test stand system in the country that could handle the F-1 program, so they had to build the test stands and everything at Edwards Air Force Base. I was in on the first few of them that we tested and, needless to say, they weren't necessarily successful from the beginning. But the problems were material problems, they weren't engineering problems. We had just not learned how to use the materials right yet, like how to weld the tubing together so you could transfer liquid oxygen and hydrogen side by side at low temperatures.

You had to remember that everything in those days operated at [very cold temperatures] and with no lubrication. So it was easy for a turbo pump that didn't work properly to cavitate [bubbles formed and immediately imploded] and just blow up. This happened on a number of occasions before we finally managed to get it to work properly. The first one only ran about twelve seconds before it cavitated and just completely destroyed itself. You are talking about transferring 220,000 gallons of cryogenics in seven minutes. It took a pretty good-sized pipeline to bring that kind of material in, and we had a lot of valves and things like that that were never designed to be able to handle that much cryogenic material. The vendors redesigned it, and the suppliers of materials were constantly revising the valve systems and the pump systems for us.

The F-1s were all test fired individually at Edwards and then they were fired at the Cape. The J-2s were taken to a place called Santa Susana, which is outside of Chatsworth, California, in the mountains. They called it "Thunder in the Santa Susanas" all the time because, if you lived anywhere in Canoga Park, your windows rattled every night.

Water is a byproduct that actually is created when the F-1s fire, and you get a flame bucket just full of water. Typically, at Edwards Air Force Base, being on the edge of the desert, you'd come out there and get ready to fire, and you'd look down there and see all these squirming things. They were rattlesnakes that would come in off the desert. They'd fall in the flame bucket, and they couldn't get out. There'd be at least three to four hundred of them down in the flame bucket and of course, you'd get fried rattlesnake.

We did do upgrading on the electrical system for the Saturn IB. At that time, we were still using vacuum tubes. We were in the transition period, going to transistorized systems. Transistorized systems were not reliable yet because of heat and vibration and one thing or another, so we were developing sequencers which would open and close the valve systems and give you the time delays you needed. So it was a hybrid system of vacuum tubes and transistors mixed together. We were designing the test boards and the support equipment to test out the whole system before it was ever brought out. It took a console that was as big as this room, just to check out the valving to be sure that they opened and closed in the proper sequence, that the timing was exact, because all you had to do was mix those fuels in the wrong proportions. They purged all the engines before they fired them, with helium because helium is colder than hydrogen or nitrogen or oxygen. So they would purge down with a tank of helium, and it would chill down the system so that it would accept the hydrogen and the oxygen. If you didn't do that, you just might as well forget it, because it wasn't going to work. You had to have a nonflammable material. Helium, of course, was inert.

We assembled the Saturn V systems, and my job was basically to teach people how to put the system together, how to make it reliable, make the electrical connections and everything so they would not go bad. We had a tremendous series of programs there, where we lost almost all our umbilicals before they were used because of shrinkage of the insulation on the lines and things like that. We went through a tremendous testing program, where we defined the various types of problems that can go wrong with a plug and cable system. At one time, we shipped almost fifty cables to the Cape, and when we went to use them, none of them worked. Brand new cables. It turns out that they had decided to put a wire braid on the outside of the cable, after the cable was built. By applying the wire braid, they pulled the pins out of the plugs because of the tension, so we ended up with about a million and a half dollars of cables that wouldn't work.

This was big cabling that is hooked to supply the electrical connections to the bird when sitting in the stand. It's long and may have as many as 200 wires to supply various phases of electrical power. These cables could be 20, 30, 40 feet long. The problem is that when you get that many connections into a given plug, a screw-in plug where they are mated and the shell is screwed on, and then that shell is sealed to keep the moisture and one thing and another out of it. There's a lot of handling of this thing, from coiling to uncoiling and coiling again, over a long period. You've seen the cables, probably, when you see the gantry move back and you see this electrical cable fall away. That cable probably only gets one use, because it's probably going to be burned in two on the bottom. That's why we used a ceramic paste that we smeared on the cable at the base where the flame would hit it, to keep it from burning in two. The ceramic paste would provide an ablative material, and that would prevent destruction of the cable. The cables can run as high as $200,000 apiece, and it is not an unlikely thing that they may go bad even before we use them, just from sitting.

After I left Marshall, my job was to go to all the sites throughout the country where they were developing parts, electrical cabling, or anything that the vendors were putting together, and see if they were doing what they were supposed to be doing, and be sure that if their people needed training, that they came in. I had the training department, which had fourteen instructors. We had about 590 students that came in on a regular basis. Every six months, they had to be requalified, and if they didn't requalify, out they went. Our biggest problem was men, because men, at least 20 percent are colorblind. So, you get a guy out there who's going to be working with nothing but colored cable, and he can be a real disaster if he's colorblind. So we gave them the Ishihara test and some things of that type, the yarn test and one thing and another, to be sure that if they were colorblind, we took them off those. That doesn't necessarily mean they were fired. It meant that we had to be sure that they were transferred to some position that didn't require color identification. Very few women are colorblind. You'll run into rare cases, but [male] field engineers, almost without exception, you would find that about 20 to 40 percent of them would be colorblind in varying degrees.

We had programs that we taught on how to go in and clean up a Saturn after it had been fired. It wasn't just a matter of taking it out and wash it off and bring it back. No, you had to go inside the tanks and wash them down. Once a man is inside that tank, he's only good for about ten to fifteen minutes with a Scott pack [breathing support system], and then he's out of air. So you had teams of people working the internals of these tanks. It was called confined entry, and we taught a program on confined entry. We did lose one man due to nitrogen. He ended up in a total 100 percent nitrogen atmosphere one day, and that was the end of that. We tried not to be careless with it.

We had a lot of programs we taught on liquid oxygen. We lost a man on the test stand because we had a liquid oxygen leak and the liquid dripped on the flooring of the test stand. The guy came along and saw it. It had built up into like a little icicle and he kicked it and it blew his leg off. He had on rubber shoes which had some oil or something on them, and oil in contact with cryogenic is just disastrous. Safety programs of that type, we taught quite a bit.

Marvin Baker joined the Air Force after finishing high school in the early 1950s. After working for twelve years in the electronics field, he was moved into the missile field. He worked on Jupiter missiles at Redstone Arsenal in Alabama. In 1963, he was assigned to Vandenberg Air Force Base in California. He did test launches at Johnston Island, a small atoll about 750 miles southwest of the Hawaiian Islands. One of the things he mentioned in his interview was the problem of cooling rocket engines. The solution of running the liquid fuel through tubes surrounding the engine was first used by Robert Goddard, who invented the liquid fuel rocket in 1926.

At Vandenberg, we were attached to the Air Defense Command. Our outfit, the Tenth ADS [Air Defense Squadron] changed the Air Defense Command to the Aerospace Defense Command. We had one launch pad there where we did all of our training and all of our checkouts. We fueled our missiles, unfueled them. With our rocket engines, we put the igniter in the rocket engine, and did gimbal tests, and the whole nine yards for launch but we never launched there. At Johnston Island, where we went to for active duty, we had six crews. I was on Project 437-A crew, which was the first US Air Force space launch crew. We had two launch sites at Johnston Island. We would spend three months out there and six months back.

We used Thor missiles. The Thor had a rocket engine of RP-1 [refined petroleum] and oxygen, and they had 150,000 pounds of thrust. We did our launches at night. Believe it or not, after a launch you could walk to the launch mount, and it was too hot to get close to. It was so hot you could see through the metal, and see shadows of people walking across the backside of it. It took somewhat like twenty-four hours to cool down, to work around. And we had it refurbished in seventy-two hours, ready to launch again.

A lot of people don't realize, but the rocket engine would have melted from the heat if the rocket engine hadn't been made of tubes. The RP-1 went down a tube, down the outside of the metal of the rocket engine and back up to the injector plate, so the liquid fuel kept the rocket engine from melting.

One of the launch sites on Johnston Island was alpha contaminated. They were doing the atmospheric nuclear tests previously, and they had a detonation on the pad, sort of like a meltdown. With a coat of paint, you can cover up alpha contamination, but it will be there for many millions of years, since the half-life of alpha is 23,000 years.

Our job was to keep inspecting and keep painting, and check the one launch mount that was hot, alpha contaminated. We took a Geiger counter and checked it out. When we left that particular launch pad, they always checked us with the little Geiger counter. This one particular crew member checked out positive, every doggone time. His fingernails and his toenails would check out too. He had come from the mountains in Colorado where uranium was discovered, and it was natural in his body. He was radioactive. So after a while they just passed him on through each time. They would say, "You all cleaned up?" and he would say, "Yeah, I am fine." Then they would pass him on through, but the rest of us had to be checked out.

Wayne Mattson earned a Bachelor's degree from the Naval Academy early in his Air Force career (there was no Air Force Academy at that time). Later, through the Air Force Institute of Technology, he earned a Bachelor of Science degree in mechanical engineering. The course work for that degree was at the University of Wyoming. He studied jet engines, rockets, and space dynamics, and then did some study toward a Master's degree. During that time, he also taught a course in astronomy and celestial navigation. He was assigned to Holloman Air Force Base in 1969.

In January of 1964, I entered the Air Force Institute of Technology and earned a Bachelor of Science degree in mechanical engineering, aeronautical option. We had a prof there who was a genius as far as space dynamics went. What we perceived to be problems that would take weeks to solve, he could solve in fifteen minutes. The first thing we found out was that it was not going to be easy to put an orbiting laboratory or space battle platform up and shoot a missile back at Earth because of the dynamics involved. There's a lot of computations. You may have to fire a retrorocket at a certain point at a certain orientation in order to hit the desired target on Earth. Lasers are a different story, but when you're dealing with physical bodies, it's not all that easy.

We were conceiving of nuclear-tipped missiles. In the event the whistle blew and the bad guys got our missile sites or we couldn't get our bombers through, we would have a battle station up there. We could fire from up above. At that time, we did not worry too much about space defense. We were thinking in terms of, if there are two battle stations up there, one of them ours, and one of them theirs, they launch one, could we have an interceptor? How would we handle that? We would get into the mathematics involved in developing when, where, and why, how much thrust from your retrorockets to get in the tail chase position to shoot him down.

I worked at Edwards Air Force Base from 1965 to '69. The Society of Experimental Test Pilots had its meeting in Los Angeles during that time. I was on the F-111 test force and our chief test pilot, Bob Parsons, was down at the SETP meeting. Monday morning, the Los Angeles paper had a horrendous headline: "Mission to Venus Possible with F-111, Says Air Force's Chief Test Pilot." Of course, we got a big charge out of that. As soon as we got our hands on Bob Parsons, we said, "What in the devil did you say to the reporter?" He said, "This reporter asked me three questions. 'Where does space begin?' I generally consider space to begin at the point where you would earn an astronaut's rating if you were flying something like the X-15 or you were going up in one of these space capsules." And then they asked him a couple more innocuous questions on the F-111 – is it pressurized, can you possibly go into space? "Well, possibly, but it's got to be pressurized," he said. We had those things plastered all over his wall. "Mission to Venus Possible." We told him, "You're going to be the first Venus astronaut." The system program office at Wright-Patterson Air Force Base thought it was very hilarious, too, fortunately.

Another one of the projects that we had within the Missile Development Division was sounding rockets. When I first got to Holloman, they told me, "You're going to have to send a batch of your troops up to Fort Churchill, Canada, for what they call PCA, Polar Cap Absorption." During the PCA, they were on alert duty up there with Nike Hydacs on the rail and there was barium in the payload. They were waiting for a solar flare, and when the solar flare went off, then they started firing. I don't remember the number of rockets, but it was around twenty to thirty. They fired them off as fast as they could, and the whole purpose was to check radiation, ionization, and the solar winds by watching the barium cloud change color and change shape. So, from the time I got to Holloman in June of '69 until the end of August, I had two crews, one up there and one back, and they'd swap.

We also fired some Nike Hydacs with barium out of White Sands. We had a very tiny launch window, about a five-minute period where we had to get that missile up while it was dark down on the ground but it was still daylight up at altitude, where the barium cloud would disperse. We were going to do this for four days running.

These were evening launches, and this was late August or early September '69. There was no big problem getting the barium up there in that launch window, but the very first time we did it, it seems our public information officer at Holloman had made a big boo-boo. He forgot to pass the word out to the newspapers that we were about to do this. Once that cloud was dispersed at altitude, you could see it probably as far east as Biloxi, Mississippi, and all the way out to the Los Angeles area. We fired that first sounding rocket, and after the debris had settled down, we all went out and laid down on the ground and watched the barium cloud shift colors and shape, very interesting. The next day, we heard that Air Force bases and Army bases all over the southern part of the United States were besieged by phone calls. One of the best was called into Carswell Air Force Base and the gist of that was, "There is a giant man in the sky and he's running after the moon. I want you to do something about it."

I have talked to some of the Navy troops who fired the Aerobees, and one of their launches led to a tightening of the safety rules down at White Sands. The Aerobee is launched from a tower, and they differentially jacked the legs of the tower for their azimuth and range. One time, they did it backwards. They fired their Aerobee and scored a direct hit on the power substation that led up to Cloudcroft!

The next day, I shot a Nike Hydac sounding rocket off, and it went up. The winds were a lot stronger than was predicted, and I wound up dumping it off Range to the west. That prompted them to add the western extension on. They made arrangements with the ranchers out in that area, so that x-number of days each month they would call it the western extension, and the people would get paid if they wanted to go to town and spend the night in a motel and eat in town. Naturally, the wives were in favor of that. But most of the ranchers would just take the money. They thought, "The chances of hitting us are very, very slim, so we're going to stay here." As it is, I never put another one off Range.

But we had one launch one day, it was after a big rainstorm in the Tularosa Basin here, and a lot of the communication lines were wet. The lines from various radars were wet and malfunctioning, and so my launch was hold-hold. Then they decided that they better send a chopper out and check that western extension. A little while later, the chopper pilot comes up and says, "Hey! There's somebody down here!" "Well, does it look like he can evacuate or didn't he want to evacuate, or what?" He said, "He can't evacuate because the road from the ranch house to the highway has a bridge across the little creek and the bridge is gone." You could see where the bridge used to be. "You better land and ask him if he was going to stay there or if he wants to be evacuated by helicopter." A little while later, the pilot came back up on the radio again and says, "Well, I can't talk to this guy because he doesn't speak my language and I don't speak his. I think he's speaking Basque, and I tried using sign language to get him in the helicopter and his message came through loud and clear, he wants nothing to do with that crazy iron bird!" So, White Sands Range Safety screams, "Hold! Hold! Hold! You can't launch until we get permission from higher authority." So I picked up the phone, talked on it a little bit and a while later, the phone rang, "Yes, sir," I said. "Some colonel, he wants to talk to Range Safety." "This is Colonel ----, you're cleared to launch; you have authority." So the safety man says to continue with our launch. What he never knew was my phone call was to one of my sergeants who made that phone call so he could continue the launch and get it out of the way.

No one was injured. The payload did not land off Range or anything like that. If you followed everything by the book exactly, you would never get your test accomplished.

There was another program, called Project PEPP, conducted for NASA at White Sands Missile Range in 1966. That's PEPP, Planetary Entry Parachute Program, the development of the parachute program for the soft landing on Mars. We would fire it up to altitude and the PEPP had a rocket motor on it, and when the dynamic pressure exactly matched that of the Martian atmosphere, the parachute was deployed. We did not have videotapes at that time. We used 16-millimeter cameras. We had a camera in the nose and we had a camera in the tail. The camera in the tail would watch the action of the parachute. On the first batch of tests, the parachutes would shred and get torn to bits and pieces. But with each successive test, NASA was able to come up with and develop a better parachute. Eventually, a parachute was developed for the landing on Mars.

When it went up, the payload went up and arced over, so it was dragging the parachute behind while it was going up. As it arcs over, the parachute was just helping to lower it back to Earth. Naturally, when it gets down to the denser atmosphere, it was large enough that it would function as a good parachute, even though it was a ribbon chute. To me, the best pictures on the 16 mm film were not necessarily the parachute, but when they would arc over and you could see the curvature of the earth and the blackness of space. That was a very beautiful scene. Then you're coming down on White Sands from many miles up, and you can see the whole Range. Of course, the parachute is spinning round and round and you get dizzy, but it's a very good bunch of photographs.

The Physical Sciences Lab would sometimes develop the payloads for SAMSO [Space and Missile Systems Organization] and put them in the sounding rockets. We had a couple of amusing instances with that. We had some female Ph.D. from the Physical Sciences Lab who had developed a payload, and it was the first time a payload had ever been developed by a female. It didn't matter if she was educated or uneducated, she was a female. So there was going to be a lot of publicity associated with that. She came out to the launch site, naturally. She was the only one that knew how to arm the payload. Before you go out on the launch pad from the blockhouse, you pass a safety checkpoint. We always had a big, husky sergeant there to make sure safety was enforced. Well, she was all dressed up, big bouffant hairdo. The safety man handed her a hard hat, which she put on her bouffant hairdo gingerly. It wouldn't have done any good if anything had fallen on her, so before she went out to the pad, the sergeant said, "Just a minute, Ma'am." He put his hand on the helmet and pushed down on it and ruined her hairdo! He said, "Okay, now you're clear to go out to the pad and arm the payload."

The next time a female designed the payload and she came out to the launch site, she was about seven months pregnant. Not the same female, a different one. White Sands had a safety man out there, a roving safety observer. She starts heading out to the pad and a few minutes later, the White Sands man came in and says, "I just threw this pregnant female off the area!" I said, "You what?!" "I just got rid of this pregnant female. She was wandering around the area." I said, "Well, you just scrubbed the launch." "What do you mean, I scrubbed the launch?" "She was the payloader! Go get her back!"

Robert Turner had a nontechnical job in the Air Force, but he had opportunities to witness some dramatic missile launches at Vandenberg Air Force Base in Santa Barbara County, California.

I joined the Air Force in March 1966. My Air Force Specialty Code was Personnel. My first assignment after tech school was Vandenberg Air Force Base in California. Coming from California, I was only a couple of hours from home. When I got my orders to Vandenberg, I said, "Oh, missiles in space!" I've always been interested in airplanes and space. That was just my cup of tea, so I was really elated about that. I spent my first four years there.

I saw a number of launches there. There was a Minuteman launch. We were about a mile away. We were out there at three o'clock in the morning, eating our box lunch. They had a loudspeaker, where we could hear the countdown. Then you would see a flash of light, then you would hear a big roar. There was a cloud. I remember this visually. The rocket went through the cloud. As it came out, it was like fireworks. They were all symmetrical, these flakes, and this horrendous, earth-shaking roar. Of course, we saw it before we heard it because of the [speed of] sound. As it went through the cloud, they detonated it because it was going off course. It was horrendous. I mean, the roar just shook all of your bones.

It was about dawn by this time. As far as you could see, there were rings of fire, where a piece of propellant would fall and make a little ring. There was nothing you could burn out there. It was just desert. I mean, just little bushes. We would go out there with our rake and canteen. We weren't allowed to touch anything, so we walked out there with our rakes with nothing do. Then, as we were coming back, about nine o'clock, the other buses were coming out to continue the fire raking. There was nothing out there to burn. That was my first experience.

You always go to a fallback area and wait for anything that happened. I always used to volunteer to go out early in the morning and see a rocket launch. That was my cup of tea. Another one was the Scout launch. We went to the fallback area. It was foggy. You couldn't even see your hand in front of you. They launched this rocket, and the sound was kept in by the fog. As it went up, it got louder and louder and louder, and it just seemed like the rocket was coming right down on your head. It was kind of scary. But it was successful.

Another one, early in the morning, before I got up, I heard this massive explosion. I said, "Well, to the parade ground." Sure enough, two minutes later, "Get into your fatigues and report to the parade ground." An Atlas had gone, they had three reentry vehicles. They usually were tests, so they weren't actually bombs, but there were three payloads. It had blown up, and we went out. They lined us up in the top of a ravine. We went down in the ravine in a big line. Down in the bottom, there was manzanita brush, which you can't even cut with an axe. We had to turn around and come back. Up and down this hill we went. We were out there all day long. This was about dawn, and we were out there until five o'clock in the afternoon, up and down the hill. You could see pieces of helium tanks and plumbing, skin of the rocket, stuff around. They told us one of them weighed about 3,000 pounds. We were looking for three re-entry vehicles. To this day, I don't know that they ever found them. Vandenberg was right on the ocean, so they might have gone into the water.

I remember one other episode. We were leaving to go home on the weekend, about Friday afternoon just at dusk. They launched a rocket, and the sun was just right. The sun had just gone down, but the glint was still there. As it went up, they had exploded it again. You saw the flame, then this swirling gold, blue, green, and red, like a modern painting, or a finger painting. It just kind of dissolved and turned. It was really a fantastic sight. The sun had to be just right, and it caught it so it was really a spectacular view.

Gerald Garing worked at the Physical Science Laboratory from 1952 until 1958. Then he went to work at Holloman Air Force Base. There was a financial incentive for people to work at Holloman. Garing recalled that the "isolated area bonus" amounted to $150 extra each month. In this excerpt, he spoke about working on the Viking program. Two unmanned Viking spacecraft were launched from Cape Canaveral, Florida, in 1975 and soft-landed on Mars in 1976.

We were trying to launch the Viking. We set up our instrumentation, and the mission had been aborted. Then they dumped all of the liquid oxygen right there on the ground. It came running out, and there was a big lake of it. It was just going along, 170 degrees below zero, eating the weeds up and everything. All of us were stepping back. We would throw stuff in there. They dumped it right on the ground! It was a pale blue, and it was evaporating real fast, but there was so much of it. I am thinking of OSHA and how things are run today. There wasn't a single person telling us "Don't do that!"

Von Braun would come out for certain Viking launches. Then we would go to the Bachelor Officer Quarters afterward for a get-together. That is where I first saw von Braun, in the same room with him.

Duke Gildenberg was involved in many aspects of early space research (see Chapter 4). In this excerpt, he first talked about launching a small missile from a balloon. The missile was used as a practice target for an F-104 fighter airplane. Then he talked about working on the Viking program. The flights he mentioned at Roswell carried an aeroshell that was designed to test parachutes that would bring the Viking landers to a soft landing on Mars. The balloon flights carried the Planetary Entry Parachute Program (PEPP) aeroshell aloft and released it. Several rocket engines on the aeroshell ignited and carried the vehicle to 160,000 feet altitude. This allowed testing of the parachute system in an atmosphere as thin as that of Mars. The tests were conducted in the summer of 1966.

We fired a little missile once from a short track on board a balloon. It was critical timing, because the F-104 would be making his approach from El Paso, and he was here in a minute or two. Then he would dive, and we had to fire our missile from the balloon. Then he would fire his Sidewinder or Falcon at it, and it would pitch up. We would have things all over the sky! The F-104 going this way, the [target] missile going this way, his missile going that way, and the balloon going up. I can remember one day, I had been working all night, so I stopped off to have a hamburger in Alamogordo. They were flying the third mission of the day. I had already worked the first two. I looked out over the pavement. People were walking by, and right over their heads, I can see this whole thing happening. The trail from the F-104, two missiles in the air, and the balloon going up, and nobody saw a thing. They were just walking. That thing was all over the sky, and nobody was even noticing it.

We did the big one, Viking. It was a successful series, launched from Roswell. That is where people saw that gondola. They had it behind closed doors just cracked open. Occasionally, somebody going by would peek in the hangar and see this [PEPP] saucer thing in there. That model they have in a couple of museums is fashioned after that. It wasn't from our release or anything; it was from seeing it back then inside the hangar doors.

When it was all over, we had three out of four successful shots. They came around with a big scroll, and you signed your name on the scroll. Then they did a microfilm picture of that, and they put that aboard the Viking Lander. So our names are on Mars. It was a little thing, but everybody got involved with it. "Hey that was great! The family name is on Mars, unbelievable!" That was great.

PEPP was the predecessor for planetary landings in general. We did that first. In fact, that was the one with the saucer-shape that wound up in the museum at White Sands. People took pictures, and they had articles in the paper, facetious articles, saying, "Hey, this is one of the flying saucers." That was exactly what it looked like.

Doyle Piland enlisted in the Army in 1957 and received extensive training in ordnance (weapons and ammunition) and missiles. The Sprint missile he mentioned in this excerpt was a two-stage, solid fuel, anti-ballistic missile.

I first came to White Sands Missile Range in 1968, and I went on to become the noncommissioned officer in charge of the White Sands Test Division. My offices were at LC-38 [launch complex 38], which was the base operation for the Sprint missile testing. The Sprint missile was assembled in a revetted [heavily protected walls] building that was made for explosive handling. It was then loaded into the cell at the Sprint site and was fired. There were a couple of things that happened, not necessarily during the time frame that I was there, but shortly before.

There was one incident where the missile was launched, came out, did a complete loop (540 degrees), and went straight into the ground. The Sprint had what they referred to as thrust-vector control valves that essentially injected a liquid into the plume of the missile that caused the thrust vector to change, and that made it do its initial dive and turn toward the target. For some reason, those valves stuck open and it just continued its initial turn.

One of the amusing things about it, we had a captain there whose last name was Walters. I don't remember his first name. He was in charge of access control, keeping people away from that area who weren't supposed to be there. He was located at a road intersection approximately a half-mile to three-quarters mile from the launch site. When describing this incident that took place, he said that the best way to describe it was, he

PEPP aeroshells were only 15 feet in diameter, so any "little green men" riding inside would have to be very small indeed. Eight of the aeroshells were flown between 1966 and 1972. None of them generated UFO reports. Courtesy of US Air Force.

was standing there looking, the missile came from 27 feet out of the ground, did a 540 degree maneuver, went back into the ground 27 feet before his mind had even registered that it had come out. That is sort of an indication of what the Sprint was. The Sprint was, at the time and may still be, the fastest accelerating object ever built by man.

The Sprint was primarily designed to intercept incoming warheads from intercontinental ballistic missiles. They used to say, although the exact numbers are classified, from the time it is launched, it has gone a mile in less than a heartbeat. It was just short of a bullet. Of course, like they do now at White Sands, there are always optics that track the missile. You can look at the film from one of the missile shots and see the missile start turning white hot from the air friction, starting at the nose and just working its way all the way back to the end of the cone. Then as the second stage motor burned out and it started slowing down, you could see it sort of turn an ash gray from the back of that cone all the way up to the nose section, as it slowed down and cooled down.

During the 1990s, I worked in electromagnetic radiation effects testing. The military, in a combat situation, will have all sorts of radio frequency emissions from everything, from the little handheld or backpack radio, to radios that are mounted on tanks, to communications vans that communicate large amounts of data to radars. Those kind of things scattered all over the battlefield. So, you have to test a very wide range of signals and modulations to make sure that this thing is immune to all of these signals. Or if it is not immune to them, at least that it is well documented and those who might be operating it or might be responsible for deployment of the entire theater understand that certain kind of equipment is not compatible with something else.

We also did testing for private industry. Specifically, we did testing for Honda, Volvo, and Chrysler Corporation. The primary function that we were doing was to make sure that automobiles did not have problems with electromagnetic radiation from any source. An example of that would be, and one of the prime things that was tested for, was to make sure that the circuitry that activated airbags would not activate the airbag as you drove by your local radio station. Or if you were driving along beside a semi-truck and he keyed the mike for his CB radio, that your airbag didn't blow up in your face, things like that. Of course, later on, more and more of an automobile was controlled by computers, and therefore, they needed to test those to make sure that the RF [radio frequency] signal from somewhere doesn't interfere with the operation of some other parts of your car as well.

Ray Martinez was born in Germany. His father was in the US Army and was stationed in Germany at the end of World War II. His mother was German. When the family moved to the United States, Martinez was eight years old and he began to learn English. He served in the Air Force in Vietnam, in munitions. His final assignment in the service was at Holloman Air Force Base. His next move was to a very different climate.

I came to Holloman in early '86 and retired from the Air Force here in January '91. I had more or less made my mind up to retire from the Air Force, and I was looking around for employment. There was an ad in the Alamogordo paper, looking for the launch officer for the Poker Flat Research Range in Alaska. I read through it and I thought, gee, this is exactly what I had been doing in the Air Force, with one exception. There, I would get to fire the rocket, whereas in the Air Force, I put them together but young officers always had the fun of firing them.

Poker Flat first started in about 1968 by some researchers from the Geophysical Institute at the University of Alaska. They wanted to shoot some rockets, and it kind of started out like a college rocket club. Not with students, but scientists and researchers, just getting together to have a good time and to do something. They got the Army involved. The Army at that time had a network of meteorological rocket launching sites, trying to develop statistics for wind patterns throughout all of the Americas. This went all the way from Alaska into Canada, a lot of the northern states, even in Michigan. They were flying Loki and Arcas missiles at that time. That kind of provided the initial seed money for the Range.

We have done a few launches for the Canadian Space Agency, where they have been studying transmission of electronic signals in space. Our highest apogee so far is 920 miles, so we are definitely in space. The space shuttle is only 250 to 275 miles, actually not very high. We've done a series of launches where we actually had vacuum containers on board the payload, and at predetermined altitudes, trapdoors would open up and suck in whatever gases that were surrounding the payload at that time. Then the doors would close. We would recover the payload for study in a laboratory, and they could determine what pollutants were at that altitude. We've done studies into ozone depletion, which is more prevalent at the South Pole, but also happens at the North Pole.

The payloads that we recover, normally, are the smaller ones, the ones that don't go terribly high. We have a locater beacon that's activated when the parachute opens. We locate them with a directional antenna with a fixed wing [airplane]. Then we go out by helicopter and either recover the whole payload or part of it. The bigger rockets like the Black Brant 12s impact out on the Arctic ice shelf and they come in ballistically. They are just a write-off.

Alaska is kind of unique, because of the majority of the state does not belong to the state or to the people. The majority of the land belongs to the Bureau of Land Management, to Native [American] corporations, to other federal entities. We have agreements with everyone whose property we fly over. The agreement states that if a rocket should land and they object to it, all they have to do is notify us and we will remove it. It's primarily the rocket motors or the stages. In the ten years I have been there, it has only happened a couple times, that anyone has objected. It would be cost prohibitive to pick them all up. The great thing about it is that they are basically inert. All you've got left in it is a little bit of liner, the seals that are in there, and then it is just a piece of steel. But on the Arctic slope, things just don't rust like they do in, say, Florida.

I think what the sounding rockets provided was a very invaluable tool for exploring a part of our world that is not accessible in any other manner. Possibly, when the hypersonic aircraft come on line, we will be able to access that portion of our outer atmosphere and inner space with aircraft. But there is nothing currently that can access that. I think the better we can understand this world we live in, the better we can take care of it.

After Clyde Wilson retired from the Army, he worked as a civilian helicopter pilot at White Sands Missile Range, where he had some interesting flight experiences.

Before I got here [WSMR], they fired a Pershing [missile] from Fort Bliss [Army post headquartered in El Paso, Texas]. It was supposed to land up in the northern part of the Range. But it allowed that it should keep going, and it landed in Creede, Colorado, which is way up in the mountains up there. They went up, the people searched and they searched. Nobody ever found it. Finally, they gave up.

Eighteen years after the firing, a hunter was up there, elk hunting, and he walked up on it. There it is on the side of a mountain, about a 45 degree slope at 10,000-foot elevation. Of course, the government, as soon as they hear about it, they have to get all excited. There is a missile lying somewhere. So they sent a bunch of EOD [Explosive Ordnance Disposal] people up there. The EOD people couldn't do anything about it, so they called us and asked for some help. We go charging up there, just like we know what we are doing, and recover it. It was quite an unusual deal. We had to use a 110-foot sling from our helicopter, and we are sitting up there at about 10,000 feet. The trees are about 100 feet tall. We have to get the EOD people as close to it as possible. Then they have to walk down from the ridge top, all the way down to where it is. Then we fly over, and we drop out some sling equipment to them. And they try to rig it up. Then we let the net out from the helicopter, and it comes down to them. They rig it up. And then we try to ease it on out without dropping it. We get up over the trees. Then we fly it over to the airport we are working out of at Creede. They had to take it apart on the ground. So we got several good, hairy missions out of it, I'll tell you. We did this in November, I think, so it was starting to get cold. This must have been about 1985.

We got out there while it was still cold, and recovered it. Interesting thing about that, the people of Creede started hearing about it. So the airport kind of filled up with all the people coming out and watching us do our thing. Some of them remembered they had helped on searching for it before. So they finally got the word it was all back. Here is this missile, and all these pieces laying down there. One of the local town's ladies said, "I've got a good idea. Let's put it back together, and then fire it back to them in the right place." That was her idea on how they ought to get even with White Sands. But the EOD people went out and rented a U-Haul truck, and they were putting it all in the back of that truck, and driving it back to White Sands and turning it in, eighteen years after that firing.

When I first got there, the Air Force had a crash up northeast of the Range, northeast of Carrizozo on the other side of the mountains. We went up there on a Sunday, bad weather. We went up there with an H-34 [helicopter] completely full of Air Force investigators that had to go up to investigate it. While we were up there and they were on the ground investigating it, we went out and we found a chute. Somebody had ejected and walked out. We searched around a while. Right before it was getting dark, it was time to leave. They had done all their investigation, and we were getting ready to come out. We had been up there all day. We were on Highway 380, going across the pass at Carrizozo and then over the open terrain on back into Holloman. As we were getting to that place where Highway 380 comes up to a crest before it goes downhill, the weather was not cooperating. It was socked in, in other words.

This guy I am with was on the controls at the time. I knew he couldn't fly on into the fog bank. So he just plopped it down on the highway, just short of the fog bank. Here we are, sitting on Highway 380 with about nine Air Force guys and all of their equipment. Everyone had on all the clothes they had, thank goodness, because it was so cold. We started taxiing. It was uphill for a while, then it was downhill. Here we are, taxiing on Highway 380, and here are these cars passing us, and boy, were they bugged eyed when they would come up! Somebody found and stopped a cop somewhere. Finally the cops came. The cops tried to guide us along so somebody wouldn't run into us. We did that for, it seemed like hours before we got down to the other side. Here we are, taxiing along, and we would come up to these wires. You couldn't taxi through the wires. So we would have to make everybody get off the aircraft. And we would pick it up to right above the wires, and then set it back down and let them get back in. Then we taxied. And we finally got it out of the [bad weather] and moved it back into Holloman after dark, on our red light for fuel. All of those people had been up there and freezing cold all day. When we got back there to land, the tower said the Security Police wanted to know if we could take them back up there to take some food to the people that were up there. And we said no, we were through for the night. That was one of my hairy missions out there.

[In July 1970, the Air Force fired an Athena missile out of Green River, Utah. It went down in Mexico with a radioactive payload.] It was so far down there it didn't work out for the helicopters. The contract people flew a twin-engine, fixed-wing down there every day, and back. We had to buy all of that dirt and everything around it at the place. We had to send a train down there, and we had to buy all of that stuff and put it on that train and come back. Our director was train commander. They brought all of that back to the United States.

Chapter 9
NASA Gets US off the Ground

Many people consider the Mercury, Gemini, and Apollo programs to have been the glory days of the United States' manned space program. The series culminated with twelve men exploring the surface of the Moon on six flights. The first phase, Project Mercury, began in late 1958 as a program to send a man into Earth orbit and return him safely.

The American and Soviet Union manned space programs were set against the backdrop of the Cold War, an emotional and scientific conflict rather than a shooting war. The two nations were engaged in a battle to demonstrate the superiority of their respective forms of government – capitalism versus communism. Some of the oral history interviewees referred to the United States' adversary as Russia rather than the Soviet Union. Because Russia was the largest country in the Soviet Union (officially, the Union of Soviet Socialist Republics, USSR) and the republic in which the Soviet capitol was located, some people used "Russians" and "Soviets" interchangeably.

The Soviet Union sprinted to an early lead in the space race, launching the first artificial satellite in October 1957. The United States accomplished a similar feat four months later, but with a satellite weighing only one-sixth as much. With objectives of reaching the Moon, both countries scored points in 1959. An unmanned American spacecraft achieved the first lunar flyby in March. The following month, the first seven American astronauts were introduced to the world. In September, an unmanned Soviet spacecraft accomplished a hard landing on the Moon. The following month, the Soviets flew an unmanned craft around the Moon and took the first photographs of its far side.

At the end of January 1961, a US rocket carried a chimpanzee named

Ham on a suborbital flight as a rehearsal for sending the first man into space. Alan Shepard took a successful suborbital flight on May 2, 1961. However, that accomplishment was overshadowed by a flight two weeks earlier that carried Soviet cosmonaut Yuri Gagarin on a one-orbit mission.

The turning point in the space race came on May 25, 1961, when President John F. Kennedy spoke to a joint session of Congress. His remarks included the following:

> [I]f we are to win the battle that is now going on around the world between freedom and tyranny, the dramatic achievements in space which occurred in recent weeks should have made clear to us all, as did the Sputnik in 1957, the impact of this adventure on the minds of men everywhere, who are attempting to make a determination of which road they should take.... Recognizing the head start obtained by the Soviets with their large rocket engines, which gives them many months of lead time, and recognizing the likelihood that they will exploit this lead for some time to come in still more impressive successes, we nevertheless are required to make new efforts on our own. For while we cannot guarantee that we shall one day be first, we can guarantee that any failure to make this effort will make us last.... I believe that this nation should commit itself to achieving the goal, before this decade is out, of landing a man on the Moon and returning him safely to the Earth.

Later that year, Project Mercury was extended to establish a basis for Projects Gemini and Apollo. Between May 1961 and May 1963, six men performed single-occupant, orbital Mercury missions. Project Gemini, using a two-man capsule, conducted ten orbital missions between March 1965 and November 1966. Those flights included essential preparations for upcoming lunar missions such as docking with other orbiting spacecraft and performing spacewalks. The Soviet space program continued as well, achieving milestones such as sending the first woman to Earth orbit and doing the first spacewalk (two and a half months before the Americans performed that maneuver).

In 1967, both countries suffered fatalities in their space programs.

On January 27, American astronauts Gus Grissom, Ed White, and Roger Chaffee died in a fire inside an Apollo capsule during a routine, nonflight equipment and procedure test session on Launch Pad 34 at Cape Canaveral, Florida. That event, which was only a rehearsal for the first manned Apollo flight, was subsequently designated Apollo 1 in honor of the three astronauts who died in it. On April 23, the parachute system failed at the end of a Soviet orbital mission, and the spacecraft's sole occupant, Vladimir Komarov, died when the spacecraft crashed into the ground.

America's manned Apollo missions began with the October 1968 flight of Apollo 7, in which the three-man crew orbited the Earth. With the craft successfully tested, Apollo 8 took three astronauts on a lunar orbit mission two months later. In Apollo 9, the lunar module, which would later be used to carry two men from lunar orbit to the Moon's surface and return them to rejoin the command/service module that continued in lunar orbit, was tested in Earth orbit. Apollo 10 carried three men to the Moon, where the lunar module separated from the command/service module and descended part of the way to the surface before returning and redocking for the return to Earth.

Between July 1969 and December 1972, Apollo missions 11, 12, 14, 15, 16, and 17 successfully took two-person teams to explore the lunar surface while the third crew member orbited the Moon. No human being has stepped out onto the Moon except the twelve Apollo astronauts on those six missions. In April 1970, the Apollo 13 was intended for a similar mission, but an onboard explosion damaged the spacecraft while it was still on its way toward the Moon. Hard work, dedication, and ingenuity on the part of the ground crew as well as the onboard crew took the craft onward, swung it around the Moon, and brought it back for a safe landing on Earth.

The interview excerpts in this chapter offer insider descriptions of the triumphs and terrors of the Mercury, Gemini, and Apollo programs.

Dr. William Douglas, an Air Force flight surgeon (a physician who is primarily responsible for the medical evaluation, certification, and treatment of military aviation personnel) served as personal physician to the original seven Project Mercury astronauts. He not only oversaw their health, but he participated with them in many of their physical examinations and training exercises. He was inducted into the International Space Hall of Fame in 1992.

As Dr. Douglas mentioned in his interview, the physical examinations of the original US astronaut candidates took place at the Lovelace Clinic in Albuquerque. Dr. William R. "Randy" Lovelace, nephew of the Lovelace Clinic's founder, joined his uncle at the clinic in 1947. They formed the Lovelace Medical Foundation, headed by the younger doctor. In 1958, Randy Lovelace was appointed head of NASA's Special Advisory Committee on Life Science, and in 1959, he and his staff administered week-long medical exams to thirty-one astronaut candidates, from which the Mercury Seven were selected. Dr. Lovelace, a long-time researcher in aviation and aerospace medicine, was inducted into the International Space Hall of Fame in 1976.

Dr. Douglas's oral history interview took place in 1990. In it, he described many of his experiences interacting with the Mercury astronauts.

I recall that before the Mercury crews had been selected, there was a short article in *Science*, the magazine of the American Association for the Advancement of Science. I think there had been a meeting someplace, and a group of psychologists had either attended it or maybe it was a psychological meeting. In any event, the reporter from *Science* interviewed some of these people and asked them what kind of person we should select for the Mercury program. One of the people interviewed said that they thought we should select Buddhist monks, being that they could put themselves in a trance-like state and not be disturbed by this terrible environment into which they were being put. Another person interviewed suggested that we select Eskimos because they had certain advantages. They were small in stature, light in weight, they lived up there in the Arctic, where they weren't too concerned about day or night because half of their time was spent without seeing the sun go down and then not seeing the sun come up. So they should not be disturbed by this cycle of the sun coming up and down every forty-five minutes. It would have been a terrible thing if we had selected people like that. There's nothing wrong with being a Buddhist or an Eskimo, but we wanted people that were trained observers, and that were scientifically oriented and could make good observations and report them back. It would have proved very little if we had sent an Eskimo up into space or a Buddhist monk who had gone into a trance. That would have not proved very much. We wanted to make sure that a functioning Western person could do this and make these observations, and so we did.

After the Sputnik crisis and the great advance of Russia into space, the country was stimulated to put an American into space, and there was a big debate as to whether the Army or the Navy or the Air Force or who should be in charge of this. The "powers that be" in the government decided that it should be a civilian effort, a peaceful, scientific, civilian, space program. They elected to give the responsibility of this to the old NACA, National Advisory Committee for Astronautics, and they changed the name of that to the National Aeronautics and Space Administration. Well, the NACA, a wonderful organization, contributed greatly to the advancement of aviation in the country. But they had no medical people at all. The life sciences were not part of their charter to investigate or to study at all.

So, when it came time to establish what was called Project Mercury, they gave this to the newly-established NASA and called it the Space Task Group. A couple hundred people were brought in from the various NACA centers to be part of this Space Task Group at Langley Air Force Base in Virginia. They needed some medical people. They went to the Department of Defense for that, and they had the first flight surgeon assigned to the space program, Dr. Stanley C. White. Stanley had been working hard, working a great deal with the Aero Med Lab at Wright-Patterson. He was familiar with the Air Force's "Man in Space Soonest" program, which I think they called MISS. [MISS was a pre-NASA, US Air Force project to put a man in space before the Soviet Union did.] They had a capsule, a mock-up of it, ready to go. Our government decided that this should be a civilian effort, so the Air Force's MISS program was abandoned and Stanley was brought down to the Space Task Group to be the initial physician there. There was also a Captain Bill Augerson from the Army (Army Flight surgeon), putting monkeys into space. We had a Navy psychologist, Dr. Bob Voas, who was assigned to develop the training program. He was a psychologist particularly skilled at developing training programs, so he was assigned to develop the training program for the Mercury astronauts.

They needed a flight surgeon, a personal physician for this group of astronauts, and I was fortunate to have been chosen to be that person. That was the nucleus of the medical staff of the Space Task Group. We had another man, Dr. Jim Henry, who was an Air Force physician. He was a very, very wonderful scientist. He'd been up at Wright-Patterson also, and he was to operate the scientific aspects of this thing. Jim Henry and a wonderful man named Otto Gauer from Germany had made a real great and wonderful contribution to cardiovascular physiology. They developed the Gauer-Henry reflex, a well-known, basic principle of cardiovascular physiology. So we had some really great talent in those days.

I did not do everything to the same degree that those astronauts did. They trained on the centrifuge and so did I, but not to the same g-levels and the hours and hours on the centrifuge. I felt this was part of my job, to know what they experienced. So I rode the centrifuge, not with the idea of ever going into space, but with the idea of knowing what they experienced and what it's like to ride on a centrifuge, so I could talk to them and discuss these things with them. They would know what I was talking about, too, and be convinced. So I did everything that they did. Came out to Lovelace Clinic and went through most of the physical examination procedures. My job was to sort of be with them when they were doing this. When they were going to have something done, if time permitted, I would have that done to me, too.

We had a wonderful, wonderful lady named Rita Rapp. Rita started out as, what we jokingly referred to as the "vampire" because when the body is stressed to a certain point, hormones from the adrenal glands and the metabolic products of these can be found in the urine and also the hormones themselves in the by-products of the blood. So Rita followed us all around and drew blood from us after having ridden the centrifuge, the heat chambers, and all these other things. One time, she was drawing baseline blood samples, hopefully that these things would all be when we were least stressed. When mine came back, the catecholamines, which was one of the things she was hunting, were very, very high. she said, "What in the world happened to you that caused you to show this great stress response?" I thought back on it and I had been interviewed by the local press, but I wasn't used to it then. It was stressful to me. So that's why my stress hormones were high at that time.

They went through stress testing procedures at Wright-Patterson, and I went up there and went through many of the tests. I went through a water survival course. They put a boilerplate [a nonfunctional mockup used for training] of the Mercury capsule out in the Gulf of Mexico, and put a person in it. You were expected to climb out of that thing in the open sea and get into a life raft, and I did that. We had desert survival. I really didn't survive on the desert as those men did, because my job was to make sure that they weren't put into physiologically hazardous situations. But I went out with them and did many of the things that they did, just not to the degree that they did.

There was one awful thing, the MASTF, Multi-Axis Spin Test Facility, which was at Lewis Lab [now Glenn Research Center, in Cleveland, Ohio], I think. It was a little, cage-like affair that you could put a subject in, and that cage was in another cage, which was in another cage, which was in another cage. And all these cages could be rotated on their own axis simultaneously. They would put a subject in there, and I can't remember if we asked him to close his eyes or to do something, and we had instruments in it also that gave rates of spin in the three axes – yaw, pitch and roll. All of these are operating simultaneously. This thing was spun up with compressed air. The subject was given a sidearm controller, a stick much like you would have in the Mercury spacecraft. After it was all spun up, he was to open up his eyes or something, I don't remember what now, and zero-out, null-out all these axes until he was stationery, which he could do by using compressed air jets by using his controller. The purpose of all this was to determine if he could, in fact, do all of this in the event that a malfunction caused the Mercury capsule to tumble in space. Whether it was best if he would null the pitch first, then the yaw, or in what order they should be done. I don't remember what order they came up with, if any. But that was a very stressful experience, and they all did it. I did it and it was kind of fun, but these things were big toys for big boys, I guess. We enjoyed all of these things. It was helpful, exciting, and lots of fun.

I think the fellows really thought I was in competition with them at the beginning. It was only after about six months of that association together that it became abundantly clear to them that I had no desire to be a competitor and I was not the stalwart figure that they were. I don't know that I could have been an astronaut. One of them came to me later and said that he had suspected that I was a competitor and treated me as such. He apologized for it, and he realized that it was not the case at all. We were very, very close friends with all seven of them, and I have remained so. Except for Gus Grissom [who died in the Apollo 1 fire in 1967], of course, and we still get together.

We had the hardest time convincing great scientists – in fact, the President's scientific advisory board – that we could do this job, that they were not going to kill people. The mere exposure to the space environment was felt by some to be a life-threatening thing. It was due to ignorance. People did not know about these things. We were frequently asked if there were ever any doubts. "Were you afraid that these people were going to have something happen to them?" Fear of the unknown. Well, there wasn't any fear of the unknown, we felt that it was known.

Astronauts, at least those men, had been quick to point out that zero-g was just one spot on the continuum from positive g's to negative g's. They had experienced all of these, including zero-g, flying their fighter airplanes. They had experienced zero-g many times, and it wasn't anything unusual or frightening.

I recall in the Wright-Patterson zero-g, parabolic airplane experience that John Glenn was seated against the wall, and I sat cross-legged in front of him. We both wanted to see whether Newton was right, I guess. I wanted to sit there and push off and float right in front of John and then start rotating my arm in front of me to see if I would rotate the counter-direction. And I was rotating, spinning my arm around, and as far as I could tell, I wasn't making any progress. So I stopped, and John said, "Go ahead, Bill, you're doing fine, keep at it." Then the warning horn sounded because we had to pull out of the dive. He grabbed me and sat me down, and it was only then that I realized that I was actually upside down facing John. We were chin-to-forehead, so to speak, and I didn't know it. It was a perfectly normal and natural thing for me. I bring out this story to point out that this zero-g thing is no problem. Everybody knows it now, but in those days, it was felt to be a great unknown. To me, it was a pleasant, wonderful experience. You could push off from one end of the airplane and fly down to the other end, turn around and fly back. A marvelous experience, wonderful experience!

The acceleration [force during launch] you got was thought to be

An airplane traveling a series of vertical parabolic arcs can produce up to 45 seconds of zero-g at the top of each arc, allowing astronauts to practice working under that condition. Here, Apollo 14 astronauts Alan Shepard (front) and Edgar Mitchell (rear) are using parabolic airplane techniques to practice moving equipment under the condition of the Moon's gravity, which is one-sixth that of the Earth's. Courtesy of NMMSH.

more than a man could stand, and one of the committees that came to us wanted us to call a halt to the whole program and then put scores and maybe tens of scores of chimpanzees on the Navy centrifuge. And spin them until they killed them and autopsied them and made sure that there wasn't anything wrong before they would give their blessing on us to proceed. Well, Stan White is a wonderful speaker, and he got up in front of this great committee. I can see him now, walking up in front of them and haranguing them to the effect that this is no great problem. He finally convinced them that it was okay. They had come with a preconceived notion that it was a terrible thing to do with people, and it wasn't at all. Those of us that were astronauts or worked intimately with them, the engineers, all of us, had no great qualms.

Of course, we were concerned. Well, you put a darned Atlas up there that had been blowing up on the pad time after time after time, and that concerned us as to whether the darned thing would blow up on the pad or not. But as far as actual acceleration of launch, the experience of being in zero-g for three orbits, the return and the accelerator forces that come with reentry, all of these things were literally of no concern to us at all. None. Zero. It just didn't bother us in the very least.

Bill Augerson and I accompanied the astronauts up to the Navy's centrifuge in Warminster, Pennsylvania. Let's give credit where credit is due, for heaven's sake. Bill said, "You know, what if something does happen on reentry and for some reason the astronaut does not survive?" All of these naysayers will first ask, "Did you do anything to determine the effects of acceleration on the human body following a period of weightlessness?" We decided that we had to do something, so we got a big tank of water, a huge bathtub, and put a subject in the bathtub and let that subject stay in the bathtub for the equivalent of three orbits. Then we gently moved him over to the centrifuge and spun him through the reentry accelerations. Nothing happened to him. He survived with no problem at all. I was the subject on that one. It was a thing we felt we had to do. We never wrote it up, and we never reported it. It was a great wonderful experiment of one subject which didn't mean a thing, but if anything had happened, we could say, yeah, we thought about it and here's the proof that we thought about it.

It wasn't courage that we did that. It was cowardice more than anything else, because we didn't want to get blamed for something. The commander of the lab wanted to be the subject, Bill Augerson wanted to be the subject, I wanted to be the subject. We really had fierce arguments as to who should be the subject. We finally ended up flipping a coin, and I lucked out and won it.

Exposure to heat, noise, vibration, and the like was mostly part of the selection program done at Aero Med Lab at Wright-Patterson. We did take them down to Redstone Arsenal, as it was known then, at Huntsville. Von Braun ignited some sort of a rocket engine, and we all walked around the supporting structure of the gantry while it was firing, to give them some experience with that, to let them know that all the noise was not that horrendous. And it wasn't. From that time on, we didn't do any training from the standpoint of noise and vibration. We had done it as part of the selection process.

We really didn't know what we were testing for. We just gave these men almost every kind of exposure to every kind of stress that one could think of that was not clearly hazardous to their life or health, just to see if there were any of them that broke down under this stress. I don't think any of them did, even the ones that were not selected.

For astronomy training, we took them to the Morehead Planetarium in North Carolina several times. The people at the Moorhead Planetarium built a little mock-up of the capsule. At least, it had the window in it with the proper dimensions of the window and had a couch in it. A view that they got through the window of this mock-up was the equivalent of what the astronaut would see through the window of his spacecraft. We couldn't move the window around as you could the spacecraft, but you could move the universe around by using the planetarium. The purpose of that was two-fold: he could recognize constellations through the window, and if he had a map of some sort that told him what the sky was like at that time of the year, he could know where he was. This was, of course, in case all else failed, and he wouldn't have any other way of knowing where he was. They did want him to be able to do some navigation, if you will, through that little window. My part was not very profound, except that I was to go with them, and when they weren't using the planetarium, I did. I had the greatest time in the world, turning that projector around. It gave you a great feeling of power to be able to manipulate the whole universe! It was a lot of fun.

[In February of 1960, a representative from the Army Ballistic Missile Agency proposed a training project in which the Mercury astronauts would experience controlled noise and vibration inside a simulated

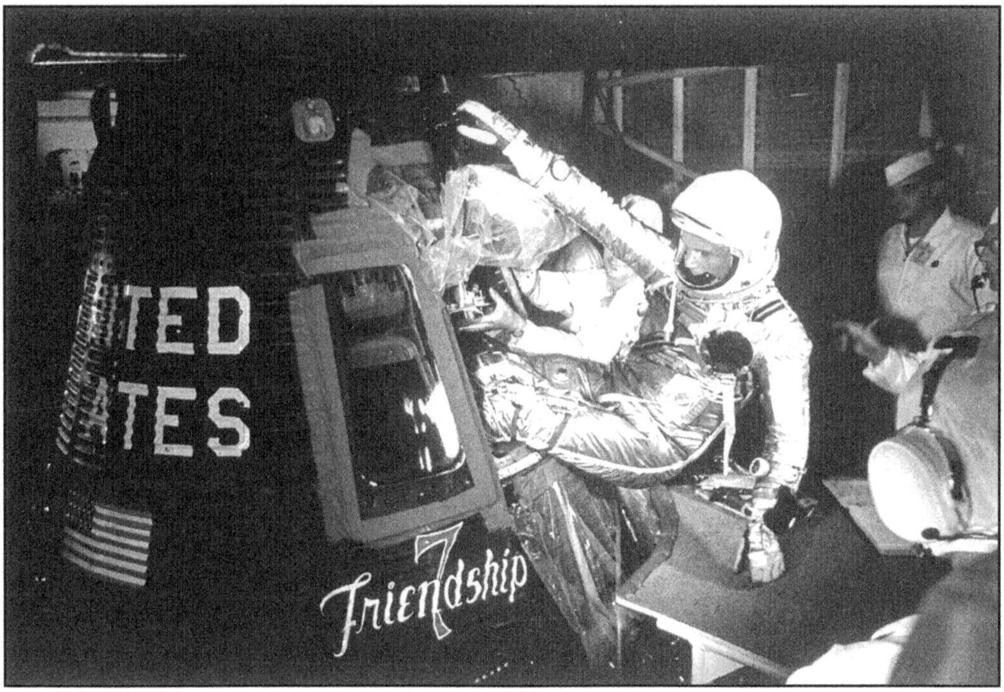

John Glenn's entry into the Mercury capsule illustrates how tight the fit was in the small, one-person spacecraft. Once inside, the astronaut was strapped to a form-fitted seat for the duration of the mission. Courtesy of NASA.

Mercury capsule mounted above a Jupiter engine being static-fired, and Dr. Douglas strongly objected to it.] I guess I thought that it was an unnecessary hazard. Rocket motors do fail, and the things do explode at test firing. We had already shown that the astronauts clearly responded properly to noise and vibration. I think as a result of this suggestion, we did take a mock-up of the Mercury capsule and put it next to one of the big wind tunnels that was called "the blow-down" tunnel at the NASA Langley Research Lab [in Hampton, Virginia]. The blow-down tunnel was very, very noisy. We put, I think, Scott Carpenter right next to this tunnel as it was blowing down and making a great noise and, of course, he survived it without any great problem at all. So there just didn't seem to be any purpose in putting a capsule on top of a Jupiter and firing it. It served no purpose. It was a hazard we could avoid.

There wasn't any psychological aspect of the training. We didn't try to change their psychology. Bob Voas was our psychologist there, but he didn't give psychotherapy, that sort of thing. He was there to develop, and put down on paper, the training program itself, which included the centrifuge and all that. We didn't feel the need for any psychological training at all. These men were chosen because they had been used to hazards and dangers, and they were brave men to begin with. But they weren't "derring-do" sorts of guys: "Send me to the front, chief, I'll win the war for you." It wasn't that kind of thing at all. They wanted to know "what are my chances, what is this machine going to do?" And sometimes they said, "No way." It was the doggone Mercury capsule. They built it with a hatch that was put on by 144 nuts, bolts, and screws, and if they had to get out of the capsule, those nuts, bolts, and screws had to be removed. Or, there was another way out. They had to remove the right side of the instrument panel and crawl out through the apex of that cone. You could push out the parachute canister and get back there and you could do it, and we did it. We did it on the ground and we did it on the ocean.

But if I could "toot my own horn" just a little bit, when I saw that thing, I said, "No way!" That's not the way to build a thing you're going to put a man into. I went to the engineers and went to the management, and I went to the astronauts, and I said, "That's silly. You don't put a person in an elevator without giving them an escape hatch. In any elevator, there's that little thing up there you can push out. If the elevator gets stuck, you can get out of there. You've got to give these people some way of getting out, and quick!" I think we didn't get that explosive hatch put on until Gus's flight. I think it was through my insistence and my admonition to the astronauts to get on my side and fight this thing that we got the quick-release hatch put on. I always felt bad about Gus Grissom and Chaffee and White in Apollo 1, that nobody fought for them for a rapid-opening hatch. They died as a result of it. Two years later, we had an Apollo spacecraft with a rapid-opening hatch on it which, had somebody fought that battle, the guys would have probably survived. Tragic thing. It was all from a lesson not learned. It didn't happen in Mercury and they never had to get out rapidly, so I guess somebody said, well, heck, it didn't happen in Mercury, so we don't need to encumber the Apollo spacecraft with such a thing. But they should have, and they did, eventually.

Each astronaut would wear [several] biosensors, at the front part of the chest, at the top and bottom of the sternum, and on either side of the chest in the mid-axillary line, which is right down the side of the chest. I thought that it was quite important that we always have those placed in the exact same spot on each man on every flight, mainly because somebody might criticize us at some time if there was a change of some sort. It might cause a critic to say that was due to misplacement of the sensors. The only way I knew that we could do it was to put little tattoo dots on the skin. I practiced on myself. I put plain old India ink and took a hypodermic needle and pricked through the dot several times. Sure enough, it left a little tattooed spot on my left forearm. I showed this to the astronauts and told them that I wanted it done, and some of them complained about that a little bit, having to mark up their bodies, I guess. I promised them that if at any time in the future, they wanted them removed, I would find a plastic surgeon that would do it for them. With that promise, they all agreed. I didn't think at the time that anybody would ever want them removed, because it would be sort of a badge of honor. If somebody ever says, "What are those little dots?" they could say, "I was a Mercury astronaut and those little dots were where the sensors were placed." It happened to me, and I've never had mine removed. They haven't had theirs removed. At least, they haven't asked me to remove them. So I had to teach a technician down at Brooks Air Force Base how to tattoo, and I had to let him do them on me first so he could get plenty of practice without making too big a spot. He did, in my case, make one bigger than necessary, so I had to tell him not to make so many skin pricks. After practicing and making four tattoos on me, he was capable enough to make them on the astronauts, and he did.

[Each astronaut also had to prepare himself personally for the stresses he would have to encounter, so each one undertook a physical fitness program suited to his own needs.] There was some agitation, at first, to hire a coach and to engage in a very involved and stereotyped physical fitness program, which astronauts didn't think they had the time or inclination to do. I agreed with them from the time standpoint, at least, and I knew that they would establish their own program tailored to themselves. They could do that better than any coach could. And they did. Most of them took up running as a method of keeping fit. I think Scott Carpenter did some other things. He liked to fence, and he liked to work out on the trampoline. He wasn't too much of a runner, but he did these other things. Interestingly enough, now in the light of present knowledge, it was the wrong thing to do. But we didn't know that then. This kind of endurance training does not suit one to withstand g-forces as well as things like weight lifting.

I can describe the preparation for a launch. There was Hangar S, which was down at the Cape, and it was in that hangar that the spacecraft was prepared for the launch. We also had my little medical office up there on the second deck, along with the astronauts' quarters and the suit room and a little dining room for them to eat. All seven of the men and I went down there, and we had weeks, at least, before the flight. That was our refuge, our place. Nobody else was allowed in there. Not quite. We had a wonderful nurse named Dolores O'Hara, everyone called her "Dee," who was there and a Sergeant Heron, a very fine Air Force master sergeant who was a medic-type person. Dee and I and Sergeant Heron and the seven astronauts was about all that were allowed in there on a routine basis. If somebody wanted to come in, they had to get permission. We came down there to prepare for the flight. I gave the physical examination before the flight, and then the physical examination after the cancellation of the flight, and again, over and over.

And we had a little humor into this thing. I think it was Scott Carpenter that was the one that used to joke with me a little. After one of the centrifuge rides, he was in the shower and I was in writing up my report on his ride. I heard a crash in the shower and I thought, "My God! What's happened to Scott?" I ran in there and there he was, lying on the floor unconscious! I turned off the water and grabbed him and about that time, he couldn't contain himself any longer and he started laughing! It was a joke! He did it to scare me.

In Hangar S, we pressurized the suit with air. We had these tanks of compressed air outside the building, and they had a chain-link fence around them. But I was concerned about sabotage of the air supply. Even with the chain-link fence around it, I had no way of knowing that, at any instant, the air supply was pure. So I asked the engineers to put a "T" in the air supply line and a little valve on the "T." And I got a little aquarium and got some guppies in it and then let the air bubble through that "T" and through an air stone in the aquarium, with the feeling that if the air was ever contaminated with cyanide or something terrible, that it would kill the guppies. So every morning, I would always check the guppies to make sure that they were all alive, and that air was bubbling through this thing continuously, weeks and months on end. The guppies all seemed to do well. John Glenn came in one day and said, "Hey, Bill, your guppies are dead!" And, oh Lord, I ran in and the guppies were fine. They had pulled another joke on me.

I had to be with them and do physical examinations and make sure that they were suited properly. I have to mention another very important person and this was Joe Schmidt, the suit technician. It was his responsibility to make sure that the space suit was going to support the life of the astronaut. He had a great responsibility on his shoulders, and he was very conscientious about the whole thing. He was also a member of the team of people that were allowed in this area in Hangar S. Joe was right with us, and when the astronauts were all suited and the suits were checked for leaks and the astronauts moved down to the transport van, Joe and I went with them out to the launch site. We both went up in the gantry with the astronauts, and Joe put them in the capsule, the spacecraft, as they preferred to call it. Strapped them in and made sure that everything was proper and just made sure that Joe never got blamed for anything. I stuck my head in at the very last thing and looked in to make sure that all the straps were fastened tightly and securely and properly, and signed off on it so that Joe would have a back-up so that nobody could ever say that Joe Schmidt missed a fastener, because Bill Douglas checked after him.

After that, Guenter Wendt, the man in charge of the white room up there, was the last guy to see the astronauts. [The "white room" was a small area at the end of a walkway extension where the astronauts made final preparations, such as putting on their helmets and detaching their portable air conditioning units, just before entering the spacecraft.] Wendt would stick his head in there, and he was the last guy to check all these fastenings and make sure they were right. Then the hatch went on, and we all left. My job, after we got them in the spacecraft, was to go down to an old blockhouse that had been assigned to us by the commander of the Cape, where we put an operating room that was just about as complete as any operating room in the finest hospital in the country at that time. We had crews of surgeons there. We had an orthopedic surgeon, a general surgeon. We had physicians that were skilled at treating burns, we had internists, we had operating room nurses. The surgical team was scrubbed and they had their caps and gowns and gloves on and were ready to operate.

We had two specialized recovery vehicles standing by for each launch. One was called a LARC, and I think that stood for "light amphibious recovery craft." The LARC was a large, boat-shaped vehicle with huge balloon tires. It could negotiate the surf and move about on the surface of the ocean, or it could move very well through the palmetto-covered areas of the dry land of the Cape. I am less sure of the name of the second vehicle. It might have been APC, "armored personnel carrier." At any rate, it was a smaller, totally enclosed, tracked, tank-like vehicle. It had been modified by covering it with a sprayed-on ablative material, which would protect it should it have to go into a fire area. It was also fitted with a grappling arm that could attach to the capsule and drag it out of a fire area. Both of these vehicles were manned with rescue people.

We had a helicopter standing by. It was our job to be airborne at the time of launch, and if there had been an abort off the pad, the helicopter would go out to wherever it landed. Of course, it never happened, so it never got used. But at each launch, I was airborne in the helicopter. When the launch was successful and the orbital insertion had been achieved, I went back to the airstrip and got into an airplane that went down to the recovery site in the Bahamas. There, again, we had another complete team of people. Mostly, by this time, they were not surgeons, but they were internists. That was the debriefing area, and we had the engineering people there to do their debriefing and the medical people there to do theirs. As you know, at the end of his spaceflight the astronaut was picked up by helicopter and put on a carrier. The carrier steamed towards the island, and when they got within range, they were brought in by helicopter. The astronaut had had one physical examination and one debriefing on the carrier, and then we repeated the whole thing again in more detail and at a more leisurely pace at Grand Bahama or Grand Turk, depending upon whether it was an Atlas or a Redstone mission.

[What's the story behind Deke Slayton not being able to fly?] First, let's set the record straight. Deke's heart problem was not a result of his centrifuge exposure. I noticed it after he had made one very low-stress exposure on the centrifuge. When we went back and looked at earlier EKG strips, it was clear that it was there before the wheel ever turned. It was not a result of his exposure.

Now, you might say, how come I didn't notice that at the beginning? Well, we were using the Navy's centrifuge at Johnsville, Pennsylvania. And prior to the astronauts' coming, I had been up there to take a short indoctrination in how to monitor a subject on the centrifuge. The Navy, in order to save EKG paper, I guess, because it is rather expensive paper, 10 to 12 inches wide and it went through a pen-type recorder and had several channels of information on it. So they tended to run the paper speed about as slow as you could get it, which really wasn't adequate for me to evaluate it clearly. You could only tell rhythm with it, you couldn't see the EKG wave pattern. So, I interpreted this change in his rhythm as being a result of his respiration. When one inspires, the heart speeds up, and when you expire, it slows down. I thought that was what it was. But this type of arrhythmia is supposed to go away with exercise, and so when his didn't, I speeded up the EKG machine and saw that there was clearly something wrong.

I asked the Navy doctor who was in another office to check it out with me and we agreed that this was auricular fibrillation, which essentially means that the auricles (the upper chambers of the heart) are fibrillating and the ventricles are not, but they respond in an irregular fashion to the impulse from the upper chambers. So the heart rate is quite irregular. But that's usually the sign of something else being wrong, so we stopped Deke from riding anymore on the centrifuge and took him down to the Navy hospital. It turned out he had a high white blood count. The Navy doctor we saw down there made the diagnosis of infectious mononucleosis and told me that occasionally infectious mononucleosis was accompanied by this arrhythmia. So since Deke was an Air Force officer, I elected to take him down to the Air Force School of Aerospace Medicine at Brooks Air Force Base, and I consulted by telephone with Dr. Larry Lamb. Dr. Lamb told me on the telephone that, yes, it was quite true and not to worry; that infectious mononucleosis is frequently accompanied by auricular fibrillation. Incidentally, auricular fibrillation in this case was intermittent; it did not occur all the time.

So we took him down to Brooks and they took Deke back into the EKG room. Larry came out and said that he wanted to tell me right away that virus infections, which mononucleosis was thought to be, are never accompanied by auricular fibrillations. Of course, I was very upset about this because he had told me earlier that it was, and not to worry. Anyhow, Deke had some fibrillation at Brooks and also had an enlarged lymph node in one of his armpits. We took him to Wilford Hall Hospital and had that removed, and it was benign.

Dr. Lamb again assured me that this was, to give it the long diagnostic word, intermittent idiopathic atrial fibrillation. Idiopathic means that nobody knows what causes it, and intermittent means, of course, that it's not continuous, and atrial fibrillation means that it's an irregular heartbeat caused by the irregular rhythm in the upper chambers of the heart. So, in retrospect, whether I did the right thing is debatable. We've got to remember that in the context of the times, we were competing with the Russians. In spite of the fact that this was a peaceful, scientific space program, we were supposed to keep all things of an unpleasant nature out of the press. In a sense, this was a classified piece of information. Since Deke was a member of the Air Force, things like that have to be reported, no question about it. It's a flying safety matter. I went to the Surgeon General's office in Washington, and the Surgeon General himself was not there. I spoke to the Deputy Surgeon General and told him the whole story, and said that I personally did not feel that this was a hazard to flying safety. He said, "I agree, and you have given me this information and I appreciate it, and I don't think you have to report it through normal channels because of the security aspect of it." So, I didn't.

Well, quite some time later, Deke was selected for a flight. Deke kept having these periods of atrial fibrillation, and in his own mind, he said, "Gosh, I've had those all my life, but since it has been called to my attention, I have been paying more attention to my heartbeat." Frequently, you would see him reach over and sort of surreptitiously cross his hands and put his left thumb on his right radial artery, taking his own pulse. And he said, "You know, when I have this, I do feel a little fluttering sensation up in my neck. I have felt that little fluttering sensation all my life from time to time, and I assumed it was a normal thing." So, very honest man that he is, no question in my mind that he had this all his life.

Deke was selected for an orbital flight, and he was in training for it. I got a call from General Roadman, who was at that time an Air Force flight surgeon and the equivalent of NASA Surgeon General. He said, "We have information that I am not privileged to discuss the source of, that John Glenn has heart trouble." John had not flown yet. John had been selected for the first orbital flight, Deke for the second, and they were both in training for their selected flights. I said, "Absolutely not! Nothing at all's wrong with John Glenn's heart." So, a day or two later, General Roadman called back and said, "This comes straight from the White House. Does any astronaut have heart trouble?" And in all honesty, I said, "No." I said that the only thing that even comes close to it was Deke Slayton's. Quite some time ago, I had reported it to him and to the Surgeon General of the Air Force. A day or two later, he called back and he said, "That's it. I've gone back to my White House source, and my White House source tells me that it is in fact Deke." So Deke was immediately grounded by NASA. I objected to that strenuously because I felt sure that Deke's condition was not detrimental to flying safety or to his ability to manage the Mercury spacecraft. As a matter of fact, he had been told by various physicians that when he had this atrial fibrillation, he should lie down and take it easy and it would go away. He found that this was not an effective treatment for it, so he decided that he would run and he would run until it went away. And in doing this, he developed a greater stamina than any of the other astronauts. He could outrun them all!

So I was absolutely convinced that there was no problem as far as flying safety was concerned. Even if he developed it in orbital flight, we would not be concerned because that was a normal thing for Deke Slayton. Jim Webb, who was the administrator of NASA, asked that a committee of Air Force Flight surgeons examine Deke and render an opinion. We had several flight surgeons from NASA examine Deke, and with the exception of one, we all agreed that it was no problem. The committee as a whole thought that it was all right to use him, but Webb said, "I don't like that answer, I want a committee of Air Force flight surgeons since Deke is in the Air Force, to come up and examine him."

So we took him to the Surgeon General's office, and I presented Deke's case to about twenty flight surgeons there. Several of them examined him carefully. At the end, we met together and came to the conclusion that this would not be a hazard to flying safety. They wrote a letter signed by the Surgeon General to Jim Webb saying that this committee had met, examined him, and had found Deke to be fit to fly an Air Force aircraft or NASA spacecraft, words to that effect, I can't remember exactly.

Again, that did not satisfy Mr. Webb, so he asked that a committee of civilian cardiologists examine him. We got a committee of three eminent civilian cardiologists to examine him, and they came up with an agreement of the diagnosis. In the recommendations section of their report, they said they recommended that if NASA had another astronaut equally competent, that they should select him. Well, that was the answer that Mr. Webb wanted, so he immediately grounded Deke. But who's to say who is equally qualified? It is entirely possible that Deke was intellectually the best qualified man. And we had already shown that there was no great physical stress to flight. The deputy administrator at the time said that [premier heart specialist] Paul Dudley White was a personal friend of his. If Paul Dudley White agrees that Deke can fly, he was sure he could convince Mr. Webb to let him fly.

I took Deke up to Paul Dudley White. I went up in advance and explained to Dr. White that whatever you do, don't come up with the answer that if you have an equally qualified astronaut, let him fly instead.

I said that puts NASA in an untenable position, and they don't know what to do then. So, sometime later, he examined Deke and came with that same response: "If you have an equally qualified astronaut, I recommend that be allowed to fly instead of Slayton." That's nearly the end of the story. In the meantime, I left NASA and Deke was promised that he could fly in Gemini because they said they just didn't want him to fly alone. Gemini is a two-man vehicle, so he could fly in Gemini. Well, they didn't let him fly in Gemini for reasons I don't know. Then he was told you can fly in Apollo – that's three man. Well, they didn't let him fly in Apollo until the very last, when they let him fly in the Apollo-Soyuz mission in which he was the docking pilot. That particular job probably required the greatest flying skill of any of the other two men's jobs because Deke had to carefully dock the two vehicles very precisely, which he did. He did a fine job on that one.

At the end of that flight, through some malfunction on landing, all of the crew were exposed to nitrogen tetroxide, I think it was – one of the gases, at any rate, that caused some concern with some lung inflammation. But to the best of my knowledge, that was the only flight in the whole series of spaceflights that we have had from Mercury through the shuttle where anybody has had any problems of any kind with relatively normal flight.

Henri Landwirth was a Belgian Jew who was in a concentration camp as a youngster. Just before he would have died, he was rescued by the American troops who came in to whatever camp it was. Henri has his serial number tattooed on his forearm, I think his mother and his father died in the camp, I'm not sure. In any event, he and his sister survived. He finally came to the United States and became the manager or the innkeeper of the Holiday Inn at Cocoa Beach, Florida. The astronauts would stay there, and Henri would keep these seven rooms vacant so that these men would have a place to stay whenever they came down there. He charged them, but you could always get a room at the Holiday Inn if you were an astronaut or if you were Bill Douglas, fortunately, because Henri always had a room for me.

Henri felt that the astronauts, these original seven men, should leave something for posterity. He convinced them that they should establish the Mercury Seven Foundation, which is a foundation devoted to giving scholarships to people in the sciences and engineering. One has to have, I think, completed the first two years of college. We do not select the individual; we select the college or the university that has a good track record in graduating engineers and scientists. We tell them that we have a certain amount of money available for a scholarship if they want to select the deserving person. There's sort of an understanding that we will continue supporting them financially through their Ph.D. if necessary.

Don Larsen arrived at Edwards Air Force Base northeast of Lancaster, California, during the Mercury program. He participated in preparations for the upcoming Gemini program.

I rolled into Edwards in 1962. Mercury was pretty well over, and they were developing Gemini. We weren't into Gemini, itself, as much as its recovery system. They had two types. They had one we called a paraglider, which was a Rogallo wing, kind of like a hang glider that would unfold. They could actually fly this thing down and land it, like at Edwards on a dry lake bed. The other recovery system, which they ultimately went with, was water recovery systems. We had to develop these parachutes, both for Apollo and Gemini. That was done, not at Edwards, but at El Centro, California.

We would go to El Centro and test the systems. We had various ways of doing it. One of the common ones, on the Rogallo wing, they'd actually tow it behind a C-47, then drop it, and it would fly down and land. The other way was for higher altitude. They would kick the thing out of the back of a C-130. The parachute would pull it out, and it would fall away. They'd build some speed, and they'd open this Rogallo wing. They had a lot of problems with them breaking up. They soon trashed that.

Then they started working with what they called ring sail parachutes, which were used on both Gemini and Apollo. We developed those using an F-4 Phantom, a test plane that actually belonged to the Navy that was stationed down at El Centro. All we did was use the airplane. They provided the crew and servicing for the plane, but we had to go down and load the package. We used a modified MB-1 Genie rocket, which was later redesignated the AIR-2A. They took the rocket off of this thing, used the front of it for their data link and electronics. They used the back of it for the parachute package. We'd load this package up on the centerline of the F-4 just like a bomb, and he'd go up and fly whatever speed they needed. The F-4 was just about the hottest thing they had at that time. It released these things, and the parachute package would come out of the back of the Genie rocket. They would recover it in different ways. They had aerial recovering they were messing around with. They were using both helicopters and the C-119. They had a thing like that back then where they recovered some reconnaissance spacecraft. They just snagged the parachute canopy. They were thinking about the possibility of airborne recovery of both Apollo and Gemini, without even doing a splashdown. A splashdown would be if they didn't catch them. They would just snag them right in the air.

The Genie rocket we used worked great. The Genie rocket, full up, weighed about 2,000 pounds, parachute and all. Well, that wasn't too much of a strain, but even with that, they had problems with C-119s. We were also using the H-19, H-21, other types of old helicopters. The helicopters worked best, but they just didn't have powerful enough helicopters yet that could actually hold this weight. This was testing concepts. It never got to Gemini or Apollo. They finally decided that aerial recovery or land recovery was too hard to do.

The ring sail parachute is just a parachute that has slots in it. They're not a continuous canopy like a personal parachute is. You can open these [slotted ones] at much higher speeds without tearing them apart or overloading the structure that's decelerating. Often, the parachute holds up fine, but you tear up the equipment that you're trying to recover. Ring sail parachutes had different ways of opening. Most of the time, they would be deployed with a drogue gun that actually forced it open, explosively. Or they would flow out a pilot chute that would then pull out the parachute, and the parachute would stay fully reefed. They had line cutters that would cut this thing in sequence, gradually opening it. It would just go out radially, *puff,* there it is. They worked with a lot of these parachutes.

The technology for them was borrowed from special weapons systems for retarding nuclear bombs. Those people were already releasing the bombs at very high speeds, and very heavy weights, and retarding them. So they just borrowed some technologies from that. It went right into the space program, and then they made it bigger. I don't have clue what a Gemini weighs, or an Apollo, but they're considerably heavier than any of the weapons we were dropping. They had to land them much slower, because the weapons landed quite hard. They had to set them down easy if a man was to survive. I've been told the difference between landing on land and water is not really that great when they are using recovery systems, because the water is very hard when you hit it with any velocity. The Russians' problem was not having access to the oceans, so they had to develop land systems. They went with parachutes and retrorockets.

While the Gemini flights were going on, Gerald Garing was working on equipment for the next phase, Apollo. The Apollo spacecraft consisted of three modules. The command module (CM) provided crew seating and operational controls for the flight. The service module (SM) contained the onboard propulsion systems, electrical power supplies, and storage space for consumable supplies during the mission. When attached together, those two modules were referred to as the CSM (command/service module). The lunar module (LM, originally called the lunar excursion module or LEM) was designed to carry two crew members from the CSM to the lunar surface. While the LM was detached, the CSM orbited the Moon. When the landing crew was ready to return to the CSM, part of the LM served as a launch platform and remained on the Moon while the remainder of the LM rendezvoused and reconnected with the CSM. The landing crew then moved into the CSM, taking the collected rock samples with them. At that point, the LM was no longer needed, and it was sent to orbit around the sun or sent back to crash into the Moon. On the return to Earth, before the spacecraft entered the atmosphere, the SM was released to burn up on reentry. The CM carried the crew to a parachute-assisted landing in the ocean.

In 1964, I went to the NASA altitude chamber at Clear Lake, near Houston. We put an actual Apollo capsule inside, well instrumented. We would run the chamber up to different altitudes and verify the functioning. The capsule itself was made for 6 psi [pounds per square inch], so the Delta P [pressure difference] across the bulkhead was like a

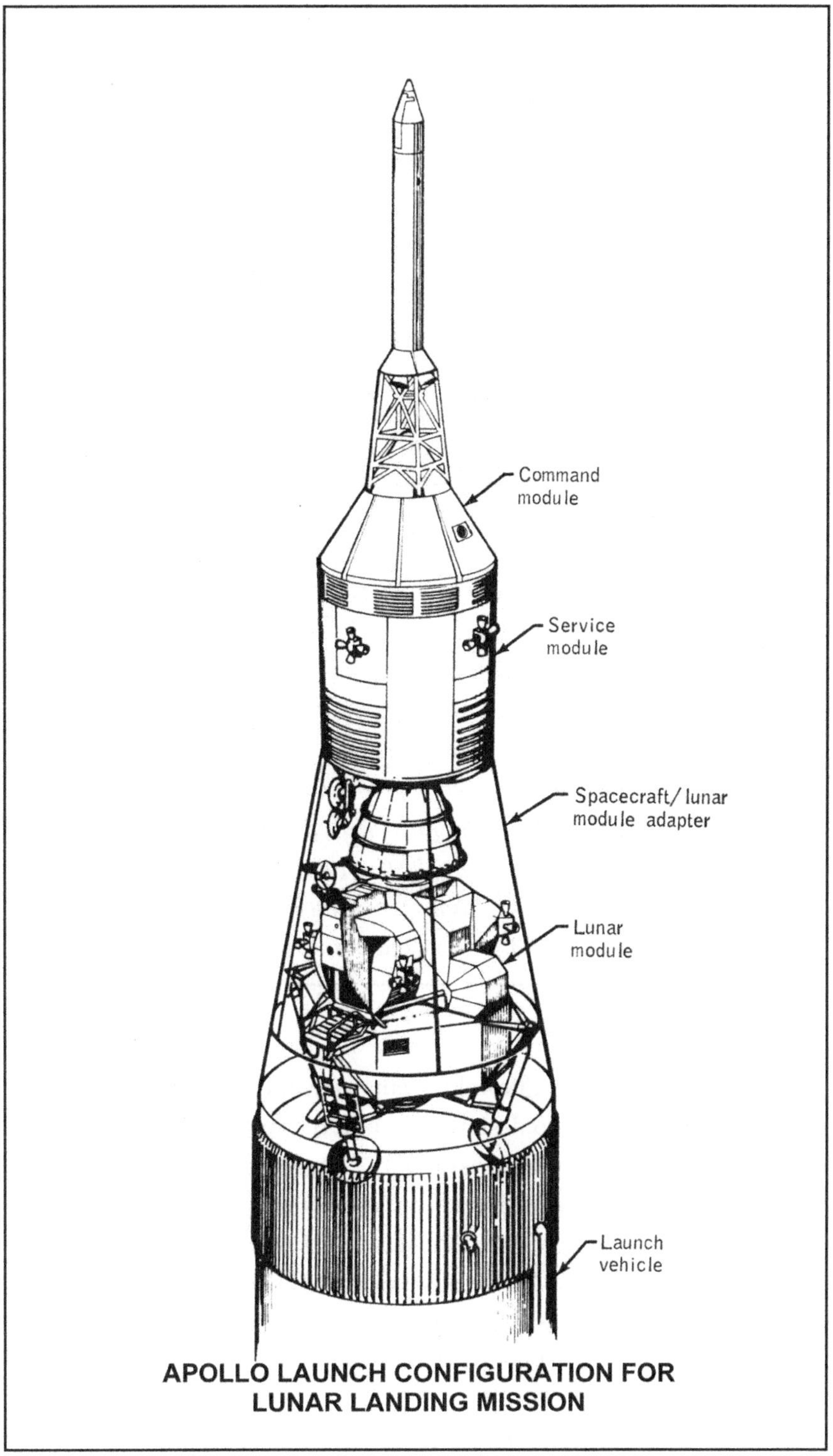

The configuration of the Apollo spacecraft during takeoff. Courtesy of NASA.

plus or minus 7 pounds [compared to normal atmospheric pressure]. We put a test vehicle in that chamber, and we verified all the parameters of the structures that allowed it to be used at 6 psi. When they first evacuated the capsule [to 6 psi], the door caved in. We came back and fixed it. So that was interesting.

We worked with Apollo 17. We were going to determine if there was water on the Moon. So we built these low-frequency antennae that opened up off the service module and would fly around the Moon and sound the Moon to see if there was permafrost and something like that.

Lew Curatolo grew up in West Virginia, one of a family of ten. His father, a coal miner, and his mother had immigrated from Sicily in the early 1900s. Curatolo was born in 1925, and after several years of military service during World War II and the Korean War, he went to work for some civilian contractors involved in government missile programs including Navaho and Titan. While working with Lockheed, he was pad leader on the Agena missile launches. Curatolo talked about Cape Kennedy and Cape Canaveral. Both were names for NASA's manned spaceflight launch facility in Florida. The name was originally Cape Canaveral; but after President Kennedy's assassination, the facility was renamed Cape Kennedy in 1963. In 1973, the original name was restored.

[In the late 1960s, Curatolo went to work for North American Aviation on its contract to design and build the Apollo program capsule. He became acquainted with several of the Paperclip Scientists, including Wernher von Braun.] Von Braun was very unexcitable. Very congenial and very interesting to talk to. He would ask some very simple, basic questions that you knew he already knew the answers to, but he would want to know if you knew. That's the type of individual he was. He really wanted to know if you knew what was going on. Dr. Debus was the director of the Kennedy Space Center, and he would make an appearance prior to launch, with the astronauts or with the back-up teams. He was very quiet, nonexcitable, but he also showed so many signs of brilliance

Mercury astronauts Gordon Cooper (left) and Gus Grissom (center) confer with Wernher von Braun in 1963. Courtesy of NMMSH.

that you'd feel stupid just being around him. That's getting a little far-fetched, but that's the type of people they were.

Dr. von Braun was a very demanding individual. He expected results, and he didn't accept any lame excuses. But he was very helpful in making the determinations of whether a decision needed to be made for the successful launches or static firings or whatever. And Dr. [Kurt] Debus was pretty much the same. I think that this was instilled in them during their days at Peenemunde [the German facility where the V-2 missile was developed during World War II]. That you can appreciate, because they were very regimented, very organized. And that ambition rubbed off on the people we worked with during the space program and the unmanned launches.

They expected the utmost. They wouldn't settle for anything second best, and that was a lesson in itself. During this realm of transition or evolution when we were going from unmanned to manned launch, there was always someone that had to coordinate the work effort and the procedures at the launch pad. You must realize that the launch pad was almost completely remote from the launch control centers or the blockhouses, as we used to call them back then. The launch director had no way of knowing what was happening at the launch complex. As a result of that, there were some problems, because there was no central point of communications.

So, the pad leader was developed by the von Brauns and the Debuses and of course, I'd like to mention some people like George Page and Chuck Gay and Paul Donnelly and [Caldwell] Johnson. Those people migrated down to Cape Kennedy from the small missile ranges like White Sands, Wallops Island [Virginia], and the Pacific Range at Kwajalein. They assembled at Cape Kennedy because they were experts and they had missile experience. The pad leader was invented by this group of people because they had to have a person out there that could have total authority like the launch director had. That's how I was appointed pad leader. At that time, I was the general supervisor for all the technicians and I was playing a dual role – pad leader and general supervisor.

A pad leader essentially is in charge of all activities that happen on the launch pad in preparation for the launch and the prelaunch activities. These activities were coordinated through the control center or blockhouse. You had to apprise the launch directors and the test conductors of all activities currently in progress and the problems encountered. The pad leader was the central point of contact for anyone that had any concern about the prelaunch activities or the launch activities of the missile, and was responsible to see that the technicians and the engineers had the proper documentation and drawings to do what we had to do in order to bring it up into launch configuration, including the ground support equipment. It was a pretty large task, and it had to be precise to make sure that we had reached the pinnacle where we said we were ready to launch. That's essentially what a pad leader normally did.

We had another gentleman that came from the von Braun team whose name was Gunter Wendt. He was from the Peenemunde era, and he was a pad leader when we were still launching Geminis. He was over with the Gemini crew, and there were only two of us at that time that were actually called pad leaders. We were the eyes and the ears and the coordination for the launch directors at the launch pad, and we had total authority. Nobody did anything without our concurrence or without us taking it back to the launch director and getting permission and concurrence from him. We had to be on our toes and make sure that we knew every incident that was happening at every moment of time on that launch complex, on that launch vehicle, and on that capsule. That's how I was assigned pad leader, which carried me on through to the Apollo program.

I got involved with the Apollo capsule boilerplate at the very early stages, about 1967. The boilerplate is a facsimile of what the real thing is – the configuration of the capsule itself, structurally and inside and out. It is an experimental piece of equipment, but it will progress into what will be the final design criteria and configuration, and that's what we were there for.

At that time, I was in charge of all the technicians and the engineers and the operations people. My job was to make sure at the time that the two boilerplate series [i.e., a mock-up of each of two capsule designs] were properly in configuration to be shipped to Cape Kennedy for checkout. It was a highly instrumented piece of equipment. So after the boilerplate was shipped from Downey, California, to Cape Kennedy or Cape Canaveral, to Pad 34, first we took it to the hangar and then we checked it out. We took the capsule to a place called the AF Hangar at Cape Canaveral, and we checked all of our instrumentation there. We stacked it to the service module, and then we took it to the pad and we did our prelaunch checkouts. The prelaunch checkouts were a lot of simulations that we had to do. At that point in time, the project began to progress on a very even keel and steady basis. We got the boilerplate off successfully and with a lot of good data which was later applied to Apollo 7.

From there, we went to Pad 34 to do the main launch. But prior to that, we had the Apollo 1, which would have been the first manned spacecraft Apollo. Everything checked out right in the vertical simulator building. We went to our vacuum chamber tests and all, and everything looked fine. I've got to say at this point that the spacecraft was very crude, in the state it was for manned launch. It wouldn't have made orbital launch. But the astronauts that were willing to take that chance were all hardened test pilots and they said, "Hey, we can live with this." So we were having a countdown simulation when the Apollo fire happened. It was very tragic and very catastrophic.

People at that point became precision conscious, that elapsed time versus real time had to be closely coordinated with precision. One second meant the difference whether Grissom, White, and Chaffee would have lived. It's a hard lesson to learn. But we recovered from the Apollo 1 catastrophe, and we got on with business. Schirra and Cunningham and

Eisele took over because they were the back-up crew, and they did a superb job. We can't be concerned about what could happen. It happened and let's get on with business. So they successfully tested Apollo 7 in another configuration, using what we learned with 1. It was a hard, bitter, ugly lesson to learn, that the way we had outfitted the boilerplate in 1 was not right. You might say it was a disaster waiting to happen, with 100 percent oxygen in the cockpit, wiring strung all along the deck of the spacecraft. You might say we did a poor job of design. And we paid the price.

When a Saturn V launched, it was the most thrilling thing I have ever experienced. My responsibility was to make certain that the crew was inserted in the capsule and were properly secured and had the proper procedures to go into the launch count. After that was completed, then we had to move the gantry away from the launch vehicle. During final count, we had to fall back to an area closer to the launch pad, because in an emergency we had to react accordingly and go back in and evacuate the crew. That was quite chilling. After you knew that they were inserted and in orbit, it was just a big lay-down. "Let's go have a drink, guys."

Dr. Rufus Hessberg was involved with many phases of space research (see Chapter 4). After he left NASA in 1979, he held a leadership position in the international Aerospace Medical Association for thirteen years.

I was a flight surgeon in the Air Force back starting in 1947. My first eight years in the Air Force were, first, training as a flight surgeon and then, secondly, as a pararescue and survival surgeon with the Air Rescue Service. I was stationed in Germany and had an opportunity to discuss my next assignment back from Germany. I indicated an interest in wanting to get in the research and development side of the Air Force program and was fortunate enough to be assigned to the Aero Medical Research Lab at Wright-Patterson. From there, after my three years at the lab at Wright-Patt, and having already started a program on the human centrifuge to look at man's tolerance for launch and reentry g's, and having determined through our research that those accelerations would be acceptable as long as one took them in the transverse vector –

chest-to-back is really what we are talking about. We were responsible, as a function of our research, for the ultimate design of the couch which was used for Mercury, Gemini, and Apollo, and launching in the couch mode to withstand the accelerations of both launch and reentry.

From 1955 until I retired from the Air Force in 1973, I was constantly involved in one way or another with some aspect of our space program. My last four years on active duty were at NASA headquarters as the director of space medicine, and I stayed on for another six years as a civilian.

[What attracted you to the problems of spaceflight?] I suppose the challenge was the main thing. It was something that no one had ever done. I think you have to realize that the research that we were doing in those early days was a combination of high-altitude, high-speed aircraft flight protection. It was almost a small step from the work we were doing for the high-performance, high-altitude aircraft to spacecraft. Man's threshold in terms of where does space start for man, physiologically, it's about 63,000 feet. You take other disciplines, and space begins a lot higher. We had been well in excess of 63,000 feet in our chambers and were well in excess of 63,000 feet in these experimental high-altitude, high-performance aircraft. The protective mechanisms that we had been working on, and our concerns about man and could he withstand the various factors involved, made it real simple to transition into looking at the same things and what would you do to put a man in space. We were already, physiologically speaking, working in the space environment. It was intriguing, it was challenging, it was stimulating.

One of the things I liked best about it was the fact that there was nobody telling you how to do things, because it hadn't been done. We were writing the textbook as we went, not having to follow a cookbook. I guess maybe it's part of my character and personality, but I thoroughly enjoyed the challenge of having to make decisions and develop research programs for something that hadn't been done before. I think many people can understand and appreciate what I'm talking about, in that it's not as much fun to just have to follow somebody's already prescribed procedures and just do what you've been told and trained, as it is for exploring and pioneering. That grabbed me and kept me enthused and motivated the whole time.

The work on the centrifuge, of course, is slow onset of acceleration and it's sustained, whereas impact work is high onset and very short term.

The work that Dr. Stapp was doing was pertinent and germane to such things as the landing of the capsules in the water and the sudden impact of that. We had to use his information, in terms of saying that the impact of the capsule in the water was going to be within the tolerance limits that he had developed at Holloman [see Chapter 5]. But we were dealing with a different physiological mechanism of the human body, when you talk about the centrifuge type of accelerations as opposed to the sudden stop on a sled, such as the Daisy track that was the workhorse at Holloman. The long track had a few runs on it with bears, chimps, and Stapp. But the real workhorse was the short track, the Daisy track.

I went from testing parachutes and jumping out of airplanes to riding centrifuges. Then when I went to the lab at Holloman, I all of a sudden found myself riding the sled on the short track. So I guess I've subjected myself to about everything you can do in the realm of flying, other than never satisfying my one ambition, and that was to actually fly in space. I never got there.

In 1963, I joined NASA and went down to the Manned Spacecraft Center as deputy of the crew systems division. We were responsible for all of the medical research and development for the future flight programs, Gemini, Apollo, and Skylab. I think the biggest challenge was probably the Apollo design, and we lost. We went hard for a mixed gas atmosphere in Apollo, and they weren't ready. We felt it would be a better physiological environment for the Apollo crew to have a little bit of nitrogen in the atmosphere and not 100 percent oxygen. But there were a lot of honest-to-God engineering problems associated with it. Not that they might not have been overcome, but the controllers for the gases [were problematic].

[The Manhigh capsule had a mixed gas atmosphere.] But in Manhigh you were dealing with an entirely different size, shape, description, and volume vehicle at the time. And of course, you always had the 100 percent oxygen button-up-your-visor situation available. We would have had to go with a whole engineering design of a lot of the Apollo plumbing, and no one felt there were controllers available that would sense the atmosphere. The Manhigh was basically a bleed type of system. It wasn't sensing and adding, it was basically starting with what was in the capsule at ground level with an oxygen flush, and then just replacing with oxygen to keep the pressure, but we weren't bleeding nitrogen into the Manhigh. It wasn't a real two-gas atmosphere. When you start going for a two gas atmosphere, you've got to be able to sense both gases. You've got to add proportionately what you need and, at the same time, maintain the pressure range that you have designed to keep the capsule at. It was pushing Apollo's, or pushing the state of the art at the time, but we sure wanted to go to a mixed gas.

Vance Brand earned degrees in aeronautical engineering and business administration. He became a pilot in the Marine Corps, and in 1960 he went to work for Lockheed as a flight test engineer and then as an experimental test pilot.

In April of '66, I was one of nineteen astronauts chosen by NASA. In life, you keep coming to Y's in the road. You go left, or you go right. When I got into the Marine Corps, I went in as an infantry officer. It was purely an emotional thing, wanting to fly, watching planes take off while I was eating breakfast. I just had to try it, and I loved it. It was an evolution then as I, fortunately, got into fighters. I loved that and got some jet experience, and then went with Lockheed. Lockheed sent me to a Navy test pilot school. I just seemed to be following a path of air achievement in aviation. To me, flying in space and going through the excitement of the ride, and getting to fly higher and faster was just plain up further into achievement in aviation. As it turned out, that impression was true. It was a great new experience. It was all emotional. I just decided I'd like to try it.

The astronaut training lasted about a year. It started out with classroom training, field trips, survival trips. Went to Iceland, The Valley of Ten Thousand Smokes in Alaska, Hawaii, all over the United States. Went to Valles Caldera here in New Mexico, and near El Paso to a feature where there used to be a maar, which is an exploding volcano. We went on those sorts of field trips for geology. We had a survival trip to Panama. Back then, Apollo was in gear, so they said it's not likely in the Arctic, but you might land unintentionally in the jungle someplace. Classwork basics, orbital mechanics, physics of the upper atmosphere, space medicine, all sorts of things that were required for a basic understanding of space science.

After the Apollo fire, I was assigned to work with engineers in the use of the materials in the spacecraft being flammable. I was in Apollo 8 doing the countdown and various things. Helping Kennedy Space Center, getting the vehicle ready for both Apollo 8 and Apollo 13, as it turned out. Al Warden was the prime command module pilot for Apollo 15, and I backed him up. He and the crew did a great job on the mission flying to the Moon and exploring Hadley Rille and returning. Unfortunately, Al didn't break a leg or get measles. I didn't get to replace him.

William Chisholm had been interested in aviation since he was in high school. After he graduated, he worked as a general laborer at Cape Canaveral from February 1951 until he was drafted into the Army in September 1952. After his service commitment, he went back the Cape, and worked as a technician with Boeing for several years. Later in his forty-one-year career with Boeing, he went into quality assurance and worked on the Apollo program.

The Apollo astronauts were really down-to-earth people. Obviously, they get excited about the fact of what job they are doing, because it was so unique. But the thing is, they were regular people. They talked to you. Asked questions. It was kind of a quickie, but the point was they were interested in what was going on. And when a flight came back, they always thanked everybody for doing a good job.

[What kind of pressures were you under with Boeing to meet deadlines with the Saturn V to try to get Americans to the Moon in time to beat the Russians?] That's an interesting question. You kind of felt like you were trying to meet the schedule, but it was always dealt with "we want to do it right." Of course, that sometimes gets to be a problem. We have had places where people sometimes thought the schedule was more important, but we always managed to stop them. We actually had a room where most of the managers sat, and they tracked all of the paperwork to see where we were at and what was going on. All of the paperwork was reviewed at least three times. It was done by the person who did it, by the lead inspector, and by the records people. Every record was checked at least three times, to make sure that everything was done right.

On every missile, somewhere on the phase while we were in the hangar, they would open up the tanks and actually go up inside and inspect them. People from Marshall Space Flight Center would come down and support some of our local mechanics and inspect inside of the tanks. The first stage used, if you will, kerosene. That's probably the easiest way to describe it. That is what we called it, kerosene and oxygen. To inspect the oxygen tanks, you actually had to build rooms made out of plastic. So that they would go in, kind of undress, go into the next one and put on the oxygen clean stuff so they could go up into the tank. Any grease is a real disaster with oxygen. They had caps with tape across their eyebrows, all of their gloves were taped to their sleeves, and all of the booties were taped to the pants, so that absolutely nothing was exposed, other than the clean suit.

I would sit at the door and inventory what went in and what went out. The two-man policy always exists. You had to have some monitor sit there and write what went in and cross it off when it came out to make sure that you didn't leave anything. That was true anytime you went on board the missile. And of course, the size of stage one, you could put a lot of people inside there.

I think one of the funniest things that happened involved a pair of clackers, if you remember what clackers are. One of the Boeing people, for some reason, had a pair of clackers in his toolbox. They were out on the first stage working on the engines, and somebody walked by and picked up the clackers and started playing with them. Needless to say, the monitors in the firing room got all upset because they kept hearing this *click, click, click, click*. Actually, nothing was wrong, but there was a lot of investigation. You know, just hearing noises that was not normal. Are you hearing things or not? You are hearing something and you're not sure what.

Before launch, it takes about five hours to fuel it, and then they had some of the cryogenics they had to put on it. I used to always go up to the top level and start walking down to make sure everything was shipshape before we left the pad. All the trashcans were removed. All the cables were tweaked up. All of the swing arms were retracted. I used to walk the whole tower just before we vacated.

Al Cornelius earned a Master's degree in communication in 1961. A few years later, while he was in the Navy, he was the officer in charge of operating an educational television station at a recruit training facility in Chicago. After his active duty ended, Cornelius worked briefly as a high school communications teacher and then as a salesman. In 1967, a casual acquaintance asked if he might be interested in an open position at NASA. Cornelius applied for the job, which became a satisfying career for him.

I was hired in the administrative department at NASA in Houston. At that time, it was called the Manned Spacecraft Center (MSC). I started off as just a buyer, you might say, and then the contract administrator, and later as contracting officer. Then I was selected to the procurement operations area. When I left there in 1991, I was chief of procurement operations.

There was a policy there, at what has now become the Johnson Space Center, that the administration wanted to make sure that even those who weren't in technical fields would feel like they were part of the mission. So on various flights, they would select someone from administration to come over. It was a real job. You were able to sit at the console. My particular job on Apollo 13 involved medical telemetry, and that kind of thing. I jumped at that opportunity right away. Of course, not having any notion that there would be the problem that there was. So at the time we had the problem, I was there in Mission Control and involved in doing what I could to expedite the process so they would have that medical information and be able to analyze it.

When I look back on it, I was involved in some very significant times. These were good times for the country. Yeah, we had our setbacks and all. But for instance, nobody had to tell us to come in on Saturday. If we needed to come in on Saturday, we came. And we looked at one of the greatest things that ever happened, technically speaking and programmatically speaking. The camaraderie was good. We understood what the job was. We understood the importance of it. And there was sort of a military aspect of it also. We would go to meetings, and guys like Gene Kranz, who was the flight director on Apollo 13, for instance, the guy that was really in charge of all that, he looked just like a general.

The job was important. When you prioritized the mission, crew safety was always paramount. That was always the most important. And well it should be. Crew safety, the lives of those astronauts comes before anything. But the mission accomplishment was second on the list. And it wasn't a matter that we had all the money we ever wanted, but the funding was there. President Kennedy had made a commitment that by the end of the decade we would be on the Moon. And we made it.

There was no foolishness there. They wanted results and they wanted people to understand the criticality of their job, the criticality of the mission that NASA had. We had a goal. We had a mission. We had our eyes on it. Whether you worked in administration as I did, or in the technical disciplines as an engineer, you understood that we had to get on with it. That was popularized in the [History Channel's] Apollo 13 movie, *Failure Is Not an Option*. We are not going to fail. We are going to get the job done. And by the way, that movie, I thought, was a very excellent representation of the actual people and events that took place.

The Apollo 13 mission was a very tense time. I can only speak for myself, but I never even considered that we weren't going to bring those guys back. It just never occurred to me. I think that says something for the confidence that most of us, all of us I hope, had in the management of that problem.

There was a feeling there that we were going to get these guys back. It's kind of like any other drama that unfolds on the stage of world history, I guess, that you do the best you can. You depend on other people to do the best they can. And then that's all you can do. You are a human being, and yes, you make mistakes. But you don't want to make them if you can avoid them. You want to do everything as professionally as you possibly can. One of the things that impressed me is that you've got to depend on other people. One person can't do the entire job. Gene Kranz couldn't have done it all by himself. He had a team of very sharp technical people who had very analytical minds, and that was so important and impressed me so much.

I participated in the return of the mission by serving as a console operator. It was not a skill, you might say, but it was a necessary job. They stressed to me that we've got to have this information. We've got to have it as soon as it comes in. The telemetry was working all of this time, and so we felt like we ought to be working too. It is hard to explain the feelings that you have like that. You are tense, but not just nervous about it. There is a tenseness, and we knew that regardless what your job was, it was important. And the end results bore that out.

It was sort of funny. President Nixon gave those of us in Mission Control the Presidential Medal of Freedom. I am very proud of that. I had an opportunity that people working in the procurement and business and administrative side of the house seldom had. We had a ceremony later on to make the various awards. I remember looking down and here was the fellow who was at MIT, who was primarily responsible for writing the program that allowed the Apollo to make the slingshot maneuver that they did to get around the Moon and head them back to Earth. And I was thinking physicists, engineers, procurement people, program administration people, everybody joined, not only on Apollo 13, but every program that came along. It was a great time to be involved in the space program of this nation. And after Sputnik back in '57, we had a little catching up to do.

It was not like playing a football game exactly, but the fact is that we needed to be where we were at that time. Not only because of the competition with the Soviet Union, but because it was the right thing to do, I think. We didn't go to the Moon because it was there. We went to the Moon because it enabled us to gather scientific data. There were a lot of spin-offs that came from that program.

James Gimbel worked with the Corporal missile's guidance system while he was in the Army in the early 1960s. That type of work eventually led to an opportunity for him to work with the Lunar Landing Training Vehicle for the Apollo program. The LLTV was an ungainly device used to train astronauts for landing the lunar module on the Moon.

The United States Army was very good to me. I enjoyed my time in there. I was on some kind of a list or something, and they watched my education. They forced me into a lot of schools that gave me the opportunity to sort of broaden my perspectives. When I went into the Army Security Agency, I started to understand a little bit more the technologies that we were trying to accomplish at the beginning of the space program. The digital technology and the digital pulse radar that we needed were coming in. That spurred a lot of interest. When I got out of the military [about 1965], I went to work for a research firm. They recruited me for that specific purpose. And it led right straight in to going to JSC [Johnson Space Center] and working on the first program, the LLTV, the Lunar Landing Training Vehicle. Other programs were winding down at the time, and the beginning of the Apollo was just coming in, so there was a lot of up-front money.

We were going to go on the Moon, and we were going to go across the lunar surface on this pogo stick. We called it the "mogo pogo." We almost killed every astronaut that got in that thing. We finally gave up. The navigation system on it was analog data; there was no digital computer in that thing. In analog, you had to have a lot more time for the processing unit to figure out what the transistors are actually telling it. We used to have an overhead crane that we attached the flight crew to. We would call up John Young and say, "John, would you come over? We have this thing ready to fly today." He would always come up with some kind of excuse not to come over, because we would slam them up against the wall or something. The thing never worked. So we mothballed it.

I was at Ellington Air Force Base in Houston. There were a lot of things happening at one time. It was hard to grab hold of something, because there were so many interesting projects going on at the time. You felt like, boy, I sure would like to be over *there*. Those guys got a good job over *there*. I stayed in the electronics area, working on the guidance control and the navigation of the Lunar Lander Trainer, and had some inputs to the module design. We did a lot of "Hey, that works better than what was originally designed!" So we had some influence on the original design of the lander. We had a chance to get into the actual flight on a daily basis. We went through that five-year period a real tight group. The flight crew was there every day. We cooked our dinner on a communal stove. We started working at twelve o'clock at night in preflight for early morning flights. The camaraderie and the closeness lent itself to keeping yourself sharp, going to school, staying up on everything.

Weber Aircraft built the escape seat for the cockpit of the Lunar Landing Training Vehicle. We knew that was a dangerous aircraft. We lost three out of five of them. When we went to Edwards to pick them up, we knew that thing was a microchip away from exploding. You have this big JP4 [jet fuel] engine running full blast. It's on a gimbel, you know, it rotates to give you some reaction against wind and velocity. We knew that that plane was very dangerous.

When you punched out [ejected], if you ever had to, you had to go through the canopy. The canopy didn't eject. You just went through it. It would break both of your shoulder bones, your legs, probably your knees.

Nicknamed the "flying bedstead," the Lunar Landing Training Vehicle was an awkward but effective training device for astronauts preparing to land on the Moon. Courtesy of NASA.

So what we did, we took the old canopy out and laminated the roof of this vehicle with Styrofoam, where the hatchet on the back of the seat would shatter it going through. But the seat pulled 12 g's instantaneously, so you know you are going to be blacked out. The human body can't go from 0 to 12 g's within half of a second. All of your blood is down in your toes. So, we knew what would happen. We had a dummy that we shot out of these seats. We called him Buck Rogers. With all of the instrumentations, we knew that limbs were going to break. Probably, the pilot would never fly again.

It was soon after that we, the insiders, knew that Neil [Armstrong] would be the guy that stepped out on the first mission to the Moon. Publicly it was not known, but inside it was known. Neil, who was one day just one of the guys, become like a china doll. We all hovered around him, make sure his lockers were right, make sure his timing and training were right. One day, you weren't that close to the guy, just another test pilot coming through. Then next day, you felt proud to know the guy. This is just human nature. You knew, historically, this guy is an explorer. This guy is going to do something. We feel that way about Columbus, and his name will be in history. One day, this guy was just a test pilot, just a regular guy, who played dominoes and cards with us. We knew this guy is going to be something. So he crawls in the cockpit [of the LLTV] and he gets up and gets out of control, and he pulls that D ring [to eject]. He busts out, and we say, "Oh God, we have killed our chief."

He came out 50 feet above the ground. Hauling. I was right underneath it. It went *boom, boom, boom*! Your mind, the human mind is slow to grasp what is happening. One of the helium bottles that supplied the pressure to the system blew. A gentleman and I were standing there at the hangar door, watching our aircraft disintegrate. Well, one of these bottles went between us. We were standing there and I sensed, more than felt, this object go between us. We look back, and it is buried inside the hangar wall. I said, "We are in big trouble." Then you start running. Well, the explosion is already over with. The flames were dying, so we ran out there. It was right by the hangar. So me and a guy named Smith ran over to a NASA photographer, who had an old, big camera. He was there every day. We didn't know the gentleman. He showed up in a van and he put his camera out, and he took pictures of every one of these flights. Now, I am sure he would be bored about this time, because we had been in operation for three years. The explosion happened right over his head.

The grass was burnt all around him. We went out there and looked where the camera had been standing. The man walked away from this. From then on, he got the telephoto lens, it must have been as long as that table, and he was back there 200 yards with a big telephoto lens. He wouldn't get anywhere near that aircraft.

The Weber seat was automatic. It had three explosions simultaneously. One of them to get you out of the aircraft, and you're pulling 12 g's so you couldn't pull a ripcord. Another explosion to open parachutes, and one that was called butt-slapper. It slapped you out of your seat. That's new technology time. Now, we see it all over the place, but at that time, Weber was way out on a limb. In fact, Weber said, "Hey, guys, I will build this for you, but no returns here. We are going to do our best to build you this seat." It had a lot to do with F-111 experimental stuff going on at the same time. The medical people told us, "Well, he is going to really be screwed up. It is going to take a lot of g forces, and it's going to erupt all of his capillaries in his arms and legs." It came out on that flight. He was exploding on the runway. The films don't really show that, but Neil was literally coming out of a fireball.

I ran out there and opened that visor. He was laying on the ground and he said, "Man what a ride!"

I roped the area off. My boss told me, "Don't let anybody in there." So I started running people off. The head of NASA was there, and I told him to get the hell out of there. He shouldn't have shown up in shorts. The next day, my boss said, "Man I wished I had never told you that." Anyway, Neil shows up a couple of days later, and he was kind of blue all over, but he survived. We were all pretty grateful.

Naturally, we went down to investigate. Flight investigations were thorough and we tracked it down to electronic failure. That kind of reemphasized that we were dealing with a bomb, and we should treat it with a little bit more respect. Overall, we did. We knew it was dangerous. Anybody that had been around aircraft knew that JP4 and hydrogen peroxide do not mix very well. We didn't change very much. We changed some of the procedures.

It was a good team. We had three LLTVs flying at one time. Maybe one would fly one day, and the next one would fly the next day. But if one of them had any malfunction, the other was pre-flown and ready to move up in the flight. You couldn't make the guys go home. They would wait to see the next vehicle, or see if anything happened. Sometimes, you had three shifts of people hanging around. They were glued to the importance of their work. We knew how important it was.

You know, we all thought, if Neil comes back from the Moon and in his debriefing said, "Boy, I could have landed that without training," we are all going down the road. It is still a contractor business. If you were not needed, you were gone. So when he came back, the first thing he said, God love him, was "If I hadn't had that training, I wouldn't have made that landing." All of a sudden we got all kinds of money. And the bosses were, "Hey, don't go off now, we've got plans for you." All of a sudden, we were really needed. And then all of a sudden, all the flight crews that had avoided flying this thing were on top of us, trying to get flight time. We were very busy.

Neil Armstrong floats beneath a parachute near the fire-engulfed LLTV. Courtesy of NASA.

The next accident probably was caused by trying to do too much at once. We wound up eventually destroying three of those things, but no one died. A lot of people got singed hair and stuff, but we improved our methods of retrieval and safety. And we were allowed to shut things down when we thought anything was wrong. Before that first crash, we were so energetic. The flight time was more important than the mission, and so we were pretty zealous about getting that thing in the air every morning.

When we landed on the lunar surface, let me tell you something, there were 7,000 people puckered up to see if that thing would actually stay on the surface instead of sinking into the sand, into the dust. I was personally sitting behind a console, and everybody white knuckled on that landing. I trained that pilot to make that landing. I knew that gentleman personally. I didn't want to see him tilt over 15 degrees. I wanted to see that sucker standing up, because, not only my career, but my interest in science was right there. If something would have happened with that lunar mission, we would still be grounded today. There would not have been another attempt. Politically, it would have been completely fiasco.

I had thirty people that I invited into this one room to see this mission. You could have heard a pin drop. It was our chance to be in history, our moment in the sun. The divorce rate in our outfit was so high, because we were so concentrating on what we were doing. The fifteen to seventeen hour days, and the weekends and the nights. The excitement and the adrenalin were so high at the time that that one landing was absolutely something! It was one of the high points in my life, that three-minute time it took to land that sucker.

I knew how much time he had. I knew how much fuel was left. [The lunar module had a 5-foot-long probe extending beneath it. When the probe touched the Moon's surface, it would signal the crew to shut down the engines so they would not kick up a dust cloud that would obscure the crew's vision during the landing.] I could see the shadow of that probe, and I was saying "Hey, that probe should be dug in right now!" Of course, you didn't say anything, because everybody in the room would have hit you with something.

I believe a lot of people were standing on street corners watching that TV. I am sure that a lot of people were excited about it, even the children were excited about it. The technology behind it and the people behind it, the unsung people, the people that got up early in the morning and took a field truck out to fuel that LLTV, and the times we had to satisfy the requirements that NASA put on us to train these people, the interaction of the human beings at the time. You gathered in little cliques. It was almost like the service, the platoon competition. It was great. There was a feeling of self-fulfillment.

We know the prolonged threat of harm to the body takes a big toll on us. During the Apollo 13 flight, there was a lot of ground people sitting there, and I was one of them, thinking this is not going to work. What they are coming up with is too far out. How are you going to tell the public that you've got to write this crew off? That's what my opinion was at the time, but I am sure glad I was wrong.

The debriefing on that particular flight is a very touchy subject. We always want to protect our own. There were some cool heads on that flight. To go into that type of stress for that long of a period must have been real tough, because there were no contingencies for that mishap. We had no scripts to go by. Those guys over in Buildings 2 and 37 were in the back rooms scratching each other's head. Man, they were really hustling to get it fixed and bring that crew back. During that time, the crew was under a tremendous amount of stress. Ground support people like myself, you know, we were inactive at the time. The experts took over and they would demand information from us, but only just on one specific subject. It was an army of people that were involved, and those managers did not have time to soothe egos. They wanted information, they wanted it fast, and they didn't care who you were.

Cool heads prevailed during that flight. We had some good chiefs, tough people, and they knew flight. They were ex-pilots. Being a test pilot, you had to know how to get out of tight spots.

Jere Daily worked for Boeing as an industrial engineer on the Minuteman missile during the 1962 Cuban missile crisis. After his time on the Minuteman, he was sent to New Orleans.

They were building the first stage of the Saturn V at the Michoud plant in New Orleans. I was in the scheduling phase of it. It was interesting because we got to see some of the manufacturing problems they ran into on the Minuteman and on the Apollo. They had to create a way to weld the sloped area of the first stage. They tried welding it, and when you go from one weld to the next, it would cause the metal to move and, I don't know what the term would be, but it would cause the weld to snap. So they came up with a method of damping it, so they could go down and weld it properly. Then they would take them up to Stennis Space Center in Mississippi to test fire them. Then eventually, they were shipped by boat to the Cape.

I was just in New Orleans for sixteen months, and Boeing transferred me to Huntsville. They also did some testing there. That was the integration contract, and that was my first contact with people from NASA. We used to prepare a concise PERT [Production, Evaluation, and Review Technique scheduling] chart for the meetings they would have, to show the major areas and how they were coming into line. We got to meet some of the NASA bosses. It was interesting there also, because we saw more of the facets of the Apollo program coming together, how many contractors that were involved, testing that had to go on. It was fascinating.

I was just in Huntsville a year, and then I was transferred to the Cape. I got down there before they fired the first Apollo. I did get to go out on the launch site, and I did get to go in the transporter crawler [that moved the launch vehicle from the assembly building to the launch pad] and see the effectiveness in that. That was very interesting, because they had to keep it level going out. They used water. They used pipes going underneath it, and they had water levels on each corner so the water would balance. It was simple, but it worked. The pathway out there is gravel. I think I remember it being 6 feet deep to absorb any change when the crawler would go out there.

We got to see the first Apollo, the F-1s [engines], the S-IC [Saturn V first stage], the 500F facility bird [Saturn V facilities integration vehicle] go out to the pad and try all of the swing arms to make sure everything fit. They actually fueled it and unfueled out there just to make sure all of the equipment worked. One of the interesting things was they had eight swing arms. I think that is the correct number. They were concerned about winds coming up and the natural damping characteristics of the Apollo vehicle itself. So one night, a Dr. Cox came in there, and they did the "Cox test." They had the 500F vehicle in the VAB, Vehicle Assembly Building. They had the platform about two feet away from the Apollo, and this was up toward the top of it. They had four guys with tennis shoes on, sitting at the edge of the platform, and they had a safety strap on so they wouldn't fall. Dr. Cox attached an accelerometer to the side of the Apollo. They had these guys push it with their feet to a certain distance, and then let it stop, to see just how long it would take it to stop. Well, the first time was okay. He said, "Let us do it again and do it a little further." So they did it the second time. There were three of us, a NASA guy, another Boeing guy, and myself, standing there observing it. We were up on another platform. During that test, the escape rocket came loose on top of the thing. Well, this being a dummy bird, it was bolted to the inside of a ring. The other Boeing guy and I took off running. The NASA guy just stood there and watched it. We thought it was going to fall, but it didn't. It just went down inside the ring and landed against the nose of the rocket. Anyway, out of that they decided they needed another swing arm. That was the one that clamped around the base of the escape rocket, just in case the wind came up before they could get it back to the VAB.

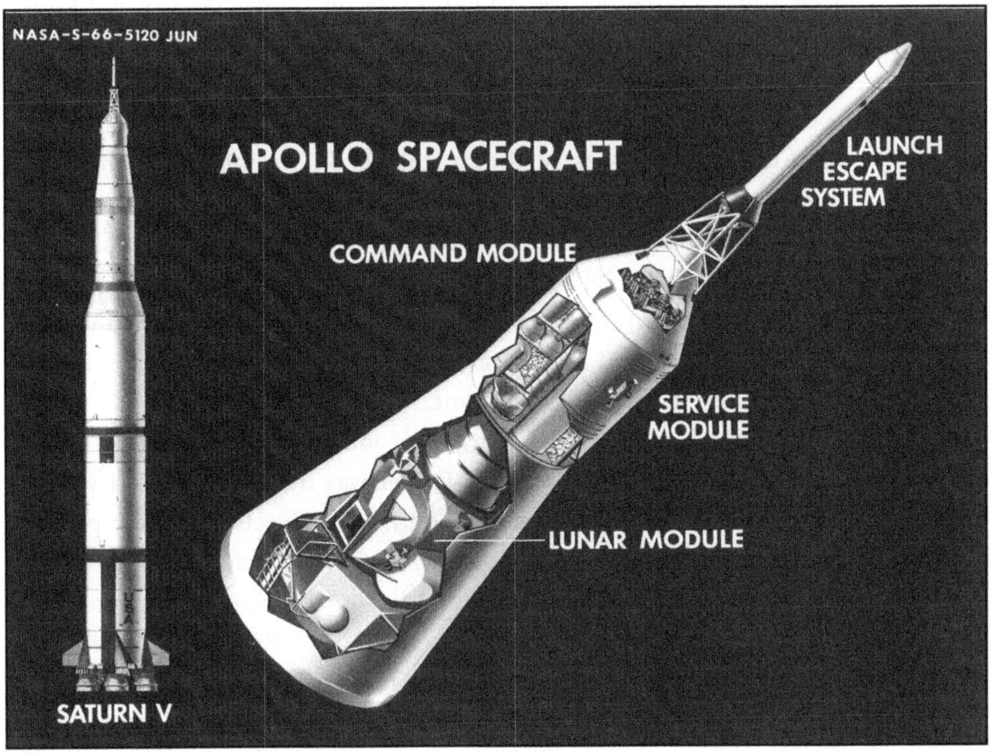

A small rocket attached to the nose of the Apollo command module was designed to pull the CM away from the rest of the launch vehicle in case of an emergency. Courtesy of NASA.

They have four launch complexes down there. We were at Launch Complex 4, where we had set up office areas. We did our scheduling in there. Launch Complex 3 was set up for the first launch. Eventually, they moved us into the VAB and set up Launch Complex 4 for launching. Boeing and NASA had a program there, that if you made a suggestion, you got an award. There were three sets of double doors into the launch complex center, and they would open one side and leave the other side locked. People would come and jam their hand against that. So, I made a suggestion that they unlatch that other door every day, and I got a ten-dollar suggestion award. That was pretty funny, I thought.

When they first started Apollo, we lived in Titusville, about twelve miles away from the launch complex. We had had a swimming pool dug the day before they launched. The soil down there is very sandy, and there was a sump pump in the bottom of the area where they dug the pit. The water would seep in, and this sump pump would pump it out gradually. Well, when they launch the Apollo, it is latched down because it is liquid fuel, and it isn't released until it reaches full thrust. Well, the earth shakes. You can hear the rumble and feel the earth shake. In fact, we had some collectible plates on the wall in our kitchen, and we had actually taken those down. We were afraid they might fall. But the interesting and funny thing that happened was that the water came up so fast in the pool that the sump pump pumped it out and dug a six-foot furrow in our lawn.

Last February [2002], I went down to see my first shuttle launch, and I expected the same thing from the shuttle. Well, it ignited and just took off, because it is solid fuel. As soon it ignites, it is full thrust and it goes. That was an interesting thing for me.

I worked at the Cape four years, and I got see all of the Apollo launches up through Apollo 11. Then I was transferred, and I worked on the 747 [airplane] for sixteen months as an electrical liaison planner at the field at Everett, Washington. They had started the lunar rover program in Huntsville, and because they didn't have a chamber to test it in zero atmosphere, they transferred it to the Boeing facility at Kent, Washington. So I moved back in there and worked on the rover program from December of '70 to May of '71. The astronauts would come in, and they would put on their gloves, and they would sit in the rover to make sure that the controls worked and they would fit very good on their hands and everything. So I got to meet some of them. A funny thing, they were in there doing a test, and they would wear a one-piece jump suit. When they finished one of their tests, the astronaut handed me the jump suit, and said, "Here, take care of this." And I did. I kept it for years. It finally wore out. I shouldn't have done that, but I did.

The lunar rover vehicle folded up and was loaded in the side of the descent stage of the lunar lander. When they would land, they would open the side of it and pull the lanyard, and it would be deployed. It was spring loaded, so as soon as they started pulling it, the wheels would pop out and the base of it would unfold. It had piano wire with titanium chevrons because they couldn't use rubber tires on the Moon. They were just wire mesh. They had a quarter-horsepower DC motor in each wheel, so you could steer it front or back or all together. That way, if one section didn't work, they could use the other section. They were trying to think of everything. Especially, you know, if an astronaut gets two miles away from the lander and the thing breaks down, it could be a big problem.

After earning a Bachelor's degree in physics and mathematics, Charles Deiterich worked on various projects including flight simulators for private aircraft and electronic components on the Minuteman missile. The second time he applied, he got a job with NASA and was involved with flight dynamics on all of the Apollo flights. [Note: Some of the following material is rather technical. Enjoy as much of it as you are comfortable with.]

I started the last day of August in 1964 with Johnson Space Center, then Manned Spacecraft Center. I was assigned to work for the flight dynamics branch on Apollo 201 and Apollo 202, which were unmanned lob shots. First, 201 was down in the Atlantic Ocean, and 202 was a lob shot into the Pacific Ocean, if you could consider that a lob shot. Mainly, the job was trajectory oriented, where we would do two things. We had to build our tool boxes, which was the control center computer requirements, where we would determine what kind of information we needed in real time to be able to control the mission. And then we would control the flight. We had to do a lot of studying with the mission planning analysis people who built the pre-mission trajectories, and figure out what was sensitive and what was not sensitive, you know, what parameter we should monitor.

Then during the flight, we would control the flight. For the unmanned flights, most of it was preprogrammed. There were a few commands that you could send, but normally we didn't do that. My job was mainly in the area of reentry and abort modes, and telling them what abort modes should be used, and that sort of stuff, depending on what conditions were present at the time of vehicle failure.

One was in the spring and one was in the summer of 1966. The 201 landed down near Ascension Island. We launched it, got up pretty high, and then as it separated from the booster, we burned the SPS engine, which was the spacecraft main propulsion engine. I think we burned it [for 184 seconds, refired it 10 seconds later for another 10 seconds,] and got it going probably close to 29,000 feet per second. I'd have to go back and check for the exact numbers on that, but it was fast. It was faster than a Gemini coming back out of Earth orbit, because we wanted to see how the heat shield would react, and we also wanted to test the engine, and test the stability system and control system. The thing had to have a sequencer that would separate the service module off, and then turn the command module around for re-entry into the atmosphere.

Then on Apollo 202, we launched it and after the S4B burn, we did an SPS burn. We did another small SPS burn, and then we got over near Australia and burned downward three or four times. One long one, and then a couple of short ones just to check the relighting of the engine. It went probably around 29,000 feet per second and landed out in the Pacific Ocean. Now, 201 did not have a platform and a guidance system, it just had some gyros to control attitude. The second flight (202) actually had a navigation platform and a guidance system in it, and it did actually steer to a target.

I was on the console when the Apollo 1 fire happened. We were doing a countdown test, and what they call a plugs-out test, where they pull the plugs out of the vehicle, and go on internal power. Everybody felt real bad about it. We tried to figure out what happened, and it kind of set us back for a while. They had a block one command module, and a block two command module. The block two was on the drawing boards, and it had better design features and a few other things. So we never flew a manned block one command module. We went over to the block two immediately, and tried to do a lot of safety things, like no longer did we put pure oxygen at 14 psi in the spacecraft. We just put normal air, and then we just replaced it with oxygen once we got the altitude and the pressure went down.

In September 1968, we were getting ready to fly Apollo 7, and I was working on Apollo 7 at the time. My boss called me up and said, "You're not going to work on 7 anymore. You're going to work with this Apollo 8. (We called it C prime.) We're going to go to the Moon, and we're going to orbit the Moon and come back, but you can't tell anybody." So it was. Everybody else seemed to know what was going on, but you can't talk about it, you know. So here we are in September, and we're going to fly in December. The ground system really wasn't ready. It didn't get ready until November. Some of the processes really weren't working. You couldn't compute anything. You could talk about it. You could look at some charts and stuff that the guys were running on the off-line, but the real-time system wasn't ready yet. We had to figure out how you can do more data, how you can do platform alignments out when there's no darkness. Normally, we align on the dark side of the Earth, when you're in the shade. On the way to the Moon, you ain't in the shade. So we had to figure out all these kind of things. What kind of information we were going to pass for a high-speed re-entry. What is the crew going to look at when they're coming, to make sure the computer's not deviating? I mean, all those things had to be done. We did that in four months.

Anytime that you have two vehicles separating, you always have issues. There's always something happening. On Apollo 8, we separated from the booster, and of course they simulated turnaround and looking to see where they would pick up the LM [lunar module], because the LM is located behind the Apollo in the booster. So they turned around and looked at it, and one of the panels didn't open all the way. They were on some kind of hinges. So in the future they actually blew the panels off, and they'd go fluttering away. Well, Frank [Borman, commander of the Apollo 8 lunar orbital mission] thought that the booster was going to run over him. He thought, it's getting closer, it's getting closer. So we ended up doing a mid-course maneuver much earlier than we had planned. We didn't think it was getting closer, but he wasn't sure. Of course, he could see better than we could. We did have tracking on it, but when you're so far out, you really can't tell small distances. So we did a maneuver and separated from it.

On Apollo 10, [Thomas] Stafford was the pilot. This is when we just flew up to the Moon, went down to about 60,000 feet, and then came back up again and did a lunar rendezvous. We were getting ready to jettison the LM. We had said, we want you to go to this attitude and separate with it behind you. He said, "No, no, I can do it okay." He got ready to jettison it, and he jettisoned right into the sun. He couldn't see it, and it was fixing to do a big burn to put it in solar orbit. Well, we got far enough away and it worked okay. Just another separation thing that you never think about, but there it was in the sun.

The other thing that happened is the pressure between the LM and the command module was not relieved, and so when it separated, it popped off. It really gave a big jolt. It went away pretty quickly instead of being able to back away from it. What he planned to do was just separate and then back away and get behind it. Well, it didn't work out that way, because it popped off and he lost it in the sun.

On Apollo 11, once you get all the rocks and stuff out and close out the LM, you really don't want to hang around it. They had another rev to go before they were going to close the hatch. Well, they come around the corner and the hatch was already closed. We go, hey, we don't really want to stay on this thing, because it's got the engine with things armed, and it's ready to do stuff. So we had to separate one rev early. We would deorbit that thing out of lunar orbit, and then we'd check the seismometers from the ground on the Moon's surface.

I was the lead guy on Apollo 11. Turned out, I was the lead on most of them. Just happened to be at the right place at the right time. While we were getting ready to fly 11, I was working with Jay Greene, who was a FDO [flight dynamics officer], and Steve Bales, who was the guidance officer, working with the crews and figuring out what kind of information we're going to pass them during flight. They want to hear what they need, but they don't want to hear what they don't need. We would work with the crews on that sort of stuff and plan how we were going to fly the mission.

Apollo 12 is the one when there was a lot of thunderstorms down there. Pete Conrad was the commander. We had a lightning strike after liftoff. Of course, the plume is ionized, so any static charge will find a place to get down to the ground, and it did hit the spacecraft. The spacecraft guidance failed. I think the IU, which was the guidance system for the booster rocket, the S4B for the Saturn V. It had a little glitch, but it didn't quit. It took us into orbit. It turns out that if it would fail under normal circumstances, you could actually switch over to the command module computer and fly on into orbit with the command module. So we had a backup way to get into orbit. The whole command module computer had to be brought back up before we could go to the Moon. A lot of people thought we were kind of crazy, because we got it back up and running.

There was a big discussion about turning computer platforms on and off on the way out. Somebody said, "You probably ought to turn them off." Somebody else said, "While they're on and running, just leave them alone." So it was always controversy of having a problem when you powered back up. There was only one platform on the Apollo. We had what we called body-mounted attitude gyros, which were strapdowns [sensors strapped to the capsule]. You could use those for attitude also. They didn't have any accelerometers in them, so you could not actually do navigation, but they could tell you which way you were pointed. If the ground had radar tracking, you could tell them how to point the vehicle to get back, so we had a backup way to get in. But if you ever got to the Moon, you couldn't land unless you had the platform. The LM had the same kind of backup system as well. So if you had failures, you couldn't go forward; but if you had failures once you got someplace, you could always get back home. You had a backup way to get out.

Of course, we did not have the TDRs [Tracking and Data Relay satellites], which gives the shuttle almost total communication coverage with the ground. Now, you have almost a hundred percent coverage. We didn't have that then. So as soon as we flew past Bermuda, we were out of sight, out of mind. We had no more tracking data. There might have been a ship down there someplace. But anyway, Jay Greene was the FDO, and I was the retro for that launch. We [i.e., the Apollo spacecraft that was orbiting the Earth before firing its rocket to propel it toward the Moon] came over Australia, and there was a network controller that's responsible for the network. When tracking comes in he says, "Hey, we've got tracking." Well, he called it. Man, he was about four minutes early. In orbit, if you are lower, you get around quicker. It sounded like we might have gotten around quicker. Except for Greene and I, everybody else was happy because they got data. All the systems guys were happy, they got telemetry data. We weren't so sure, because we wondered, "Is he high enough? Is he in orbit, or is he coming in?" Well, it turns out that they had multipath, which is a reflection off the atmosphere, or something. They got the signal early. We were concerned there for a little bit.

On Apollo 13, we were fifty-some hours out when the dadgum thing broke. Worst place in the world it could happen to you. Well, I guess if you didn't have the LM, it would be worse. So we didn't have any way to come straight back. If we'd have had the service module engine, we probably could have come straight back, but we were not sure what we had. It turns out that in the early Apollo missions, when we did what we called translunar injection with the S4B, we would put ourselves on a course that we would go out, and do a figure eight around the Moon, and come back close to re-entry condition. It wouldn't take much tweaking of the trajectory to get in the entry corridor. The corridor ran from like 5.5 degrees to 7.5 degrees. Anything outside of that, you would either skip out or pull too many g's.

Anyway, as we got further and further into Apollo, we wanted to go to some landing sites that we didn't have the performance to get to doing free return trajectory. So we would do what we call a hybrid. We'd get out there part way, and we would do a little maneuver which would optimize the lunar landing stuff, but it would take you out of corridor. It was after that that the thing failed, so we were not even going to free return trajectory. The first thing we wanted to do is get back on a free return trajectory, and then the cheapest maneuver put us on Africa. I said, "Why don't we do just a little bit more, and push us over in the Indian Ocean? At least we'll land on water." The command module could land on land if there wasn't much wind blowing, but you really didn't want to do that. You didn't really want a land landing with the command module. We landed a bunch of them out here in the desert. So anyway, we did a little bit more of a maneuver to put it in the Indian Ocean, knowing full well if worst came to worst we could always get them out of the water. Hopefully, we could do something to tweak it up later on. We were really kind of scrambling at the time. So, we did that maneuver. Of course, we ran out of battery power and that sort of stuff, so we tried to keep everything to a minimum. I forget exactly when we shut the guidance system down, but eventually we shut the guidance. We shut the whole LM down and just barely kept the life support system going.

We came around the Moon. We did what we call a PC plus 2, which was pericynthion (which was the back side of the Moon) plus two hours. We did a burn which would speed the trajectory up, and actually moved us from the Indian Ocean over to the Pacific Ocean, and shaved, I don't know, ten to twelve hours off the landing time, plus it got us out closer to where the normal recovery ships were, which is what we wanted to do, too.

Then we were coming in, and, oh, there was a hurricane out there someplace, but we decided to ignore that. It turns out it was a good thing.

The recovery guys wanted us to move. I said, "Are you sure it's going to be there?" It turns out if we'd have moved, we'd have put them right in the middle of the hurricane. We'd have been dead, because we'd have had to bring a lot of stuff up [i.e., power a lot of equipment up] and do a big maneuver. We could move the target with a burn if you had to. Or we could use lift during entry for up to a thousand mile change. I always kind of had that in my back pocket. We didn't do that, but we did have to tweak up the corridor because the LM was boiling water. We use the water boiler for cooling. It was venting a little bit, and it was continually changing the trajectory because it is very sensitive. So we had to do another burn.

Back in Apollo 8, we found you can actually look at the Earth and find the terminator, which is the boundary between the dark side and light side, and it points mostly north-south. Using that, you can actually control your attitude and do a burn. Well, that's what we did with the LM on Apollo 13. They got the coast in the window lined up. It was a little bit out of plane, but you know you're not going to change the plane very much. We had enough that we could actually control the burn. I think Fred [Haise, LM pilot] had one controller, Jim Lovell [commander aboard Apollo 13] had the other controller, and Jack Swigert [CM pilot] timed it, because it was a timed burn. They lit the thing off with the descent engine and did a maneuver just by looking at the terminator on the Earth.

I was the lead retro, which is the return-to-Earth guy. We had four teams, four shifts, but I was the lead guy. Once the explosion occurred, we dropped down to three teams. I was basically responsible for the trajectory planning and recommendations for all of the return-to-Earth maneuvers, including reentry and vehicle separation activities. Of course, the flight dynamics team discussed these plans with our backroom people and then presented them to the Johnson Space Center management and the flight directors.

There were several new techniques that we had to develop to overcome situations that had never existed before. For example, for a normal entry, the command module/service module separation was controlled by the command module and thrusters on the service module. However, for this case, the service module was totally dead. I had remembered from Apollo 10 that when they jettisoned the LM from the command/service module, they forgot to depressurize the tunnel between the command module and LM. So when they jettisoned the LM, the air pressure vigorously forced the two vehicles apart. In fact, the LM popped right into the sun and the crew lost sight of it. I remembered this situation and decided we could use this air pressure to separate the LM from the command module. So, instead of normally bleeding the air out of the tunnel, we left some air pressure in it. First, we had to jettison the service module from the LM/command module combination. We used the LM thrusters to maneuver away from the service module. The LM thrusted forward, the service module was jettisoned, and then the LM thrusted backwards by the same amount, thus putting it back on its original trajectory, with the service module moving away.

Then later on, we used the LM to put us in the correct attitude. We vacated the LM and separated the command module from the LM by using the air pressure between the two when the docking ring was jettisoned. We had no way to use the LM thrusters for separation once the crew had gotten into the command module. This worked out real well.

Another item was the fact that there was a lot of debris floating around outside of the spacecraft. These particles were confusing for the optics when trying to do star sightings to align the gyro platform for attitude determination. We needed something that would be very easy to recognize through the telescope. I figured that the sun and Moon would be easy to see and since they were in the computer catalog of stars, we decided to use them. Although not as accurate as using stars, they certainly were very easy for the crew to recognize and were good enough to get us home. It turns out the crew had no problem using the sun and Moon to do their alignments. Once they had the initial alignment, they could then fine-tune it using stars acquired by the computer.

The fact that the computer and inertial platform were turned off for most of the trip home was unsettling to a lot of folks. We needed a way to physically check that we were on the right trajectory, as well as provide an external cue that the command module would be in the right attitude for reentry. Looking at the Earth-Moon geometry, you always enter the atmosphere on the side opposite the Moon or at least partially around the Earth from the direction to the Moon. I decided that the crew could watch the Moon on the scribed line in the window and hold that attitude until the Moon set and then track the horizon on the same scribe line to atmospheric reentry. They were heads down towards the Earth. And then to physically tie the trajectory to something they could monitor, I predicted the time of Moon set. Thus, the crew watched the Moon and their clock, and exactly as predicted, the Moon set on time and the crew watched the horizon and tracked it to reentry. It turns out that this was such a good cue we added Moonset check for the rest of the Apollo missions on their real-time reentry maneuver data.

One of the things that we were really concerned about is the command module computer had been down. Its platform had been turned off, so it had gotten cold. Electronic things drift when they get cold, so the systems guys had to measure the drift and put in a drift compensation value so it would torque itself back to keep it correct. They did a really good job, even the accelerometers. Turns out that [Apollo] 13 was probably the closest we got to where we were aiming as anything. We didn't aim for the aircraft carrier. If the ship captain were on the IP [impact point], and got hit, I mean.... So he always stood 5 miles off the target point. But as far as where we were aiming, we were within less than a mile.

The other thing on 13 was that we had a radioactive cask on the LM which was used to power the ALSEP, the lunar surface experiments. It was a radioactive thermo generator, which generated electricity. With the AEC [Atomic Energy Commission] guys, there were always radiation concerns. There was a real concern about where this cask was going to go. Prior to flight, I don't know, it was just premonition or whatever, but they wanted to talk about how we were going to get rid of it. I guess because on Apollo 12 we had a LM cask, and they didn't have a plan. As you do more flights, you take care of more and more things. So we were starting

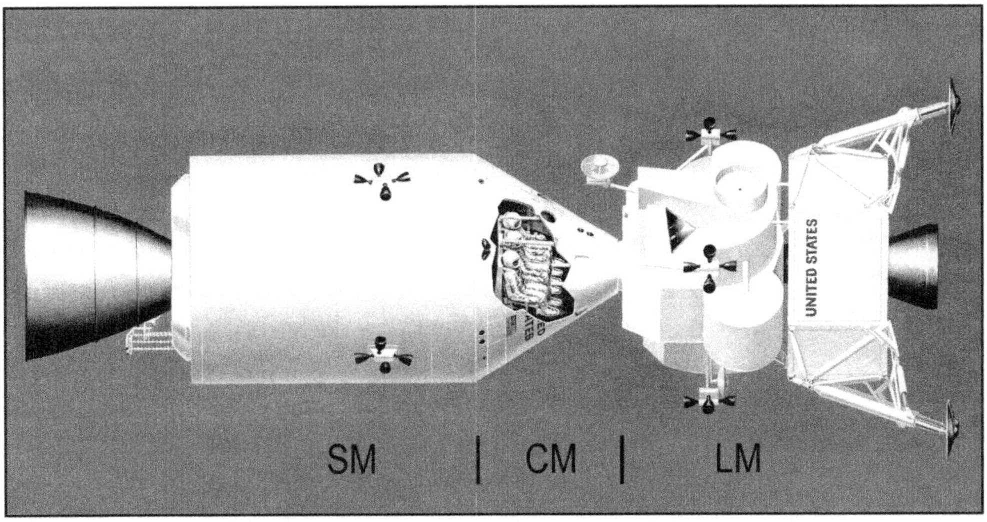

This is the configuration of the command module (CM), service module (SM), and lunar module (LM) after redocking. Courtesy of NASA.

to take care of more things, and one of them was what do you do with this cask if we have to let it reenter? It was a thing that would survive reentry no sweat, but they just didn't want it coming down in a populated area. So we got the guys who did the trajectory, and they told us what different L over Ds [lift over drag] the cask could have. I could change the value of L over D in the ground computer. So after the crew landed, I ran the aerodynamics and gave the AEC two places it might be. I actually also had done that long before we got back. They went out, and sniffed with their airplanes, and they didn't see it, but they did track something at one of the points we gave them. So it's probably in the Mariana Trench. We put it in deep, deep water.

When 13 successfully splashed down, there was a big celebration. In fact, somebody always sent cigars to the control center. So we all lit up cigars. But you know, that was the first flight that we saw the command module on drogue chute. The drogue chutes came out at about 25,000 feet, stabilized, more or less, floated down a little bit, and the mains came out at about 10,000 feet. We actually had a picture of them on drogues in the control center from the aircraft carrier.

[After leaving low-Earth orbit, the astronauts separate the CSM (combined command and service modules), turn it around, and then steer its nose into the LM's docking collar. With the two modules linked up, the LM can be pulled from its attachment to the Saturn rocket's third stage, the S4B that was refired for the escape from low-Earth orbit.] I think it was on Apollo 14, we had a hard time docking with the LM when we tried to pull it out of the S4B. We kept banging, and the docking device would not latch. It took them a couple of hours. Finally, they got the latches to engage, and they separated the LM from the S4B. They had what's called a probe, which was a thing that would fit in the docking collar between the two. Once you got the two docked together, you could take the probe out. You had to take the probe out to crawl between the vehicles. Normally, we left the probe with the LM. On 14, the guys wanted to see what was wrong with the probe. They brought the probe back, which meant we had to find a place to stow it. Of course, the retro's [retrofire officers] were always worried about mass properties. How much does it weigh? Where's the CG [center of gravity]?

We were worried about several things. One thing to worry about was the L over D, which was a function of where the CG was inside of the vehicle. The further the CG was up the Z axis, the higher the L over D. If it got down to the center, you wouldn't have any L over D at all.

The other thing to worry about was, we were responsible for how we point the engine bell. The engine bell has to point through the CG to keep from having startup attitude transients. So we were worried about where the fuel was, and where the water was, how much water, where the crew had stowed things. We had to work through all that stuff. We had some processors that we could compute the CG. We'd get with the flight activities officer, who was the ground support for the crew. They worked all the stowing list with the crew. So we'd get with them, and they'd tell us where everything was stowed. All this stuff was well documented. We had tables where you could put every little piece in there. We would then figure out what the CG is, and then we'd know how to point the engine bell. And we'd also know what the L over D was for reentry, because part of our computer processors required the L over D for reentry. Of course, we had to know how much rocks they got, and where they put the rocks. And this probe weighed a couple hundred pounds. So it changed the L over D. We had to find a niche. The FAO, flight activity officer, had to figure out a place to put it. So they did that. Then they would offload a lot of garbage and stuff it into the LM when they were getting ready to jettison it. So we had to keep track of all that sort of things.

The fuel tanks in the service module had like 40,000 pounds of fuel in them. So as the fuel went down, the CG would change. The fuel in the service module affected the CG of the CSM. L over D was only important to the command module, which was all that we worried about for entry, as the service module would be jettisoned and burn up anyway. So we had tables that would track the CG. We had tables where there was three guys, or one man with CSM because two guys were in the LM. That was a confusing issue when we were flying 13, because we were using the LM to push the CSM. We had foreseen that we could do that.

Apollo 15 was pretty non-controversial, and 16 was too. I don't remember any big things happening with 17. I did feel sorry for poor Joe Engle on 17 because he was originally selected as LM pilot, and it was Jack Schmitt who finally flew as LMP. It was the last flight, and they wanted to get a geologist up there. So Jack took Joe's place and flew the mission.

We never reused the Apollo capsules. The heat shield is an ablative material, which means it evaporates. As it evaporates, it cools. The tiles on the shuttle don't evaporate or anything, so they're reusable. I suppose they could have refurbished them, but they never did. It probably would have cost them more to refurbish than it would to build.

Chapter 10
Paul Haney

Paul Haney paid his way through journalism school by working nights for the Associated Press. After earning his Bachelor's degree in 1945, he worked for newspapers in Pennsylvania and Tennessee before joining the Washington [DC] Evening Star in 1954.
In his 1992 oral history interview, he talked about his early fascination with the newspaper industry:

> *My mother's father was the editor of a paper in Mobile [Alabama].... I met him only once, when I was 5 years old and attending kindergarten down there in Mobile. He took me into the print shop – I'll never forget this – he sat me down on his lap in front of a linotype machine. He took my left hand and dragged it down the [second] row of keys, which, if you know anything about printing, spells S-H-R-D-L-U. That had a tremendous impact on me. That was one of the most dazzling machines I had ever seen at age five (this was 1933), and I can still feel that moment. I was just overwhelmed with technology.*

Haney combined his fascinations with news reporting and technology when he joined NASA three months after it was formed. Between 1958 and 1969, he served in various public information posts at Cape Canaveral in Florida and the Manned Spaceflight Center in Houston. He ultimately becoming known as the "Voice of Mission Control." His NASA service spanned the entire Mercury and Gemini programs and the first three manned Apollo missions. After

Paul Haney was inducted into the International Space Hall of Fame in 1993. Courtesy of NMMSH.

that, he worked for a British television network and continued to report on the NASA programs. His behind-the-scene stories include the Apollo 1 fire and all of the Apollo flights.

The first real recognition of space that I had was in 1955, when the [President Dwight] Eisenhower government announced that there would be a National Geophysical Year in 1957 and the US would take part in that. For some reason, I handled the story on the copy desk of the *Star*. I had to write a two-line headline for a late edition, and I remember that I wrote "The US to Orbit 'Moonlet' Around the Earth." Some of the old-timers said, "Gee, that Haney, he can really understand science." That's what happens at newspapers. You could go out and handle a story on the gold standard, and you immediately became a gold expert. Suddenly, I was the bright shining star of science at the *Star*.

In 1957 when Sputnik went up, I was assigned to do some special coverage for the *Star*. I had four or five guys, and we were running around town making phone calls and trying to shake down all the big science types and get statements, and we did "man on the street" interviews. The advent of the space program [was one of the year's biggest events]. People's jaws dropped and their eyes glossed over. I'm not overstating it. The fastest you could go in an airplane in 1957 was around Mach 2 or 3 – maybe 1,200 to 1,800 miles an hour. All of a sudden, even to play the game in space, you had to go 18,000 miles an hour to be in orbit!

By 1957, there were ongoing discussions in Congress about how the US had "been had" by this god-awful Russian bear [i.e., the Soviet Union]. How did they get so far ahead of us? Suddenly, we were examining our arithmetic programs. The schools were deficient, and mothers were deficient, and fathers were deficient. As a result, we created "new math," and I'm not sure anybody ever understood the "old math." We decided that this was terribly serious. The hearings leading to NASA started in the summer of 1957, and the Eisenhower administration took a great deal of the heat. Lyndon Johnson was Majority Leader of the Senate, and he took a very aggressive posture at the hearings. Johnson made Eisenhower look good. Eisenhower had real troubles. His second term was a downhill thing.

It was a typical Eisenhower staff thing. There was a space council, but you never saw these people, and I wonder if they ever met. It was populated by various figureheads on the White House staff, and NASA was trying to struggle along. But the hardware we were launching was largely Army hardware that had been willed to us. It was like trying to raise somebody else's kids in a second marriage, and it wasn't easy. Probably the greatest technological advance that Eisenhower identified with was the Jeep. He decided that nothing was ever going to eclipse that, but space was an interesting toy. Ike [Eisenhower] went to Huntsville when we dedicated the Marshall Space Flight Center in honor of his old general buddy. He may have gone to the Cape once, but I'm not sure that he did.

[Former President Harry] Truman was pretty much the same. The Cape was started during the Truman years. It was called the Army Long Range Proving Ground because Truman was an artillery officer in World War I. In 1947, they began testing the air-breathing missiles there. Then the Air Force took it over, and the missile business took off in the late '50s. I remember Truman came to NASA one day and I was flattered. [NASA Administrator James] Webb called me in to brief Truman on the Mercury program. Typical of Truman, he took one look at the Mercury capsule which I was talking about, and he nudged Webb, sitting next to him. "Tell me, young man," said Truman, "how does a fellow take a pee in something like that?" He didn't wait for an answer. He just burst out laughing.

I don't know what I was doing the day [President John] Kennedy was inaugurated, but I remember it snowed like hell the night before. There were 20 inches of snow in town, and it was a real struggle to get the snow shoveled into the river. Typical Kennedy, he showed up in street clothes! Robert Frost was there reading a poem, and the lectern caught on fire [during the invocation]. It was curious, the tenor of the town. You loved Ike [President Eisenhower] because he was a grandfather figure. He was a nice, gray-sky kind of a person. What a different feeling it was, all of a sudden, to have this young President Kennedy. He was in his early forties. He and Jackie were having babies. Washington looks at itself in terms of the personality of the President. It's always been that way and always will be, I suspect.

To Kennedy, space was like the Holy Grail. Twenty days after Shepard flew [America's first, suborbital spaceflight], Kennedy went before a joint session of Congress and proposed the Apollo program. I don't think there were but two or three people in NASA who knew what he was going to say. We thought he was going to propose a beefed-up Mercury program and perhaps go along with what we called "the two-man Mercury," which became Gemini. But we were, for the most part, thunderstruck at this idea. He said the nation should land a man on the Moon and return him safely to Earth before this decade was out. That was May 25, 1961, and we had flown only the first of six manned Mercury flights.

I knew I had signed on for quite an interesting voyage. It was going to last ten years for me. All of a sudden, the whole establishment began to pick up. The space budget was more than doubling each year. Lyndon Johnson, the Vice President, was responsible for space. One of the things he did was hire Jim Webb to be the administrator of NASA in 1961. Webb worked for Kerr-McGee as executive vice president and understood the oil patch. He worked for Truman as Director of the Bureau of the Budget and later went to the State Department. He was Number 3 at State under Dean Acheson. He was involved in the tremendous decisions of the late '40s on China. Webb loved to harken back to those great days of the late '40s when he worked for Truman.

In any case, in 1958, they decided to set up a space agency. To everybody's amazement, there was a lot of competition for it, in the military particularly. That just knocked everybody flat. They took bits and pieces out of the Air Force, Army, and Navy and grafted them into a civilian agency called NASA. In a previous life, NASA was the National Advisory Committee for Aeronautics, NACA. With NACA, there was no central authority. They purposely had no driving administrator. They had some very distinguished staff. [Charles] Lindbergh was a director of NACA for years. They were quiet, they covered their asses tremendously, and yet I think they were very effective in their way. One difference between NACA and NASA was that NACA had no test directive. NACA got great ideas on airfoils, surfaces and laminar flow, what makes lift, and "Coke bottle airplanes" [an aircraft design in which the fuselage narrowed in the area where the wings were attached, to reduce aerodynamic drag]. The whole X-series came out of these brilliant minds at NACA, but when they'd get up to the test point, NACA turned to the Air Force or the Navy or industry.

Poor old NACA never quite scored in the public relations area. That dawned on me when I went to work for NASA just three months after NASA started. That was a year after Sputnik went up. I always felt it was kind of like another football team showing up for the game a year late, while the other team was running up and down the field making touchdowns. We actually got into business October 4, 1958. Sputnik went up October 4, 1957. The agency let it be known they needed some information people. I was at the *Star*. To me, it was like Columbus saying "We're going to sail west and we need a cabin boy. Would you like to join us?" Here I was, an assistant city editor of the best paper in the United States. Without hesitation, I walked away from the *Star*. I had been with the *Star* for four years, and we had won three consecutive Pulitzers. I don't think any desk has done that before or since. It was a tremendous sense of accomplishment. But NASA and this space possibility, I couldn't understand anybody holding back. One of the things I had to do first was recruit one of the service PIOs [public information officers]. I hired a guy named Shorty [John] Powers, who got a lot of identity [as the "Voice of Mercury Control"]. When I left the *Star*, I remember the editor of the *Star*, Newbold [Newbie] Noyes saying, "Paul, that's the dumbest thing I ever heard, you walking away from the *Star*. I'll tell you what, I'll keep the seat warm and open." I used to see Newbie at some of those big dinners that they have in Washington. I'd say, "Newbie, I don't know, I think I'll stay another year or so." It was fun.

All those satellite events in 1959 and the first weather satellite in 1960. The first camera flown in something called Tiros [the first successful low-Earth orbital weather satellite]. The first picture they took was of Montreal and of the St. Lawrence River harbor. The second picture was of Communist China. It was sharp and showed much detail of where the Yangtze River flows into the sea. Tiros had a [low-resolution] camera that took a picture 100 miles on a side. Well, the 35-mile [high-resolution camera] was so sharp from that first go-round that we let the CIA dictate to us that the camera was broken and we had nothing to release. We released some fuzzy pictures on the 100-mile scale, and they were pictures of oceans and North Africa and things that were not very controversial. But the boys over at Langley, their eyes were rockin' with the definition. We could see stripes in a parking lot in San Diego. I remember that picture very well. You could see the markers where the cars were to park at the Naval Station in San Diego. All of that was fascinating to me as an information function.

I remember Echo I [the first passive communications satellite experiment]. It was a [metallized] helium-filled balloon and people could see it, particularly if you were in a darkened area and the light was still hitting it. Here we were, just three or four people in the NASA information office. I can remember people wanted to see this so badly. It became an international story. I couldn't get the projections from the engineers on when the thing was going to be over a given city. I am not a mathematician, but I remember doing my own projections for about three or four orbits. We moved them on the wire and they were page-one stories all over the world. We had calls from Delhi, India, and from Bombay and China and everywhere else, even in Russia. It got so big that all of the major newspapers in the world carried little boxes on page one, "Echo Sightings." In fact, they were ultimately incorporated into the weather box.

Echo was a great space demonstration for NASA. It was also remarkable how the first communication satellite worked so well. I can remember standing on a hill in New Jersey and hearing the first words when they came through the local lab there. We got to do everything first, mistakes and all. For the most part, the public was awestruck. The Mercury thing was the first project approved on the first day of NASA. That was the Number One Project. It was, again, generated a lot by the Cold War and the fact that we knew the Russians were playing around with those things as well, and we wanted so desperately to beat them in space. We never admitted to the people how intense that was, but it came out in so many ways. I remember the Russians had a long string of successes when they hit the Moon in '59 [a hard landing with an unmanned spacecraft]. That was the time that [Soviet Premier Nikita] Khrushchev landed in L.A. He went out to a Hollywood set, I remember, and got ahold of some movie star and told her her dresses were too short, and if the United States wanted to do as well on the Moon, they should pay more attention to morals. This guy was really crafty. They were just running away with us in every area of space. Now the Cold War has come to an end, and I still wonder how they could have done so well in space accomplishments. I can't quite resolve that in my own mind. We wind up with all the marbles, I guess. But they had a very productive space program.

I guess it was late Mercury, but I remember a Russian spacecraft going over the Cape one day. They called us in the simulation [facility] and said, "Hi, there, this is Voskhod." Voskhod was their two-man spacecraft. The message was taunting, to say the least. The guy doing the talking spoke fairly good English. He said, "Well, be sure and do your lessons and if you succeed, you can come up here and fly with us. We're having a beautiful day and we can see Pad 19. We can see the control center." I remember he identified these things at the Cape. "We are approximately over Tallahassee, how do you say 'Tallahassee'?" It was a kick in the gut. CapCom [capsule communicator at Mission Control] said something like, "Roger, Voskhod, have a nice flight." They came back, "Oh, we will. We have lots of them." You know, they were really sticking it to us – the Red Baron kind of stuff during World War I. I remember sitting there listening to that message, and I really felt like a scalded ape. I thought that was dirty and yet, if things had been reversed, I am sure we would have done the same thing.

Chronologically, the space program was incredible. It doubled every year in the infant years. I think it started at around $180 million in 1959, which was the first year. There was some kind of supplement that drove it up to about $200 million. These were Eisenhower budgets. Then, in 1960, the last Eisenhower budget went to $400 million. Kennedy came in. Everybody thinks of him as a "gung-ho" space President, but he scared the daylights out of us as a candidate. He went around the country wondering if the Mercury program should have an Atlas or a Titan under it, the kind we used in the Gemini program. Kennedy really terrified us because it showed that he didn't have much footing in space. But he certainly came around. He became a quick convert to the real space effort. In the spring of '61, [Soviet cosmonaut Yuri] Gagarin flew in orbit, and about three weeks later, Shepard flew about 300 miles downrange. Some equated those two as pretty similar events.

I talked about Eisenhower being kind of a gray-sky thing. NASA, in its beginnings, was similar. It was dominated by old bulls who came out of the NACA world. The "Vatican" was Langley Field. All the great things had come from Langley Field. In fact, Samuel P. Langley was "Mr. Aviation." The Wright brothers, in the view of the science establishment in Washington of the early 1900s, were a couple of second-class bicycle builders from Dayton, Ohio, and they didn't know doodly about the principles of flight. The establishment still thought the world and all of Langley. Langley Field in Virginia was the Edwards of its day. All great things aeronautic came out of Langley Field. When NACA was created in 1915, it shared a hangar with the US Army Flying Signal Corps at Langley Field, 160 miles south of Washington, which tells you something.

The old bulls, as I call them, metamorphosed from NACA to NASA. I remember a fellow named Henry Reid. I visited with him six months after I went to work for NASA. I went down to Langley and was ushered into the presence of this great fellow who was the director of the Langley Research Center. Harry Coleman, who became the managing editor of *Av[iation] Week*, was with me. We had just gone to work for NASA, and Reid sat us down and he said Project Mercury was idiotic. He said he was doing everything he could to kill to Mercury Project, then based at Langley. I couldn't believe that man. I wish I had a tape of the conversation. There were others who felt the same way. Vannevar Bush was a big science charger at Yale before and during the war. In 1948, he was one of the chief science advisors of the US government. He said it was ridiculous, this is a quote, "to think that the US or any other country would ever use a missile as a means of war." This was after World War II, [a few] years after the A-3s and the buzz bombs hit London.

When Kennedy came in, all of a sudden, there was a sea change as we got into the Apollo program. Senator [Clinton] Anderson from New Mexico was very much a part of this. He was a power in the Senate. I remember just after the Shepard flight, the scene in the Oval Office. Anderson was there, and [Oklahoma Senator Robert] Kerr and Johnson. Kennedy was sitting in the rocking chair and wouldn't look up. He kept rocking. All seven astronauts were there, along with Bob Gilruth [the first director of NASA's Manned Spacecraft Center], seated on two long couches. Kennedy kept pushing these guys in informal questioning on the impact of what a man could do versus a machine in space. Glenn was probably the most eloquent in responding. Glenn was very close to the Kennedys. He'd go up to Hyannis on weekends. He taught Jackie how to water ski, as a matter of fact. Glenn emphasized that a man could reach over and turn a dial, whereas a machine, unless properly programmed, could fail. That scene sticks in my mind.

Twenty days later, Kennedy went up to the Hill and made his pitch, "I think this nation should put a man on the Moon before this decade is out." Jim Webb was aware that the plan was rattling around in one or two other people in the administration, but he wasn't sure. That was strictly a one-man decision, and no task force worked up a bunch of alternatives, which is what so often happens when you go fight a war or make any great national policy move. That was John Kennedy, pretty much on his own, and we still don't know how many people he talked to! It was absolutely amazing. This conversation I am talking about took place on May 8, 1961.

After the Apollo announcement, there were huge real estate deals. NASA bought 90,000 acres north of the Cape, which was six times the size of the old Air Force Cape at 15,000 acres. NASA, this young kid on the block, a product of a forced marriage of a bunch of component parts, grew to about nine feet tall overnight. The real estate deals were very prudent politically. The big rocket to loft Apollo would have to have a testing place. Why not in Mississippi, just up the river from the old Michoud plant in New Orleans? Michoud would become the center of production for the Saturn's lower stages. Huntsville, in nearby Alabama, was in charge of booster design and production. The Tombigbee River could be used to barge the elements of the Saturn into Huntsville. I never understood why we were going to test all these things, and why we didn't do it like the Air Force did and build them in California and fly them into the Cape and shoot them. But, again, this was politics working.

At White Sands, we had to put a NASA center on the other side of the mountain, which was an extension of another center. We decided to do the center in Houston. Before that, all the manned stuff was going to be moved from Langley to Goddard, close to Washington. All of a sudden, it got bigger than Washington, so it was decided to go to Houston. Before we finished siting new functions in 1962, the so-called "space crescent" was created, stretching from California to Florida. The crescent included test facilities in New Mexico, a new space center in Houston, and production in Oklahoma.

These were giant steps, suddenly taken. But none was as big as the mode selection [for lunar launch] in '62. It became a big controversy within the scientific community, whether you fly direct ascent, lunar-orbital rendezvous, or Earth-orbital rendezvous. [The direct ascent mode would require a very large rocket to launch straight to the Moon. The lunar-orbital rendezvous scenario would use a relatively smaller rocket to carry the spacecraft into Earth orbit, and then fire again to launch toward the Moon. After completion of the engine burn, the rocket would be cast off and only the spacecraft and lunar lander would continue to the Moon. The Earth-orbital rendezvous alternative would launch two of the relatively smaller rockets into Earth orbit. One would carry a cargo of fuel, and the other would carry the spacecraft and astronauts. While in Earth orbit, the rocket carrying the Apollo capsule would be refueled by the other rocket so it could propel itself toward the Moon.]

Von Braun and the people in Huntsville wanted to go Earth-orbital rendezvous. The people at Langley (what eventually became the Houston center) wanted to do lunar-orbital rendezvous. It was the best idea that ever came out of Langley. Others were for direct ascent. You would build a rocket about as big as New Mexico to get there. It got very cumbersome. Anyway, the phase selection took a year. I can remember writing the final release when we made that decision. Joe Shea, the head of the Apollo program, had to brief all the various elements, including the White House. We were on Lafayette Square, at the Dolly Madison house. Finally, in August of '62, the mode selection (lunar-orbital rendezvous) was made, sixteen months after the decision to go to the Moon was announced.

On the Mercury flights, we flew monkeys in several of the Little Joe flights, three of them, as I remember, out of Wallops Island, Virginia, which was near Washington. They were a lot of fun but were conducted in secrecy. Again, those were Eisenhower rules. They really hurt. At the launch site, we operated on the basis that you could do a launch announcement only when there was fire in the tail of that bird – nothing prior to ignition. It was always a canned announcement and it said, I can still remember, "The Air Force (or NASA) today launched successfully a (fill-in-the-blank) type rocket, and it was last seen going downrange." This boilerplate said nothing and gave me fits. If the rocket didn't ignite, sometimes we said nothing. At the Cape, the problem was the Air Force owned the real estate. They wouldn't approve the releases unless it was favorable to the Air Force launching crew. They always worked their PR games. It got better when we got our own real estate and started launching our rockets.

During the Eisenhower years, the model public information program was the Russian program. If it didn't work, we said the thing flotched. We never explained why. That was the Air Force and the Army and the Navy way. You said that something that was launched today was a canned experiment. One paragraph, three lines long. I can almost remember it. And if it worked, that's all you put out. If it didn't work, you said, "Due to mechanical difficulties, the XYZ booster experiment failed today," and you didn't even describe where it failed. First stage, second stage, third, and I don't give a damn if this thing blew everything in the Cape up, you never deviated from that statement. You can imagine how this grated on this old newspaperman's bones! We couldn't even announce that a flight was coming up next week or next month. We were saddled with a quarterly presentation that in the next three quarters, we will fly three Mercury missions. This was the Eisenhower approach, and that was a great breakaway, because the Russians weren't even doing that. The Russians never announced anything in advance, and the Eisenhower administration thought that it was a pretty damn smart way to go.

I knew that it just had to change, and I hoped that Kennedy would feel like that, and I hoped that we could change it. And, by god, we did! I wrote the change orders for the Mercury program, and I remember we were at the Cape just before Shepard flew and that's where the change came in.

That was an enormous change when the Kennedy administration came in. The way it timed out was that the Shepard flight was May 5, and I went for the change and presented to [T. Keith] Glennan [the first NASA administrator] and his group in December. All the engineering people, all of the people in headquarters and in Houston voted "no" on my proposal, which was that we would announce flights forty-eight hours in advance and open the gates to live coverage. This just blew their minds. But I knew damn good and well these things were going to get covered some way, whether we did it or not. And Glennan, God love him, was the only one who voted with me and he said, "I'm not going to put this policy change aside, I will simply put it on the desk for the next administrator." He said, "Frankly, I favor it, but I appreciate all the no's around the table. I can't outvote you all, but I do have the power to keep it alive." And he did.

The new administrator, Jim Webb, came in February. Webb looked at it and said, "How does the White House feel about this?" I said, "Well, I expect to get a report from there soon. It's over there now." I had presented it to [President Kennedy's White House Press Secretary] Pierre Salinger earlier that week. I went to the Cape to set up the press quarters for Shepard. On the second or third of May, I got a call, and it was Mary Lincoln, who was Kennedy's secretary. I remembered because she didn't call very often. She said, "This is Mrs. Lincoln. The President would like a word with the senior person in charge." I looked around and I didn't see anybody that outranked me, and I said, "Well, that would be me, Paul Haney." Kennedy had a habit of doing this. He would not call the administrative chief of an agency, he would call down in the bowels of the ship. He is one of the few Presidents who did it, and he did it for effect. He knew what he was doing, and instead of calling Gilruth or Webb or whoever, he would call the guy on the line.

Mrs. Lincoln said, "Would you hold, please? He's talking to the Secretary of State in Paris (or something like that) and he'll be right with you." Then Salinger picked up the phone. He said, "Paul, he's wondering," and I could hear Kennedy talking to Salinger, "frankly, the boss is wondering how good is that escape rocket?" And then there was some laughter on the other end. Fortunately, I knew the answer. I said, "Well, Pierre, it's been tested and it's worked fifty-seven out of fifty-eight times up until right now. And the one flotch was excusable because of lightning or rain or something at the test site." I don't remember why the hell I remembered that. I heard Salinger say, "Hell, it's worked fifty-seven out of fifty-eight times." Then Pierre said, "You can tell how his mind's running, he's wondering how do you escape from something like this if it doesn't work on the pad?" I said, "It'll work. That's the surest component we've got." And he said, "Okay, you've got a 'go.' Do it!"

I remember I just threw the phone up in the air. I don't know when I was ever so exhilarated by a decision. I said, "God damn! We got it!" That changed the whole format, and it opened the windows to the public information officer on the line calling out the progress of the flight. I mean, it was radical in its departure. It opened the program up and made us responsible for not just the good things, but the bad things as well. Kennedy understood that. The program changed dramatically.

Hell, when we got into the Mercury flights, all three networks would get on and stay with the flight as long as it lasted. Sometimes they would be on for six to eight hours. Then they'd come back in the evening with the recovery film. We would feed it into one of the networks. We organized a national pool, which wasn't just television. We organized the word pool as well. All of them could share the pool product. You could sit there, as a writer, and look at the tape and the television. It was one hell of an effort, and I was very proud of it. My single biggest contribution.

I never had any "Walter Mitty-ish" [a mild person with a vivid fantasy] ideas. I never wanted to fly on a spacecraft. Hell, I didn't want to fly in an airplane, except that it's a lot faster to get somewhere, and maybe that's what set me apart. Most of the rest of the people in the program, doctors and lawyers, most of them secretly wanted to be an astronaut. I never wanted to be an astronaut! I still don't want to! Some of my best friends were astronauts, but they appreciated that of me, that I didn't have any little secret desires to go wear parachutes or get hemorrhoids from sitting on a parachute pack for twelve hours a day. Simulators are dirty, smelly things, and you are sitting there smelling somebody else's body odor. It's just not my cup of tea. I did it in order to understand how the dials worked and what all the levers were. You had to do it to understand it.

The Mercury Seven were closer than any other group of astronauts have ever been. They were really close, and they were aware of their significant position and selection. I don't know of a group of people that handled the public adulation any better than they did. These guys were like the Beatles. They were as well known or better known than the Beatles in that era. Tom Wolfe's book [*The Right Stuff*] was a wonderful book. That was a sorry movie, however. The book, to me, conveyed the right stuff. All of that happened. I wrote Wolfe a fan letter after it came out, and he wrote me a page-and-a-half letter back, which kind of amazed me.

That was exactly the book that I wanted to write from Day One, before I got into the program. And Tom Wolfe, another newspaper guy from the other Washington paper, wound up writing it.

As individuals, the Mercury Seven generally got on extremely well. There were certainly some differences. I thought John Glenn and Wally Schirra were going to blow each other away, along about Wally's flight. They got into one hell of a public pissing contest, and I was trying to negotiate between them. There are always going to be cliques. Essentially, Shepard, Grissom, Slayton, and Schirra represented the pilot's pilots. John Glenn was pretty much "Mr. Independence" and stood by himself. Gordon Cooper was the kid of the group. Scott Carpenter was kind of flaky; you could never tell what planet he was on. He would get terribly interested in flying and then, one time, he decided he couldn't fly anymore and he didn't fly for quite a while. Then he got interested in the undersea thing. They called themselves the aquanauts, I guess. He took a leave of absence from the NASA program. Scott had a wide range of interests and is a fascinating guy. Just to sit down and talk to somebody, Scott Carpenter is still, to this day, one of my favorites because he has such range. Scott got to fly because Slayton got bounced out of a flight at the last minute.

The Air Force threw Deke [Slayton] away like a dirty piece of paper. Finally, it went to Dr. [Stanley] White for decision. Dr. White was "Mr. Heart," a big muckety-muck. He was called in and he said no, he wouldn't let somebody fly with an atrial fibrillation like that. Deke was shot down there in 1962, and then they really didn't know what the hell to do with him for a long time. Finally, the Air Force said, "If you're not flying, you're no longer TDY [temporary duty assignment] to NASA," and they cut Deke off at one point. That's when NASA hired him as a civilian to head the crew program, because of all his background. He was made assistant director for flight crew, and that was a nice bit of coverage by NASA. He became part of the senior staff in Houston, of which I was one.

Something similar happened to Shepard with his inner ear business. The Navy began to threaten him. "If you can't fly airplanes, you're not on our rolls," they said. Al went off on his own and found a physician in Los Angeles who specialized in inner ear problems. The guy did an operation on Al and brought him around when nobody else in the medical establishment would do it. It was Meniere's Syndrome, which affects your balance mechanism. John Glenn fell in the bathtub and wrecked his inner ear, which knocked him out of space and everything else for a while. But he managed to come around, and he sort of conquered that on his own, like good Marines are supposed to.

The older astronauts were left to fend an awful lot on their own. Leo DeOrsey [attorney for the Mercury Seven] was the first outside adviser who was a Washington attorney. He had managed a number of Hollywood show biz and New York personalities like Arthur Godfrey. DeOrsey wound up owning the Washington Redskins. DeOrsey was truly significant in that he worked out a lot of the policies for the Seven, and outside interests. He got them a personal insurance policy, which was really an inside contentious point, particularly with Webb. Webb said, "Look, if anybody ever gets hurt or killed, there will be an immediate $100,000 payment to the family, a special bill that would sail through Congress." It was just an understanding that a special bill would go screaming through Congress and, frankly, the pilots listened to all of this and said, yes, but that's just kind of a wish, a hope. One of the things DeOrsey worked out was a joint insurance effort from the private sector.

When Gus [Grissom, who died in the Apollo 1 fire] got killed, Betty Grissom had $100,000 in her possession within thirty-six hours of his death. It took something like four years to get the US government to pay off. On the last day of the North American [Aviation] contract on Apollo, which came along, I think, December 31, 1972, Betty got $400,000 and the widows of the other two got $250,000 each. There wasn't any negotiation. It was "here's a check," and it was the last day on North American's billing to NASA, put in as a purchase order. Checks were cut and they were sent along. It took that long after a 1967 fire.

Even though the Seven had some disputes among themselves, I have seen it happen again and again, they would retire from a room, sit down and arrive at a position. They would come back and announce their decision, "Five of us want to do it this way and two of us...." That's what I admired. Frequently, I was the carrier of the position back to the other room. So, there was an involvement, and there was a certain closeness.

I went to Panama for two weeks of jungle training in 1964, and that was interesting. I went with Pete Conrad. He was den mother for the new fourteen [astronauts]. The site was south of the Canal. The guy who ran the program said he knew where he put us, to stay there and don't wander around. Don't go to Colombia, which was only a few miles away. I took along a transistor radio, which I don't think we were supposed to do, but I wanted to hear a little news. I felt cut off. We could get a news program in the morning and evening, and the rest of it was Panamanian rock and roll. One night, this guy that ran the program came back to visit us. We were just trying to get along with our little hammocks, chopping wood, and eating berries, and existing by catching fish in streams. I remember how dark the jungle was. All that heat, and none of the sunlight was penetrating. This guy, who ran the program at Albrook Air Force Base, heard this noise and he pulled his machete. It was my transistor he heard. He thought some group was sneaking up on us.

The night before they came to pick us up, Pete and I got to wandering around. No more than a hundred yards from base camp, we ran into some Choco Indians, who are indigenous to that jungle. They were most entertaining little guys, even though we had no language. We did signs and so forth. I don't think any of them were over five feet tall. They could go up a tree faster than I don't know what. They were very outgoing, very expansive, and they showed us around their little community. They invited us to dinner, which was nice of them, I thought. They were having boa constrictor for dinner. That's right! Boa constrictor! They had chopped this snake up like cordwood. Each piece was about twelve or fourteen inches long. They had these sitting around the fire, and every once in a while, they would turn these pieces of meat. It was the most delicious meat that I have ever tasted! It tasted exactly like chicken breast. It was light and beautifully done. It got to be a standard joke when we got back to Houston at a dinner party or something, Conrad would look at me and say, "This chicken is wonderful. It tastes just like boa constrictor!" I haven't had any boa constrictor since.

There were some horrendous snakes down there. I can't remember the name of the one which is among the four or five most poisonous in the world. The story goes that you are dead before you hit the ground. We found one of those snakes and killed it and used it for bait to catch some fish. We had some great meals in addition to the boa constrictor. We had broiled fish. The leader of the expedition showed up on the appointed day, picked us up, and we went back to town. We had one night of liberty in Panama City before we came home.

On a geology training field trip, we were walking in and out of the Grand Canyon with a bunch of Harvard geologists leading the way. Invariably, they would stop about every 300 yards and look at a big rock pile and say, "What do you think's been going on here?" Geologists just keep asking those questions. The toughest part is the going down, as opposed to the coming out, because you have your brakes on all day and those muscles in back of the legs get a little tired. Al Shepard and I were in the same team. We got to the bottom near the Phantom Ranch, where we were going to stay overnight. We had about 200 yards to go, and we were both really dragging ass. Shepard looked at me and said, "I'll beat you to the tent." We sprinted and he touched out about a half yard in front of me. There was a tremendous lot of this childish excess of competition, hiking, swimming, beer drinking, and the likes.

The Mercury program wrapped up in the summer of 1963. Gordon Cooper's flight was May of 1963. I moved to Houston in the fall of 1963, and we did a wrap-up conference on the Mercury program. Gemini was well advanced, having been announced in Houston on December 7, 1961, at a Houston Chamber of Commerce meeting. To me, it was interesting because Gemini was just called "the two-man program." It didn't have a name at that point.

There was a big change and maturation there [with the Gemini astronauts]. The first Seven, as I indicated, were just kind of off on a cloud to themselves. The Second Nine, as we called them, came in early '62 and went into about a two-year period of training. They were way in the back of the bus. I mean, they just weren't allowed to come up front and associate with the Seven. It was almost like 3A teams visiting the Big Team, and the Big Team sure as hell let these new junior members know. I mean, they did not welcome them with open arms, and they didn't go out of their way to do it in public. You had to work in the same hall to realize what was going on. There was very definitely a caste system there, and it was remarkable as time went on.

After 1963, there weren't any more Mercury flights. Because there was one person in the Mercury spacecraft, it was a lot easier to identify with a single person. When the Gemini flights started with two people, it was a lot harder for people to remember the name of the person in the other seat. By the time Apollo showed up with three people, it was quite a crowd, and it just became "the crew." By the time the third group of astronauts came aboard in '63, the New Fourteen, as they were called, the second group and the New Fourteen made friends quickly. But the older heads still seemed to maintain their distance. The recognition factor was like two different worlds.

The Second Nine were a different breed of cat. They were more "Yes, sir. No, sir," and salute people. There weren't more Academy people, but

Ed White just looked like he had been extruded from a West Point cookie cutter. I mean everything he did, he even ate square meals. We used to joke that he probably made love square, you know. One of his great lines was, "When I stepped out of Gemini 4, I felt red, white, and blue all over." He made that as a quote, and I tried to talk him out of it. *Life* magazine tried to talk him out of it, but he wouldn't change it. And for all of his wonderful qualities, he was a great athlete and almost made the Olympic team in 1952 [he was back-up in the 400 meter hurdles]. Incredible runner. He had a heartbeat of around 52 just sitting idly. He had so much going for him. His daddy was an Air Force general, and he knew the whole military existence. But he saluted everything and everybody.

When White and [James] McDivitt went up in Gemini 4, that was really a grabber. White's EVA [extravehicular activity, i.e., spacewalk] was really the big accomplishment. Ed had done West Point, and then had gone to Michigan later and got a Master's degree in aeronautics. McDivitt was a very bright aeronautical engineer, and that's why he was sitting in the left seat, I guess. But then that let White be the guy to go out and do the first EVA, which was a wonderful accomplishment, one of the three big goals of Gemini. What mystery surrounded that event. We weren't supposed to let anybody know it was going to happen. NASA got so secretive. Actually, the NASA establishment wanted to spring it in flight. If the hatch opened and the systems looked all right, we were suddenly going to say, "Oh, there goes Ed White, and he's going to get out and walk around." I railed against this for months. I said it was stupid and if something happened, it will look like it was totally unplanned. But I got voted down. Shouted down! That's when we had to sign a book promising to say nothing. EVA practices were held in total secrecy. It was the damndest mess, but it came off because the AP [Associated Press] guy at the Cape found out there was a strong possibility of EVA on the flight.

The EVA was actually targeted for Gemini 6. [Thomas] Stafford would have done it. Then it looked like it could be ready as early as 4. We did it early because we wanted to do something ahead of the Russians and actually beat them on an EVA. That's why it was pushed up so far. But it worked out beautifully. We finally did announce it three days before the flight, after it had leaked out pretty good. We went through three press conferences. We did a press conference at eight o'clock in the morning in Houston, we did one at ten o'clock at the Cape, and then we were back in Houston for something else that afternoon, just to take care of the EVA. Earlier, Bob Seamans, who was then the deputy administrator, came down to the Cape for the buy-off session on the mission. He said, "What EVA?" I just sat there and kept my mouth shut. Chris Kraft started explaining about the EVA, and then Seamans looked at me and said, "Paul, there hasn't been anything announced on this." I said, "No, there hasn't." And he said, "Well, I don't think that's kosher. I think it should be announced before the flight." I just sat there and looked at the tablet. Kraft made some noises and said, "Well, we've had input to that effect," or something like that, and I let him explain it all. After the meeting, we did as Seamans suggested. This was strictly NASA, not the military, playing security games.

On Gemini 4, they had a hard time closing the hatch. Actually, it wasn't secured until about the next rev. They had to take the damn thing apart, but McDivitt was enough of an engineer and at that point, the first EVA, we didn't know how cold it got on the other side. We weren't too sure about the heating in the suits and things like that. So, they sort of pulled the hatch down but they could not secure it. Then when they got around to the sun side again they both, quick like a bunny, took the door apart. They dropped a couple of screws, which floated off. Then they took some screws out of something else and they finally figured out a way to secure the thing. None of that came out at that time. Everybody had their foot in my mouth because they were afraid I was going to say something. This would be a sign of failure, you know. We weren't into talking about failures much.

Gemini 8 was the first Gemini docking experiment. The dock was performed over South Africa. Neil Armstrong and Dave Scott were the pilots. As soon as they locked up with the Agena, which had been independently launched, the whole huge affair started tumbling. Because everything had been all right up to there, it was assumed it was the Agena's fault. It turned out it was a thruster stuck open in the Gemini spacecraft. It was stuck open and thrusting, so it was providing the impetus to take this thing end-over-end, about 25 to 30 feet long in combination with the Agena, which was the second stage of an Atlas. It was our first real in-flight emergency. We were in Earth orbit, so we were only picking up the crew every twenty-five minutes. They docked over Madagascar, and they checked in over an Indian Ocean ship about twenty minutes later. Neil Armstrong's voice came on over the line, and we could see it on the telemetry and wondered if there might be a mistake here. Neil said, "We're revving up at a pretty high rate and I can't turn anything off." I remember that's exactly what he said. They were spinning up at incredible rates. Had it gone on much longer, they would have blacked out.

They were told along about Australia, I guess, to get off the Agena. As soon as they did, they could see it wasn't the Agena's fault. Of course, they had imparted this motion to the Agena and they could then identify that it was a stuck spacecraft thruster. They had to use the reentry ring. It stopped the stack thruster. They had to use half of the reentry ring gas to stop that motion. They did get it stabilized by California on that rev, and they were directed to come on in on that rev. They were to fire their reentry rockets over east Africa. They landed about 180 miles off the coast of Red China. There was a lot of speculation: Would they clear the coast of China? There were people who were going to call the Red Chinese, who we weren't exactly friendly with, and say, "Hey, if some guy named Armstrong from Ohio comes down, please give us a call and treat him right until we can get there." We fired up an airplane, and Kraft and I were going to go to Ellington Air Force Base and get on that plane and make a fast trip to the China shore. The reentry was quite successful. They splashed down 180 miles off the coast, landing beside a US destroyer that just happened to be in the neighborhood. The destroyer picked them up.

Gus Grissom, who had the second manned Mercury flight, had gone through a terrible time with NASA and Washington because that spacecraft, Liberty Bell 7, sank. Gus went to his grave defending what he did in his Mercury flight. Gemini 3 got a little controversial, too. John Young, who was sitting in the right seat, absolutely idolized Gus Grissom as a pilot and as a man. In the course of each day of the simulations, Gus, a creature of habit, would invariably say, "Gee, I wish I had a corned beef sandwich along about now." And John would always agree. Schirra or somebody would get in a car and go down to Cocoa Beach to a place called Wolfey's and get some corned beef sandwiches, the standard New York kind of corned beef. So, in the second orbit of the flight, almost like they had rehearsed it in the sim [simulation], Gus turned to John and said, "Damn, I sure would like a corned beef on rye along about now," at which point John reached in one of his flight suit leg pockets and pulled out a corned beef sandwich on rye and said, "Just happen to have one with me, Commander." They hadn't bothered to log it on the flight.

Jim Webb, the administrator, really raised hell about that. Webb was an old Marine Corps pilot himself, who had learned to fly the mail between St. Louis and Washington in the early 1920s. He really woodshedded those guys over that. In fact, the Senate Space Committee held hearings to deny them promotions. It had gotten to be a fairly routine custom after you flew, you got the next stripe, a promotion for the military guys on TDY to NASA. Webb went up and testified that he didn't think a promotion was in order because of their reckless behavior in taking that sandwich. I mean, it was really bad news, but it went away after we got into the program.

The thing about Gemini that I can't get over was the rapidity of the flights. We flew ten flights in nineteen months, including three launches in ten days. I don't think the shuttle has ever equaled that frequency, which is almost incredible since all those flights were in 1965 and 1966. Turning them around at six to eight week intervals, I still don't understand how we did it. The whole program got very businesslike, and it was almost like working an airline terminal or something because the turnaround was so fast. We were trying to show that we could do that. The Gemini 9 crew was killed in a plane crash. That was the first time we lost a crew, and the flight still got off on time. I guess there just wasn't much time for any hokey-pokey. Somebody in another crew got a little too much beer in them and got a ticket, which we managed to get shushed up. Nobody wound up in jail, let's put it that way. That was an extremely active period.

There were some interesting political happenings. The first rendezvous in the Gemini program, which was one of its primary objectives, occurred over Saigon, probably by accident. I will never forget that. Right over Saigon in December 1965, Gemini 6. That was Schirra and [Frank] Borman. Gemini 6 was the quick part, which just went up and rendezvoused with 7. Seven was there for two weeks. You can imagine what it smelled like when they opened that hatch! The crew had been sitting there in a space the size of the front seat of a Volkswagen [Beetle] for two weeks. Anybody that's driven from L.A. to New York nonstop in a Volkswagen can identify. Lots of people were doing that in the '60s. Borman had little brain sensors taped to his head. He looked like he had escaped from the local nuthouse.

The heavy political stuff came down in Apollo 8, which was the first circumlunar mission. I had now worked for NASA for nearly ten years, and I was specifically told for the first time to make no reference to a reconnaissance spy ship that the Koreans had just taken off the coast of North Korea, along with the whole crew aboard. That was the *Pueblo*. The

Air Force general that was TDY to Apollo came over to me that morning, and he said, "Just delete any reference you might have in mind regarding the *Pueblo* for the rest of the duration of the mission." That's the only time I was ever censored as far as the news went. I, of course, said, "Aye, aye, sir." That was December of '68, and by then the Vietnam thing was really going full bore. The Tet Offensive was right around then, as a matter of fact, and that war didn't get any hotter than that.

I don't ever remember a time in my life when I was as busy as those two years, because in addition to Gemini, we were interweaving Apollo tests. Apollo was enormously more complex than anything in Mercury and Gemini. I noticed early on in blockhouses at the Cape, the crews from Huntsville [i.e., Paperclip Scientists, see Chapter 3] would cheer and break out the beer and sing some German songs after launch, because when the booster got into the sky, they were done. They wanted to hear whether the bird got to Grand Turk [Island, a tracking station in the Caribbean] on time and whether it was going into orbit, but at that point, they were done and it was party time. They'd go down to Cocoa Beach and get sloshed and eat too much. For us in Houston, the mission was just beginning. In the early days of Gemini, we would stay at the Cape through launch and then get in a fast plane and get out to Houston in about an hour and a half, one orbit.

The 1965 and 1966 time period is almost a blur for me. Interestingly, my marriage came apart in the middle of 1966 and the divorce became final on January 27, 1967, which was the day of the Apollo fire. I used to tell people that the day started very positively; the judge signed the decree at 9:00 in the morning on a Friday after a bunch of slips. The fire was that afternoon at 6:31 p.m. I didn't go to bed again until the following Tuesday at the Georgetown Inn in Washington. I was up for something like seventy-two hours. I remember going to the Cape and going through some things, and we went back to Houston and then we went to Washington.

I was working the afternoon of the Apollo 1 fire. We were doing a simulation. I was in and out of the simulation all day. I was across the street from the center in Houston, and Chris Kraft called. How he ever found me, I'll never know. I was at a friend's house. Kraft said, "You better get your ass back over here. We've got real trouble." I went over to the control center and sure enough, we did have real trouble. We put out a release fairly quickly, and then we had a very long night and a long weekend. It was a Friday night. It was just one of those god-awful accidents. When you ignite pure oxygen at 5 pounds per square inch, it's like using a cigarette lighter in a gas tank. It was horrible to listen to. I can still hear the voices. Gus Grissom didn't say a thing, but Ed White did. White was the communicator. Roger Chaffee screamed.

An oxygen line was ignited and became an acetylene torch. It was like a welder's torch. It cut into Chaffee's neck. The torch broke loose and started snaking all over the cabin, whipping back and forth. Within forty-five seconds, the cabin pressure went up, and it burst at about 30 to 31 pounds psi. It burst right along the bottom of the cone shape and then it became ambient air inside. The fire was out, but the destruction was done and the guys were dead. It was about three weeks before launch.

I remember the fire was a very difficult time. But we got through it somehow. A lot of tears. I remember discovering that Shepard could cry. We shared a room at the Georgetown Inn for the funerals. I remember we were having a drink after the burial ceremony for Gus in Arlington. Al, with tears running down his face, said, "Dammit, I hate those empty slot fly-overs." He couldn't bring himself to say he missed Gus. We got back to work in a hell of a hurry. We changed directors. Joe Shea was relieved almost immediately as Apollo program manager, and it really had a tremendous impact on Joe and a lot of other people. As so often happens at accidents of that kind, there was a tremendous feeling within the program that we just ought to close up and say nothing to anybody, not respond. I didn't feel that way and neither did Joe Shea. Some idiotic orders coming down from Washington said nobody was to say anything to anybody. That idiocy reigned for a while, and I have seen it happen in other cases.

About a week after the fire, Shea and I decided that we really owed it to some of the people that were covering the program very carefully to do an off-the-record background or briefing kind of thing. So we went up to New York and went around the circuit, the *Times* and the networks and the news magazines. We hit all the major ones and then went down to Washington and did some more. It was appreciated by the people who got the briefings. George Low was brought in almost immediately as program manager. He was already in Houston as deputy center director.

Low had been one of the first people on the Mercury program. He started with NACA in the '50s. Nobody ever worked harder at the space business than George Low. I remember I wanted to get a Mercury press kit approved. George looked at his calendar and he said, "How about 4 a.m.

next Thursday morning at my house in Bethesda?" And that's when it was done. He said, "That's all the time I have." The guy was sleeping three hours a night, and this was before Mercury flew. After the fire, I remember Low telling me he was putting in ninety hours a week at the office and taking work home. I don't know of anybody who worked that hard, year in and year out. Yet George always managed to get up in the morning four and five days a week and run five miles before he went to work, whether he was in Washington or the Cape. He was in remarkably good shape. After Apollo, he went back to Washington as acting administrator of NASA for a couple of years. He set everything up for the shuttle program to get started.

Low and I had many, many really bitter battles. One of the longest and most protracted was about whether there would be a television camera aboard the Apollo command module. I had been fighting for it from the early Mercury days. We never got any television in Mercury. We got none out of Gemini. There never was quite enough weight allowance for that. So I started fighting early for the Apollo thing and just announced to anybody that would listen that we weren't going to take "no" for an answer in Apollo. RCA had a black-and-white camera that weighed about three pounds. Westinghouse had one in color, and we still had a battle royale over onboard television. Unfortunately, when we came down to it, we were a couple of flights away when you really have to commit on hardware and requirements. Shea was with me. He thought there should be a camera aboard. I remember a guy named Lee was the LM program manager and he fought like hell for TV on the LM, but Low came out against it and announced it wouldn't be on Apollo, period. Washington people were lining up behind Low. That was the one time that I really went out on a limb and specifically leaked the fact that it didn't look like there was going to be a TV camera aboard. I leaked it to NBC and *Time* magazine.

We were up and flying eighteen months after the fire, which at the time seemed an extraordinary delay. The investigation dragged on for better than a year. But by October of 1969, we were ready to fly. The first Apollo Earth-orbital mission was a ten-day affair, just to test the hardware in Earth orbit. Apollo 7 got to be the most contentious flight I ever remember. It was mostly that Wally Schirra got his back up. He was a next-door neighbor of Gus Grissom. They were really dedicated, so you had the feeling that Wally was flying this for Gus, and he was going to do it right. He was going to do it his way, and he wasn't going to take a lot of crap from a lot of young engineers. Plus the fact that Schirra really got pretty pushy and obnoxious. He had a hell of a bad cold about three weeks before the flight. The whole crew, I was involved and a couple of other people, went duck hunting down in Florida. It was cold, and we were wading around and slithering around in not enough clothing. I remember that Wally and I came down with a cold before the flight. I could hardly talk during the mission and neither could Schirra. We found we were sneezing at the same time, and I got to be the control cold because they could tell from what fever I was running how Schirra was doing. I'm sure the cold was part of Schirra's aggravation. He got snippy as hell and one of the crew members, Donn Eisele, also got snippy. Walt Cunningham was also there, and he would do whatever the big guys did.

I can remember Schirra raising hell with Mission Control over some kind of a measurement they were given to make. This was unheard of in all other manned NASA missions. Schirra had hardly finished his criticism when Donn Eisele chimed in, "Yeah, I'd like to talk to the joker who thought up this experiment." You never had anything quite like that, where the crew was taking hide off the flight controllers, and the controllers didn't appreciate it one damn bit. They finally called Deke Slayton, the boss, to chew the crew's ass a bit and tell them to talk nice, but Deke wouldn't do it. I know that Schirra knew what was going on. After talking politely to Slayton, Wally said, "Well, Deke, you've got to do what you've gotta do." Slayton couldn't bring himself to reprimand the crew. Astronauts always insisted that they were the capsule communicators and nobody else, but *nobody* else, talked to the crew directly during a mission. There is no way around it. Just an interesting bit of protocol. But that particular flight is remembered for bruised feelings and the cold was part of it. Fortunately, our colds broke about a day or two before the mission ended, and everybody was in a happier frame of mind after the 10.3-day mission was over.

Still, Schirra didn't exactly apologize, but he didn't go on trying to beat up on the flight controllers. Also, the statement was made, "That son-of-a-bitch will never fly again in anything NASA makes, nor will any of the rest of those people." I heard that more than once rattling around the halls of Houston and in Washington, and it turned out to be quite accurate. In fact, Wally left the program a few months later. He was certainly at a point where he might have fit into some long-term lunar plans. There were six lunar missions and all the pilots wanted one. But as soon as Schirra found out it wasn't going to be, he bailed out in quick time. Eisele and Cunningham followed suit fairly quickly. Eisele was of the second group and Cunningham was third group. Each of them flew only once and that was that.

The second flight jumped out of sequence. We wound up doing the Christmas Apollo 8 flight. We went to the Moon a little early. It was a hell of a lot riskier than it appeared, because if what had happened on 13 had happened on 8, we would have a dead crew in orbit around the Moon for Christmas, and they wouldn't have gotten out of that orbit. The lunar module acted as a lifeboat in 13 when the cryopumps failed and blew up. If that had happened on Apollo 8, where there was only the command module, there was no other way to get home, no extra oxygen to breathe.

Apollo 8 was a very gutsy mission. We used to call that mission "real-time flight planning," and it all was decided about ninety days before it flew. It was one of those back-of-the-envelope developments. It was John Yardley, who worked for McDonnell Douglas, who suggested it, even though it was North American hardware. Anyway, George Low, of all people, "Mr. Conservative," jumped on that idea. Low sold it up the line in about seventy-two hours. It was like Michael Jordan inventing a new shot to win the NBA title. Apollo 8 had enormous impact. The whole world was watching because it was Christmas Eve. I can still hear [Jim] Lovell describing, "The Moon looks gray" and "It looks like plaster of Paris." He really did a tremendous job. There were a lot of PR surprises. I didn't know the crew was going to read from the book of Genesis, although everybody assumed that I did. I knew about an hour and a half before they did it that they were going to read some Biblical quotations. It worked out all right, too. It was primarily Borman's idea. He was quite a church person. The other two crewmen didn't object.

Al Shepard, as crew chief, put little airline flasks of brandy in the Christmas dinner. When Borman found them, he sequestered all the brandy. He was going to be "Mr. West Point" and not allow any loose happenings on his shift. I got a hell of a kick out of that. Borman never said anything about it, but after the flight, he brought those three little flasks of brandy back and gave them to Bob Gilruth. They sat on the Gilruth's mantelpiece for years down in Dickinson, Texas. We used to refer to it as "out of this world brandy." Lunar orbital brandy! I don't know what ever happened to them, but don't you know, they would be worth a little something. Shepard is better remembered for a six-iron sand shot from the lunar surface during his Apollo 14 visit. The fire was really a hell of a lot of baggage to carry around, but Apollo 8 seemed to cleanse it.

The December 1968 Apollo 8 flight is well documented, and it was one of the highest points emotionally, I think, for everybody associated with the program. It somehow washed away all the guilt left over from the fire. It did so exactly two years after the fire. It had a tremendous cleansing effect because it worked so well. The mission came out of nowhere. It was like kids in a huddle drawing football plays on the ground.

Apollo 9 was Jim McDivitt, Dave Scott, and Rusty Schweickart. That was kind of a "back to business," ho-hum space test. They flew the LM for the first time, but it was strictly Earth orbit. Dave Scott, I think, knew more about Apollo than the people who designed it. He carried it all around in his head. He literally knew every wire and the strength of every wire, the capacity of every wire. He had the command module learned quickly, and he went on and learned the LM just for good measure. It was absolutely incredible – when people wouldn't believe a drawing, they would ask Dave. He had that kind of a computer mind. That flight was in early April of '69, and Schweickart was supposed to do a bunch of tests to find out why so many astronauts got airsick or spacesick. Borman was sick on Apollo 8 but he hid the fact, went to great length to conceal it. He did a data dump into [a tracking station on] Ascension Island, knowing full well that the station wasn't manned that day, but that somebody would come around and pick up the instrumentation about twenty-four hours later. He was throwing up all over the place at lunar commit. Lovell later went along with Borman and said, "Yeah, I wasn't feeling so good either."

Anyway, in Apollo 9, it was one of the first times they thought they ought to start looking at some airsick questions. So they put a bunch of tests on Schweickart. Wouldn't you know it, they picked somebody who was totally, helplessly, and incurably spacesick prone. Schweickart was one of those people who, if he flew a thousand times, he'd get sick a thousand times. The real question is why didn't the doctors know that? They did a little EVA, and Rusty managed to barf all over his helmet. We had to do a lot of covering on that flight. Finally, we had to announce that Rusty wasn't feeling too well. That's a whole other book unto itself.

Apollo 9 was very worthwhile. Then 10 came on, and that was kind of gamey. That was about the time I left. I did all the preflight on the crew and went right up to the flight, then I left the program. I actually worked it in London for ITN [Independent Television News] and subsequent flights. That was a good crew, [Eugene] Cernan and [Thomas] Stafford had some ups and downs in a lunar orbit mission. I remember a stream of computer directions went in and told the engines to fire back at the lunar surface when they were supposed to go in the other direction. Cernan distinguished himself for saying something like, "Holy shit!" They almost washed his mouth out with soap because he reacted as anybody would.

I left in May 1969, and I had done preflight things with the Apollo 11 crew. One was a rock-hunting trip out in El Paso. I also remember doing a pre-flight interview with the 11 crew as part of the ITN crew. I was walking both sides of the street. The best thing about 11 was it happened on my forty-first birthday. When the 1201 emergency alarm was sounded, I was just listening to that as they came down [for lunar landing]. I knew that was an abort parameter and, boy, I took a bunch of breaths. Then I remember it was [Buzz] Aldrin saying, "Just ignore that." He said that very quickly. And he did one hell of a job calling the landings. If you listen to that tape, he was busy. Charlie Duke was capsule communicator. After the landing, he said something like, "We're all about to turn blue down here," which was a pretty honest reaction. They had something like three or four seconds worth of fuel left in the descent tank. You couldn't cut it any closer. But Neil seemed to know what he was about, kicking up the dust and everything. That was really hairy, I thought. What if the lift-off rocket doesn't work in the LM? There was no backup. There was just one way to get out of there. It had to work, and that was the piece that was developed at White Sands, Las Cruces. They fired that thing up how many hundreds of times, and it worked real well. It had one hell of a success record. It worked six times on the Moon, too.

I did all the Apollo 10 through 17 flights from London. ITN and I were going to try to do Apollo 13 with just a minimal effort. Of course, that was the one that blew up fifty-eight hours into the flight. Along about the sixtieth hour of that mission, I was on a plane to London. By the time I got to London, we knew it was just going to be a figure-eight around the Moon and back. That was decided in the intervening eight to ten hours after the explosion. I didn't get any sleep that week, either, because we were on the air virtually the whole time I was in London. Then they wanted me to cover the splash-down, so I went to Hawaii and then back to Houston. I wondered what country I was in.

I have to tell you about the Queen [of England] and her story during

Apollo 12. ITN moved into a new building, and you know, the British are so good on ceremony. The Changing of the Guard is just a fantastic piece of show business. They do it so well, and they do it every four hours or something. People line up for blocks to watch it, and it's worth watching. It was announced that the Queen herself was going to come around and dedicate the new ITN building. I noticed everybody was getting a little bit nervous, and lights were falling, and the tea was cold, and it was obvious that they were getting a little clanked because the Queen was coming. The day before the event, which happened during the 12 mission, one of the copy girls simulated the Queen, and we did a walk-through, the whole damn thing. The Queen was to come in and look through cameras and snip ribbons, and then she was to come on the set. We were to explain the lunar toys to her, and then afterwards there were twenty people, and I was one of them for some reason, designated to go up and meet with the Queen informally in the board room. It was all very informal, yet we stood on little duct-taped X's with our name written on them, all very informal.

So I am standing there on my X, and the Queen walked into the board room. She walked right over to me and asked, "Mr. Haney, tell me, how is Mr. Neil Armstrong's health?" I said, "Just fine, ma'am, as far as I know. I saw him a few weeks ago in Houston." I didn't attempt to tell her any Aggie jokes or anything as I might have, but then she said something about the weather and how much she enjoyed the lunar toys. She wandered around and said a couple of things to other people. About a week later, the mission's over and I'm back in Houston. The very day I got back, I ran into Armstrong at lunch. I went up to him and said, "Hey, a lady in London was asking me about your health the other day." As I framed the question, his face got white. "That was the Queen," he said. I felt like I had been harpooned. "What the hell was going on?" I asked. Neil said, "Oh, Jesus, you won't believe this. Sit down."

Then he explained that after the 11 mission, NASA and the State Department had laid on thirty-five countries in forty days. Talk about jet lag. They used Air Force One. I think the first stop was in Bermuda, then maybe through the Canaries. London was about the fifth stop, the first major city in Europe. Well, somewhere along the line, Neil got a virus. By the time he got to London, he had such a cold and was so hoarse that he couldn't talk. The plan was that London was an overnight, about thirty-six hours, and he was going to stay in bed and get shot full of B-1, or whatever it is that makes you healthy again, and then pick up the trip to Paris and do all the other things. Well, they got to London and Jan Armstrong, his wife, found out that it wasn't just a trip to Number Ten Downing Street, they were invited to the palace. She told Neil to get his ass out of that bed, she was going to the palace, no matter what. He agreed, so they went over at four o'clock for tea in the family room at the palace.

The Queen understood that he couldn't talk. They sat around and sipped white wine and ate crumpets and made small talk with the grandchildren and the grand-princes. On the way out, Neil said, he felt he ought to say something like "thanks" to the Queen for her hospitality. He figured out something to say in a loud whisper and he got up close to say it. When he tried, he coughed and spit right in the Queen's eyes. Hit her square! He pulled back and so did she. Then he came up to say "I'm sorry" and hit her again with another cough! At that point, she was cowering. The lady-in-waiting gave her a bath towel. Now, six to eight months later, she comes up to me knowing full well that this question is going to go back to Houston. You know, "How is Mr. Neil Armstrong's health?" because nobody else was involved but Neil and her. It was a real *gotcha*, and here he is, recounting it to me, which flushed it out. Extraordinary happening.

Apollo started out to be a very long orbit of the Earth. It was not a lunar landing device at first, and the apogee would have been about 200,000 miles, which would have put it within twenty or thirty thousand miles of the Moon on a close approach and would come in fairly low, a 200 or 300 mile Earth orbit on the other side. It would have taken a couple of weeks to do one orbit in those early designs. After May of 1961, Apollo assumed the landing portion. The programs did run concurrently, which made Gemini all that much tougher because you did your homework on Apollo on weekends. We did several launches, the Saturn 1B and the first Saturn Vs were fired in 1964 and 1965, in off-weeks when something with Gemini wasn't burning the barn down. In briefings from our friends at Huntsville, we were told that Saturn V had engine-out capability. That is, there were five engines in the first stage, and it could accomplish the mission with four. It was on the first or second booster-only kind of test shot that we learned we could fly on three engines. That came as something of a surprise to me. I was, as usual, "Charlie Expert," trying to tell the world what was happening, and I usually wound up learning a lot more than I told the world. It was always a learning curve.

Chapter 11
Walter Cunningham

"Apollo 1" was NASA's designation for the first manned flight of the Apollo program. After a fire killed three astronauts in a training accident (see Chapter 9), however, that designation was retired. Following several unmanned rocket tests, the first manned flight of the program was designated "Apollo 7." It was a 4.5-million-mile shakedown flight for the command/service module (CSM). Launched on October 11, 1968, it completed 163 orbits of the Earth during its eleven-day mission. During the Mercury and Gemini missions, the astronauts had to remain in their seats throughout their flights. The Apollo capsule was larger, and the three-man crew had room to get out of their seats and move around the cabin. Along with Mission Commander Wally Schirra and Command Module Pilot Donn Eisele, Walter Cunningham flew as the Apollo 7 lunar module pilot. No lunar module (LM) was attached to the CSM on this flight, but the crew practiced maneuvers including a simulated LM rendezvous and docking. Cunningham was inducted into the International Space Hall of Fame in 1983. His interview for the NMMSH oral history program was conducted in February 2006.

I was a half-time employee at the Rand Corporation while I was working on a Ph.D. in physics at UCLA. This was about 1960. In 1959 they selected the Mercury astronauts. I read the background of the Mercury astronauts, and I wasn't terribly impressed. They all had a good, decent flying background. But they had all been through test pilot school for what it is worth, plus or minus. Their academic credentials were not

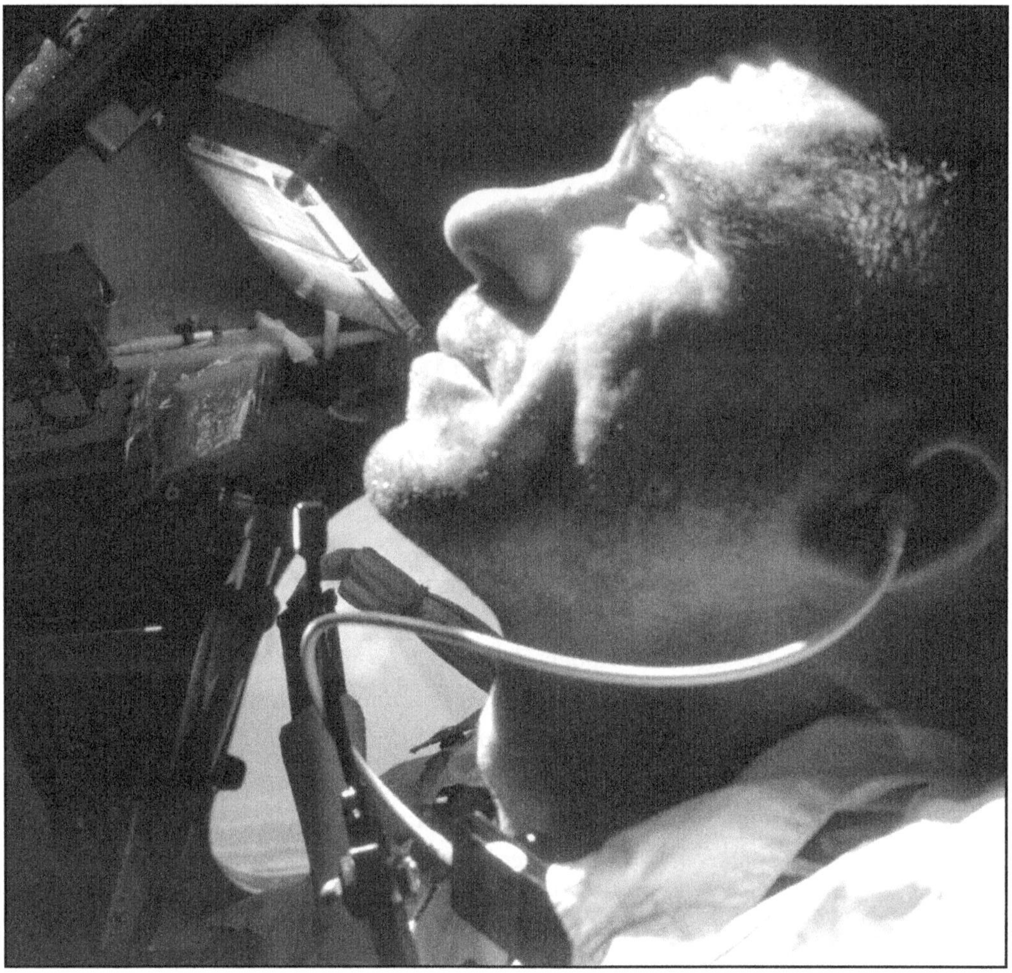

Walter Cunningham during the Apollo 7 Earth-orbit mission. Courtesy of NASA.

all that terrific, I didn't think. But I envied them because they had the opportunity to do things that nobody else was going to be able to do, flying higher, farther, faster. It was something I read about and thought, boy they are really lucky. Those are fortunate guys.

Then in 1961, I found myself driving from Van Nuys, California, over the Santa Monica Mountains to Santa Monica, where I worked at the Rand Corporation, starting about seven o'clock in the morning. So it was a little before ten in Florida. It was the morning of Alan Shepard's flight, which was a little fifteen-minute suborbital mission downrange. As I was listening to the count, I was up on Mulholland Drive. I will never forget, I pulled over to the side of the road. Stopped and listened to the count.

They got down to three, two, one, liftoff, and I heard this voice screaming out loud as it could be, "You lucky son-of-a-bitch." I realized it was me. I am not given to this kind of spontaneous outburst, and I looked around to see if anyone heard me. And I finally went on to work.

That was May 5, 1961. It was only later that I realized that probably was where I changed from just envying these guys to deciding that's something I wanted to do. The only qualification that I didn't have that they were still looking at was having gone to either the Air Force or the Navy test pilot school. I thought I had decent flying credentials and a very good academic background. I already had a Master's degree and I was working on a doctorate in physics, but they didn't think I had enough qualifications. I just had to accept it. I thought, well, maybe after I get my doctorate, maybe they will be interested in me.

Then NASA dropped their requirement of being a graduate of test pilot school. I remember in the spring or summer of '63, I was working at the Rand Corporation in their astronomy department, and the assistant head of the department was a man by the name of Gerhardt Schilling. Gerhardt had just spent a year on loan to NASA at headquarters. I think it was after John Glenn's flight [America's first manned orbital flight], I was chatting with Gerhardt. I told him, "Now, that is something I would really like to do. I would like to go fly in space."

I had no romanticized idea about it. I wanted to do it because it was hard to do, and they were selecting people to do it. I think that's what drove me more than anything else. It wasn't like I had an insatiable urge to do this. It is kind of strange as I think back on it because, at the time, I certainly wasn't very introspective about any of these kind of things. Gerhardt said, "Well, if you are really interested, I'll make a phone call to NASA." I wasn't even aware of the significance of that. He made a phone call two days later, and he called and told them what my background was. I had 1,500 hours in jets, my education, what-have-you. I got a call two days later from someone that said, "Are you serious?" I said, "Well, yes. Next year I should be getting my doctorate, and I would really like to do this." They said, "Well, we are not selecting any astronauts next year, but we are selecting a group right now. Would you like us to include your application?" I said yeah. So it was a case of now or never, almost.

They sent me piles of paper. I did all of the paperwork. I found out later there were, like, 770 qualified applicants. They were just starting to place a little bit more emphasis on academic background, physical sciences education. And as I had learned, they had dropped this requirement to be a graduate of test pilot school. So I filled out the papers, sent them in, and went about my business. I was attending summer school classes and working full time at the Rand Corporation. Every couple of weeks, I would get a call from NASA of the character of, "Well, you are still in this so far, are you still interested?" And I always said yes, and went on about my business. Went to my two weeks summer drill that year [in the Marine Corps Reserves]. I remember that I probably got seventy-five, eighty-five hours [of flight time] during those two weeks, because I flew anything I could, anytime I could. I just liked flying.

Finally I got notified to come down and take the physical at the School of Aerospace Medicine at San Antonio. That was first contact I had with anyone in person having anything to do with the program. There were thirty-four of us left by the time I got down there. I looked around and for the first time saw what my competition was. It was probably the first time that I had any doubts. I thought, hey, I might not be able to do this. There were some tremendously well-qualified people, and most of them were active duty. I was a reservist. And out of that thirty-four, twenty-eight had passed the physical screening. We knew they were going to select, I think they said twelve to fifteen, something like that. We didn't really know. It was all a mystery to all of us.

Depending on when you came in, it could be seven to eight days to do all of this physical testing, and they came in in groups. I probably met eight or ten others that were going through the physical the same time I was. And I began to realize just how long the odds were against me coming in. Several of those people I had met, this was their second or third time to go through this process, just to try to get selected. And I was just a young, I think I was a captain in the Marine Corps by then, full of all of my own capabilities and not realizing where it fit with all of these other guys.

Anyhow, I passed the physical, and we went down for personal interviews in Houston. When we went down there, I could see guys [astronauts] like Dick Gordon, Jim Lovell, Pete Conrad, and everybody seemed to know everybody. I was the only guy that didn't know a soul. Never flown with any of them. I was beginning to think this is an old boys club, and I'm just not one of the boys. We took a lot of other tests, psychological tests. I did very well at the School of Aerospace Medicine, and I think that was the only reason I was still there. I could see a lot of these other people had some very good flying credentials.

I was one of the fourteen that they selected. On October 14 of that year [1963], they announced it. I learned in the past year, quite by accident, that I was number fourteen. Paul Haney was a public affairs guy at JSC at the time. I sent Paul an e-mail, and he sent back and said, "You knew, of course, that you were number fourteen, didn't you?" And I said, no, I had no idea. But recently, and from reading the Neil Armstrong biography *The First Man*, I began to realize I had to be one of the longest long shots you can imagine. Because I was a reservist. I wasn't a regular officer. NASA was looking at me as a civilian because I wasn't on active duty, and I think they wanted to have some civilians. They had Neil Armstrong and Elliot See, until he got killed. And in our group, there was myself and Rusty Schweickart, who had been an Air National Guard pilot but he had resigned his commission. So I was the only one that was an active reservist. From the beginning, they always classified me as a civilian. I was a little bit odd because I was a scientist in their mind, with a physics degree. I was a non-test pilot. Half of our group that were selected were non-test pilots. I had never flown with any of them. I had never applied before. So it was quite a good fortune that I ended up squeaking in.

My story is in a book I wrote called *The All American Boys*. It is on the human side of the space program. It tells it like it is, or was. Maybe it will get back there again, I don't know. But the first year was essentially collective training to get us all on the same flight plan, if you will, because we had a variety of backgrounds. We had at least one Ph.D., Buzz Aldrin. There were some astrophysicists. I was a physicist. NASA needed to get us all up to the same level in things like orbital mechanics, astrodynamics, those kind of subjects. We had to become familiar with the program, the various pieces of hardware. We had to get to know the players. One of our early meetings, I remember, was with Wernher von Braun at Huntsville. We went to the California sites, and we went down to the Cape and saw some unmanned launches. I think they were some of the early Saturn I launches.

I remember the first speech we had after we were selected. We didn't know what to do when we got there. The guys that knew people before probably had some idea. But the rest of us, we show up, and there is going to be a press conference to announce us. We had never had a press conference. I mean, what do we do? We were worried. Paul Haney was the guy running the press conference, and I remember Paul saying,

"Don't worry about it. They won't ask any tough questions. [More like] 'Were you a Boy Scout?' or 'What is your religious preference?'" So before we went into this press conference, I went up to Paul, because I was a little bit concerned about that "What is your religious preference?" And I could see that Rusty Schweickart, the other civilian selected, was up there wanting to talk to Paul too. His concern was the same thing. I had not been religious since I was about fifteen or sixteen years old. I thought, how am I going to answer this question? Paul said, well, tell them what your mother's religion is. He wasn't terribly concerned about it.

When they asked the question, we were in alphabetical order, but they started at the bottom of the alphabet. When they got to Rusty, he had a brilliant answer. That left me not much room to maneuver. He says I have no preference. They got up to me I said, well, my mother has always been part of the Lutheran church. So we went on. We got past that sticky

Geology was one of the training programs the Apollo astronauts had to complete. Here, Cunningham is shown on a 1964 geology field trip at the Philmont Scout Ranch in northern New Mexico. Courtesy of NMMSH.

wicket. But it was really my first experience at just how uninformed and simplistic the media really is to this day, in everything.

I spent my fifth anniversary [with NASA] in the air on Apollo 7. The training was quite demanding, quite frustrating. My responsibilities around the office were electric power systems, environmental control systems, and the likes. I did a lot of work on the Gemini program, with their fuel cells and systems. I did a little bit of simulator training on Gemini, but I never was tapped for a Gemini flight.

If you read my book, several of us were very concerned because we didn't seem to know what was going on. Bill Anders, Alan Bean, Rusty Schweickart, and myself, we felt kind of like outsiders that didn't know what was happening. I remember going in and talking to Wally Schirra one time because I think he was kind of head of our section. What is going to happen? When are we going to fly? He says, "Don't worry about it, Walt." He obviously knew something I didn't know. A short while later, though, we had the flight crew assignments. Flight crew assignments changed around a lot in those days. Sometimes by physical problems the guys had. More frequently, because somebody got killed.

We were originally the Apollo 2 crew. The Apollo 2 mission was in many ways redundant to the Apollo 1 mission. For reasons of time, money, and what-have-you, they cancelled our mission. They put us as the backup on the Apollo 1 mission. Being backup meant it could be several years before we would rotate into prime. Then a couple of months later, they had the Apollo 1 fire. So that was cancelled. Three weeks later our crew, Wally Schirra, Donn Eisele, and I, inherited what would have been the Apollo 1 mission but subsequently came to be designated Apollo 7.

We were basically an insensitive bunch. I can only recall one emotional feeling about Apollo 7. I was kind of emotional that nothing happened to delay the launch. I mean, other than that, I did my job. I felt like nobody in the world knew as much about the Apollo spacecraft, overall, as I did at that time. I knew we had specialists that knew a lot of things about it. But if the other two guys had died right off the pad, I felt like I could do the mission. Everybody asked, "What were your thoughts, because the last one didn't even get to the pad before it burned up?" That never entered my head as a potential problem. I had been on the fire investigation board. I've always known that you've got to expect loses.

By that time, I was thirty-six years old. I can remember that about the time I was thirty-one or thirty-two is the first time I can remember of having the thought, "I have had a marvelous life. If I die tomorrow I will feel fulfilled." So I always felt like everything after that period of time in my life was just something extra and special, and I appreciated it.

I had two thoughts I can remember on the pad. One was, "Boy, I hope they don't have any reason to scrub this." Because if they scrubbed that flight, there is always a chance that you will not be on the next one. I didn't realize how much of a long shot that would have been at that time, but that is what we were concerned about. You wanted to get above 50 miles high so you could be an official astronaut. And the other thought was, "If this mission fails, it is not going to fail because of me." I am sure the other guys, especially the rookies like me felt the same way – don't screw up.

Donn and I had kind of got a bum deal there, because we were the only guys that ever flew with a commander who did not spend some time trying to look out for his crew afterward. Wally was always kind of about Wally. Other guys had big egos too. All of us probably did. But the other

The Apollo 7 crew (left to right): Donn Eisele, Walter Schirra, and Walter Cunningham. Courtesy of NMMSH.

guys always tended to look out for the guys that flew with them, and it helped their careers after that. We were not fortunate enough to be in that position, but we had a wonderful mission.

I think it was an extremely important mission. In the meantime, it gets lost in the noise. People hardly ever ask about Apollo 7 or Apollo 9, because they were the two Earth-orbital missions in the Apollo program. Today, you run into people who think the Apollo program pretty well started with Apollo 11.

We were very much aware on Apollo 7 that we weren't sure this country could stand two disasters in a row. So we were a little bit concerned for the program that we not have any kind of a problem. But we all knew that we [America] were landing on the Moon. It was just a question which mission was going to make it.

I have never been considered a very diplomatic guy, but on that mission I was proud of the fact that I was busy kind of keeping things on an even keel diplomatically. My normal reaction would have been "to hell with you." But when you are in a world that is this [a few feet] wide, and there are three of you, and you've got eleven days to go, you kind of do what is necessary to keep that atmosphere as compatible as you can.

Wally didn't care, because he had already known that he was going to be leaving the space program. As a matter of fact, when we came back in and landed at Ellington [Field Joint Reserve Base, near Houston], instead of going back over to the office or even to his home, he went down to the NBC station and announced his retirement. Wally's ass was in a jam, because he had had a fight with Deke [Slayton]. I can remember, when we came back into the astronaut office at the Cape, Deke took Wally aside, and they had it out kind of privately for about twenty minutes or so. I'm sure that Wally felt justified. To this day, he defends his behavior, even though he knows I certainly disagreed with it at the time.

Most of the time, motivated men will put the mission first, as we did. I kind of laugh inside a little bit when I see them talking about all of the psychological aspects in picking these crews. I still have a copy of an article that was written when they announced the Apollo 7 crew. In it they talk about the happy-go-lucky commander Wally Schirra. I think they describe me with a look of a "dyspeptic falcon," whatever that was, kind of the intense Walter Cunningham. And then the in-between Donn Eisele. The whole tone of the article was kind of like we were psychologically matched for this thing. Nothing could be further from the truth. I mean,

Deke didn't mess with those kinds of head jobs at all [Deke Slayton picked the crew members for each mission]. He put three people there, as he said, "When I selected them I figured they could do the job." Eisele wasn't even originally on our crew. He was really supposed to be on the Apollo 1 crew. He mentioned, once afterward, he would have been dead if he stayed there. But he had a shoulder out of joint. So all of this talk about psychological matching and everything, I guess I'm a Stone Age guy about it. But I think it will drive things as you go forward, because that is what our society has come to.

Being in space was everything I expected. I had no surprises in space. I think I would be surprised if any of the other guys had surprises in space. We never talked, at least, I never talked to anyone else about it. It was kind of against the code to go and, say, ask Wally, "Wally, what is it like up there?" I mean, you just didn't do that. You didn't want to betray that you were even curious. Because after all, you were just going to go do it. So there were little things like that you just didn't ask about. I was a physicist. I expected that a drop of water would turn into a sphere and things like this. And you pick up bits and pieces from other folks. I was a little concerned as to whether I would feel sick. At the time we weren't sure, but in retrospect, about 25 percent of the guys did. Nobody in our crew felt sick from space. But I remember both myself and Donn performed a little bit of our own experiment as we got up there because we weren't feeling bad. Your head feels full, which is natural. The blood doesn't have gravity pushing it back to the heart. I can remember going through various head motions and things like this.

We were in the first spacecraft where we were free to move around, and we did some tumbling things. I don't remember any special caution about it. But I think that it is kind of almost luck, who it affects and who it doesn't. I had my own theory about that at the time, a theory I shared with Story Musgrave [see Chapter 14] one time. Story had the same kind of theory. It was a visually oriented thing. Stop and think about it. Down here, we thought of the spacecraft as the heat shield is the bottom, you lay on your back above the heat shield. In space, though, in zero gravity, you are floating around. So you've got an axis in yourself, body-centered axis. You've got the axis of the spacecraft, and if you are floating, then the instrument panel is here and you are here. If you look out the window, you've got the Earth-centered axis system. And it's sitting over here. Nothing is exactly where it ought to be.

I took the egocentric approach: I'm here and anything else is someplace from me. So it never seemed strange. The Earth was over there, tilted this way, even though it was us that was tilted. And the same thing was true with Story. He tended to look at it egocentrically with our own axis system, and referenced everything else to us. It's still a mystery they are working on, but it is interesting.

We did simulated rendezvous and docking with our booster the S4B, which is the third stage of the Saturn rocket. We separated from the spacecraft and then re-rendezvoused later, separated from the booster, and re-rendezvoused later with the booster. We were able to practice the re-rendezvous techniques that we were using. It was a way of finding out early on if we had problems that we had to fix in the rendezvous techniques.

If we had just launched the first mission and gone to the Moon and landed on the Moon, think of all the risks along the way because of all the unknowns. Maybe they worked and maybe they didn't. But we ironed out all kinds of procedures and stuff before we got to Apollo 11. So you've got these other missions in between. What you do is you bite away at the risk. You incrementally reduce the risk so that the next mission takes on incrementally more risks to make incrementally more gain. But part of the risk, you put behind you. It's like if you look at the risk of catastrophic failure analysis for a mission, a shuttle mission for example. If they say it is one in 258, a little over half of the risk is during boost. So as soon as you've got boost behind you, half of the risk is already gone, theoretically. I mean, we all know that. So you incrementally reduce these risks.

People ask me many, many times, was I afraid during the mission because of all the things that could happen to go wrong? And my answer is not very romantic. You see, we were only at risk a total of a couple of minutes during the whole mission. If you are in orbit, you don't spend any time worrying about whether the parachutes are going to open. Because they don't have any immediate impact on your life until just before it's supposed to open, or the drogue chutes, or a canister, or the engine during retrofire. I once figured out there might be just a total of a couple of minutes during the eleven-day flight of Apollo 7 that you had any reason to have any anxiety.

I think that there were a lot of good reasons to do what we did, but it was to fight the Cold War that we did it. It is the intangible benefits that impress me more. In 1961 when Kennedy made that speech, he threw down the technological gauntlet to the Soviets. It had to do with which country had the best way of life, the best system, the best economy, the best technologists. And then we went ahead and did the impossible. Can you imagine how impossible it was to say you are going to land a man on the Moon in the next nine years? I mean, you've got to be out of your head. The Soviets started just shortly after that on their own lunar program. They spent a lot of money on that program. They assigned crews. They developed spacesuits for it. They developed boosters, the N-1 rocket. I think they had four flights. The longest one lasted a hundred and twelve seconds. They didn't cancel their whole program officially until 1971, although they gave up on it pretty much in 1969.

But the Soviets never forgot that the Americans, when they were faced with a tremendous project, especially one that was technology based, that we could pretty much do what we said we were going to do. The Soviets remembered that a lot more than the American people did. I think it was in 1984, when President [Ronald] Reagan met [Soviet President Mikhail] Gorbachev in Reykjavik, Iceland. Remember, they took that little walk together? President Reagan threw down the technology gauntlet again. He told Gorbachev we are going to develop an anti-ballistic missile system that is going to render all of your missiles just absolutely impotent. The Soviet Union had spent so much money developing that system. They spent a whole lot more on their military and space program than we did as a percentage of their Gross Domestic Product.

Gorbachev heard that, and he remembered what we had said and done before. He believed it, even though many Americans wanted to call it Star Wars and laughed. I happen to believe that program is feasible, incidentally, although a lot of Americans don't. And he believed, in fact, that we would do that. After spending all that money, they either had to spend a whole lot more money to develop the same thing, or they had to change their ways. I think that that was one of the key things that happened in the Cold War to defeat the Soviets. We showed we could do what we wanted to back in 1960s, and it was an impossible thing for us to do.

I am pretty simple minded. I have always felt like if you did a good job it was going to be recognized, and you would get an appropriate reward for it. Even to this day, I am not politically astute enough. I didn't recognize the politics that was being played all the time. I felt like it was okay to say honestly what was on your mind and not have to shade it, or not have to look out for your back from this guy or that guy. There was a tremendous bit of politics that was going on in the office. And even at the time, I still didn't want to believe it. I am a romantic about it.

I felt like I had a great academic background. I felt like my flying credentials, or my flying capability, was as good as anybody I had ever run into. There is nothing that I've heard of any pilot doing that I didn't think I could do with the right training background. And I thought that would be enough to carry me. That was insignificant compared to who you knew, who you championed with, what politics was played. I mean, just unbelievably. And to this day I still retain a bit of that, I don't want to call it innocence, but at my age that's an admission of stupidity. (Laughs)

Chapter 12
Gene Kranz

Eugene (Gene) Kranz was directly involved with Mission Control for the Mercury, Gemini, and Apollo flights. He progressed from procedures officer at Cape Canaveral to flight director at the Manned Spacecraft Center in Houston. He was a flight director for Apollo 11, the first manned Moon landing, and was lead flight director for Apollo 13, a manned lunar mission that suffered a crippling explosion on the way to the Moon but was brought safely back to Earth. Kranz, a major figure in the popular 1995 movie Apollo 13, *was portrayed by actor Ed Harris.*

An April 2010 Smithsonian magazine article explained the story behind Kranz's trademark vests. His wife, Marta, said, "Gene wanted some kind of symbol for his team to rally around. I suggested a vest." She began sewing vests for him, using white for him to wear during spaceflights because his was the "white team," and other colors for splashdowns and other occasions. "I started wearing a vest during Gemini 4, and it was an immediate hit," Gene Kranz said. "From then on, I put on a new vest on the first shift of every mission."

Kranz was inducted into the International Space Hall of Fame in 2012. When he went to Alamogordo for the induction ceremony, he was returning to familiar territory, as he explained at the beginning of his interview:

I was at Holloman for a couple of years in the late '50s, so I'm familiar with the winds and the dust storms. We used to love to have the dust storms come in because we could get inside the hangar, and we would work all of the engineering problems that we had been putting off, putting off, putting off.

I was working for McDonnell at that time. I had just come back from Korea, and I was working as a flight test engineer on one of the early B-52s. Our job was to adapt various weapon systems to the B-52 air frame. At the same time, we had a variety of secondary tests. The Quail [decoy] missile that we were developing used a JD-5 engine, which was used in the T-38. We had a trapeze we would swing down out of the B-52's bomb bay [to launch the Quail]. It was an incredibly successful test program. It was one that brought me into what I would say was the "hands on" side of the job, and it really helped me as I moved into the space program just a few years later.

As we were getting ready to wrap up the program at Holloman, I saw an advertisement in *Aviation Week* magazine that said they were forming a Space Task Group. I thought that sounded pretty cool. That might be the next adventure. I had been fortunate all through my life that I was doing what I loved every minute of the day, every second. I thought, gee, let's take a look at space. I wanted to go down to the Cape, because I

Gene Kranz in Mission Control during the Apollo 16 launch on April 16, 1972. Courtesy of NASA.

thought the launch operations down at the Cape was the place to go. My wife said, now, Gene, you've got to get more education. You ought to go back to Virginia. You ought to work at headquarters down there and go to some of these great schools they have in Virginia. So I ended up out at the Langley Research Center, which was headquarters of the Space Task Group. I arrived, and they had a pool of about six to eight young people who were still forming up. They had about a hundred people total in the Space Task Group. Some people were going to engineering, and some people were going into recovery operations, doing the testing of the water systems and figuring out how to get them aboard the aircraft carriers or the destroyers. Some were going down to launch operations, down at the Cape. I was just sitting in the bull pen.

One day, about two weeks after I had been on board, a gentleman comes up. I knew him by sight, but I didn't know him by name. He said, "I'm Chris Kraft. You are working for me now." And then he said words I'll never forget. He says, "I want you to go down to the Cape and write a countdown, write some missions rules. When you're through, give me a call, and I'll come down and we'll launch." I had been two weeks on the job. All I had seen were the various manuals for building the spacecraft. I didn't have a clue about big rockets. The spacecraft was a virtual unknown. And it was just, "Okay." Military background: You salute Chris Kraft, and you go down and figure out how to write a countdown.

This brought me into my first meeting with the astronauts. Literally, I was so green at this new business that I hopped aboard NASA's mini-airline. It was East Coast Airlines. We hopped aboard the airplane and got down there. I didn't know whether the Cape was north or south of our landing field, once I got down there. I was standing sort of in the door of the aircraft, getting out, trying to figure out what to do next. Up drives this convertible, a beautiful red thing, and all of a sudden one of the guys who had been on the aircraft with me walks down there. He looks at me, wondering what to do. He says, "Where are you going?" I'm going out to the Cape. He says hop aboard. This was my first introduction to an astronaut. I found that out very quickly as we were going through the gate, and the air policeman at the gate sort of gave this man a high ball. All of a sudden, we must be hitting close to 70 or 80 miles an hour. The guy reaches over and says, "Hey, I'm Gordo Cooper." We shake hands, and that was my introduction. Gordo became the guy I was always rooting for to get the first flight or do this or do that. He was absolutely a joy. He was probably the most passionate and enthusiastic. All of the other guys were steely-eyed missile men in the original Mercury Seven. But Gordo was a guy who just loved every aspect of his job and he just enthused it. He gave you the same enthusiasm he had.

We were very fortunate, because at this time the space community was relatively small, and they recognized how far we had to go. So, when I first started getting into the business of writing a countdown, obviously you go to the blockhouse. I had heard about a blockhouse. This is where everybody is encased in a cement facility, up near the launch pad, so they can see what is going on there. The guys do the countdown and push the button.

I had my first meeting with Tom O'Malley. He was the Atlas test conductor at this time. I'm writing a countdown. What are the kinds of things that we ought to work together on? Obviously communications, obviously we are going to bring a command system. Basically, you are going to be sending various commands off to it. We are going to have telemetry. Tom sort of guided me through this.

Then Tom says, "The next guy you ought to talk to is range safety." I said, "Who?" He says, "That is the guy that has these vertical wires out there, and as we launch they've got to make sure that we don't tilt back toward the launch pad or towards the population. And they've got the ability to blow up the rocket." So I said, okay. I went and met the range safety guy, and I took some time taking a look at the various observers they had. They had observers situated around the launch complex, and they had a little grid. This grid gave them the indicator as to whether this spacecraft was going vertical. If it was not going vertical, but tipping back towards a launch population area, these range safety officers would tell Mercury control, "Hey, we are deviating the trajectory. Do you see it?" Mercury Control would say, "Yeah, we are seeing it. You've got twenty seconds. If you've got a crewman on board, you had better get him off." It was that kind of a relationship.

After getting to know a little bit about range safety, then I had to figure out something about the spacecraft. So now I go over to Hangar S and talk to the people on the spacecraft. Who are the various test conductors? This was a process of finally seeing these pieces come together, writing them down. Then the next part of the job was to develop what we called mission rules. Mission rules, in the early days, were very similar to what we use in aircraft and what they have in commercial flights today. You have a minimum equipment list. These are the things you've got to have working in order to say we're "go" for launch.

That was the early part. But once you've launched, what do you got to have in order to keep going the path you want to go, as opposed to aborting? You have conscious decisions you have to make all through powered flight. So this was the next part of the job, but the most interesting one. It came to me as sort of a shock. I expected to see a fully professional team of people show up there for the countdown. But it was sort of like, remember in sandlot softball, where you would have two teams divide up, and you would have one guy get the bat and throw it to the other guy. He would get it and you would keep going up the top, and the guy that got first to the top of the bat got the first choice of the players. Well, that was sort of how we built the launch teams.

Chris Kraft, at that time, had only three guys that were full-time controllers. Two were trajectory, plus myself, plus himself. That was the team. Everybody else had to be borrowed. So he would go over to engineering, and he would try to convince two guys to come over from engineering. And he would say, okay, we need a doctor here, because pretty soon we are going to launch a man. So we want this doctor to become part of the team. Then you had to go to the Air Force and get somebody that was capable of talking about the range facilities that we had. So you get this team coming together for the first time. This was before you had any real simulators. So, you would run a paper countdown. You would go through this thing, and you would learn to say "Roger, go, go, go." Or you would say, "Okay, hold on, I'm going to check." It was sort of like just play acting. All of a sudden, it was time for our first [unmanned] launch, and our first launch, Mercury-Atlas 1, blew up shortly after liftoff. So that was the end of that mission.

What we called the "four-inch flight" was a[n unmanned] Mercury-Redstone. That was the one that was really a stunner. That brought me right back to some of the things that I had learned in flight testing at Holloman. We were launching our first Redstone rocket, and we only had two televisions in Mission Control, one at the flight director's console, and one down at the spacecraft communicator. That was an astronaut down there. Why they gave those two guys televisions and nobody else, I don't know. I guess they didn't want us being distracted by things that were moving. They wanted us to look at meters and dials. So, we finally got to the point to where we sent the firing command, and we saw a great big billow of smoke come up. It really wasn't smoke. It was basically just steam boiling up all around there. All of a sudden, something takes off and boom! It is going like a shot. We had never seen a rocket launch before. Chris Kraft is looking at this thing on TV, and he said, "Did you see that?" I said, "Yeah!" And we were both looking at that thing, and all of a sudden the television camera pans down and the rocket is still on the launch pad. We hadn't launched the rocket. We had launched the escape tower.

Well, this started what turned out to be a not-too-swift day. The next thing we see come out, we see a great big wad of stuff come out, and it is a parachute. Then comes out some chaff. They had strips of the aluminum foil they used from a standpoint of tracking and then dye markers that came out of the top of this rocket. Then the parachute starts to inflate, and we said, my gosh, if we've got the wind coming off the ocean, this thing is going to tip over and really be a mess. So then it is "What we do about it?" And there was a lot of discussion.

In the meantime, to add further to this confusion, our rocket engineers weren't speaking to us, they were speaking German to the people in the blockhouse, so we had no clue what they had seen and what they had done. We finally concluded we have a live rocket on the launch pad with no control over it. So somebody says, "What are we going to do?" Well, we've got to reconnect the umbilical [a cable that supplied electrical power to the launch vehicle]. Pretty soon, we got the information from the launch pad that the umbilical had been pyrotechnically severed. There was no way we could reconnect it. Then the discussion is "go out and get a cherry picker and some pruning shears, and cut the risers on these parachutes before it blows over." But nobody wanted to get that close to a live rocket. Then, somebody says, "Why don't we shoot holes in the tank with a rifle?" So they were going out to security to get a rifle and see if that's the way we want to go.

Well, at this time we wrote the first true, real rule in mission control. "If you don't know what to do, don't do nothing." And that served us very well, because we sat there for the best part of about ten hours as we watched the batteries deplete. The way they had designed the system was, if you lost all power internal to the rocket, all of the valving would go to a safe position and it would vent the pressurized tanks and everything else, and then you could approach it safely. This is what we decided to do. This was probably our greatest learning mission that we had in the very early days, because we found out that we absolutely knew nothing about the systems we were working with.

Sum substance of our knowledge of our spacecraft was what was called a pocket checklist. It was a checklist that was about six to eight inches high and probably about five inches wide, and it fit into the flight suit of the astronauts. There was nothing about the booster in there. All of a sudden, we said we've got to get a heck of a lot smarter. Now, one thing that goes back to Holloman, whenever we had a flight test going on with an aircraft, every flight test engineer worth his salt has got some integrated schematics; he's got the drawings the thing was made from. If there is a problem with the test, you want to give them some straight answers, because range time and flight time is very expensive. So, basically, you had to be the guy with the answer. We did this by developing a series of integrated schematics of the system that we were flying. They weren't the greatest. A lot of them were hand drawn on the back of just a piece of paper, or you would sit down with the engineers or the mechanics and figure out how things worked. But this led me to start in the direction that says that from now on, this team is going to have knowledge of the systems they are launching and are going to fly. So this is sort of the beginning of a very steep learning curve as we went through Project Mercury.

Remember the comedian Bill Dana? He did the Jose Jimenez, the reluctant astronaut stories. Whenever it got boring in the rocket's countdown and things were getting sort of tense, while somebody was working the problem, the launch director would bring up one of the Bill Dana segments in there. And we would just listen, you know. I think back, I won't say how casual it was, but basically it was a very personal, emotional. We had an attachment to each other that was truly a brotherhood in those days, so this was the way that we had done business. We knew the risks. It was high risk, but we also knew that you had to live with it and you had to power down periodically. Today, if we played some of these kinds of the tricks over at the launch pad, the people would be fired. We've lost our sense of judgment in the kinds of people that managed incredible risks in spaceflight. But in those days, we were all what they would call cowboys today. But we weren't cowboys. We were thorough professionals.

"Stars and Stripes Forever" sort of became my moniker from very early in my life. My father died when I was seven, and my mother had three kids to raise. It was right after the Depression, with no insurance. So she turned our house into a boarding house. This was during World

War II. We lived right next to the USO [United Service Organizations], so the USO would send the GIs over to the house. These people who lived there became my heroes. From that day forward, I just had an intense love for country, and patriotism, and respect for the kinds of people willing to risk and sacrifice. So all through my career, as I then went into the military, "Stars and Stripes" became part of the day. When these eight-track cassette recorders came into play, I had one put in my small Mercury [automobile]. And right off the bat, it was stolen. So this time I riveted that sucker in, so nobody is going to take this! We had six kids, and I would drive them to work or to school. The first thing they would hear every morning was "Stars and Stripes Forever." They made a deal that I had the radio and tape player in the morning and they had it in the afternoon coming home.

All through my career, every meeting I ever had in Mission Control where we were talking about where we were going, every staff meeting I had, I always started off with "Stars and Stripes Forever." In fact, as I was approaching the end of my career, I moved into the headquarters building at Johnson Space Center, and I picked the office which was directly opposite the flagpole. So right out that window I had the American flag. Every morning, I kicked off our teleconference with all my people, half hour before work started, with "Stars and Stripes Forever." That was part of my life. I love it.

The challenges we were facing in 1961 were many. It was to build this team, to acquire the capabilities and the skills to launch the rockets we had, principally the Atlases, which we were flying at that time. And to continue the drive forward. The challenge was that the medical community had a very dominant part in the early Mercury program. Those of us who had worked in aviation had no doubts about man's ability to live and work and perform in space. Therefore, we found it sort of difficult to have to work to a set of programs and projects, and move forward in what we call incremental steps, when we feel that we are capable of much greater steps.

About a month before President Kennedy made his speech, we had blown up our second Atlas rocket. We immediately turned around and started the preparation for Alan Shepard [our first manned spaceflight]. So it was sort of a not-so-good feeling. We had had a major failure in one of our launches, and now we were getting ready to launch Alan Shepard. Al's flight came off very well. I believe about nine days later is when we first heard that Kennedy had made the "go to the Moon" speech. The words still ring in all of our memories: "We choose to go to the Moon, we choose to go to the Moon in this decade." And this word *decade* was the key word. Eventually, we believed, we were on a track that would have taken us into space. But we were looking at a much more leisurely path than what President Kennedy had set up for us.

There were two types of people in the program at that time. There were those people who I would equate to Cold War warriors. These were the people who recognized the early challenges the Soviet Union had given us. And myself, personally, I had seen the Russian MiGs flying the other side of the DMZ [demilitarized zone] when I was over in Korea. So to me it was a question of, just for our own national security, to rise to the challenge that President Kennedy gave us and to defeat the Soviets in this thing which was called the space race. There was another group of much younger people, fresh out of college, who saw this as the dream mission, the opportunity of a lifetime. They did it principally because of the challenge. So there were two discretely different frames of mind as we continued to go forward in the program, and particularly as we had to meet certain milestones.

What was interesting was our knowledge of what the Soviets were doing. We had a book where they would put all of the newspaper clippings and the headlines from the newspapers. Once a week, they would pass this book around to us so we could see what our competitors, the Russians, were doing. At the working level that I was at, and I believe up at the levels above me, we had no direct intelligence on what the Soviet Union was doing. But the real challenge was one of winning the space race for the Cold War warriors, and winning the space race principally as a result of the technical challenge that we faced. Two different approaches to doing the job.

There was an enormous amount of frustration because every time you'd go home at night, you would see a newspaper article that basically would cite the Soviet triumphs and our failures. It was this kind of a relationship that the media was building within the country. When are we going to catch up? When are we going to set our own American space records? Alan Shepard's launch, I think, sort of tempered that almost antagonism for a period of time. It was the launch of John Glenn [America's first manned orbital flight] that really caused this entire approach from the media to become much more positive of the work we were doing. The John Glenn flight, I think, did many things, but in particular, it gave the people of the Space Task Group the confidence that we were moving in the right direction. And that we had "the right stuff."

It was in very rapid succession that we put up each one of the remaining missions, the Mercury-Atlas 6, 7, 8, 9, and proceeded to continue the development of and the proof to the medical community of man's ability to perform in space. The greatest challenge we faced in that era was with Scott Carpenter, because Scott, once he got airborne, was completely enamored of not only the environment, but basically he tried to track down these snowflakes that John Glenn had reported. [During his first orbital flight, Glenn had reported seeing "a big mass of some very small particles, that are brilliantly lit up like they're luminescent." These particles were later referred to as *snowflakes* or *fireflies*.] Scottie was the fill-in for Deke Slayton. Deke was scheduled to fly that mission. It was a late change in missions that brought Scottie in there, so he really did not have the benefit of the training that he should have had.

But the key thing is, the training was very rudimentary. There was no spacecraft simulator except, basically, a mockup that existed. It was right adjacent to our control room down at the Cape. There was no way for the crewmen in that mock-up to interact with the thirteen tracking stations that we had around the world. So all of this occurred in real time. The first time a spacecraft communicated, got a chance to talk to John Glenn or any of the astronauts, was the first time they appeared over the site.

Scottie, really, I don't think had all of the benefits of the training. He was cruising along on orbit in dual modes, both automatic attitude control and manual attitude control. He was hosing the propellant. We kept advising Scottie to get off the use of propellant. Quit doing whatever you are doing. He would never explain it to us. Finally, he was basically out of both forms of propellant, both systems, as he approached the Hawaii site. And he was still trying to track these snowflakes. It was only through a very alert controller out there that started reading off the pre-orbit check list to him, and even after the spacecraft had crossed beyond the radio line of sight, this controller kept reading. Fortunately, we had sufficient aberration communications that Scottie was able to finally get on track and start pulling his spacecraft back to center.

When he got to the California site, he was virtually out of fuel. Again, we had a great controller, Ted White, there and Alan Shepard. They coached him into getting into an attitude where he could fire the retrorockets and have a reentry. Now we were going to have a reentry different from what we had planned. That was probably the closest to an accident that we had throughout the entire Mercury program.

That led to a major change in the training of all crewmen who would fly Mercury thereafter. We provided more of the classroom and direct hands-on instruction for these folks. By this time, within the control teams, we had decided that we needed better information in the consoles. So we started developing integrated schematics. We started writing the procedures we used. In the early flights, we were at the mercy of the contractors delivering the right information to us. By the time we got through Mercury-Atlas 6 [John Glenn's first flight], we knew that we needed better information for the control teams as well as the crews for training purposes. So we started to move in an entirely different direction. At the same time, we moved in the direction of developing a professional flight control team. We realized by the end of the Mercury program that we couldn't use part-time players. In Mercury we would borrow from the engineering organizations, medical organizations, range control organizations, and put this team together. Quite frankly, for instance, in the Glenn mission – and this was true in all missions – we would have a training session, and then at the next training we had different people in there. So there were no full-time professionals. By the time we finished Mercury, we were on track and really understanding the nature of the work in the space environment.

So did the crew members. The original Seven were all the steely-eyed missile men. They were classical test pilots. They had to come to grips with what is the role of this group on the ground? Who is supposedly in charge of my mission? That was not an easy mountain to get over. But very quickly, they learned that up in orbit they were basically blind. They did not know even whether they were in orbit at the time they got engine cutoff. It was totally coming from the ground and the ground tracking systems to let them know they were in orbit, and they were going to be there for one orbit and two orbits, and so on. So we would coach them through this process. All of a sudden, the crew side of the system, the astronauts' side, started to move in toward the ground system. So by the time we moved into the Gemini program, we both better understood the nature of the relationship that we had to have in order to succeed in this new business called space.

We used a lot of the communication techniques that they used in the Navy, where they would skip our regular signal off the ionosphere, then back down to a station. We would communicate to Australia through several skips. My call sign was Double Fox Brass One, and as it went around the world you would have a brass two, brass three, brass four, to get over to Australia. Then the response would come back in the same fashion across the Pacific and then back to home. So you may have communication that had been relayed three or four times before it got to the station and once or twice before it got back. The answer you got back was almost unintelligible. We spent a lot of time developing and learning what we called speed printing. We would teach the controllers to print up to twenty to thirty words per minute. It always had to be printed, so the teletype operators could take that, print out the message, send it to the site. And hopefully it would get there before the spacecraft arrived.

It was the same thing when the spacecraft would leave the site, and we would hope that the message from that site would get to the next one down the line. So we were working with an awful lot of impediments, from the standpoint of communications to computing technology. But at the same time, I think this was probably the way we should have grown into space. Because it forced the focus upon the team, and the team coming together and establishing the set of directions, the policies that we needed. No one individual in Mission Control, whether it be flight director or any of the flight controllers, had all the answers. So we had to recognize that the answers had to come from an integrated team because we had a complex spacecraft, a complex mission, and we were working with, at that time, what we might consider simple but complex new technology.

As we moved into the Gemini program, this is where we finally came to grips with our inability. The majority of the Mercury people were technical dinosaurs, and now we came to grips with this new thing called computers. They put a 4 kilobyte computer onboard the Gemini spacecraft, and we had no computer experience. So then this took us to go to the colleges and the universities of our nation, and we brought in young people who had been working with computers in laboratories. The Army at Fort Bliss was a major provider of controllers, particularly in the training role, because of the ground-to-air missile programs that they had been working on. So we now brought in the younger group of people, and from that day forward, we started benefitting from the young people that would come in. As we moved forward in the technologies needed for space, these people would bring it to us. At the same time, we were growing in experience and leadership and, in particular, continuing to understand the nature of the missions that we had to fly.

In the Mercury and Gemini transition, there were two very important lessons we learned. The first lesson we learned was the need for a professional space operations team in Mission Control. The other thing that was very important was to reduce some of the ambiguities on who is the leader of this team. When we had the problem with John Glenn [during the Mercury 6 orbital flight, a faulty sensor incorrectly indicated that the spacecraft's heat shield and landing bag were no longer locked in position], the control room was flooded with program managers, people from the contractors. Everybody had a different answer, but nobody understood the problem. They moved in different directions. Walt Williams [launch operations director for the Mercury program] was basically the dominant personality, and with all of the ambiguities, he just laid down the law and said we are going to reenter with the retrorocket package attached for John Glenn. But the bad news was, we hadn't even told John what our problem was as we had gone through the second and the third orbit. It was only when he was coming in over California that we told him that he would have to do this. And he said, "Why?" The answer was, "Well, the flight director told us so." So we had to come up with an answer that we would give to John Glenn during reentry. We had already started reentry when we told him why we wanted him to come in. So this basically dictated that the role of the flight director is probably the simplest job description in America today. One sentence. It says the flight director may take any actions necessary for crew safety and mission success. That's it. And for over five decades, that position description has been observed, and we have had quality of leadership within the ranks of the mission control people.

From the media, we were getting blessed. We didn't have any Fox or CNN. Literally, in the media, we had space and science professionals. And they were good. They spent their time. They got their homework. They would come down in the Mission Control. They would understand the mission. They understood some of the physics of space operations. They understood the risks associated with high technology.

One of the benefits that we had as we moved into the early space programs was the directions that came down from headquarters were enforced universally in Mission Control, that this program would be conducted in the open. We weren't going to be like the Soviet Union. Everything that was going on, the media would be briefed on. Each flight director, after every shift of every mission, would have a press conference, and you'd have three press conferences a day. This is how information got out to the media. At the same time, the media was listening to what was happening in Mission Control. And they understood what was going on in Mission Control. If anything, I think the media acted as a stimulus for us to do our job even better and to do it more complete.

By Apollo, we were truly in a race. We had transitioned very quickly. In fact, during the late Gemini program, we had two mission control rooms in our control center, one on the third floor and one on the second floor. We would be actually flying a Gemini mission on one floor and we would be training for an Apollo mission on the other floor. We were overlapping. The preparation that we had for what we were doing on Apollo was not as complete or as rigorous as it had to be, due to the other duties in support of Gemini.

As we were approaching the Apollo 1 mission, we were roughly about thirty days out, and the launch date had been slipped many times. I was the flight director. Three flight directors for that mission were Chris Kraft, myself, John Hodge. I had the white team. We ran two tests in sequence, what we called the plugs-in test and the plugs-out test. Both were countdown, dress rehearsals for a launch. We were about one month out from the launch. The day before, I had run the plugs-in test and we staggered. We had communications problems that were unbelievable. The test procedures just didn't hang together. But by the time we had finished that test, I had managed to patch around the majority of the communications problems we were having with the launch team. We moved directly from that test into the plugs-out test, which is a full-blown dress rehearsal, including the final minutes of the countdown where we transition from external power on the launch pad to internal power in the spacecraft.

I had kicked off the countdown that day and handed it over to Chris Kraft around noontime, and he continued the countdown. But I remained in place until he got into the actual countdown or the crew was involved. And shortly after the crew moved into the spacecraft, Gus Grissom had reported, "The spacecraft has just got an odor inside, and I don't understand what it is, but it is horrible." So basically, he reported a noxious odor inside the spacecraft.

Communications problems continued. Our test procedures just didn't hang together, and we struggled through that afternoon. Frequently calling a countdown hold, sort out problems, continue on, hold, sort out problems, continue on. Until finally at 6:27 that afternoon, we again went into a hold before we transferred from external power on the launch pad to internal power in the spacecraft. Four minutes after entering that countdown hold, the control team was startled by screams coming from the crew. And they listened to the crew as they died. With their death came anger, because we knew we were responsible for America's first space disaster.

Throughout that late afternoon and evening and into the night, we kept trying to find the cause. What was the reason for this? We secured Mission Control. We couldn't communicate out. Various experts came in. We saved all the recordings, all the records we had had for the launch. And then, we have a watering hole when times get bad, as well as good times, the Singing Wheel. We went over and had a few beers that night, and many of the controllers had to call their wives to come and take them home. This was a very difficult time for the team. Myself, Hodge, and Kraft had worked in flight tests. We had worked through the Mercury and Gemini programs. We fully understood the risks of the business. But we had a bunch of young kids now who for the first time in their life had faced death, and they faced the fact that in some way they were responsible for this.

I had flown fighter aircraft. I had worked out at Holloman [Air Force Base]. During my time at Holloman, I had a little placard that showed a bi-wing airplane, a World War I type of airplane, hanging in the top of a tree. And it, basically, indicated that unlike the sea, aviation is completely intolerant of incapacity or neglect. I thought about those words, and that airplane up in a tree. I was chief of the flight control division. I had the responsibility of all of the controllers and the flight directors. I started off the talk that day and indicated that there was no way to pass the buck to anyone else. We had to assume total and complete responsibility for the failure that we had had. Because we knew we were not ready, we knew the procedures weren't ready. Basically, we were drinking from a fire hose and we were behind the power curve.

Then I finished with the fact that from now on, this team in Mission Control will be known by two words: *tough* and *competent*. *Tough* meaning that we will never again shirk from our responsibilities because we are forever accountable for what we do, or in the case of Apollo 1 and our crew, what we had failed to do. *Competent* meaning we will never take anything for granted. We'll never stop learning. From now on, the teams in Mission Control will be perfect. And those words set the stage for the work that we would do in the following years that took our nation to the surface of the Moon. That addressed the pause we had during Apollo 12, when in the launch phase, we were hit twice by lightning. We had Apollo 13, in which we had a disastrous explosion 200,000 miles out in space, and the odds of bringing the crew back successfully were very limited, very dicey, but the team pulled this thing off. So for the rest of our professional career we lived by those two words.

We designed a badge, the emblem of the mission control team. The price for admission to our ranks were discipline, morale, toughness, competence, commitment, teamwork. Because as a team we must never fail. So, I believe the team came away from that massive failure in Apollo 1 with a totally different resolve that we would never again be accountable for such a failure. From that day on, we would make sure that it would never happen again. And this held true. All the way through the early years into the space shuttle program, those words served us well.

I believe the key to our success in spaceflight, in the early years as well as today, is the quality of the training that is given to the members of the integrated control team, both crews as well as ourselves in Mission Control. I had the responsibility in later years for both crew training as well as the controllers' training. We had this person, we looked at him as a cop, we called him Simulation Supervisor, SimSup. The SimSup would observe our preparation for the mission. We spent about 60 percent of our total mission preparation time developing what we called mission rules. Mission rules represent the risks versus risks and risks versus gains strategy that we would use if we encountered problems on the journey to the Moon or whatever our objective is. These debates on mission rules would go on to late at night. The instructors would be sitting and listening in there. They would recognize when we just made an arbitrary decision so we could go home. Or we made a decision, and they just suspected that we didn't have a good technical base for it. Or there was a disagreement on how the crew wanted to do it versus how the team wanted it done. They would set up these very complex training scenarios that would address these issues that had come up in training.

At the same time, the instructors knew the controllers intimately. They knew the people who were, maybe, having problems at home. They knew the people who were, maybe, party animals. So the team's approach was to basically address the problem, actually focus the problem on the team and force the team to solve problems where they were inadequate in either knowledge or personality or membership or whatever it was. And during this process we would be recorded, and then we would debrief.

We learned the expression "ruthless honesty." Basically, it was honesty at all cost. You will always identify whatever deficiencies that you had. This came to be very helpful when we got into the mission environment, because we were used to stepping up and identifying if we had a problem, or if we didn't know. This was the first thing. We didn't try to guess. It was just "I don't know, but I will find out, and I'll find out quickly." The concept of ruthless honesty characterized the team. For people who couldn't live in that environment, basically, they elected not to remain a member of the team.

The training period was roughly about three months in duration. We started off with the launch phase part of the mission. In the launch, you had many different abort modes as you were going uphill, and this was generally where we started off working with the prime crew. We had had a good background. By the time we got to Apollo 11, we had been through launches for 7, 8, 9, 10. So the team was relatively up on the ball. It was really a question of letting the team bring the crew on board and getting this relationship that we needed to be successful.

Launch phase is extremely dynamic. Decisions have to be made very rapidly, and once you have made the decision, it is irrevocable. You've got to pick a pack. We then moved into the training for the Earth-orbital phase, the translunar-injection phase, and right on down the line. We went through this phase mainly because the software development people were continuing to develop the stuff that we needed for the lunar landing, both on board the spacecraft as well as within Mission Control. So we started off doing this stuff that we knew how to do and had the software available to do, while the development people were still bringing along the simulators and the stuff we needed for lunar landing.

When we got to the lunar phase of the mission, my SimSup was a guy by the name of Dick Koos. He was one of the Army guys that came to us from Fort Bliss. Again, he was ruthless in his identification of the team. He worked with the team. He knew every team member. As we went through our training process, somewhere along the line, he developed the feeling that we were a bit too cocky. Generally, the last training session we have is considered graduation exercises. It was a relatively simple set of problems where both ourselves and the crew would leave with a good feeling about ourselves. Well, Dick Koos felt that we weren't at that point in the program yet. So he gave us a problem that brought us to our knees. This was associated with a series of computer program alarms. We thought we had responded properly. The flight director always starts off the debriefing. He is the first one to practice this ruthless honesty. He has to set the standard and basically identify that we had serious program alarms and that it is the judgment of my guidance consultant that we should abort, terminate the landing attempt. Then the SimSup came aboard. And he says, "What was your second cue? Because you in your own mission rules say you are required two cues. You had one cue, what telemetry was telling you. Where was your second cue?" And we didn't have an answer.

Then we recognized that we darn well better go back and study these computer program alarms and have a better feeling as to what was happening in the guts of the computer when we got these alarms. We expected something dramatic to be happening, but it wasn't. There was some aberration occurring internal to the computer. By this time, the crew of Apollo 11 had deployed to the Cape. So we got together with the Draper Lab, and we went through the computer program listings. We identified what we considered were the secondary cues if we saw these program alarms. We listed these things for all possible alarms that might occur and went over them with the backup crew the following day. The thing that was almost incredible about this was the situation our training people gave us for our final simulation was exactly the situation we saw during landing [of Apollo 11]. If you want to talk about "somebody up there is looking out for you," you sort of get that feeling. A lot of times, I've wondered if we had seen these alarms, would we have aborted? And I tend to believe not. Because in training you're more prone to take aggressive actions, where in the real mission you will back off just a couple of notches, and you go through things. But basically it is very interesting.

The actual landing and the descent was extremely difficult. The real initial problem that we faced was a communication problem, where we couldn't talk directly to Eagle [the Apollo 11 lunar landing module]. So we had to relay communications through Mike Collins [the command module pilot, who was orbiting the Moon]. We normally worked with about a three-second time delay from communications to processing within the computer complex. But once we started adding Mike Collins in the loop, we add another two seconds into it. Three to five seconds doesn't sound like very long, but when you are hurtling toward the surface of the Moon, you can go an awful lot of distance. If you are not proceeding in the proper trajectory, you can get into trouble real quickly. Anyway, the communications problem forced us to download, and Charley Duke, as the space capsule communicator, was absolutely beautiful. We started changing orientation of the spacecraft. About this time, my trajectory controller, Jay Greene, comes in and he says "Flight, I don't know what's happening here, but we are out a radial component of velocity and we are halfway to our abort limit." Now, *abort* is one word that is not used casually in Mission Control. When he said those words, everybody else sort of perked up. Then he said, "I'll continue to watch it."

At the same time, we were going through a series of checks in the spacecraft, getting ready for the Go/No Go, to start powered descent. And we continued to have dropouts. We had a minor electrical problem that we resolved very quickly. Periodically, we would have good communications and no good communications, and we only had one more opportunity to go down to the Moon. We had a mission role which was totally on the back of the flight director. There is no crash recorder on board the lunar module. So the question is "Do I have enough data to know what is going on, so if we have a disaster we can figure out what happened and fix it?" I make the decision that we have enough communications, both telemetry as well as the information from the crew, to continue the descent to give the crew the Go for the start of powered descent. About this time, we recognized that we were going to be landing long. There was some kind of aberration to the trajectory that was taking us further downrange than we had expected it to be. This is just a casual comment that we were going to be landing near the toe of the footprint.

We start the engine and go down and start our descent. We are actually throttling up, we are actually braking. We are reducing energy for the orbit so the Moon's gravity is bringing us down toward the surface. As we start the descent, we updated the lunar module's knowledge of its altitude above the lunar surface, based on the ground. But now we have to allow the landing radar in the lunar module to update that computer because we need very accurate measurements. If there is too great a difference between what we put in from the ground and what the actual is, the very dumb computer will try to solve that instantly. During training, we occasionally get too great of a difference between what we gave and what is given, and the spacecraft will just turn upside down and drive toward the surface and then try to recover and start doing whip-a-dos down there. So we are waiting for this.

This is the point where we start having these computer program alarms. So now we've got the team working in two directions. We've got to get this [new data] saved in the computer, but the computer is going belly-up on us. That is where this training really paid off, because we had nailed down these computer program alarms and what they could do. We had two types, what we called a bailout, where the computer is too busy to get all the work done and it is now doing only high priority things – updating guidance, navigation, firing jets, and this kind of stuff. And this is the kind of stuff my guidance controller can look at and see happening. These were the secondary cues we were looking for. At the same time, if these bailouts occur in rapid succession, very likely you are going to go into what we call P00 do [pronounced "pooh do"], do program zero zero, or the computers will go into hold and missions will have no choice but to abort.

So we are now starting to talk the crew down. Communications have opened up a bit, but we know we are going to be landing long. We are, basically, GCA. If you are familiar with Ground Controlled Approach, which you use in airplanes, you're telling the crew to stay off, do this way or that way, right on down the line as we get down close to the surface. We now are starting to look at the fuel. In the back room, I have a controller that has two stop watches, and he is looking at a single analog trace. In the LM fuel tank, they had a capacitance measuring probe that could only go down to the bottom of the cylindrical portion of the tank. We have no measurement down in the bottom of the tank. It is like driving your car with a gas gauge reading empty. But we knew from testing that at an average throttle setting of 30 percent we had two minutes to fly. So now the crew is throttling above 30 percent, below 30 percent, above 30 percent, up and down. So we've got a guy in back room looking at an analog trace, and he is trying to mentally estimate how many seconds above 30 percent we are with two stop watches. He is relaying this information to a controller in the front room that is telling us we have 60 seconds to go, 45 seconds to go, 30 seconds to go.

Now my team gets real quiet in Mission Control because there are only two people talking. It is my controller giving fuel, my spacecraft communicator, and then if need be, I will talk, which is "land" or "abort," one or the other. Those are the only two decisions. So we are quiet in the room. We get down to time to say, "15 seconds," and we realize the crew is going through engine shutdown. They are down on the surface. It took a few seconds to recognize that particular event. Then the recognition seeps into the people in the viewing room, where we've got Senators, Congressmen, program engineers, and everybody else. And they start cheering, applauding, stomping their feet.

At the same time, I have to start going through a series of stay, no-stay decisions. My team can't celebrate. We now have to say within two minutes if it's safe to remain on the Moon. We give the crew a stay at T plus two. The stay times were set up so that if we lifted off at that time, we could go into an orbit and have enough fuel to get back to the command module. Then we go through another stay at eight minutes, and again we give the crew a stay. Then we go to another stay at two hours, and we then give the crew the Go for the T plus two hour stay.

Then the team starts to hand over the process to the next team. We actually have two teams in Mission Control at that time, because the doctors felt that we ought to schedule a crew sleep period there. I don't think anybody ever believed that the crew was going to land on the Moon and they are going to want to go to sleep for a while. They are going to want to get out of the spacecraft and start the exploration. So I have two teams in there. One team that would handle the sleep, and another one that would handle the EVA [extravehicular activity]. So we now have three teams in Mission Control, milling around while we make the decision and wait for the crew's call. So, basically, that was it. I went over and did a press conference. Then coming back, I finally had a chance to power down and just start my own mental celebration. We had just landed on the Moon.

Once you got on the surface, the crew had an awful lot of work they intended to do. It has always been a trouble for the crew. They always want to take a look at the Earth out there. They wanted to do this, and you have to hurry them along. They are getting behind. You know why they are getting behind, because they are taking a break and starting to enjoy this business of being on the Moon for the first time. This was true of all missions, and it's really cost us. "Okay, guys, let's get with it. Come on, let's get going back on track."

The landing of Apollo 11 on the Moon was very important because I think very few people recognized how young that team was. I was the oldest guy in the room that day at thirty-seven. My team members' average age was twenty-six. The majority were roughly four years out of college, and this was something that had never been done before. We were going to take an American to the Moon. We were very well trained for it. But it was also important to let people know that whatever happened that day, the thing that was interesting, and it was very stark, throughout every training session, we would go into what we called a battle short, where we physically block all power breakers going into the building. Then we would lock the control doors. And when we locked those doors, they will not be opened until we had either landed on the Moon, we've crashed, or we have aborted the landing.

There is only one of those outcomes that is good. I think it was very important for my team to recognize that I would stand behind every decision they would make. I understood how difficult, I understood the risks that were involved, I understood the fallacies of the people that were in there. But as a team, we must not fail. This turned out to be very important. One of my controllers expressed it in a documentary. He said those words were exactly what he needed. In fact, he says, "Kranz also added another set of words to that. He says, 'I'll stand behind every decision that you make. We came into this room as a team and we will leave as a team.'" He basically indicated that was what gave him the conviction to stand up and make tough decisions.

[The communications problems and the potential overshoot of the landing site in Apollo 11 had to be solved before the next mission.] Mission Control, and the team that we had supporting us, was very aggressive in reaching out to new challenges that were out there. We went off to Draper Lab and we said, "Gee, is there some way, based on our trajectory observations, that we can put an incremental delay into the start of the descent? Or either advance or delay this thing?" We called them a "Noun 69" update. And doggone, they came back and said yes. It's a relatively simple process. Do these computations you want to provide, on the ground. You voice them up to the crew. They can enter this in the computer. It's going to take you back to the point where you really want to target for the landing. Apollo 12 landed about a football field from an old Surveyor spacecraft. That proved that we could put it down exactly where we wanted to put it.

The real key in Apollo 12 was a lightning strike. That just about ended that mission. We got hit twice by lightning right at liftoff, and we had three crewmen inside a dead spacecraft thundering up towards orbit. If the reactant valves had closed, that mission was over because the fuel cells would starve and there was no way to restart the fuel cells in there. Again, this is a team that has been really incredible. A young controller by the name of John Aaron recognized some aberration that he saw from a test that had been accomplished a year earlier. We didn't know that lightning had struck. But in a test he had seen a year earlier, this aberration had occurred in his displays, and he recognized he had to go over to a backup subset on our telemetry system. So he made the call to the flight director. He said, "Flight, have the crew go SCE to AUX." There was this pregnant silence. Everybody is saying what is SCE to AUX? So the CapCom dutifully relays it to the crew, and then on board the spacecraft, on the onboard tape, Pete Conrad's saying, "What is SCE to AUX?" We'd had to voice this up several times until Alan Bean, recognizing that each one of the crewmen had a segment of display panels, realized, ah ha, there's a [Signal Conditioning Equipment] switch right down here at the bottom of the panel. So he throws the switch to AUX, our telemetry comes right back, and the team can get back to work.

This was a two-minute period of time where we thought we had lost the whole thing. And with the lightning strike, we delayed injection to the Moon so we could check the spacecraft out as thoroughly as possible. There is no way to do the kind of checkout you would have liked to when you had that kind of a glitch in the system. But a very gutsy move: We decided we'd go for it.

Probably the first fifteen minutes of that whole thing [the problem with Apollo 13] was the most difficult time for me, because on the ship, just prior to the time [Jim] Lovell called down, "Houston, we've got a problem," we'd already had two electrical glitches. We'd had a problem associated with the instrumentation in oxygen tank two, and we had an electrical transient that shifted the beam width of our communications antenna. Electrical glitches are not unusual. You see them, generally, every mission or every other mission. So as soon as that problem occurred, I thought, okay, we've got another electrical glitch in there. We'll stabilize the situation as it is, and then we'll start doing the troubleshooting during the crew rest period. Then about four or five minutes later, one of my controllers comes to me and he says, "Flight, the crew mentioned they had a very large bang associated with it." Then I remembered an almost identical thing, because he continued, "It looks like some of our jet valves may have closed." I had an identical problem happen to me on the Apollo 9 mission, where we undocked and just the shock of firing the retention devices had closed off some of the valves, just a transient shock. I now moved into what I would say is the second phase of thinking about Apollo 13, which is proceed a hell of a lot more carefully because something is going on here we don't quite understand. The third step was the final moment, I think it was about twelve minutes after the explosion, when Lovell looks outside of the hatch, and he says, "Hey, Houston, I see some kind of a gaseous substance out there. I believe we are venting something." And this immediately triggered survival.

As I went through those ten to twelve minutes, it was just frustration, because we couldn't wrap our arm around it. All of a sudden, the direction was clear. Basically, this is where I call the controllers and say, "Okay, flight controllers, settle down, quit your guessing. Let's start working this problem. Electrical bus A is still good. Don't do anything to screw it up. The lunar mod is attached. We are going to use that as our life boat. Now get me some backup people in here and get me more computing and communications resources." So that was the point where, all of a sudden, the issue became one of survival. The focus was clear, and now the team could get to work.

I could do a lot of speaking with Lovell. He said, "My wife is watching TV, but nobody in America is watching it." [Lovell's wife was aware of the problem because she could listen to communications between Houston and the Apollo 13 crew on a "squawk box" in her home.] On one of the networks, they had the *Dick Cavett Show*, which was a live show, and Lovell couldn't understand why the network wasn't carrying it. On another one they had a rerun of *I Love Lucy*, and then he said, the third network, at least in Houston, was showing the baseball game. So in Mission Control, all the players were watching baseball. This is sort of the situation. It had become humdrum. We had been to the Moon twice, so, what's the big deal about going there again?

The only reporting that was going on was that which was coming from Mission Control. So, the media weren't in the action. We had our first press conference roughly six hours after the accident, at which time we had established provisions for them to come back into Mission Control. When we designed the mission control room, we had a little booth over in the left-hand corner of my room that was all glassed in, and we had enough room for three reporters. We had one from the newspapers, one from radio, and another one from TV. They were all chosen from pools of people that were out there. Also in the same little room, they had one of the media people from JSC [Johnson Space Center]. These people now started reporting live, and all of the communications that were going on were patched right into them. So we were living in a fish bowl.

One of the things, to me, was the kind of respect we had for the media. I set up my work plan. I was lead flight director. So it was my job to get the team on a rotation sequence, work the problem, and show back on console only when we had answers. So I moved my team down to the second floor. At the same time, whenever I needed a nap, or some kind of a break, I would go into the viewing room on the second floor. This was one that was identical to the one we had on the third floor. We had all of the communications piped in, so I could listen to the communications. It was cold as an ice box, and if I wanted to grab a few minutes' rest, this is where I would go. But at the same time, this is where the media went. I was right there. We were at opposite ends of the room. They wouldn't come over and bother you, disturb you. They would leave you go. They had the opportunity for one-on-one interview questions. But this was the kind of relationship and respect they had. This was also the place where I had teams scattered all over doing certain things. If I wasn't in one of the rooms working, they knew where to find me. So this acted as sort of my node when I wanted to get away and get some time to think, or in particular when I wanted to talk to Kraft or one of the other flight directors. We would have sort of a meeting down in that room where we had time to talk things over to see which way we were going.

Go back to Apollo 11. When the team's given a problem, they always worked best. Basically, that is the nature of the team. They feel very comfortable. They get into action. They are not observers of something that is going on. They are participants in what's going on. This problem that we had on 13 was so encompassing, I set up three basic teams. And there were many subsets of teams. There was one led by John Aaron, and John was managing all of the resources available in the spacecraft. These were power, water, and these things were consumables. Arnie Aldridge was one of my branch chiefs. He had been a controller, worked in the Mercury program out at a tracking station. Basically, he would integrate the procedures. Then I had another guy by the name of Bill Peters, who was a very capable guy but he wasn't a good communicator. But I liked the way he thought, and I gave him the responsibility for listening to what was going on and seeing if there were any crew survival issues related to the path that we would take.

These teams would report in. Remember, this was the days before word processing and all this kind of stuff. Everything had to done by hand, printed by hand, drawings we needed, actually done by hand. It was really interesting to see these markups. But again, go back to the basic discipline of the team. As this Arnie Aldridge guy is working the procedures, you could look up in the upper right-hand corner of every checklist page that was coming up, and you could see a date and a time. The previous time that you had seen this package would be scratched out, and a new time put in there. Every page had this. This started from a technique we used back in Mercury, where as we would get ready to lift off, I would have the controllers hold up the copy of their mission rules. I could look around the room, and we'd change colors every new edition we came out with. If we were looking for a yellow cover sheet and somebody didn't have a yellow cover sheet, they went and got the proper updates. So this is the kind of thing that got the job done, without all of these major communications and all of these new technologies you've got. It was this kind of a background that allowed us to incorporate the new, young people very easily and very comfortably.

At the time of Apollo 13, the Mission Control team was Super Bowl championship quality. We took the challenges of that mission, and we met them head on. We made the difficult decisions. The team worked together. There were many, many, many players. It wasn't just the flight directors. I had Arnie Aldridge doing the procedures and John Aaron doing the resources. I had people addressing survival issues. We had astronauts coming on to work the simulators. We got word in from our contractors all over the United States. We got people at the University of Toronto giving us ideas as to separation techniques that we could use to minimize perturbations to the orbit. It was really amazing to see this team come together and wrap their arms around that crew.

We were extremely fortunate because we had a very professional media in those days who understood the risks of the business. They gave us the benefit of the doubt every time we came to talk to them and give them the status. They were actually in the control room. We were operating under the observation of the media – television, written word, and the radio. We lived in that fish bowl for the entire period of time of Apollo 13. But we came out at the far end. We saved that crew.

When I found out that Jim Lovell was writing a book on Apollo 13, I thought about all of the other missions when we had had close calls. I was sort of surprised that this had become a big deal and so many people didn't know about it. It completely changed the direction of my life, and many other people in there. All of a sudden, I got the opportunity to write a book [*Failure Is Not an Option*]. I got out on the lecture circuit. I have the opportunity to talk to various museums and kids in school. If we are going to move forward as a nation, these are going to be the generation that we have to inspire to move forward.

[Before the 1995 movie *Apollo 13* was filmed,] I had finished a very long interview with Japanese public broadcasting TV on site. This was two days, and it was probably the most intense interview I've ever done in my life. The Japanese had set this thing up so that they had the pictures, the events. They had the recordings, right on down the line. Literally, by the time I had finished the second day, I felt exactly as I did when Apollo 13 was over. It was really tough. So, Ed Harris [the actor who played Kranz in *Apollo 13*] got this set of video tapes, and he studied them. He decided that the only way that he'd be able to fulfill the duties of this part was to essentially get the members of the Mission Control team to coach him every step of the way. So he hired Gerry Griffin, one of my flight directors on Apollo 13. And he hired Jerry Bostick, who was my trajectory controller. Both of these guys were key members on the team. It was funny that Griffin and Bostick would come back after they had done the filming and say, "Boy, we've got you nailed. Wait until you see this!" They'd sort of pull my chain, but they wouldn't say what it was.

I didn't realize how distracting it was to my team, but apparently this really bugged them: I destroyed government ball point pens. I was one of these guys who was perpetually picking up pens. Every shift I would begin, my secretary would put a brand new box of pens on my desk. I would pick one up, and I would just [fiddle with it] when I was thinking, and all of a sudden the thing would fall apart. And I couldn't find all of the parts, so I would get a new pen. Well, they got Ed Harris, you know, playing with this pen.

But the one thing that was interesting is that the flight director's console is real plain Jane. It has only clocks, two TVs, and a status panel. So, whenever I got ready, or I was really thinking about something that had to be done, getting ready to make up my mind, I would get away from the console and I would pace. I had a headset cord that was very long. So I could go to the far corners of Mission Control Center, pacing behind the consoles, and it always looked like I was praying. But, basically, getting away from all of these distractions and start listening very intently to what people were telling you. This is when somebody would be going through one of these long sequence of events things that happened. And you want to get totally focused. So they had Ed Harris pacing. But at the same time, he would very abruptly stop, sit down and issue orders. Basically, Ed Harris, I mean he's just got that absolutely perfect, the way he would look like he was praying. So there were several characteristics in there.

The one thing that never happened is that Ed Harris sort of loses his temper when the thermal reentry procedures don't come in, and he sort of gets mad, and he yells at some of the people in the room, and he kicks a waste basket. That never happened, because you never lose control of the situation. And you never let people know you've lost control, that you're angry. It's total focus, and maintain the poise you need to keep people going.

[After Apollo 17, the government cancelled the Apollo program.] This was a very difficult time. This was particularly true of the people on my trajectory team. This trajectory team had pioneered rendezvous, they had pioneered dockings, they had pioneered maneuvers in the vicinity of the Moon, they had pioneered the landings. They had figured out how to get off. If you had an abort, they had figured out what are the options they've got going. All of the sudden, they got the spacecraft [space shuttle] that grinds away in orbit and goes around, and around, and around, and around. At the same, we were in the process of losing people because the shuttle program was building up. Chris Kraft said, "Kranz, you've got some good people in there, but I want them in engineering."

So I had to find some way to protect my most valuable assets, which were the people. One of the trajectory guys, one of the most hardnosed of all of the people taking this, was a guy by the name of John Llewellyn. I sat down there and I said, "John, how would you like to fly airplanes?" He had never been an airplane crew member in his entire life. I said, "We've got some airplanes out at Ellington Field that are doing Earth resources work all over the globe. They are doing air sampling in Alaska, and down in Antarctica. We are doing crop surveys. How would like to become one of these scientific equipment operators?" This guy is wondering what is going on. But, basically, it was to the point of finding a role for one of these trajectory guys. Then I said, "When the Skylab program comes along, I want you to manage their resources experiments. So, you are going to work the ground troops side of the system. Then when the Skylab starts, you are going to be working the instruments that are providing the measurements." So, this was trying to find logical roles for very dynamic people.

I did lose a significant number of people. There were good jobs. There were several technology generators, the Department of Energy was one, the Department of Transportation was another. I lost several people going to the Department of Transportation. They were going to build high-speed trains. We lost some of the good people, and we kept the core that we needed to. For the systems guys, their job was the same whether it is around the Moon or whether it's around Earth orbit. The systems job is to make sure the system is operating properly.

Skylab turned into, probably for me, one of the most interesting missions that we ever faced. We started off with one set of procedures, but by the time we ended up at midpoint to the end of the mission, we had rewritten the book several times. We had to learn dynamics of control moment gyros that we had in the system. All of the science technologies. Going back to the end of Apollo, I sent several guys to various observatories, because we had sun study experiments. We had medical experiments, and a couple of kids went over and did pre-med school. Basically, we would send the people out to acquire the skills so we could communicate effectively with the scientists that would be coming on board in the program. It was a major plan change, but it was very dynamic. Good people, if you can give them a challenge to learn something new, will most always step up to that kind of responsibility.

I believe that our nation becomes easily distracted. And I think this is one of the cases now. If you go back into the time when we were flying the early space program, as you moved toward Apollo you had the environmental movement start. As we saw in some pictures that were taken in Gemini, you had, basically, the pro and against Vietnam challenges that came out there, started the challenges associated with sending Peace Corps people around the world. And the Civil Rights movement was starting in there, in addition to space. So you had many dynamic things going on. People got out of their comfort zones and were involved. This was sort of the environment that we found ourselves in, where a lot of these activities sort of came into collision with each other.

In particular, I think that the dominant activity was the war in Vietnam. This was not only costing money, it was costing lives, and it basically captured the media. Maybe we were victims of our own success.

If we had lost a crew on the Moon, maybe we would have continued on for some period of time. But NASA, unfortunately, does not have a history of effective long-range planning, and in particular, expressing and bringing the American people on board into the programs. The American people came on board during the early space program because it was a race with the Russians. Our future in space now is going to depend, not upon power and prestige, but upon does it make economic sense? What are the reasons to send people into space?

There were NASA spin-off books, but we market them very poorly. NASA does not pay enough attention to getting professional marketers. Go back into the early people we worked with in the media. Some of our best publicity agents were Jules Bergman and other people that worked for the networks. But look where the networks are going. Do they have science journalists? No! They're all generalists in there right now. So, we have to find someone to articulate this message a hell of a lot better than the way we do it right now. If we are to find the future, we have to find the articulator, point one. Point two is we have to get a leader in NASA. When we had Mike Griffin, we finally had the leader we needed. He had the respect of the technical community, he had the respect of the political community, he had the ability to make difficult decisions. We had a change in administration and he was gone, and they cancelled the programs. I think that we need to see if we can find some way, it can never be done totally, but get politics out of the leadership of NASA.

We were blessed. I talked about the media, but I haven't talked about the quality of the leadership that we had, starting at the very top. We had administrators within the agencies, in one case, a pure politician who knew how to get along with Congress. But he had an incredible technical staff. We had guys like George Low in there, who to a great extent, I look at sort of as the guy that truly got us up to the Moon. We had Dr. [Robert] Gilruth, Chris Kraft, and Deke Slayton, so we had that leadership component. But the thing that was most important, these leaders gave us the job. There comes some time in a career where you have to build the team that will succeed, and then turn them loose to do the job. And that's a tough thing to do nowadays.

Chapter 13
Flying High

The International Space Station (ISS), a cooperative venture among fifteen nations, was constructed on orbit between 1998 and 2011. It has been continuously occupied since October 2000 by at least three people at a time. Like other major accomplishments in space technology, it was preceded by a series of earlier projects that provided the information and experience to make it possible. One American and eight Soviet space stations were sent into orbit and occupied for various periods of time between 1971 and 1998.

The United States' Skylab program consisted of a single-unit space station launched in May 1973. Three crews of NASA astronauts performed scientific experiments and observations on Skylab in 1973 and 1974. It remained on orbit until July 1979, when it reentered the atmosphere and broke up, raining 500 pieces of debris on the Earth. Most if its orbit was over water, but NASA predicted the chances of a piece hitting a human being were 1 in 152. The July 16, 1979, issue of *Time* magazine carried an article describing the impending reentry breakup. Americans across the country reacted with humor. An Atlanta radio station offered listeners chances to win a T-shirt featuring a bull's-eye and the words "I'm an Official WQXI-AM 79 Skylab Target." Competing newspapers in San Francisco offered $10,000 for a piece of Skylab debris or up to $200,000 for personal or property damage from space station chunks. Although NASA aimed the reentering vehicle at open ocean, debris did land in Western Australia, without human injury. One Australian citizen flew to San Francisco to collect the $10,000 prize, and the Shire of Esperance community issued a tongue-in-cheek $400 fine to NASA for littering.

The Soviet Union's Salyut program consisted of four scientific research space stations and two military reconnaissance space stations occupied between 1971 and 1986. Each was staffed for periods ranging from fifteen to 816 days. All of them were assembled on the ground and launched to orbit as single units. However, the last one, Salyut 7, docked with an expansion module to prove the modular construction capability. That expanded craft was unoccupied when it reentered the atmosphere in 1991, scattering debris over Argentina.

In July 1975, the United States and the Soviet Union conducted a joint test project, Apollo-Soyuz, in which an Apollo command module docked with a Soyuz spacecraft. Three American astronauts and two Soviet cosmonauts interacted in the combined craft, conducting joint and separate experiments during the nine-day, 148-orbit mission.

Based on the successful expansion of Salyut 7, the Soviet Union began construction of the modular space station Mir. It launched the first component in February 1986. Two modules had been added by 1991, when the Soviet Union collapsed. One cosmonaut remained on Mir during that political transition. Russia assumed ownership of the space station, and in 1993 the United States and Russia reached an agreement allowing the US space shuttle to visit Mir [see Chapter 14 for more information about the space shuttle program]. In 2001, the unoccupied Mir reentered the atmosphere, its debris dropping into the South Pacific Ocean.

Charles Deiterich continued his involvement with the Apollo command/service module (CSM, see Chapter 9) when it was used for access to Skylab. He described problems with the launch of the unmanned Skylab work station on May 14, 1973. Skylab, the first American space station, was launched without occupants. Crews would arrive later, docking their Apollo CSM with the Skylab module. Between May 1973 and February 1974, three 3-man crews occupied Skylab for durations of 28, 59, and 84 days. Throughout its six-year lifespan, Skylab experienced several problems. The first one occurred on the first launch, the unmanned Skylab module.

I was on when they launched the first one. We launched the work station itself with the micrometeorite shield panel blown off and had

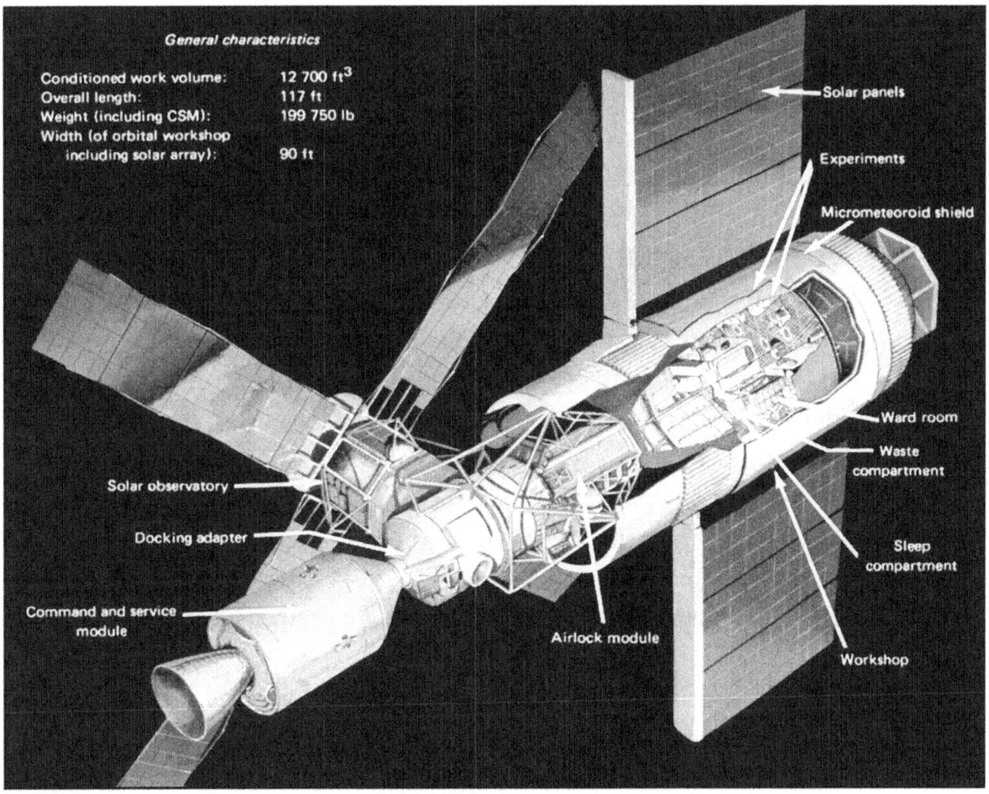

Skylab was a large cylinder (right) with room for astronauts to move around and work. Crews and supplies were transported in the Apollo CSM (lower left). Courtesy of NASA.

problems with the solar panel deploy. So we slipped the next day's launch [with the first crew that would occupy Skylab]. We had to slip that for, I guess, a week or two. Because once you miss a day, the plane of the orbit changes. You only get opportunities with good lighting every week or so, or you're trying to go to the wrong plane. So we had to deal with that. Plus we had to figure out how to put a screen over the work station to keep the thing from getting too hot. The solar panels didn't deploy, and then I guess the meteorite shield was torn off. So we had to figure out a way for them to deploy a parasol, you know a big umbrella over it. [Skylab 2 launched on May 25, 1973. The crew repaired the work station and stayed aboard it for twenty-eight days.] [Pete] Conrad was the commander on that one. We got there, and they went in. It was hot inside. It was 126 degrees in the work station. They had a little hole [i.e., a small airlock] where they were going to put experiments through. Well, they put this parasol thing that the guys developed out the hole and opened it up. Also, there was a cable that was wrapped around one of the solar panels, and they managed to get it cut loose, and deploy the panel. Joe Kerwin and Pete Conrad were out doing it.

[Deiterich was also involved with the Apollo-Soyuz mission.] I was a flight dynamics officer (FDO, pronounced fido) for that. I actually got to go to Russia for almost two weeks, where we did some joint training with the Russians. That was back when we were still having the Cold War. We'd get in a bus in Moscow, and go out to their Space Science Center, whatever they call it. I got a cold, a sore throat, and went to the doctor at the US embassy to get some penicillin or something, so I wasn't with the rest of the guys. This time, I had skipped the bus, and I had to ride on the subway out there. Well, I got on the wrong subway. Some guy comes in and says, "I don't think you're on the right subway." He had to be my KGB [Soviet security agency] guy who was taking care of me. I figured no matter what I did, I wasn't going to get in trouble because I had some KGB guy taking care of me. So he left, I got on the right subway, and got there okay.

Some of the Americans learned Russian, but I didn't. I represented the mission operation trajectory guys over there, and they had interpreters. In fact, one of their interpreters had an Australian accent. Evidently, he had been raised in Australia. I guess his dad must have been a diplomat or something over there, because he had an Australian accent when he spoke English. He was Boris something or another. We were over there the last part of September, so the weather wasn't too cold. The days were still twelve-hour days. Wasn't like it was in winter. It was pretty fun to work there. We got there the first morning, and there were some coat racks. It was in September, so I took my coat off. This Russian said, "Don't take your coat off." They don't turn the heat on until October. It was cold, so we all wore our coats. The hotel where I stayed had heat in it. We were staying at the Rossiya Hotel, which was right off the Red Square, which was kind of neat. I could look out the window and see Red Square.

One of the things that I did, we had this UVA, ultraviolet tracking experiment. We actually had to fly [the Apollo] around the Soyuz. I was the guy that computed the maneuvers for the flying around. We had to work with them to make sure that they knew what we were doing, so that we didn't run into each other. They worried about every little bitty thing. They wanted every "t" crossed, and every "i" dotted. They were much more rigorous than I thought they would be. Their equipment was much bigger than our stuff. They had bigger tubes in their radios. Like it was ten years behind ours.

We studied how their systems worked, and then they studied how our systems work. They probably learned more than we did. Like, we had a cryogenic oxygen for breathing. They used potassium chlorate, and they would heat it to drive the oxygen off of it. They had a big pressure gauge in there. If it moved they'd come down, because they didn't have any makeup oxygen. I think on our mission, they did have some oxygen bottles.

The ground crew could send all the commands to the Soviet spacecraft. The [onboard] crew could almost do nothing, and the ground could control most everything. They had some sort of programmer thing that would set up different programs. Our guys called it a jukebox. You know, our guys, the commander's responsible for the spacecraft.

Astronaut Vance Brand was inducted into the International Space Hall of Fame in 1996. He did not get to fly in the Apollo lunar program (see Chapter 9), but he did participate in the last Apollo mission. In his oral history interview, Brand described his experience with the Apollo-Soyuz international space venture. He began, however, talking about his work on a potential rescue of Skylab's first crew in 1973. Two equipment failures on the Apollo CSM had occurred, one during docking with the space station and the other six days later.

Skylab was a great experience, because we crewmen were getting to work with the engineers and trying and come up with fixes of things that were breaking. They had to put a parasol up so the vehicle would not get too hot in the crew living area. They saved the mission that way. The first crew to go up had to do an EVA, cut away the structure, and do what they could to get this parasol up. From that point on, though, even after the first crew went up and came back, there were things breaking, or wearing out, re-servicing the Freon lines, changing electrical switches. There were just all sorts of things up there. On the ground, we were busy trying to figure out how to make all this work.

At one point, there were two thruster failures on the second crew's vehicle that went up there, and we thought we might have to rescue them. Don Lind and I were told to get ready to go up and rescue the second crew. We practiced for a month. We helped people figure out how to put extra couches in [the Apollo CM] so you could carry more crewmen in the back. Then two people could go up and carry three back. Some of them would fit under the regular couches. That vehicle that was having thruster problems was to be up there two months, so during the first month we figured out how to rescue them.

Then they said, "Now let's figure out not how to rescue them, but how to sort of Band-Aid their vehicle so that it could come back on its own." So I spent the next two or three weeks working with some people to work that out. In fact, the procedures were worked out, and they came back. We didn't get to rescue them. Don said that was one of the biggest disappointments of his life, that he didn't get to do that.

In 1975, I was command module pilot for the Apollo-Soyuz Test Project. It was well worth the wait, I'll say! Simulators get you used to the idea of what launch will be like and what being in space will be like. But when you actually go up after training for so long, you just savor each phase. You savor, but also after you check off each thing that worked well, you kind of give a little sigh of relief too. People have asked if you're not real nervous at liftoff. Deke Slayton and I talked about it later, and decided that our emotion was relief. You're relieved when the hold-downs release and you know that you're not going to get stopped from doing the mission. I just thoroughly enjoyed that flight.

It was a very nominal [i.e., normal] flight until the very end. We had so many backup plans for if we had rendezvous problems and everything. We didn't ever have to use those plans. There was one thing at the very end which was not nominal. A switch was out of position. The automatic landing switch was not on, so as it turned out, I got to do something I had practiced a lot, which was put everything out manually. First getting rid of the apex cover, and putting out the drogue chute, and putting out the main chutes, and all of that. An unexpected thing that happened in that sequence was that we had a cabin vent which was used to equalize pressure, as we were nominally a third of an atmosphere in the spacecraft. As you get into the thick atmosphere, then you have to have air come in from the outside. We had firings of RCS [reaction control system] engines that contaminated that area. We'd gotten a dose of nitrogen tetroxide on the descent on the last part when we returned to Earth. So after we had

The Apollo CSM (left) about to dock with the Soyuz spacecraft (right). Courtesy of NASA.

landed on the water and tipped over like command modules did about half the time, then righted ourselves, we were still breathing that stuff so we had to get on oxygen masks. I was closest to the vent, so I passed out. Then the oxygen mask was on my face when I came to again. That put us in the hospital for observation, and treatment was very good. In two weeks I was jogging, and I don't think I've ever had any trouble from it since.

[What was it like to actually meet some Soviet cosmonauts? Was there an immediate camaraderie, or did you have to overcome a little bit of the political prejudice that was prevalent at the time?] You couldn't avoid it. You'd lived through the first part of the Cold War, and you had preconceived ideas. I'm sure the Russians thought we were probably dangerous, and we probably thought they were dangerous. But when we got together, especially when people are in the same business like pilots, and meeting pilots from another country, or engineers working with engineers, et cetera. When we all got together, it fairly rapidly became apparent that there were these other people that had families, that got up in the morning, and the men shaved like we do. You know, here were other human beings. So I think that trust was established.

There were things that helped, I think. For example, everyone was highly motivated to pull this off on our side and on the Russian side. We had a super team that had just come off of Apollo and Skylab, very capable people, and we had very capable leaders in the space program at that time that handled human relations well with the Russians, I thought. The people on the Russian side did well too. The people running the country, the Secretary of the Soviet Union on down, were wishing for success, I think. That's a very authoritarian country, and when the guy at the top says, "We want this to be a success," then it just flows. The people from the top to the bottom work hard to make it such, although perhaps in some scenarios, it might be kind of dangerous for a Russian at the time to be dealing with an American, because of security and all that. They seemed to have kind of a mind-set that said, "You can go ahead and deal with these Americans."

We got into language training. It's been twenty-one years, so I probably don't remember much Russian now. But at the time, we studied a lot, which taught us a lot about them, their culture, kind of what makes them tick. They studied English. They probably learned the same things about us, and we toured things in each other's countries. Before the mission, we got down to Tashkent, Leningrad, and other places. So there was a foundation established for understanding. I think when the crews became quite comfortable with each other, that was noticed by others in the program, and it helped with trusting people, engineers, everyone.

In Russia, the vodka was very smooth. It was very powerful as well. Back then, especially, there was the toasting tradition. I think there still is, but I haven't been there lately. It was quite a good way to improve your Russian, because we would try to make toasts in the other guy's language. I found that there was a pecking order. The senior Russian always started the toasting, then the senior American, and then you went down the ladder. I was always down there a little ways, I don't know, four or five maybe. The first part of my meal, I would eat, trying to think of what would be good to talk about. It's not just to your health. It's a three to five minute speech. Then I would try to converse, and at the same time think about how I would say that in Russian. Then finally it would come to you. About one time in five you'd get blown out of the whole scene by the fact that a person just ahead of you picked your topic.

We flew their simulators. We had classroom training on the Soviet systems, though not in great depth. We practiced technical communication, things we had to say. We went through procedures like docking, things like that, speaking the other guy's language. We got used to the way of doing things in space, ate their food that they used in space.

It was good training. We had about three 3-week training trips over there, and they had about the same number over here.

After Apollo 17 (see Chapter 9), William Chisholm was involved in the Skylab project at Cape Canaveral. He worked for Boeing, which was responsible for the launch tower and fuel supplies for the vehicle that would launch Skylab in 1973.

The High Bay at the Cape was designed to hold four launch missiles, and I think we had three. So one was modified to take the Saturn I-B, one was modified to take the Skylab, and one was being used for the Saturn IV-B. So all three of these were there at the one time.

Fear of heights is a big problem because the whole tower is metal mesh. When you are on the 320-foot level and look down, you can see all the way to zero because it is all open. And when you are on the swing arms, they are all open. These were the arms that carried everything out from the missile to the tower. We had the white room on swing arm nine, which was where the astronauts entered the capsule. We also had several more for carrying the fuel and electronics and stuff out to the missile. They used a lot of inert gas to prevent fires. You had to go out on each of the swing arms and verify that you had positive flow of nitrogen, which would make it a reduced oxygen area. On swing arm nine, the white room, you always had to go outside the arm to check the flow. That was always thrilling, to think about 320 feet in the air you are standing on the outside of all the cages. You out there, just you and your belt.

Skylab was really more hectic than the Apollo Moon landing programs. We had less people, for one thing. As soon as one crew-carrying Saturn launched, the next Saturn was a rescue vehicle for Skylab. Since you only had one umbilical tower, you had to immediately rebuild it and get that missile that was partly built up on it. Then start your testing around the clock until you were ready to go to the pad, because it was the rescue vehicle. The idea was that you had the rescue vehicle ready as soon as possible, and then it was just sit there until it was needed. It was a lot of "hurry up and wait" on that program. It was kind of frustrating, twelve hours a day, seven days a week, and then nothing. Well, nothing but routine maintenance. Go out and make sure everything is okay. But basically, you just sat around waiting for something to happen, and hoping that it didn't. Then when the previous mission was completed, and they were safe on the ground, you reconfigured the capsule.

Dr. Valeri Polyakov holds the record for the longest-duration spaceflight, spending fourteen and a half months aboard the Russian space station Mir in 1994-95. He had previously spent eight months aboard Mir in 1988-89. His total time in space, 678.69 days, ranks third behind cosmonauts Sergei Krikalev (804 days in six missions) and Sergei Avdeyev (747 days in three missions). Polyakov was inducted into the International Space Hall of Fame in 1996.

[What prompted you to want to go into space?] The first flight of the man in space, Yuri Gagarin, and the following flights of Russian cosmonauts and American astronauts, and also, especially, the first of the flight of the doctor Boris Yegorov [the first physician to fly in space, he spent twenty-four hours in sixteen Earth orbits aboard Voskhod 1 in October 1964]. I was never introduced, but I saw him several times in different meetings. The first meeting was on the Red Square, when Gagarin was reporting his ride. I was a student at that time. I am strongset about Gagarin, that he was the one who pulled everybody to space.

I went to medical college in Moscow. I graduated from it in 1965. Later, my work in the USSR Ministry of Public Health was pretty short, but it was in the area which is supervising the space medicine. After that, I started my postgraduate work in medical biological problems. There I became the specialist in space medicine.

[Polyakov was accepted as a cosmonaut in 1972.] It would be pretty hard to tell about the differences between these two training programs because I am not really familiar with the American program of training. But I will tell about the Russian. At first as a specialist in space medicine, I had to realize that there is a big problem in the space medicine. After that, to be a cosmonaut physician, I had to go through training in the best clinics of Russia in different locations. After that, being a cosmonaut, I went through a general training program. It includes knowledge of

Polyakov volunteered for his 437-day Mir mission to prove that a person could withstand a trip to Mars and still be able to walk and work at the end of that voyage. Courtesy of NMMSH.

the rockets and spacecrafts and space navigation, life support systems, systems of medical support on the deck of the spacecraft, being able to live and work in space, theoretically and on the simulator. As a civilian, I had to go through the pilot training. I had to fly by pairs and also fly by myself. Also, we had to be trained to jump with the parachute.

The first I was in the training was in '79 and '80. But we were only backup and we didn't go that time. My first flight was in 1988. I was twice as backup cosmonaut, in '80 and in '83, but my first flight was in '88. I was in training for several years, and I also talked to a lot of my colleagues who already went in there, so I was pretty much aware of the impressions and feelings. The biggest impression from the first flight was pretty much normal. I was ready for them, theoretically. But it was a real interesting job for me. I felt very good during the whole flight, on the line, during the time on the orbit and during the landing. It can be explained because I am a specialist in space medicine. I know how to use the methods of control, all of the things that protect you, how to use contrameasures the best.

[The capsule landed on solid ground, using a parachute.] We have expression "soft landing." Soft landing is, you can feel the impact. It is kind of a hard impact on the ground. But a lot depends on the way the cosmonaut feels during the going-down period, because a lot of times, you can pretty much guess, seasick, airsick, and that determines how cosmonauts feel during the landing and afterward. And also depends on the condition on the health, on the shape of the cosmonaut when he comes back. Since every time I came back, I was in good physical shape, I don't have problems with getting motion, sea, or airsick. But behind all of this stands really hard work on the deck of space station to keep yourself in good physical shape.

I pretty much didn't have any extraordinary feelings about [being in space for a total of almost two years]. First, that my desire was to stay so long. Well, of course, with permission of my supervisors in government, and also I knew in my head I had to stay so much to complete the program of the flight.

I am convinced that human can fly to Mars, work on the surface of Mars, and come back safely. Because with the flights that I performed, they created a medical biology of the model of the flight to Mars, and the model was really successful. After the first flight, which was eight months long, and after the second one, which was fourteen and a half months long, I was able to come out of the capsule by myself, to walk around, to undress, to dress, to do pretty much everything. And be conscious of everything. That was pretty much the goal of the flight. I had to show that it is possible to preserve your ability to function after being in space for such a long time. But the gravity on Mars is .37 [of the Earth's], and since I was able to stand up and walk on the Earth wearing a space suit, it shows that human will be able to stand up and walk on Mars.

[To keep Mir in orbit, we had to fire the rocket motors] once a month. That is, just to kind of change the orientation to lift it out just a little. But once every two or three months, when the cargo ship comes in, they lift the station up. The stationary orbit of Mir is 400 kilometers [250 miles]. But when it lowers to 380, then with the cargo ship, they kind of kick it up to 400 kilometers, sometimes 420. But 420 kilometers is not really good because of the radiation. Even though the difference in the height is not really big, about 20 kilometers [12 miles], the dose of radiation is rising more than 1.5. [Radiation is] measured in roentgens. Usually, the dose that a cosmonaut gets in one month is about one. And the radiation by itself is pretty much combined. It is gamma, and protons, neutrons, and heavy particles from the sun. When you sleep at night, if the particle hits your eye, you can see the flash. Like sparks in the eye when somebody hits you in the head.

I spent about six months in orbit with Yelena Kondakova [in 1994-1995 on Mir]. I was pleasantly surprised, there is pretty much no difference [between men and women living in space] . She came back in really good shape. She stayed in pretty good shape. The women can be proud that in the spaceflights, there is no discrimination. I have a joke that women participating in the spaceflight program is the highest trait of emancipation.

A short period of time after the landing, there is registration of the cosmonauts being higher [i.e., taller], and I had about five centimeters [two inches]. There were times it was registered at about eight centimeters [three inches]. But it is not the true growing. When we walk on the Earth with the gravitation, our spine is not stretched completely, it is curved. When we walk in space, it stretches out and also the distance between the discs is stretched out in the spine. That can be a real problem, because when you come back you have to fit yourself in the chair that was made for your size as you were coming up. Sometimes, it is kind of hard to fit yourself in your chair. But if you wear a special suit, it is called penguin, it kind of presses you down. In that case you will have less of a problem. Probably, for Sergei Krikalev it was kind of hard because he was kind of tall and he is thin. So probably he grew out a lot in there.

I probably will not fly in space anymore. The reason is that there are only three cosmonauts who are physicians, and there have been four flights with physicians on board. Two of those flights, I was in the crews. In this century, there will not be one in space, and for a shorter one I don't want to go. Probably my wife will allow me to go, but my granddaughter will not. When I came back from that first flight, even though she was real little girl, she asked me, "Why did you fly so long? I missed you!" And after the second flight she said, "That's it, grandfather, I will not let you go anywhere anymore."

Norman Thagard earned Bachelor's and Master's degrees in engineering science before entering medical school and earning an M.D. degree. A science fiction fan since his childhood, he had also dreamed of becoming an astronaut. In 1978, he achieved that dream and began a space career that included being the first American to work aboard the Mir space station. Thagard was inducted into the International Space Hall of Fame in 1996.

I served as the cosmonaut/researcher for the Russian Mir-18 mission in 1995. That was the nice culmination of my NASA career. I got to learn Russian; in fact, was paid to learn Russian. Got to live in Star City [the cosmonaut training facility in Russia], which a few years before had been a closed city, a secret air force installation. Training as a regular member of a Russian crew, launch on a Russian rocket from the Baikonur cosmodrome in central Asia, and then spend four months in space. I loved the flying on the shuttle, but on the other hand, the shuttle is not a facility that allows you stay in space for longer than a couple of weeks. In fact, my longest flight had been only eight days. So I had never really, as a result of my shuttle flights, felt I had done more than just visited space. But with four months on the Mir station, I really did finally feel like I had lived in space. It was a nice experience. And the Mir space station was a very nice facility for doing a lot of things you would want to do in space. I had twenty-eight experiments, I think, on that flight, most of which were looking at physiological effects on humans of being in space that long.

The ride [on the Russian rocket] is very similar [to the shuttle]. The main difference, I think, is that the space shuttles are two-stage vehicles, essentially, whereas the Soyuz is a three-stage vehicle. The g-forces are about the same maximum, only about 3 g's. It's certainly smaller than the shuttle. It is, with some modifications and changes over the years, not all that much different, perhaps, than the vehicle that Yuri Gagarin rode as the first human in space in 1961. On the other hand, it serves its purpose adequately, which is to get crews safely to and from the Mir station. I enjoyed it. I thought it was nice to fly something other. In fact, one of the big attractions that the whole mission had for me was this opportunity to fly on another vehicle other than the shuttle. I was the first American in history to do that.

This first mission, like Apollo-Soyuz, had a large diplomatic component to it. My first and foremost job was simply to get along and make sure, from the crew interaction standpoint, that the mission was a success. And from that standpoint, it was great. Of course, the Russians don't launch crews unless they are pretty sure they are compatible in the first place. So they certainly did observe us closely during training. But Valodya [Vladimir Dezhurov], Gennady [Strekalov], and I, as far as I could tell, were pretty much very compatible people. We certainly didn't have any interpersonal difficulties of any note during the whole course of the mission.

From the science standpoint, what we as Americans wanted to do on that flight was take advantage of this opportunity to have someone in space for that long a period of time. Knowing that we were going to put the International Space Station up, and want to keep folks up there for six months, a year or more, knowing that we want to send folks off to Mars, ultimately, we wanted to get this early jump on looking at long-term physiological effects on humans in space. Obviously, we designed my science program around that. The experiments wanted to look at things like loss of bone mineral, cardiovascular deconditioning, muscle atrophy, the anemia that occurs. We wanted to characterize the Mir environment, looking at what kind of bacteria and fungi grow in there. What sort of radiation do you get when you are on that station? And what is the microgravity level, the vibrations on the Mir station? So the twenty-eight experiments that were on there for me to do, we were looking at those things.

The training in the US and Russian programs is really pretty close. You have your classroom instructions, your simulators. The simulators can range from the very rudimentary to the very sophisticated, high-fidelity simulators. So all of that was very much alike. Maybe that's because that's the way you have to train if you want to be successful in spaceflight. One difference is that the Russians formally test their crews, which we do not do in the US program. Toward that end, when you would complete a course and your instructor felt that you had completed it, then you would have to face a panel of experts who would fire questions at you in Russian. And you would have to answer those questions in Russian. In other words, you took an oral final exam. It involved understanding technical questions in Russian and being able to give adequate answers to those technical questions.

Clearly, we wanted to look at this whole business of international crews. How does that work? How do you train for that? Especially in this situation, where the missions are going to be long duration. Based on our experience on Mir-18, it looked real good from any standpoint that you want to look at. The crew interactions, the training, the science results from the experiments that were performed, all of that, I thought, went extremely well. I think we learned a lot from that as Americans.

We had the Skylab experience. Here was an opportunity after twenty years to do it once again. I think it augurs well for these complex international cooperative programs in space. So, I don't see any reason at all why the International Station Alpha – I don't guess we don't call it Alpha anymore – the International Space Station program won't be a complete success. It's clear to me that you can do it.

Chapter 14
Shuttling Pioneers

The Apollo-Soyuz mission described in Chapter 13 was the last flight of the Apollo spacecraft. After that, NASA developed the space shuttle, the first reusable vehicle that was capable of orbital flight as well as docking with another spacecraft in orbit. It was launched vertically using its three main engines and two solid-fuel booster rockets, that detached shortly after liftoff. At the end of its mission, the space shuttle would leave orbit and perform a horizontal landing without power.

NASA used five operational space shuttle vehicles in the program. Enterprise, the first shuttle to enter service, had no engines and was used only for glide test flights and landings. Columbia, the first operational shuttle, performed its first mission in 1981. Challenger, which had been built as a test shuttle, was upgraded to become a fully operational vehicle and flew its first mission in 1983. Discovery and Atlantis flew their first missions in 1984 and 1985, respectively. On January 28, 1986, a little more than a minute after its tenth liftoff, Challenger broke apart, killing the seven occupants. References to Challenger in this chapter refer to that event. The entire shuttle fleet was grounded during an extensive investigation period, and the next flight took place on September 29, 1988. Endeavor, which was built to replace Challenger, flew its first mission in May 1992. Another shuttle disaster occurred on February 1, 2003, when Columbia broke up while reentering the atmosphere, killing all seven occupants. After a total of 135 flights, the space shuttle program ended with the July 8-21, 2011, mission of Atlantis.

The normal landing sites for the space shuttles were Kennedy Space Center at Cape Canaveral, Florida, and Edwards Air Force Base near Lancaster, California. NASA's White Sands Test Facility in southern

New Mexico, a backup landing site, saw only one shuttle landing, when Columbia touched down there in March 1982. Originally called Northrup Strip, the White Sands shuttle training and alternate landing site was renamed White Sands Space Harbor after the Columbia landing. When a shuttle landed in California or New Mexico, it was returned to Florida attached to the top of a specially modified Boeing 747 airplane. As described in this chapter, several additional landing sites located around the world were prepared for a possible abort situation on every shuttle flight.

Each space shuttle could operate as an independent orbital vehicle. In fact, between November 1983 and April 1998, twenty-two missions carried a Spacelab module in its cargo bay. Spacelab, which was developed by the European Space Agency, provided a laboratory environment for scientific experiments. Beginning in March 1989, shuttles also carried components of the International Space Station (ISS) to orbit as well as carrying cargo and astronauts between the ISS and Earth.

Gerald Garing was involved in many phases of America's manned space programs (see Chapters 3, 8, and 9). Like thousands of others, he worked on the hardware and ground support that made spaceflight possible.

When we finished the Apollo project, we had written all of these specifications and people were building the hardware. They were manufacturing highfalutin wire, making good this and that. So we were thinking for the shuttle, we could just go out and buy the stuff. About two years into the shuttle program, in the 1980s anyway, we went out to buy something and it wasn't there. "Well, that guy quit. Yeah, he talked about it all the time. He just retired." So we couldn't find anything. We had to go back and resurrect a lot of technology that was developed for Apollo so it could be reused with the shuttle.

When I went to the shuttle program, I went to structures. We had an "iron bird" [a full-scale mock-up of the shuttle] that had all the hard points and everything in there. You could develop a line run of pipes or wires, or find out where everything fit. We called it tooling. It was to the nearest ten-thousandth of an inch. In California, that was hard to maintain because of earthquakes and stuff. But we would keep this big, huge structure on a giant slab. It looked somewhat like the shuttle. We developed all of the interfaces, so all the systems like the bulkheads and everything had a certain space. They would draw it up, and they would actually go down and put it in the iron bird. It was in Downey in [North American Rockwell's] Building One. The iron bird saved a lot of time in the sense that the guy could come out and cut his own pipe and see what the space looked like. We even installed mock-up shuttle engines on it. We found a lot of things there, like how can you get to tighten the bolts? How can you get this to fit? So in developing the space, we also developed the procedure. These people would have to know, "How long am I going to take, to take this out? How long to put it in?" When you put it in, "How can I get a wrench in there?" It turned out to be a real asset for the company and for the whole program.

Robert Turner enlisted in the Air Force in 1966. His first assignment was at Vandenberg Air Force Base near Santa Barbara, California. In Chapter 8, he described some of the missile launches he saw there. In his interview excerpt in this chapter, he talked about his involvement with the space shuttle program in California and New Mexico.

About 1973, I came to Edwards Air Force Base. I spent the next three years there. That was probably my favorite assignment. I was personnel officer, but one of my additional duties, a volunteer thing, was as a tour guide for the Information Office. At that time, they had the B-1 being tested, the F-15, and F-14, the A-10, the YC-14, YC-15, and the space shuttle Enterprise. They were all there at that time, so I got to take tours to see those. I knew all the numbers and all the statistics.

The space shuttle Enterprise was the one that was built for the approach and landing tests, not to go into space, but full size. They were taking it up on top of the 747, and dropping it off to test the approach and landing and the gliding. The Junior Officers Council, which I was a member of, went on a field trip to Palmdale, where they built the shuttles.

The day we were down there, we got to walk on the scaffolding where they were building the Enterprise. We got to walk around and see the insides of it, because they were just building it. Then they built a four-million-dollar road from Palmdale to Edwards to bring it out. In fact, they had to take down some telephone poles and bring it through town because the wings were so wide.

They would drive the shuttle into a big gantry and then lift it up. The 747 would go in underneath it, and then they would drop it down on top of the 747 for loading. The very first time it ever left the ground, I was there. They loaded it on the 747. At eight o'clock in the morning, they said they would take it up for the first flight, and at exactly eight o'clock the wheels started turning on the 747. I know one of the guys that went to Boeing in Seattle. They had to reinforce the 747 where you could sit the shuttle on the aircraft, the two little winglets on the horizontal stabilizers, and also the reinforcement of the body. The plane they took was a 747 bought from American Airlines. In fact, when they got it, the red, white, and blue logo of the American Airlines was on the side of it. You could see where they had taken letters off. They paid fifteen million dollars for the Enterprise. That was their first shuttle training aircraft. They did their approach and landing tests, where they would drop it off and glide it in. That was all done there at Edwards Air Force Base.

[In 1981, Turner was assigned to Holloman Air Force Base, near White Sands Missile Range in New Mexico.] I was here when STS-3 [third flight of the Space Transportation System (shuttle), in this case Columbia] came to be landing here. I remember coming into town and seeing the railroad cars on the side of the railroad that I had never seen used before. It was the track that goes out to Holloman and to White Sands, but it hadn't been used until recently. The shuttle was going to land here, so all of this equipment and people had to come in from Cape Canaveral to support the shuttle landing. There were no rooms on Base. I guess the closest rooms you could find were in Albuquerque, because these NASA people came in in force and took all of the rooms in El Paso, Ruidoso, Las Cruces, and Alamogordo. It was really exciting.

We were told we couldn't get close to it, so I didn't even attempt to go out. I remember watching it in one of the offices where I worked. We saw it on TV coming in live. It was about 5 miles from where we were standing and watching it come in and land. Come to find out later, they did have viewing sites out there where people could actually hear the sonic boom

Space shuttle Columbia (center right) about to land at White Sands on March 30, 1982, accompanied by two T-38 chase planes. Courtesy of NASA.

and watch it land. I was very upset about that. But a couple of days later, they had buses going out to where it was on White Sands Space Harbor. When I saw it, they were lifting it up on top of the 747. Then when it took off to go back to Cape Canaveral, it went right over our house. It was kind of neat to see that. That was quite an exciting time when I came here.

Just after the third shuttle landed [at White Sands in 1982], I was part of the shuttle landing support team. When a shuttle was in the air, we would go out to the West Area at Holloman and be on standby in the command control center. If there was any accident, we would be notified to go and do whatever it was we were supposed to do. The first time I went out to Northrup Strip, which was what it was called then (after the shuttle landed, it was called White Sands Space Harbor), I went out with the medics. They had an ambulance, and I rode out with the medics to the site. They had this big B-52 forebody mockup which resembled the shuttle. They would have it on a flatbed truck and they would take it out to the runway, run down the runway with that truck, and the Army

An extensively modified Boeing 747 airplane lifted off from White Sands, carrying the space shuttle on its back after the 1982 landing. Courtesy of NASA.

helicopter rescue people would come in and hover over it. The pararescue men would go in the top and simulate rescuing the shuttle astronauts in case there was something wrong. There were three different kinds of egress. One was where there was no accident, one with live victims, and one with injured victims. We did those in different ways.

My dad lives in Florida during the winter, and I had an opportunity to go down there once to visit him. I planned the trip around the launch date of STS-61, which happened to be the first servicing mission of the Hubble Space Telescope. I got a pass from NASA to go down there. My dad and I went out, and it was supposed to go about four o'clock in the morning. We left at about twelve o'clock, not knowing how long it was going to take us and how to find the place. We got there about one o'clock or something. I remember we had three hours to wait. Got in there, and at the last minute the weather prevented the launch. They postponed it for twenty-four hours.

So we went home and slept all day, and next morning it was the same scheduled launch time. We knew where to go, so we didn't leave quite as early. We took a picnic lunch. We went to the NASA causeway, and waited for a couple hours anyway. As we were driving in to the site, we were stopped. This helicopter came in about 20 feet off the ground with the spotlights and all kinds of commotion. What it was, the astronauts were going to the launch pad in their van. The helicopter was going ahead right over them with the floodlight, lighting the way. It was in the dark of the night. So that was quite interesting, that the astronauts drove right past us going on their way. We were there before they even got to the launch pad.

They had loudspeakers where we were. We could hear them when the astronauts got in. We could hear the checks between them and Houston, all of the radio checks, and all of their comm [communications] checks and everything. It was quite interesting to me. I knew some of their codes, and some of the things they were saying. Getting ready, they put their visors down and hook up their oxygen. We could hear the astronauts talking to the blockhouse. It was really interesting.

We were probably about 3 to 5 miles away. You could see off in the distance, it was just like a Christmas tree. It was all lit up. You couldn't miss what it was. The whole pad was just flooded with light. And of course, all of the work being done at the last minute. They did the countdown. The weather was fine, and everything was "go." They took off on time. It just lit up the whole night sky. It went up and it kind of veered away from us. As it goes, it kind of curves toward the horizon to go into orbit. It surprised me that it went so fast. As it went on into the horizon, you could see the lights of the three main engines on the glow. Then it kind of went out of sight, and we left.

Another time I went down there was for STS-80. I called and I said I would like to get a pass, and I told them who I was. They said, "We don't have any more passes for the causeway, but we have the static test road, which is a press special site closer up." I said that's great.

They have different color passes for different sites. So as we were going, people were being directed off to the side, but we just kept going straight on. We were about 3 miles from the launch site, in a really good site. There weren't a lot of people there. This was during the middle of the day. You could hear the countdown. They've got loudspeakers there, and you could hear what was going on. They were having trouble with the hydrogen leaking. I thought, "Uh oh, that may be cause for a scrub."

They got it fixed just in time. They were just a couple minutes late.

And the thing took off. You could see the big puff of smoke. It hides the whole launch site. All of a sudden, this little white arrow thing starts rising up out of the smoke. Then it goes up, and as it goes up you can hear this really crackling sound. I guess it's probably of the sound coming back. All of a sudden, it gets louder and louder. You can feel, hear the earth shaking. And the SRBs [solid rocket boosters] come off and you see them floating down. Then it veers off to the horizon, and you see the light of the three main engines.

After working on the Apollo and Apollo-Soyuz programs (see Chapters 9 and 13), Charles Deiterich became involved with glide testing the space shuttle prototype. The STA he mentioned was a Gulfstream II airplane that had been modified to mimic the flight characteristics of the shuttle as it approached for a landing. Space shuttles landed with no engine power, like a glider aircraft. Landing was particularly crucial because the astronaut flying the shuttle could not wave off a landing and circle around to try again. Shuttle pilots and commanders (who actually landed the shuttle) trained in simulators and in the STA.

We did ALT, Approach and Landing Tests, where we released the shuttle off the back of a 747. It was a small group of people. The Enterprise, which was the name of the shuttle we did on the 747 for the approach and landing tests, didn't have fuel cells in it. It didn't have cryogenics in it. It used bottled gas for oxygen and hydrogen on the fuel cells. So it was not a full-out spacecraft. It couldn't fly in orbit. It was not as heavy as the ones that we fly now. It was neat from the point of view that there was not very many people working on ALT, so we got to make some pretty significant decisions about how we're going to fly it. I was out talking to the L.A. center guys, and I would talk to the China Lake guys, and I'd talk to the guys over at Edwards. It was really kind of neat, because I put together what they call a "mission and operations plan," which defines all the flights and what we're going to do with them. Then I actually got to build the flights themselves and got to control them.

We had a couple of captive-active flights, where we actually had the shuttle tied to the 747, and did some systems tests but never really separated. We had to make sure that thing worked right. Then, the shuttle had ejection seats in it. It had them up through, I guess, the fourth orbital flight.

In 1980, we did some shuttle stuff out at NASA's White Sands Test Facility. Bo [Karol] Bobko, one of the astronauts, came across an English version of some approach lights. They're called PAPIs, Precision Approach Path Indicators, and they have four lights. They're angled like at 18 degrees, 19 degrees, 20 degrees, and 22 degrees. They have a red filter and a white filter in them. If you get below the aim angle, you'll see a red; and if you're above it, you'll see white. If you're right on the glide slope, you'll see two red ones and two white ones. He wanted to try that for the shuttle, because at first, we thought it would come in at like a 22-degree angle. Now, I guess we're down to about 18 degrees. We tested the lights out here on at the Northrup Strip with the T-38s and then the shuttle training aircraft.

Joe Engle and Dick Truly were one crew, and Fred Haise and Gordon Fullerton the other crew on the ALT. They would fly in the STA [Shuttle Training Aircraft]. I got to go out and fly two flights with each one of them. I flew one morning for about two hours with Engle, and then I flew about two hours in the afternoon with Truly. That was really neat! It was kind of funny, when you're coming down, they just simulated from 25,000 feet down, so you're going around 290 knots [about 335 miles per hour], which is really pretty fast. You're coming downhill. At about 2,000 feet, they started pulling out and they simulate a touchdown. On the Gulfstream II, they never really touch down. They get to about 30 feet, which is the pilot eye height at landing, and then they go around. When they start down, of course, the Gulfstream was really slick, so it's going to accelerate like crazy here, at 24 degrees glide slope. So they put the engines in reverse and the plane would shake. They put the main gear down. Then when they get down where they normally put the gear down, they put the nose gear down. Of course, they never did touch down.

When you're coming down, the astronaut is in the left-hand seat and the training instructor in the right-hand seat, and there's no place for me to sit or stand. There's a sim [simulation] engineer sitting between them. So I'm standing behind them, leaning against the bulkhead, hanging on. The sim engineer's shoulder harness was there and he wasn't using it, so

NASA had a fleet of four STAs, Gulfstream II airplanes that were modified to handle like a space shuttle during approach and landing. Courtesy of NASA.

I would grab ahold of it. You can't see the horizon. All you can see is the ground coming, and it's coming up, and then he slowly pulls the nose up. Once he comes around the corner, we were cleaning up the Gulfstream II to go around. Bobko was flying chase plane in a T-38 and somebody else was out there, and they were right there beside us, it was kind of neat. When he went around, that Gulfstream actually accelerated faster than the T-38 did, believe it or not. Anyway, we went back up and made another approach. We made about ten in the morning and ten in the afternoon.

Northrup Strip was just gypsum. In fact, one of the engineers who was kind of coordinating the landing stuff out there, we drove up, and he was out there with a shovel in hand filling the holes.

Another thing that we did, we were getting ready to do night landings on shuttle flight 8, and we came out and tested another kind of light arrangement. We just had one solid line of red lights, and then we had one white light at an altitude on one-and-a-half-degree slope. When they come down, they would move that bar down until the ball would center on the bar, and then they would know they were right on the one-and-a-half-degree glide slope. We still use that. We did that out at Holloman in February 1980, I think. It was cold! We picked a moonlit night, because our safety people said if these guys are diving at the ground at that speed, you know, they don't start pulling out until they get to about 2,000 feet, and they kind of come around the corner. Well, at 2,000 feet, you're coming down at about 200 feet per second, so you're only ten seconds from running into the ground. So they wanted to make sure that the guys could see the ground.

During the Apollo era, James Gimbel had worked on the lunar lander training vehicle (see Chapter 9). After that, he continued to work for a contractor at Ellington Air Force Base in Houston. His next assignment was to help develop training procedures for shuttle astronauts. An experienced diver, he also worked underwater (simulating weightlessness) to evaluate equipment and procedures that would be used in space shuttles on orbit.

When the Apollo program ceased to exist, I was transferred over to the main part of the Air Base, and we were servicing all of the NASA aircraft. I was research and development laboratory manager, and it was a lot of open-ended research. There was a NASA contract to look into the STA program, which was the Shuttle Training Aircraft, based on a common aircraft.

We outfitted the Gulfstream [airplane], but that was just part of the laboratory's assignment. We were doing a lot of research on NASA aircraft and NASA's advanced systems. The laboratory was actually part of the Earth resources program that NASA had going at the time. I managed that laboratory for quite some time, and then I wanted to do something else. I went over to Northrop Company, because they had an opportunity to do the things that I thought were more interesting. At the time you are a young man and you want to rub shoulders with astronauts, I guess. So I managed the simulator for the remote manipulator arm in Building 9A, in NASA GSC [Goddard Spaceflight Center in Maryland]. It worked so well we made it a trainer, and it is still a trainer to this day. I was an engineer there.

Later, I managed the underwater training facility for a year. At that time, the shuttle was in operation. We hadn't flown yet, but all the trainers were up and operational. So for reasons unbeknown to me, I was assigned to help train the EVA part of the missions. The WET-F [Water Emerging Test Facility] started out in Building 5 of JSC [Johnson Space Center]. It moved to Building 260 with a big above-ground tank, and then we eventually took the old centrifuge building and put a big tank in that. We moved the centrifuge out of there, pulled the motors out, dug a big hole, and put a water tank in there. We could get the whole shuttle mock-up inside the tank. [A space shuttle was 122 feet long and 57 feet tall, with a wingspan of 78 feet.] We needed to do end-to-end studies, like get from the cockpit into the airlock, dress, then go out and do work, the whole end-to-end study. Before, on the smaller tanks, we were just doing smaller portions and then tying all of this together. The new facility was a big expense to NASA. They questioned the longevity and the use of it. But when it got into operation, it became overnight one of the godsends we had to do testing with. It is a huge tank, and the water is perfectly clear. It's just beautiful water.

To deploy a satellite that has to be manually done by EVA, eighteen months before that is done, we are visiting the companies that are building the satellites. We are building a cradle in the shuttle. We are testing, and testing, and testing, till the operation of going into the airlock, putting on a suit, pre-breathing, opening the hatchway, going out and performing your function, and then coming back in is a scenario that has been done so many times that every little second is accounted for. We see on TV about how great they did! They came out and did their job and went back in. But there are thousands of hours of people like me behind that scene, developing those procedures and getting the tools right.

We work underwater a lot. A long time ago, we decided that we needed camera lights. They needed to run off of batteries. Well, an engineer, without thinking, went to Sears and Roebuck, got him some Die Hard [wet-cell car] batteries and put them in a big box, and put that box down in an aluminum box and down in the water. They plugged into it and they used camera lights. Well, batteries gas off with hydrogen. So, effectively, it is a good-sized bomb. We pulled the canister out after a dive, after a full day's work. A gentleman reached over to bleed the bleed valve, and it exploded and blew his head off. It killed him and harmed a few other people. After the investigation we find, oh yeah, duh! So we go back and change things. It doesn't reverse the loss. But we carry on from there. There are a lot of incidents where we blew it in NASA. Most of them were pure accidents based on cutting some corner to be more efficient, but that is the way you progress, I guess.

Under water, you feel your weight in the suit from a perspective away from you. You look like you are floating around, doing your job. But you have an up and down, left and right. You've sensed that. Your senses are doing their job, and you are governing your work activity based on those senses. You get very tired, because a suit in water tends to be a little bit stiffer than it does in zero gravity. After four or five hours, you start making errors, the human factoring errors we have always been worried about.

As I was going through my career, I kept looking at the human aspect, because I was designing hardware that had to be used by human hands in space. There was no up, down, and sideways. You are working in zero gravity, or microgravity as we called it. The human factors aspect of this was nonexistent. There are only certain trainers in the world that will allow you any time at all in microgravity. Only the flight will allow you the true feeling of microgravity. In water, you still weigh your weight inside that suit, even though your suit is drifting in the water.

Rockwell built most of the hand tools, originally. Their engineers would come to us and say, "Would you evaluate this tool for us?" We would set up training hardware and get in the water tanks or get in the aircraft and get suited up, and go through the dynamic tests. We had instrumentation everywhere, all of those accelerometers and pressure transducers. And we would feed this information back to the design groups.

These underwater tests are very serious. They cost a lot of money to run. You've got twenty to thirty support people that get that facility ready. You've got people on consoles all over the site. Your TV people, contractors. They are all up and running. You've got safety divers, and you've got suit people, at least three for a subject. We did some pricing at one time, and to do a wet-out test was like $27,000, just for the labor alone, not for the facility. So when you are going into a test, you are responsible for a lot of people and a lot of money. And it is taxpayer's money. It is not NASA's money. It belongs to the taxpayer. You feel an obligation to do what you're supposed to do.

As suit subjects, and I was one for many years, you had an obligation do your homework, your lesson plan, a minute-by-minute plan that you go by. You are actually developing a plan. That is your job. So they put you in there and give you all the hand tools, and you start doing the job. At the same time, you recommend to all the contractors standing by, how they could facilitate a little better operation. So there is an onus on you to get your job done and be professional about it. You don't want to go off on a vocal tangent that means somebody is going to spend a lot more taxpayer's money re-inventing the Crescent wrench. You want to make do with what you've got but be realistic about it. If it doesn't work in the glove hand, ditch it and get something that does work. But if it's functional and it needs some modification, tell them about it. We went with Snap-on tools, but Rockwell put some large handles on it and some ruler type of thing. We put a tool caddy on our chest. It plugs right into the suit. We can un-Velcro the tool we need, take it out and work with it. It can't fly off because it has this little pulley arrangement to it. That's pretty simple, but it took two years of experiments and operation in those wet-off tanks to prove that that would work. Sometimes, you have a mirror on your arm. You have to put your arm out, look into this shiny mirror back and un-Velcro the tool you need, which is a screwdriver, a hammer, or whatever it is. You can't really see it from in the suit. There is a lot of human factoring involved.

One time, I was working underwater and I needed to go to the bottom. The tank is like 37 foot deep. If you need to go down for a reason, it is no sense of climbing down. Have one of the divers take you down. One of the divers grabbed me and took me down too quick, and ruptured a sinus. Now you're working with this blood and stuff in your helmet. And the question is, well, do I cancel the test, or do I keep working? You know, there is a lot of money and people involved in that. Another thing is, you are supposed to be a hero-type understudy doing this stuff. So there's a lot of pressure on you to keep doing your job.

I wasn't very happy with that safety diver that took me down too quick. The suit adjusts its delta p [change in pressure], but it takes a long time to do it. There is an actual mechanical spring that has to actuate. If you take the suit down too quick, that mechanical spring is popped off too quick, and you wind up over-pressurizing the suit or under-pressurizing immediately. I think he knew that. I believe it was just a human error, and he took me down too quick.

The cameras are always on you, and the medical team is always watching you. So I finished off that experiment. God love the suit techs, they are good people. They know what they are doing. So one of the suit techs came on line. You have channels you can get on to, where the suit techs can talk to you and nobody else can hear you, or the medical can talk to you and it is nobody else's business. The suit tech came on line and said, "Jim, what is that in front of your face?" "Well, I think I've got a nosebleed." You know, I am kind of covering up. I know I've got a sinus that just exploded. The suit tech ended the test early, but I completed one section of it. They were very gracious people. As soon as I came out of the water, they grabbed me and took me to the dressing room, and the medical people never saw what was going on. They will scratch you. You get hurt once, you may not fly again. So the tendency is to kind of cover up. They had to clean that suit, and I am sure that was a terrible job, because the suit had to be turned around [to be used again]. That was an old A7-LB suit from the Apollo days.

I almost got fired once. The old A7-LBs were being taken out of service and being put in bonded storage and given away to museums. We didn't have any shuttle suits, so we were doing our shuttle training in A7-LB suits. We couldn't replace the components on them. If they failed, we would have to chuck the whole suit. We knew we only had about three more suits left. If they failed, we were out of business.

Well, one day, the blue umbilical connector on a suit broke. Some of the pins broke off. There were none in storage. We called the vendor, and they said, "We will make that plug for you, but it will cost you $17,000 for one." I said, "No, that's way too much. My budget is not going to carry that." If I turned that into NASA saying I want $17,000 just to buy one connector, NASA is going to say, "Take a hike!" So I went over in Building 2, and there, inside a glass cage on display for the museum, was an A7-LB. I knew all the guards, so that night we went over there, and I was in the process of taking that connector off of that suit when a strange new guard walked in. Boy, I tell you what, there was a lot of explaining to do! Finally, I think one of the astronauts, probably John Young, made a comment at a meeting I wasn't present in. He said, "Hell, at least he is doing something. Don't fire him. Promote him." So I got out of that trouble, but the people that ran the museum didn't think that was very funny. I was replacing the connector with that old broken one.

The reason I was chosen for a test subject was because I am very honest, but I am not insulting. I don't want to be insulting, but when you get out of eight hours of suit and you've got ridge marks all over you, and you are hot, and you've got pinch marks, and your hands fall asleep in the gloves, you have to politely let the contractor know that after eight hours in zero gravity in this suit, you are not going to EVA for a couple of days. So, what if you get a contingency within those two or three days? You've got to go out. You've got a hole you have to patch. Then we will go out and tell the contractor the bad news or the good news. Well, it does put you on a spot. You are certainly offered the opportunity at times to say good things that you shouldn't say.

The astronaut society has completely changed from the old original crew. The old original crew, you have to understand was ex-Marine, or active Marine, or Army, or Navy test pilots. You give them a pair of pliers and some bailing wire and they can fix anything. They never complained. They knew as much as you did. You are trying to teach them how to do their job, and you're teaching somebody that has a learning curve of thirty minutes. You are trying to explain to them what you want them to do in space. It really requires you to do a lot of homework. You can't make a mistake. Those old guys can do anything. I never was so impressed with people in my life as that original crew. We no longer have those requirements. We don't put our flight crews through survival training. We don't demand that they come out of a background of being a pilot or some tough Marine. We have system specialists now that come out of MIT, that come out of major aerospace corporations that were trained, flew, and then go back to their research.

The present flight crew differs completely from the old flight crew. You are never going to find a person as competent as Neil Armstrong or Pete Conrad or Roger Chaffee. These guys have been replaced with the modern scientists in our aerospace programs. Training the specialist today is a far easier job, in some ways, than it used to be, because you know the mission requirements three or four years ahead of time. You develop training hardware for them. You develop the scenario, the timelines. You know quite a bit about what they are going to be doing on that mission. It is just a process of bringing through people that have been selected to go on certain flights to do certain research programs.

My career started with developing the expertise of these super jockeys, these test pilots that I admired so much. And then later on when you are training a person for one mission who is out of some university, who is politically connected and got an experiment on a shuttle, you don't feel obligated so much to teach them the basic training of the shuttle flights. You just teach them their specific script.

And it is a script. Everything that goes up is scripted to the second. When the electric razor goes, that power drain is registered and computerized and it goes against the long-term mission scenario. Everything, the water, the breathing air, and especially the power, is scripted into a second-by-second script. There is no secret that there are down times and up times. When you get up in the middle of the night when you are supposed to be sleeping, and you use power, you upset the whole fire spectrum. You start turning on the radio, speaking to your wife, all of that power has to be put back into the system. There are a dozen engineers sitting behind a console going crazy when somebody pulls out something and plugs it in that wasn't on the script. This affects all the computers, all of the people. You call in people overtime if somebody wants to play country music on a plugged-in player up there. I mean, it really does bother a lot of people.

If you took a special tool that weighed seventeen pounds on every flight, that means back there in the shuttle cargo bay, the residual weight the manifest people have to deal with is all the way down. Then someone comes up with, "I've got to have this, maybe they can scrounge a little bit based on re-entry characteristics of the aircraft." The shuttle is an old plane, and they have to increase this margin a little bit over a period of time because of the landing gear. You can't land with what you take off with. You've got to dump it somewhere. There is always a chance of being overloaded on touchdown. You don't go around. You're dead-stick landing. You have one shot at it.

We did a lot of what-if scenarios. One of the scenarios is, "What if the whole crew is in the craft, and one man is doing EVA? Something comes through and wipes out everybody on board. A big enough hole inside of the craft and everybody dies." You've got the shuttle out there. The doors are open. You've got a man out there with nobody back in the craft. Can he get back in the cockpit, close those big doors, and possibly land this sucker in a space suit? There is a certain astronaut out there, I am not going to mention his name, that is bigger than the rest of the guys. I am approximately his size and mass at this age. So the scenario was, "Jim, can you come in, crawl up in the cockpit, and get into the pilot seat, and operate the controls in the shuttle suit?" "Oh, okay." Come to find out, the hatchway is going down into the crew compartment. This is back of the pilot seat. The cockpit is upside down. I am trying to come through this hatchway with the largest suit. We are trying to come through here and get into the seat, but I get stuck. The largest suit cannot go through that hatchway. So Rockwell gets a contract to enlarge this hatchway immediately. This was done before the next flights.

Those scenarios are called what-if scenarios. Those meetings were fun. You get into a meeting, and you've got the flight people there, the engineers, the contractors, and you go the script for the flight. Somebody says, "What if the lanyard breaks?" "What if the doors don't shut?" My job was to go, "Okay, take the what-ifs that statistically might happen, go out, build mock-ups, and test the what-if problems out." And give written reports back saying, "Statistically, the average of this happening on a flight is something-to-one." Then you go back in and say, "Here is a fix, here is a go-around, and here is what the cost is."

Richard Finch had worked for government contractors in the missile programs since 1961. He had a broad educational background, including welding technology, aeronautical engineering, and business. The O-ring he mentioned in this interview excerpt was a rubber pressure seal in a joint on one of the booster rockets that launched the space shuttle.

I was hired at NASA as a troubleshooter on the shuttle program. On one job, they had two aborted launches from the shuttle Discovery. Discovery was sitting on the launch pad. They would light the shuttle engines, and when they were all up to 98 percent power, then they would light the solid rocket boosters. Two shuttle engines lit and the other one wouldn't light. So they said, "Go find out why." We knew it had to do with the valves that didn't open to turn the fuel on. They are hydraulic, so I was supposed to go and inspect the hydraulic system. But it was very obvious what the problem was. The red hydraulic fluid barrels were sitting out beside the buildings. It rains every day in Florida, and they weren't sitting on their sides. They were sitting upright. I went over one of them and turned the cap, and it was loose. So I said, you have got to put them on their side and put them under a tarp, at least. It took about three weeks, but they drained the fluid on the space shuttle and put in fresh fluid out of fresh barrels.

When the Challenger blew up, my office was in the OPF [orbital processing facility] on the second floor. As I walked up the steps to my office, I could reach out and pat the wing of the shuttle. Of course, everybody that worked in the OPF had the bunny suit, and the caps, and the moccasins with the sticky feet on them so you don't track anything in or out. The day of the explosion, we had a fire alarm in the OPF. It turned out to be some miscue that caused the fire alarm. But it was cold. I was leaning up against the sunny side of a van, and somebody finally said, "Come around and get inside so you don't freeze." I think it was probably 28 degrees. There was a wind.

The first thing they did when the Challenger blew up was they said, "Go get your paperwork on the Challenger. We had one engineer who said, "I know what it was. It was the O-rings." That day, he said, "It was the O-rings." But with NASA, you never ever jump to a major conclusion instantly. So what happened was, when they were stacking the booster, they used to stack it with a two-point lifting beam, and it egg-shapes it. It probably gouged the O-rings, in addition to the weather, because 28 degrees won't freeze an O-ring. Not by itself. It will harden them. So it was a combination of things. My direct boss was the first man ever to go out and dive and recover a booster on first launch. He was called over to where they were stacking them in the VAB [Vehicle Assembly Building] to check to see what happened. He recommended a four-point lifting beam. That made sense. If they picked it up at four points it wouldn't egg-shape the booster. So they put me to work on that because I am a welder.

Then they said, "While we are down, we need to inspect the other three shuttles." So they put me to work as the inspection writer: "Open this hatch and inspect certain nuts and bolts...." We found multiple instances of loose hardware in all of the other three shuttles, none of which had ever caused a failure. Like, the crew compartment is actually shut inside the shuttle with a space that you could crawl around. They have it mounted with struts, and those struts were only finger tight. Now, it would have been a rattle first before it came loose, but still it would have been very embarrassing.

Vance Brand was inducted into the International Space Hall of Fame in 1996. Courtesy of NMMSH.

Vance Brand was an Apollo astronaut who flew on the Apollo-Soyuz Test Project (see Chapters 9 and 13). Then he became involved in the space shuttle program and was commander of the first fully operational shuttle flight. The four shuttle test flights he mentioned and his own first shuttle flight all used Columbia. Brand's third, and last, shuttle flight, also aboard the Columbia, was the sixth night launch of a shuttle and the fourth night landing at Edwards Air Force Base. Night landings at Kennedy Space Center began three years later, in 1993.

I was the commander of STS-5 [Space Transportation System, or shuttle, flight number five, in November 1982], the first operational flight of the space shuttle. Per the plan, the first four shuttle flights would have two crewmen, with ejection seats, that would do testing. Then you would have STS-5, the flight we were on, which would have four crewmen and two mission specialists, and you'd do something useful, like put the first two US satellites to be deployed from the shuttle in space.

We had over fifty flight tests [for vehicle performance] on STS-5. One of the more interesting was, actually, I got to fly at Mach 18 doing a roller coaster, which was really enjoyable, testing it to run. Our motto was "We deliver." We had a lot of fun with that. I got to land the shuttle for my first time, which was in trouble. We had a slight overcast at Edwards to land. It was a thin overcast at about 18,000 feet. So a lot of interesting things came up on it.

We had two simulators that we used to learn to land the shuttle. One was a mission training ground simulation in Houston. The other was a Gulfstream II with a variable stability that optimized the program just for the shuttle handling qualities. It's not something where you could tweak a dial and simulate some other airplane. The right seat had the instructor and the Gulfstream controls. The left seat had shuttle controls and the student. You never landed with it. All the gear was out while you would level off at about 30 feet, and go around, and try another landing. You could come down from as high as 30-something thousand feet.

The shuttle and the Gulfstream II training aircraft had nearly identical handling qualities, so it was a good simulation. Up and away, it flies like a fighter plane, sort of. When you get down low, it's different. It's still laterally in roll and everything, but it's very sensitive in pitch. It has some features that make it different. For one thing, the pilot is way out ahead of the center of gravity. And for another thing, when you want to put the nose down, you put the stick forward, and the vehicle goes up and then pitches down. It takes a fair amount of training to get used to that, and also, you don't want to put a lot of control on the stick that's unnecessary. You want to sort of get set up, and then leave it where it is if you can. You're a glider, so you don't have throttles, but you do have a speed brake [a split rudder on the shuttle's rear fin], which could be used automatically or manually.

You put out the landing gear below 500 feet. Most people put it out around 300 feet, and the reason is, the lift-over-drag ratio is better than a brick, but it isn't real high. A lot of airplanes have a lift-over-drag ratio, when they're landing, of maybe up between six and ten. A glider maybe fifty or forty. The shuttle is four and a half at best, with the gear up. When you put the gear down, then the drag increases, and you're going to be slowing down faster. So that's why you hold the gear.

On STS-3, they had a test which required them to leave it on autopilot until about, I forget, 300 or 400 or 500 feet, and then take over manually and land. There was a lot of turbulence and tremendous wind at the time. When the pilot took over manually, he hadn't had a chance to get used to the stick. I think the gear got down even a little later than usual, because of the upset that they had. That's what tests are for: you learn. After that, we started taking over manually at about 50,000 feet.

The Apollo-Soyuz flight wasn't all that different from the space shuttle. It had staging, about two minutes and fifteen or twenty seconds, which is similar to the way it is in the space shuttle. The first stage on both the Saturn and the space shuttle were very rough, very noisy. You get bounced around a lot. You get a sort of a kick in the pants when you lift off, and you have to raise your voice to talk over the intercom to your fellow crewmen. The g's pick up as the vehicle gets lighter, as you approach a couple of minutes of the flight.

Staging is when the first stage S1B comes off and you go to S4B on Apollo. Similarly, when the solid rocket boosters come off on shuttle after burning out, you continue on your three main engines. In both vehicles, things seem to smooth out. The noise came down. At the very instant of staging, you were aware of pyrotechnics, and a bang, something leaving, but then being on the upper stage is more like electric drive. It was just very smooth. Then as you're approaching orbit insertion, you have a buildup of g's. The only thing different there is the shuttle then throttles back so you don't exceed 3 g's. It did not reach as many g's as Saturn did, I guess, at that point. So the ride up was very similar between the two vehicles.

It doesn't make much difference when you're going up, but when you're coming back in the shuttle, you have the impression that you're in an airplane. It's a fantastic sight coming back in the space shuttle. Because of the friction of the atmosphere, the vehicle is in a plasma sheath around it. It appears to someone inside that you're inside a furnace. You can look out through the top window and see all sorts of turbulence and structure and what appears to be fire. After you do the deorbit burn, it takes an hour until your landing. About half of that time, you're above the atmosphere. So you're in the atmosphere for probably close to thirty minutes, and of that time, you have the fire starting out at about Mach 22 maybe, and it goes away around Mach 10. So you're probably seeing it for ten to twenty minutes.

Knowing that you're going to make the landing at the end, not just going to plop down on the water on parachutes, it's a whole different feel to the reentry. So I like the shuttle better for that reason.

Also, you can see more and better from the shuttle than you could through Apollo. Apollo, you've got a little hatch window about six inches across, as I recall, and two small windows that were used for docking or things like that. But the shuttle, in comparison, seems to have a wall-to-wall window seat. It's really nice when you're viewing the ground, and that sort of thing.

I was also commander of space shuttle mission 41-B in February of '84. That was a crew of five, and the interesting thing about that mission was we were testing out EVA equipment that would be used to repair a satellite one or two missions later. STS-41-C repaired the satellite in orbit. So, the Martin [Marietta] backpack [MMU, or manned maneuvering unit] was tested for the first time out of the payload bay in the orbiting space shuttle. Bruce McCandless and Bob Stewart each went out untethered about a hundred yards away from the spacecraft, which sort of had all of us on edge, because we didn't want to lose them up there. I didn't know how I'd ever explain that to their wives.

If the manned maneuvering unit had failed, we planned to get the crewmen in by translating over to him, flying him in to the payload bay, and having him grab on to something. Unexpectedly, we got a test on that. At one point, we had Bruce working in the payload bay, and an aluminum foot restraint that's perhaps about a foot or eighteen inches across got away from him. He was trying to reposition it somewhere, and it got away from him and started flying out into space. I happened to be back at the controls, and I maneuvered the spacecraft over to fly into it. If that had been a person, it would have been a piece of cake.

It was extremely interesting, though, watching that [the MMU EVA]. They went out about at the speed of maybe one or two miles an hour on the wing of the spacecraft, stopped so they would not get into the orbital effects, which would make it harder for them to get back without expending a lot of propellant. They came back. They also worked at the end of the Canadian arm. So a lot of things were tried out.

That was a really enjoyable trip. We had a real good deployment system, but we did lose two satellites. This was not the fault of the crew. For some reason, two satellites that we deployed on this mission, just as another crew had deployed on STS-5, the boosters on the satellites did not ignite to send them to higher orbit. This occurred about forty-five minutes after they were deployed. A timer started the solid rocket booster itself. No one understood why that happened and why there was apparently a flaw in one of these boosters that caused them to burn out prematurely. It wasn't that they didn't start. They started, but they burned out prematurely. We felt terrible about that, but just elated on the success of the EVA tests work.

In December of 1990, I commanded STS-35. I was signed up to fly on a couple of crews just before Challenger [the 1986 disaster]. One was a life sciences mission, and one was called Atlas, which was an atmospheric science mission and solar science mission. Challenger upset all of that, and for two or three years I had a group of about thirty astronauts that were all working on different aspects of how you fix Challenger. They were making inputs to the engineers and managers on how it should all work. We all spent a long time doing that, and forgot about flying.

Finally after Challenger, the vehicle was fixed, and I guess it was STS-26 that started things off. I eventually got rescheduled on STS-35. This was an astronomy mission with a crew of seven and quite a few tons of telescopes that would be taken up and then brought back to Earth. We had ultraviolet telescopes, three kinds, and X-ray telescopes. These were to look at the hot events that were happening in the universe, explosive things, dust clouds, various things. Very specialized astronomy. We had four Ph.D. astronomers on board; two were mission specialists, two were payload specialists that really knew the equipment and what they were supposed to observe. Three of us who were not Ph.D. astronomers kept the ship running.

By this time, I thought that shuttle operations had become somewhat routine. You should never get complacent with the shuttle, but it was a little more routine. The fun part of it, at least in my opinion, is the guy in the left seat [Commander Brand himself] was getting to do a night launch and a night landing, and training for that. That's quite different, especially the landing. You're depending on lights. You're looking at lights just before landing more than anything else.

I was fifty-nine when I flew my last flight. I love flying, and I would have loved to have flown in space until I was ninety.

Norman Thagard, a medical doctor and NASA astronaut, described in Chapter 13 his experience as the first American to serve on the Russian Mir space station. Prior to that 1995 international mission, he flew on four space shuttle flights. The first two flights were on Challenger, and the next two were on Atlantis and Discovery. Some of the space shuttle missions were flown out of numerical sequence. They were numbered based on initial schedules rather than the actual launch order. For example, Thagard flew on STS-51-B in 1985, on STS-30 in 1989, and STS-42 in 1992. On some of his missions, he studied space motion sickness, or space adaptation syndrome, a condition that can cause nausea, visual illusions, and disorientation.

I was born in July 1943. Even when I was a kid, I was interested in science fiction. In fact, science fiction was really cranking up in those days. There were a couple of magazines that sold on the newsstands, *Analog*, I think, and *Fantasy*. I just felt that if there ever were such a thing as an astronaut, I wanted to be one for sure.

As a senior in medical school, I applied to the astronaut program. I got interviewed during the course of my internship, and then accepted in January '78. So I finished my internship, and then I immediately went to NASA as an astronaut candidate.

Norman Thagard spent a total of 140 days in space during four shuttle flights and his 115-day mission on the space station Mir. Courtesy of NMMSH.

You can't just be assigned to spaceflight when you first get there. There has to be a candidacy period that lasts about a year. The training is what you might expect. The first part of it really involves a familiarization with NASA itself. Toward that end, they sent us around to visit the various NASA centers, a couple of them in California – Ames and Dryden. We saw the Rockwell facility, where they were putting the shuttle together. We visited the center in Cleveland, and we went to the Kennedy Space Center. In fact, we got to see a rocket launch there. And there were academic things. You had to study things like orbital mechanics, astronomy, astrophysics, a little bit about aviation, since we were going to be flying in airplanes and ultimately on a spacecraft.

I first flew aboard space shuttle STS-7 in June of 1983. It was very exhilarating. Obviously, when you get to a thing, it is never exactly what you might have imagined, because you fantasize. You probably idealize and make things somewhat different than the real experience. But the real experience was pretty darn incredible too. The excitement of launching on a rocket, and the thrill of actually getting into weightlessness, and looking out and seeing how the Earth looks a couple of hundred miles up in the sky are very awe inspiring. It may not have been exactly what I had had in my mind's eye, but it lived up to expectations nonetheless.

I remember when the main engines cut off. You are in weightlessness but you are still strapped in, so you don't sense that right away. I held a pencil, let it go, and it sat there and floated in midair. I had to laugh because I thought that was a pretty strange sort of situation.

On the fifth flight of the space shuttle, there had been a spacewalk planned. But one of the astronauts had a touch of space motion sickness, and they decided to postpone that spacewalk. That made the issue of space motion sickness take on a new importance, a new sense of urgency, if we were going to have to postpone spacewalks as a result. So the decision was made to put a medical doctor on the seventh shuttle flight and on the eighth shuttle flight. I was lucky enough at the time to be the support crew member for the seventh shuttle crew that was already in training. As a medical doctor and already familiar with that flight, it was logical to assign me to fly, to study this business of space motion sickness. Most of the experiments that I performed were actually looking at space adaptation syndrome, space motion sickness.

I had read a compendium of all of the medical results, not only from Skylab, but there was a bit of history in there about things that had been looked at in the various programs since the Mercury program and even before, I suppose. I was already somewhat familiar with it. There was a whole slate of experiments that were devised specifically to look at space adaptation syndrome on STS-7, and those I performed.

Since you can easily suffer from space motion sickness as a result of some of the tests that you would do, you do put yourself at risk for that. I had made a conscious decision before I flew that even should I suffer from space motion sickness, I would take no medication, because I didn't want to cloud the results of some of the experiments I was doing. It turned out I did suffer from space motion sickness, and I did not take any medication. So in that sense, it was a bit of a sacrifice. I probably would have felt a lot better than I did, had I taken something for it. But I think we got good results off that flight, and it was worth making that kind of effort on that occasion, I thought.

On that particular flight, there was a woman on board, Sally Ride. In fact, that was the flight of the first American woman to fly in space. None of my experiments involved Sally as a test subject. But I did have female crew members on four of the five flights I flew, and if there are differences between men and women experiencing it, I certainly can't tell. Some women, as some men, suffer from space motion sickness and some don't. The only area I would think that would single out women over men would be this business of bone loss in long duration spaceflight. With women more susceptible to osteoporosis, I think you would worry more about the bone loss, perhaps, than you would in men. Apart from that, I am not aware of any differences.

On STS-7, we did the first retrievals of an object in space. We took a satellite that was carried up in the payload bay of the shuttle, and it had its own attitude control system. When we got into space, John Fabian and Sally used the robot arm to lift it out of the shuttle bay and release it. So it was a free flyer. Basically, it was another satellite up there with us. We used it as a test article. We fired rockets at it to see how they affected it. In other words, a proximity operations study, looking forward to the time when we would go up and retrieve things like the Hubble Space Telescope or the Solar Max satellite. At the end of that, we flew back close to it, and sort of took turns capturing it. When it became my turn, Sally made an input to the satellite that caused it to start a slow rotation. Then, while it was doing the slow rotation, I used the arm to go up and grab it. I guess I was the first person to grab a rotating payload.

When you were doing things like operating the robot arm, you had to make sure that you were anchored firmly. We had foot loops, for instance. And you've also got some hand grips around the controls for operating the robot arm, that you can sort of hold on to a little bit while you are making input. Many things are more difficult in weightlessness than they are here on Earth, because of that aspect of it.

I also flew on STS-51-B in April and May 1985. That was a Spacelab mission, and for the first time on a spacecraft with humans, we had rats and squirrel monkeys. It was basically a test of the cage system, the handling facility that would keep those animals healthy. We really didn't do any experiments on the animals while we were up there. Our job was just to feed them and to observe them. The system worked fairly well. There were some problems that hopefully were corrected before subsequent flights of the handling facility.

We also did a number of other experiments, crystal growth for instance. There were supposed to be some celestial camera photographs, but we had a problem with the airlock that would have enabled us to put the camera out to space. So that experiment, largely, was not successful. Crystal growth experiments were very successful. There was another one that looked at geophysical fluid flow, looking at planetary atmospheric flows, which you can do in a weightless environment, an experiment which you couldn't do on Earth. The reason we take a lot of those experiments to space is that they either can be done better there or that is the only place that you can do those experiments.

I also served as the flight engineer on that flight, so I had a dual role. On ascent and entry, I was the flight engineer [assisting the commander and pilot with ground communications, and calling out checkpoints], and while in orbit I was one of the mission specialists helping perform the experiments we had back in the module.

Doctor Bill Thornton, who was my crewmate on that flight, was doing many of those experiments that I performed on STS-7 looking at space motion sickness. He was doing some pharmacologic stuff, and my help with that was I helped draw some blood on him and inject some things into him while we were on that flight. But my main role on that flight was not looking at space motion sickness, just generally the other experiments we had on board.

STS-30 [May 1989], which deployed the Magellan spacecraft to map Venus, was a nice flight because we all knew that we were going to be launching a science probe that would have some significance. In fact, what that probe did very successfully was radar map most of the surface of the planet Venus. Venus is covered by visually impenetrable clouds, but they can be penetrated by radar of certain frequencies. The radar can actually map the surface and give you, via computer enhancement, a picture that looks almost identical to a picture that would have been taken by camera, had the camera been able to see through the clouds. So almost the whole surface of the planet Venus is now mapped, and that crew of which I was a member was the one that launched it on its way to Venus. I think we all felt a sense of accomplishment with that mission, because that's the sort of thing you like to think that you're doing as an astronaut.

In January of 1992, I served as payload commander aboard STS-42, another Spacelab mission, a very ambitious one, I think. Fifty-five experiments from, if I am not mistaken, ten or twelve different countries, including Japan, a number of countries in Europe, Canada, and the like. We grew plants on that flight. We had some fruit flies on that flight. It included the Bio Rack, which is a very nice facility for doing biological experiments. It was developed by the European Space Agency. We also did some more crystal growing, and I participated in that. Because I was payload commander and responsible for the operational laboratory in space, that was an especially rewarding flight for me. I thought everybody did well, and we got a lot of science data. I would say that we got well over 90 percent of all that we had intended to get on that mission, which is pretty good, given all of the things that can go wrong on a microgravity laboratory flight. The Spacelab is a nice facility for doing science in space.

Richard (Mike) Mullane was a flight test engineer in the Air Force. He became a NASA astronaut in 1979 and flew on three space shuttle missions between 1984 and 1990. Mullane was inducted into the International Space Hall of Fame in 1993.

I have always and forever been interested in flight, astronomy, science, space. I built rockets back in the late '50s and early '60s, when I was a kid. I built rockets in my garage, tested them out on the deserts of Albuquerque. I had always had an interest in flying, and I really tried to get into the Air Force Academy. I was unable to do so. It's a political appointment; you apply through your Congressman and Senators. My Congressman had a slot to West Point, so I was able to take that, fully intending to take my commission in the Air Force.

I tried to get into pilot training. My eyesight was too bad for that and actually, it took a little bit of an effort for me to get into navigator training because of my eyesight. But I was able to get a position in navigator training in 1968 and graduated from that program. In 1973, I went to graduate school at Wright-Patterson Air Force Base, Ohio. I graduated

Mike Mullane completed a total of 236 orbits of the Earth during his three shuttle flights. Courtesy of NMMSH.

from there in 1975 and went to the Flight Test Engineers course at the Air Force Test Pilot School at Edwards Air Force Base. Pilots and non-pilots both attend the same academics. The fly-in training is obviously different between the pilots and the engineers. That was a key event in my career because I think that, more than anything else, made me a little bit unique for the applicants that subsequently applied to the astronaut program, and it enhanced my abilities to compete for that program. After graduating from that Flight Test Engineer course, I went to Eglin Air Force Base in Florida and was flying in many different aircraft there, primarily the F-4

Phantom and the right-seat F-111s, doing weapons testing there. It was there at Eglin that I applied for the astronaut program.

Up until the shuttle program, [nearly] all of the astronauts that we have ever had were military test pilots. The shuttle program was the first program that said half of the astronauts will be pilots and half of them will be non-pilots. As soon as they said that, I put in my application for the mission specialist position and was fortunate enough to be selected in 1978.

All the requirement called for was a Bachelor of Science degree and three years of experience in whatever that field was, to be eligible for mission specialist position. I had a Master's [in aeronautical engineering], I had my flying background, and Flight Test Engineers school, so I put in an application. I was fortunate enough to be called down for an interview. They probably called 250 or so down for a physical and an interview at the Johnson Space Center. I went down there for that interview and the physical, and it was from that process that they then selected thirty-five. I was fortunate enough to be one of those thirty-five.

The training program is less structured than what I had experienced in the military. In the military, you go into a program that's so many months long. You enter, not knowing anything about an airplane, and you graduate nine months later being an expert on that airplane. NASA's was a little bit less structured than that, but more thorough. You start out with a lot of lectures by prominent scientists in various fields, but the group of astronauts that comes in for the shuttle program is such a varied background. You have military test pilots, who all their life have done nothing but fly airplanes, and then you have post-doctoral students from college who have never been in an airplane, and people somewhere in between.

We had a lot of lectures at the very beginning by people in oceanography, meteorology, geology. We went on a geology field trip up in the Taos area of New Mexico. They are trying to impart on you a sense of how all of the sciences are affected by spaceflight and how spaceflight can enhance those sciences. So you get a several month classroom-type atmosphere of these briefings, trips like the geology field trip, and planetarium work. After that, you end up in a particular job associated somehow with preparing a veteran crew for a mission. We have a lot of simulators that require crew support, software verification. You will end up in a simulator. That isn't a formal training program, but when you are sitting in that simulator, you are learning about the shuttle and studying the various systems. So it's kind of a self-taught type of program. It's not like you have an instructor standing over you. You get plenty of opportunities to ask questions of engineers and answer them, but it's more of a self-taught type of thing, being exposed and having the opportunity to fly the simulators.

We do a lot of simulators, and we do a lot of support of payloads that are being manufactured to fly on the shuttle. You'll go to a factory and be the astronaut point of contact for developing the human interface with that satellite. You will go over and validate software for the remote arm or procedures for the remote arm. They will rotate you around through this, so you are not just sitting in one. Maybe you'll do one of our avionics integrations lab simulators for six months, and then you'll go over to the remote arm and vice-versa. People will be continually rotated through working on the space suit and training in the procedures. So after several years of that, you've been exposed to all the shuttle program and all of the supporting aspects of it. Mission Control is another place, working in Mission Control as CapCom [Capsule Communicator]. So, by the end of four years of that, you are an expert in the manned space program and in the shuttle, and then you are prepared for flight assignments. Around the four year to five year point is when they will start assigning a person to a flight.

I don't think that there was anything that was uniquely difficult in the training program, other than waiting for five years. Or in my case, we were the first shuttle group, and the shuttle program didn't develop as quickly as we thought when we were hired. So I flew about six and a half years after I was assigned as an astronaut, and that was about in the middle of my group. The earliest people flew at about five and a half years, and some of them went maybe seven years before they actually flew. You felt like you were prepared much earlier, and the delay was very frustrating for somebody that's dreamed all their life of flying into space. The idea of waiting and waiting and waiting and training and training, after awhile it gets very frustrating.

A mission specialist astronaut is a unique position to the shuttle program, and probably I can explain it best by explaining what the pilots do. The pilot astronauts are primarily responsible for flying the vehicle into orbit, and returning it from orbit, and for controlling its attitude and pointing on orbit. The mission specialist job really encompasses everything else, which is the on-orbit operations. If you are operating the remote manipulating arm, it will be a mission specialist that will do that, either deploying a satellite or retrieving a satellite. If there's a spacewalk, it will be a mission specialist doing the spacewalk. If there are onboard experiments, the mission specialists are primarily the ones responsible for the on-board experiments. If there is a satellite, one of these spinning satellites that you have seen in the early shuttle program, those were always mission specialists that were responsible for checking those satellites out and deploying them. So whatever the mission is on orbit is primarily a mission specialist's responsibility.

My first flight assignment was STS-12 [which was later rescheduled as STS-41-D]. It was the first flight of the orbiter Discovery, and the primary payload was three communications satellites. Two of them were what we referred to as PAM satellites, propulsion assist modules. Those are the type that are spinning and you see them pop out of the shuttle. And we had this much larger, new type of satellite called Syncom that rolls out of the payload bay and gets a spin from the actual deployment mechanism. And we had a large solar array made by Lockheed Corporation. It was a solar array that extended 115 feet out of the payload bay. On that particular mission, the solar cells were massed simulators. We were testing it structurally, rather than electrically. There were a few active solar cells that they tested, but mostly it was a dynamic test to extend this thing out, fire the orbiter steering rockets, and see how, dynamically, this thing was affected. Inside the cabin, we had a pharmaceutical laboratory experiment called CFES, continuous flow electrophoresis system. McDonnell Douglas and Johnson and Johnson were partners in that.

The ascent in a space shuttle is the scariest thing I have ever done in my life. At the same time, it was a dream come true. When you are sitting out on the launch pad before liftoff, your soul is simultaneously occupied with gut fear and boundless joy. You are really terrified about this ascent and at the same time, it's something you have dreamed about for so long that it is the most wonderful experience waiting out there for you. In fact, I once mentioned that to an audience, and a mother in the group said that that's the way she felt when she was in labor. So, maybe all of the mothers out there know, closer than they think, what it's like to be an astronaut preparing for the launch.

It was a dream come true for me, but very frightening. And this was before Challenger. But we all had a sense, certainly in the astronaut office, the military flyers had a very keen sense of the hazards associated with this vehicle. No escape system on it. Launch is a particularly terrifying event because you don't feel that you have any control over the vehicle. It's not like in an airplane, where you can always feel that you have ultimate control of it, and if something happens, like you run out of oxygen, you can always eject from it. The shuttle, you don't have that sense. You are a prisoner of the machine. You feel like you are strapped into this thing, and you are a total prisoner, and it absolutely has to fly exactly right. There's a few options there if you lose engine power or you have an abort capability, but you don't feel that you have the control you do in an airplane. That's why I say it's exceptionally terrifying to sit there and go through this. But at the same time, I want to emphasize that it's also a very joyful experience for those of us who dreamed of doing this all our lives.

My second and third missions, STS-27 and STS-36, were both Defense Department missions, so I cannot talk at all about the operations. I can talk about the orbit that we were in on my final flight, which was aboard Atlantis. The inclination and altitude that we flew were unique to the space shuttle program. In fact, unique to the manned space program. The inclination refers to the tilt of the orbit to the equator. No American astronaut has ever flown in higher inclination orbit. Most of the missions that we fly out of Kennedy are what we call low-inclination orbits that only cover a little area around the equator. A lot of people have the misimpression that when you are in a shuttle you see the world as a basketball or marble, and they tend to think of the Apollo pictures taken from the Moon, which is wrong. We operate in a very low orbit and we are very close to the Earth. You only see the part of the world that your inclination allows you to see, and in most of the missions it's close around the equator. You don't even fly over the United States, for example, except the extreme southern part of Texas and Florida.

On my last mission, we flew a 62-degree mission, which means that we got to see the world from 62 degrees north latitude to 62 degrees south latitude. We also flew at the lowest altitude that an orbiter has flown at, 110 miles. So that was a very unique experience, seeing the world. We got to see the world up into the Arctic areas, up in the 62 degrees north latitude, and you are seeing close to 1,000 miles slant range higher, and then down in the Antarctic area. In fact, we were the first crew to see the Antarctic at all. That was really exciting, to see so much of the world, to see the northern and southern lights, for example. They were absolutely fantastic. That was an exciting mission. I wouldn't trade that for anything. It's a little frustrating to fly missions and not be able to talk about what you did on them, but I wouldn't trade it for anything after flying that high inclination and in being low. That was the other thing. You got to see a lot, and your detail was a lot better than being up several hundred miles.

Another sight from that mission that was spectacular was flying over the city lights of the United States. We had many night orbits that carried us across the eastern seaboard of the United States, and seeing the city lights from Key West all the way up into Canada. It was clear, just magnificent. You don't have a sense when you are looking at those city lights that you are 110 miles away. It's more like you are in an extremely high-flying airplane and the lights show up very, very good. You could see Long Island, Manhattan, the Hudson River. Absolutely beautiful.

Another sight, going back to STS-27, my second mission. It was a high-inclination mission. Not quite as high as 62 degrees, but still high inclination. One of the unique memories I have of that was flying over Albuquerque, my home, in the daylight and looking down and seeing the deserts near Albuquerque, where as a kid I used to fire these rockets off. It is hard to describe the sense of having done that, having dreamed as a kid firing these rockets off the deserts of New Mexico and now, here I am in this rocket that's 240 miles up. It just sends your heart into another adrenal sprint!

I think that probably the most anxious moment I have ever had in my life, well, "anxious" is probably an understatement of the century, probably "sheer terror" is more accurate. It was on my first mission, the first flight of Discovery. Our first launch attempt was June 24, 1984. I ultimately flew in August. So what happened in June that delayed us for two months? Remember, this was the twelfth flight of the shuttle program, so we had eleven missions that counted down, the engines ignited, and they lifted off into Earth orbit. So everybody sort of has that mind-set that when the engines ignite, you lift off and fly into orbit. Well, on this particular mission, we counted down and got to when the engine sequence starts at six seconds. So, you're sitting in this cockpit and as you might imagine, you are pretty tense. The engines start and you get these strange sensations you've never heard before, a loud wrenching growl in the cockpit that is violently shaking. We had T minus 6 seconds, engines start, vehicle shaking, and then at about T minus 3 seconds, everything stops and we get a warning tone in the cockpit and a warning light indicating our engines have all shut down. That was terrifying because that had never happened before. We were totally unprepared for it. The system was really unprepared for it. You could tell immediately in the headset that the launch control team were confused. There was just a lot of confusion.

After being in an organization that was very, very professional and had covered all problems, and all of a sudden you don't know what's going on, and it's obvious that the people on the ground don't know what's going on, and then they start talking about a fire. There was some hydrogen gas leaking and they had a fire going. You're in this cockpit and you have visions. What is going on out there? It was very terrifying for several minutes there while they sorted out exactly what had happened. A valve malfunctioned. The liquid engines are started early, because you can shut those down if there is something wrong. And there is a lot of redundancy in the engines and a lot of redundancy in the entire vehicle. One of the valves that operate when the engine starts to sequence did not move at the proper speed, and the computer sensed that and did exactly what they had programmed it to do. The computers are saying this doesn't look right to us, so shut them down and we'll take a look at them. The computer and the software worked perfectly. Absolutely perfectly. There was no real threat to our lives, but we didn't know that at the time. It was quite harrowing to sit there and listen to that and wonder what in the world is going on out there!

There was another time, too, on STS-27. They always filmed launches with these long-range cameras. They have a lot more cameras than the ones that you see on TV, and after missions launch and while the vehicle is safely in orbit, the engineers will play these things back and look to see if anything funny happened. On STS-27, they looked at the early part of the ascent, and they saw something fall off of the solid rocket booster. They saw something come flying off and were a little bit curious about what that could have been. They asked us to use the remote arm (in fact, I got to do this because I was trained in the remote arm) to look underneath the belly of the space shuttle to see if you could see anything up there. We looked under there and saw a lot of damage to the heat tiles, a lot of damage. You start thinking, uh-oh, what is this going to do on entry? It's no threat on orbit, but what is this going to do on entry? Entry is normally not a very threatening event. You don't have this feeling like ascent. You are more like in an airplane. Entry is much more benign than ascent. But that particular entry was one where your heart rate was up a little. You wonder, am I melting underneath here? We get back, and the tile damage is quite severe. One tile was completely gone and a lot of them were gouged, but there was no burnthrough of any metals, so we were ultimately okay.

They had changed some type of the manufacturing process for the very tip of the solid booster. I don't remember exactly why. But it was not a good process, and a piece broke off. It shot-gunned the belly of the space shuttle with this spray of material, and those tiles on the belly of the space shuttle are very fragile. They are like Styrofoam. You can stick your finger right through them, so it doesn't take much. And if you're going a couple hundred miles an hour up through the atmosphere and something sprays across it like that, it's going to cause severe damage, which it did. I forget how many they ended up having to change out, several hundred tiles before it could fly again.

Dean Schaaf, and electrical engineer and pilot, served the last three and a half years of his Navy career at abort landing sites for space shuttle flights. When he retired from the Navy in 1991, he went to work for NASA, still in support of manned spaceflights. In this excerpt of his oral history interview, he described the contingency landing site facilities and shuttle flight training operations.

We've got four shuttle contingency landing sites that NASA has used for several years. Some of these were picked up after Challenger. The actual [contingency] landing site for Challenger, when the accident happened, was Maron, Spain. We used to use Dakar, Senegal, as a site then. They activated the International Airport at Casablanca, Morocco, just a few weeks prior to that launch, so we would have a site there. But we never continued to use Casablanca. It was too busy an airport to use as a landing site. A team of NASA experts and astronauts searched for another site, and they found Ben Guerir, Morocco, which was about midway between Casablanca and Marrakesh. It was a SAC [Strategic Air Command] base abandoned in 1962 by the United States Air Force, and the runway was 12,800 feet.

These four sites that I talked about all have shuttle-unique landing aids. We've got the PAPI [precision approach path indicator] lights, at 6,500 and 7,500 feet prior to the threshold. We've got ball/bar lights for the one-and-a-half degree glide slope for the landing. We've got microwave landing systems. We've got long-range navaid [navigational aid] to find the area. And we've got a numerous amount of generators that we have to maintain. There is not power at these places, so they have to be generator powered. There are communications packages, satellite comm. We've got three INMARSAT (International Maritime Satellite) satellites that we set up at these sites, along with UHF, VHF comms. We've got xenon lights, which are one billion candlepower each, searchlights that we put at the end of the runway. They use this at Kennedy Space Center and Edwards for the end-of-mission landings also. If they do a night landing, they have these lights set up so they essentially light up the whole runway. So we have those at the sites, and all the ground support equipment.

Three of the sites have barrier systems, what we call SOAS, Shuttle Orbiter Arresting Systems. Following Challenger, it was determined that we needed a barrier on any runway that was less than 12,800 feet. This barrier is a massive system, copied after the barrier that we would erect on an aircraft carrier for a crash landing. The net is basically 300 foot wide, 26 foot high at the center point, and is supported by 45-foot stanchions, and then connected to energy absorbers by some pretty massive equipment. We've put it 200 feet into the overrun. If the shuttle moves into that barrier doing a hundred knots, it will stop in 850 feet, doing very minimal damage to the shuttle. It is designed so it won't crush the payload bay. The retarding force is a big bundle that flips up after the nose gear passes over, then it catches the main landing gear and the forward edge of the wings.

One of the last things they used the Enterprise for was, when we flew it to Dulles for the Smithsonian Institute, they towed the Enterprise through this barrier and saw how it would play out across the leading edge of the wings and everything. I've seen pictures of that in some shuttle history books. They also built a miniature shuttle, which was about 13 feet long. It was in a warehouse in Philadelphia, and they had a miniature barrier, and they fired that across this warehouse using a rocket launcher and saw how it was caught in this barrier. That was an interesting test.

To set up a contingency site, we leave Kennedy Space Center. The team is made up of about fifteen United Space Alliance personnel, the manager and his technicians. I am the NASA ground operations manager, the NASA man in charge. There is a DOD [Department of

Defense] contingent, three weather guys, normally, that do our weather forecasting and launch balloons for us. Their manager, they called him the DFC, Deployed Force Commander out of DDMS [DOD Manned Space Flight Support Office]. We normally take a security manager with us from Space Gateway Support, another contractor. Xenon support if it is a night landing. We will need three guys in each one of the sites from the xenon shop at Kennedy.

Two days before launch, DOD brings a C-130 in to support. That C-130 is used for SAR [search and rescue] medevac missions. On that C-130, there are eight crew members, nine para-rescue specialists, PJs. There are airdrop Zodiac rafts on that aircraft that they could launch mid-ocean if the astronauts have to bail out, which they have the capability since Challenger to do that. If they come up short of a landing site overseas and they have to bail out, we could go out and pick them up. There are also two positions on board, a nurse and a med tech with about 2,500 pounds of medical equipment, because there are no hospitals at these areas.

The first three or four launches with the shuttle, they had bailout capability. They had ejection seats on them, but they don't now. Now, they use the pole. That was redesigned after Challenger. If something happens and they are going up hill and they can get control of the shuttle, a controlled glide, so to speak, they can blow the side hatch. The pole telescopes out 10 feet, and just as quick, they can hook on to one of the clips on there. They roll out and it keeps from hitting the wing as they go out. They've got their survival equipment on them. They've got a one-man life raft on there and ten minutes of breathing air. They would bail out, and then we know where to send the C-130. The C-130s in Patrick [Air Force Base at Cape Canaveral] are on alert to do that during the launch. Out at the TAL [Transoceanic Abort Landing] sites, the highest probability area at launch, having to do that bailout is within 200 miles of the launch site, and within 50 miles of land sight. So we cover that with C-130s on both sides.

At the TAL site, they've got the C-130 at launch minus two days. We also bring a C-21 aircraft in. That is the Air Force version of a Lear jet. An astronaut arrives about two days prior to launch, and he uses that C-21 for a couple things. He gets airborne. He checks out our landing aids, that everything is set up as far as intensity and everything. We set them up for operation maintenance instruction (OMIs). But just a good feeling, visibility to see the haze that might be around there at our African sites and this kind of thing. He gets airborne about an hour and a half to an hour prior to the launch, and he is airborne through the launch at each of the three sites. We tie all of those in through UHF radio, back to our comm centers in the buildings and back to Johnson Space Center.

So those astronauts over there are talking to the weather CapCom back in the mission control center, sitting right next to the CapCom, the Capsule Communicator, who will be talking to the shuttle and feeding the flight director information. We will not launch at Kennedy unless there is good weather at one of the TAL sites. We have to have one good TAL site for them to launch. It's happened three times in the history of the program where we've scrubbed a launch because there is bad weather overseas but perfect at the Cape.

If you add all of these up, you have about twenty-five contractor and NASA personnel and thirty-five DOD at each TAL site. Plus about fifty host nationals, whether they are Moroccans, Gambians, or Spaniards, that furnish the fire suppression support. The internal rescue is done by DOD-trained personnel or guys from Kennedy Space Center. If we have a crash landing on the runway and the astronauts can't get out on their own, we can send this team of guys in with breathing air and rescue them out of either the top hatch or the side hatch, however they need to. We have got the docs there and the C-130 to treat them on site and fly them either to the trauma center in Seville, Spain, or Ramstein, Germany. So that's the whole TAL scenario.

A normal trip, if everything goes on clockwork, we leave seven days prior to launch, we launch, it takes us two days to pack and to stow at the site, and then the return trip home. So a ten- to fourteen-day trip is a normal trip. When we get week delays and this kind of things, then we sit there, because the break-even point is like nine days. If you have a delay that is less than nine days, you stay in place. If you have one more than that, you come home, because it's cheaper to fly you home and send a team back.

The shuttle landing at White Sands is really a secondary mission to what they do. They exist to provide training for the astronauts. The shuttle training aircraft, the Gulfstream that is half shuttle and half aircraft, they have four of those at El Paso. The astronauts will get in their T-38s, fly to El Paso, get in the STA, the Shuttle Training Aircraft, and come fly their training missions at White Sands. There are two primary runways, and then they've got a third runway, and then the opposite runway from that. The third runway is their training site for the TAL runway. Going the opposite direction on that runway is ELS, Emergency Landing Site, anywhere in the world.

The other runway is made just like Edwards, and then they've got one made up just like Kennedy Space Center. The width of the markings and everything are identical to the 300-foot wide runway for Edwards and 8-foot markings. Kennedy only has 4-foot markings, and all of this is out there on the gypsum. They used petroleum-based tar to mark off the lines. They've got reflectors out there on the runway that would reflect the xenon lights and show them the edge lights. That is how we do it at the sites overseas. Once we turn those xenon lights on at night, we put reflectors every hundred feet down the runway. We also have, since you are on a lake bed, a portable approach lighting system. There are no approach lights out there like a normal airport.

At two of our sites overseas where we don't have those, we use flashlights. We put these flashlights on about a foot-square, half-inch plywood and tie-wrap fluorescent flashlights. We put them in a 3,000-foot string out there, and then a pattern on the underrun. We used to do the same thing at White Sands, but it was taking so many D cell batteries! (Laughs) They have a system now, it is still DC powered, but it is little LED lights. It doesn't take a lot of money to maintain, but it does exactly what the astronauts need to train on.

Normally, those STAs will make nine or ten approaches. Their approach is at 19 or 20 degrees dive at the PAPI light. Then at about 2,000 feet, the transition to one-and-a-half degree glide slope, lining up on the ball/bar that sits on the left side of the runway. They will come down, and when they get to 30 feet [eye height] off the runway, that puts them at about the same height they would be in a shuttle, if the shuttle was touching down. At that time, they call "touchdown," and then they go around and do it again, so they never actually touch down when they are doing their flights.

I've had the opportunity to fly in one of those. I flew in the training flights at Kennedy. We went out and did ten approaches on the runway there. And, boy, you have to brace yourself behind the pilot, because you are really diving!

Before they fly, the commander will have done a thousand landings in this kind of thing. Of course, they use simulators too. The simulators are good, but they can't quite simulate the feel. The Gulfstream has been modified. They drop the landing gear. They've got modifications on there to make it handle just like a shuttle, and the speed brakes dirty it up so it can fly at that angle. When they call "touchdown," the instructor on the right-hand side takes over. It is a regular aircraft now. He goes around and they set up and do another one. They do this day in and day out. That is the primary mission of the White Sands Space Harbor.

Rick Berry was a helicopter pilot who began working as a civilian for the Army at White Sands Missile Range in 1986.

Being as the White Sands Space Harbor is an abort landing facility for the shuttle, we do simulated abort or landings out here of various categories. We will go out there with five or six helicopters, and crews, and all of the facilities. They have a great big, B-52 type body they call "buff." They drag this thing around, and all of a sudden, poof, we get a call, "The shuttle just crashed off the field." We have our helicopters set up, and one called "air boss" launches immediately and goes and locates it. Then he directs all the other aircraft and the ground people to it. In each of our Hueys [helicopters], we generally have a couple of litters, and we also have what we called the "flight docs" and the medics in the aircraft. When the ground people get there, one of our aircraft is called the "fire aircraft," which was one I normally flew on. We have all the ladders in there. We go over to the base camp, pick up about four or five firemen, take them out. They get out with their ladders and all of their fire extinguishers and everything else, rescue the people, bring them back to the aircraft, load them on the aircraft, and we supposedly fly them off to whatever appropriate facility they need to go to.

After being heavily involved in the Apollo program (see Chapter 9), Lew Curatolo retired. Before long, however, he became bored with retirement, and a former colleague asked him to work for NASA following the 1986 Challenger disaster. In 1988, Curatolo went to work at NASA's White Sands Space Harbor, where he prepared the facility for a potential landing during shuttle flights.

White Sands is targeted for an alternate landing site for the shuttle and in the event of any emergency, depending on what orbit they're in. White Sands is an ideal landing spot for the shuttle, which was proven earlier because of bad weather at Edwards several years ago [i.e., in 1982]. The astronauts seem to think that that's a very wise choice as an alternate landing spot. The opportunity came up for me to be stationed at White Sands as safety engineer and quality assurance engineer to receive the shuttle in the event it does land here. I have to make the safety assessment if the shuttle lands here, to make sure that there are no toxics and no hazards before we evacuate the crew, and that the spacecraft is not in any danger of any explosions or anything. I am the first to go to the spacecraft and visit the crew and make sure that everything is in the proper order for them to evacuate the spacecraft when they land, and that there is no hazard that will jeopardize their life or the hardware and or ground crews that we have here to support this effort.

I wear a several hats here and I like it. It's very comfortable. It's not a pressure cooker like we had at the Cape. Of course, if we ever do have a landing here, it will change.

Every time the shuttle flies, we have to gear up just as if the shuttle was going to land here. We go through the launch count, and we go through the orbit count, and we go through the landing sequences just like anybody else all over the world. We're reading off the same book that the astronauts read off and the launch director reads, all in concert, and we respond to the commands. It's interesting. We're better prepared than they were in 1982, because they realized that they were ill-prepared and it gets real complicated. The launch is complicated, but the landing is almost as complicated because there are so many things, and the chain of events that have to take place to prepare the spacecraft to be flown back to Cape Kennedy.

When it lands, then we have what we call "wheel stop," and then the astronauts acknowledge everything is okay. Then we take over as the ground crew, the recovery crew. We have a convoy commander, and we deploy all this equipment we need to provide special air conditioning and provide a special coolant for the electronics and hardware. We have to prepare to make sure he didn't blow a tire, because that's very hazardous, and that there are no toxic materials (propellants) being vented off, that there are no explosives that are armed and ready to go off.

The first thing that happens is that my team does the safety assessment. We go around to the thrusters and the main engine with a special instrumentation and special safety gear, and we take an assessment to see that the toxic levels are within limits and they are not where they could do any damage to the spacecraft or to the crew or to the ground crew. It takes us about forty-five minutes. Then we call the convoy commander and we say, "You have a 'go' for a downgrade." We start letting in the stair truck and the GSE [ground support equipment], and then we bring in the guys to open the hatch. If there's a dire emergency, we have to let him know to blow the hatch with explosives and we go in and get him, and then it's a different problem.

So, the chain of events would normally be that we'd go in and do the safety assessment to make sure that there are no toxics, that there are no hydrogen leaks, that there is no fire, and the brakes are all right. The brakes have a beryllium drum that is highly toxic and can cause you harm. And you have Freon that is highly toxic, and you have hypergolics and monomethylhydrazine and N2O4. So you are sitting there with a loaded bomb, so to speak, and you want to make damn sure. These guys have been in orbit for over a week, let's not wipe them out here on the ground. That's a call I have to make and advise the convoy commander we are in a configuration to exit the crew. Once this is done, then the ground crews move in and remove the tires and the brakes, and then you take it down to the service pad and evacuate all the hypergolics and all the cryogenics. Then we prepare to fly it back to Cape Kennedy on a 747. This takes, as a whole, maybe five to seven days.

It's a twenty-four-hour operation, and we want to get it into the air back to the Cape as quickly as possible. It's not just a handful of people. There's hundreds of people involved in this thing.

Story Musgrave grew up in an intellectually barren home (see Chapter 1), but as an adult, he developed a zeal for knowledge. He became a medical doctor after earning two Bachelor's degrees and three Master's degrees in a variety of fields. When NASA began recruiting astronaut candidates with scientific skills rather than test pilot training, he applied and was accepted in 1967. During his space career, he was involved in the Skylab program and flew six space shuttle missions. Dr. Musgrave was inducted into the International Space Hall of Fame in 1995.

The Marine Corps exposed me to books and visualization. Just going out to the airplane was not quite enough, and I started studying the manuals of how to maintain airplanes and started studying drawings about how the systems work. I also got an appreciation that I wasn't going very far without an education. I wanted to fly then, but without even a high school diploma I could not fly. That was a limiting factor in the Marine Corps, so in a pragmatic kind of way I said, "I need to shape up a little bit here; this is going to be a limitation later on." I got a little different view of what college was from some of the pilots. It was different than the constrictions of high school. I went on to college. I didn't know what to study, I had no idea. I had to kind of beg my way into college. I studied math because I like numbers and was good with numbers. I went to work for Eastman Kodak. That was the very earliest days of building mathematic models of business problems.

From there, it was graduate school at UCLA, mathematical models and computers. Then I got interested in the nervous system. How does the brain work? How does the brain do the same functions that I now understand in the computer? I deal in technology, but then I'm always trying to get the human aspect into it. I understood the nervous system, and while I was pursuing those things at graduate school at UCLA, Sputnik went up around that time. So it wasn't like, "Did you always want to do space?" No!

I read things in the nervous system, communications theory, design for the brain and those things. Decided I wanted to study the brain. That was a major leap. I didn't know if I'd do a Ph.D. in physiology or an M.D., but I didn't want to limit myself. An M.D. can do both research and

Musgrave's first name is Franklin, but he is known by his middle name, Story. He was the first person to fly six missions on the space shuttle. Courtesy of NMMSH.

practice, so I said, "That's it." I didn't have the pre-med, so I had to go and take pre-med. I did that at Marietta, southeastern Ohio.

When I left the Marine Corps, I took up flying light airplanes again, got all my ratings including airline transport rating, went on to a commercial flight instructor. I took up parachuting, too, free-fall parachuting. As soon as NASA expressed an interest in flying people other than military test pilots, then I said, "There it is." That was the bolt of lightning that says there's my call. I had the education. I had the airplanes, the flying, the parachuting. I had nature. I had my ability with machinery. You cannot be a good space worker unless you're grounded in machinery and tools.

As soon as NASA expressed interest in flying scientists, I left the traditional clinical training path and got a Master's in physiology. All my research was in aerospace physiology.

In 1967, I was selected as a scientist astronaut with NASA. The selection process was very different from what we called "mission specialist" in 1978. You had to have a doctorate. You had to write essays as to what kinds of research you would do in space. You wrote proposals as to what kind of scientific research you wanted to do, and you got evaluated on that basis. You got evaluated by the National Academy of Science on the kinds of research you had done already. It was a little different selection process. They reduced the thousands of applications down to seventy of us. Then they handed the seventy of us to NASA to complete the process and eventually, eleven of us were chosen. I was unbelievably excited about the job.

We went to NASA in Houston and got the academic training, all the different sciences that are basic to spaceflight – weather, meteorology, astronomy, geology, astrophysics, aerodynamics, all of the basic sciences and basic engineering. We, as a class, had a physician, myself. We had astronomers, engineers. We all came from different backgrounds, so you had to slant those sciences at the right level so we could all comprehend.

Then we went off into the Air Force for a year. I went to Reese Air Force Base in Lubbock, Texas, and did training in a T-41, which is a Cessna Skyhawk; a T-37, a Cessna twin-engine jet; and then a T-38, which I would fly with NASA so to make an easy transition. That was a marvelous year. Just a delightful year to just give myself over to the Air Force.

From there, I came back to NASA in Houston and got to more pragmatic stuff. We started into spacecraft systems, which in those days were Apollo systems. We started getting our feet wet in the Skylab program, which we were going to do research and development on. Then we started getting some Apollo simulators, that kind of stuff. Did that for a year and a half or so, and then I got assigned to be on a backup crew for the first human Skylab mission. That was a three-year effort, to develop Skylab and to train for it.

We were much closer to launching a Mars mission in 1967 than we are in '97. Much closer. We were just years from doing it in '67. Now, that sounds strange, but when Kennedy in '61 gave his very famous speech and marching orders on space, they were going to move. Less than a year later, Saturn lifted off. That's the way we did business then. Today, it would be twelve years.

We invented project management back then. Basic project management is you need to follow the decision process, not just hardware. You need to identify decisions which need to be made up front, because a lot of decisions cascade into every other decision. And you cannot make that decision later on, or you have to go back to the drawing boards on all elements of that project. So we identified all decisions that had to be made early, and they were up front. The decisions which were autonomous and only affected that element, you made them later on. We had extraordinary leadership. The entire world looked at us not only for our technology, not only for our integration and harnessing of this vast number of people and companies, and how we pulled them together and got them to work in unison. But, we were looked upon as the best project managers in the world. Nowadays that is absolutely not true. We don't have that anymore, industry has it. Politics is the number one reason it changed.

When I joined NASA I was working to go to Mars. I was not nuts. It was totally reasonable. Kennedy says, "Go to the Moon." Within a year, a Saturn launches. We're coming from absolutely nowhere. We don't have the buildings. We don't have rockets. We don't have anything, and he says go to the Moon. We had only had one summer of flight, one twelve-minute, suborbital flight. In eight years there, we did four programs. For me to think that if I'm going to have a thirty or thirty-five year career in the space program, in the mid-'60s, it was totally reasonable to think that I could make Mars. And I was a physician. Who would you want to send along more, not only for the medical aspects, but to study the possibilities of Martian life? It was totally reasonable.

It didn't happen, and I don't regret it either. I don't regret things. The only thing I could regret is if an opportunity presented itself and I did not pursue the opportunity. Whether I had one flight or the six I had, that's not what's important.

The first time you go up in the space shuttle, you're coming from a familiar field of the aviation world, with the aviation experiences. You're coming from the hundreds of hours of simulator time, and some of the mentality says you are looking for recognizing familiar experiences. You're looking for, is this thing flying the way the simulator flew? You're looking for grasping the differences between your expectations and what is there. Yes, your eyes are so big you can hardly get them in your head, and it's the raw excitement that you have, even though it is grounded very much in the training you've had. It's not unlike going to a spectacular ride at a theme park in which you have some expectations. You see other people do it and you see what it does before you get on, and now you're wondering what it's like to be on the thing. You are incredibly excited, the anticipation of what's to come!

You lift off at one and a half g's. You build up to a little over two, go back down to one and a half, and build up to three for the last couple of minutes. It's not like the 3 g's you're used to doing in an airplane. It's much harsher. You're lying on your back and you've got the oxygen bottles that are hanging over the sides and their weight at 3 g's. They are really heavy. They're tugging on your harness. They're tugging on the straps. You're in a seat that's not designed to have a suit in it, because all the seats in the shuttle were designed to be shirtsleeve from the very beginning. I did two shirtsleeve launches, and then after Challenger we went to [space] suits.

Now, the magic of free fall and zero g, you can never anticipate what that's going to be like. We have created and evolved with this force between us and Earth. It's like part of our constitution. And now it's not there, so a lot of magical things happen then, and you're amazed. Then I started learning. I really want to have this experience. I want to give myself over to it. That's an attitude a lot of astronauts do not take. They want to bring their Earth-based self up there and maintain the Earth-based self. That's a basic mentality when you go to combat: the enemy's not going to affect me. I am going to stay myself. It doesn't have to be that way. The human being's an incredible parallel processor. You can be just as disciplined with checklists and procedures and with technology, and at the same time have an experience. You know when to be creative and when not to be.

I got up there and said, wow, this is subtle stuff and you have to work on it. Spaceflight does not come to you. You've got to go get it. I had three missions where I never saw the stars. Don't feel bad, that's the average. If you want to see the stars, you have to decide to see the stars. If we're sitting here in this room and you've got the lights on, you're not going to see stars. You've got to take your time, dark it out, get all the lights off. Say, "Hey we're taking a dark pass." You get a forty-minute dark pass to see the stars. So I started right off saying, wow, spaceflight is subtle. There's an incredible number of experiences out there, but I've got to get subtle. I have to surrender to this. I've got to give myself over to it and let it talk to me, and let spaceflight affect me. Now, I can have a real experience. And so you embark upon that world.

You come back home and you write journals. You talk to yourself. You examine what that experience was, and now you try and work on the next one. You try and play the spaceflight experience you had against the humanities. You study psychology, literature, pHi Losophy, all that kind of stuff. Always in those classes, taking notes saying, "What does this mean to spaceflight? How do I bring Aristotle into spaceflight?" You work on the experience you did, and you try to flesh out what it really is. Then you go up there more sensitive, more attuned, and you go up there a better explorer in terms of the human experience. It is amazing how you get better flight after flight after flight. People say, "Did you get bored after all those?" And I'm thinking, "Ha ha!" First of all, if I was bored I wouldn't keep doing it. But it's amazing how much more rewarding every flight becomes, because you become a real explorer, an inner explorer, and you become sensitive. You become subtle.

You get to know the Earth incredibly well, and you look for more and more subtle things: the zero g, the free fall, the orientation, what you're going to see in the sky. You become much better at looking at the sky. You become much better at the whole experience. You can fly under the Earth, not just over the Earth. It's kind of esoterical. Why not? You're not over the Earth, you're in space. You're not over it; you can fly under it too.

When you close your eyes, every other body sensor you have tells you you're in free fall. Your cardiovascular system is pumping blood; it doesn't have to pump uphill. All the strain across these tendons tells you, "I'm in free fall." If you step off a diving board here, if you go on a roller coaster, when you leave a trampoline, that's free fall. The second you leave that trampoline, you're in a free-fall condition. When you close your eyes in the shuttle, you ought to have that sensation. I closed my eyes and nothing happened. Think about it, my perception is that I'm just kind of floating around here in the shuttle. I says, "That isn't right. What is going on here, and what am I perceiving? Am I perceiving that I'm floating or am I perceiving that I'm falling?" Finally, I really don't think that I'm sensing what I ought to be. So I closed my eyes and pretended I was stepping off a ledge on a very familiar mountain in Colorado. And I achieved free fall! It took four flights before I ever experienced free fall.

Anytime I wanted, I could turn this free fall sensation on. It told me the distinct difference between sensing a float and sensing a fall. Then I got even better. My fifth flight, just as the engines started throttling down, I closed my eyes so my eyes could not take over and dictate to the other sensors the wrong message. You close your eyes to engine shutdown, and then you're in a delicious free fall forever. So you just fall through the dark forever, and it is absolutely marvelous.

We have a genetic component that also tells us that gravitational vector; it determines our geometry. But we do need to start procreating in space – animals, plants, and everything else. And study and get that dividing line between what are the contributions of genetics verses environment. Space is a neat place to do that. Down here on Earth, it's very hard to get to.

I had six shuttle flights, and the last one [STS-80 in 1996] was important. It may have been significant because this is my last shot at it, but I don't think so. I think it was that I had zero responsibility [on reentry]. Different people on the crew had different responsibilities. Some were responsible for launch, some were responsible for entry. In terms of mission specialist, I had no responsibility for the entry. I just got to see my last one, so I got to sit back and watch. Hey! Man, I'm going to have an experience here. So, in terms of ecstasy, that entry was the highest moment I had on six spaceflights. I had a chance to step back and watch a team work. To watch a team work, to see how they do an entry, to see the traditions of naval aviation. I just kind of reveled in watching that team work and mission control. I thought, man, this is what performance is all about, this is what spaceflight is all about. In a way, I'm very happy that the highest ecstasy was a point at which I was not a player in the game. I was nothing. I was just taking it in.

I had a chance to really look at the fire out the back window at the start of reentry, and it goes all the way down to about Mach 10. No one has gotten any kind of details to report about that. I'm telling you, it is unbelievable that the shuttle stays together with that amount of fire. You are astounded at the amount of fire that is out that back window. Now, in terms of physics, every single ounce of energy that goes into launch has got to be taken out in the heat of reentry. You look out that back window, and something is organizing that flame. It is not just flame like a fireplace or some other thing. It is organized in a disc. I cannot tell you how big this disc is. It's a circular disc, and it rotates something like an antenna on an AWACS [Airborne Warning and Control System] airplane.

It is unbelievable. Now, I don't know any of what's going on. All I can do is tell you the visual of what I see. I caught it on video too, the first time that it had been videoed out the back window. I can only think that shocks are organizing the flame. So you have a disc so far back, and I don't really know how far back because I don't know how big the disc was. You get this organized structure up there, and all of the fire comes back and feeds this structure. And then you get these horrendous pulses of blue lightning coming off. No one's spoken about this before. This blue lightning comes off the nose, and it goes right in and it feeds this disc up there.

I'm standing up. I have no cooling. I'm standing up on my parachute, my oxygen bottles on. You can't stand up during reentry, but I did. I made it. My buddies kept turning and saying, "You're an animal, Musgrave, you're an animal." Ha ha! What an animal! I'm looking at this stuff, and every time a rocket went off, it would blow the whole structure away. The whole structure would go zap! And then it would form again. What I really couldn't believe was that like an aurora, right exactly down the middle of the vehicle, the colors change from one side to the other. I had green on one side and pink on the other, just like an aurora phenomenon. I couldn't believe any of that, but I had a monitor, and I stick it in front of [mission specialist Thomas] Jones's face. I'm just talking about this stuff, and they were in disbelief too. But I would say, "Hey look, here's the monitor," because my video camera was feeding the monitor.

It is Fourth of July all the time in the shuttle. There is stuff everywhere, and it's rotating. It gets out so far, and it pulses at you. The light goes on and off; it rotates in the sunlight. You get ice chips that shatter, and all of a sudden they break, and they go in all kinds of directions. We know that even bits of grease, or this or that, makes an incredible light show out there.

Andrew Gaffney was a medical doctor who worked on a variety of projects as a visiting senior scientist with NASA's Life Sciences Division in 1987-89. Because of his expertise in cardiology, he was selected as a payload specialist on the first Spacelab Life Sciences mission in June 1991. STS-40, the eleventh launch of space shuttle Columbia, was a nine-day mission that carried the Spacelab module for Spacelab Life Sciences 1 (SLS-1). It was the fifth Spacelab mission, but the first dedicated solely to biology. STS-40 was also the first spaceflight that included three women crew members. Dr. Gaffney was inducted into the International Space Hall of Fame in 1996.

After medical school, I got a fellowship in cardiology from '75 to '77 at the University of Texas, Southwestern Medical Center in Dallas. When I finished my fellowship, the person I was working with had been doing research for NASA. The year I finished my fellowship, NASA had one of its first requests for proposals for space shuttle and Spacelab medical experience. So in July of 1978, Dr. Gunnar Blomqvist, a professor at Southwestern, and I turned in a proposal to do a series of medical experiments examining cardiovascular functions in weightlessness and then study the adaptation process on return to Earth's gravity. That went in as what NASA called "Announcement of Opportunity, 1978," and some 400 proposals from around the world were received. Ours was selected as one of the major experiments on what was initially called Spacelab 4, SL-4. From '78 to '83, we did definitional studies.

In the original announcement of opportunity was fine print saying if there were not crew members available from the regular astronaut corps, trained and capable of performing these experiments, they would consider flying scientists to perform the experiments. It was forced on them by the National Academy of Sciences, I think, much as happened in the Apollo program. It also turns out that a very well-known scientist at NASA, Hans Mark, who was the associate administrator at NASA, had insisted that there be a scientists' program aboard the shuttle. So in the spring of 1983, they announced that they had decided that they would have scientists fly the mission. I thought about it, and it seemed like the perfect deal because, unlike the career astronauts, who basically gave up

During the STS-40 mission, Andrew Gaffney collected medical data on not only the astronauts, but on thirty rodents and thousands of tiny jellyfish. Courtesy of NMMSH.

their medical careers, I could keep mine. I could stay a cardiologist and fly in the space program. That was kind of a win-win deal.

I applied, went through the selection process, and was picked as one of four candidates to fly on SL-4. Strictly by coincidence, things I had done in Dallas and in Copenhagen, lung physiology, muscle physiology, exercise physiology, cardiovascular physiology were a huge part of the flight. If I had designed a career, I couldn't have overlapped better with what wound up on that mission. So I was pretty comfortable that I would be selected.

They ultimately picked two of us. Then the second person developed a medical problem less than a year before the flight. What a tragedy! So they put a substitute in. I continued to be part of the crew, and ultimately flew. What was, on paper, a two-year contract to go to NASA, train, fly, do the experiments, and come back, lasted from January 1984 until September 1991. One never knows where fate will take you.

There was a tremendous gain in knowledge from the Skylab program. If you think about the focus, the early life science efforts were really aimed at survival and function. First, could a human survive in space? And then, were there showstoppers that would keep them from functioning satisfactorily? The operational questions. Virtually all of the experiments that were done prior to Skylab were operational in nature. They were done on a time-available basis. No emphasis or priority given to life sciences. It was strictly add-on, whereas in the Skylab program, the equipment was phenomenal, unparalleled. Space available, they flew physicians and scientists. You had excellent laboratory equipment, you had professional scientific crew, and you had the space and the time. You got the long-duration exposure.

I think that most of what we did following that, and the Soviets did, came right out of the Skylab book. That was really the first major gain in medical physiology in space. The problem was, though, it sort of stopped. Skylab was done in the '70s. So people knew that data, hypothesized a lot of things to fill in the gaps. When you don't have any data, you can argue until the cows come home, but you don't really know. I think we had gone on so long that we kind of thought we knew a lot of this stuff because we had said it so often, and there was nothing to challenge us that it wasn't correct.

There was a whole series of hypotheses that I think turned out to be, delightfully, incorrect. The best thing that can happen to you doing an experiment is to get something you didn't expect. It is sort of boring when you know what you are going to get, you do it, get it, and on to the next thing. It is really exciting when you do it right, confident of the data, and the data is totally different from what you thought. I think that probably best characterizes the Spacelab life sciences results. There were a lot of surprises, delightful ones.

This is an interesting, I think, untold story that at a time when [President] Jimmy Carter refused to allow our athletes to go to Moscow to participate in the [1980] Olympics, NASA and the Institute of Bio-Medical Problems, the Russian counterpart of Life Sciences, were doing human and animal collaborative experiments and exchanges on a regular and continuing basis.

When you get the actual working people who do the same thing together, whether they are Soviet or American, they generally can get the job done. They kind of put aside the politics, and in our case there is a universality of medicine. Human physiology is basic to mankind. Whether you are a Communist or a capitalist American, you still have the same cardiovascular function and pulmonary function, and you can cut through the crap. There were still games. There were things that they wouldn't tell us and sometimes held back. But by and large, I think we published everything we knew. There were no secrets, basically. The whole point was to discover, to do it in a way that is publishable, testable. So we exchanged a lot.

Probably one of the more exciting times, just after the Challenger disaster, there was a life sciences meeting in Houston. I had gotten, through back channels, a note from Oleg Atkov, who was a cardiologist who flew 237 days in space at one time. He had flown the first echocardiograph in space, performed exercise testing, had been the control subject for a noncountermeasure flight, which was pretty gutsy. And he had sent a note through a visiting professor at the University of Texas that he was going to be at this meeting, and could we meet? So I introduced myself, and he said, "Find me after the session." Of course, the Russians were involved, and there was not a lot of time during the session.

But we went back to his room and spent about two and a half hours sharing cardiovascular physiology thoughts and approaches. I told him about our experiments. He told me about his. He told me stuff that they hadn't published. We became friends, and to this day are friends. He has been out to my home and stayed with us. He has visited our university at Vanderbilt, and I have actually managed to visit his facilities in Moscow. So we had a pretty good exchange.

There were other things unpublished, where the Soviets had a kidney stone problem. If you think about it, kidney stones come from an underlying propensity to form stones, concentrated urine, high protein diet, and high calcium. Those are all things that make you more likely to form a stone, and they are all typical of spaceflight. They, in fact, had a crew member form a kidney stone, and they had to abort a mission. They made up some story about it. But what is not known is that we provided the protocols for screening the cosmonauts to identify stone formers and to encourage them to maintain a calcium-hydrated state, perhaps modify diet. That was done by getting the best guy in the US, flying him to New York, meeting with their docs, all secretly, no publicity, to share and to try to maintain the health of people who are doing risky things generally for the common good.

I think the Russians provided us a long series of opportunities on their biosatellites. We got a lot of animal data that we would not have gotten, and they got a lot of technical help. So it was a win-win. I think mankind benefitted from a collaborative effort. It was quiet and unpublicized, but very effective.

To do the training [for spaceflight], you try to, at least intellectually, develop a mind-set on how you do things when there is no gravity. It turned to have been amazingly effective. The other thing, which is probably more remarkable, is the rapidity with which one adapts to the weightless environment. My Russian friends had told me that the learning curve is probably flat after about twenty-eight days. I think that probably after about a week, it is already starting to get pretty flat. The first day, you are sick. The second and third day, you're feeling better, but it's still not normal. You still are being a little careful about moving around, particularly head movements. You are still not automatic and totally adept. You're forcing things back into Earth reference, vertical, up, down, and those sorts of things, and it is just not comfortable doing things upside down. If you're a scuba diver, you do the same thing. If there is something very close to the bottom of the water or low on a rack, the tendency is get close to the floor and work on it, rather than just flipping upside down.

Within about a week, I remember, we had one period where we weren't scheduled, where we actually could play around and just have some fun. I suddenly realized that I was standing on the ceiling. I am standing on the ceiling in a laboratory whose mock-up I worked in for seven years, and it wasn't upside down anymore. I deliberately went over onto the side of one of the instrument panels and just stood there and looked down at it, and it seemed okay. I didn't think I was standing on the wall. It was like in less than a week, I had internalized vertical, and it didn't make any difference.

I would think to myself, here I am floating in space, around the Earth, without gravity, or in free fall, and feeling perfectly normal. Somehow, I thought I shouldn't feel normal, but I did. I felt like I belonged there.

I can tell you a story about it. When I was at NASA headquarters prior to the flight, I was on the committee with Sally Ride [a physicist who in 1983 became America's first woman in space] to look at a Mars mission. One of the big questions for these long-duration missions is whether you have artificial gravity with these countermeasures. I remember thinking at the time that this just wasn't very rational. If we could handle it engineering-wise, it would be a lot easier to spin the thing up and have an artificial gravity. Sally said, "You don't understand, but you will." I accepted it at face value. I respect her. She is a bright lady. So I kind of filed it away. And I was sitting up in the flight deck, looking out over the tail of the shuttle at the OMS [orbital maneuvering system] pods, and had this incredible sense, first of all, bonding with a piece of equipment, which was strange. I mean, this was our vehicle, and I was floating there. I remembered what Sally said, and I knew exactly what she meant.

It is so nice, just floating there. In some sense, it converts you almost to a spiritual quality, rather than this physical thing that has weight. You know, it has to be dealt with, just like it isn't there. That feeling, I am sure, is what most people feel when they are in space and would resist aggressively an attempt by engineers to pose an artificial gravity when you are lucky enough to be in a place where you don't have it.

It's not an organized religion thing. We use a phrase in medicine, an out-of-body experience, if you don't have the appropriate receptive feedback. I actually did some playing around with this. I was taught that to go sleep, you can do progressive relaxation techniques. You teach patients to do it. I have found it really helpful with sleeping in motels 250 nights a year for several years. It just helps. So I was trying to get to sleep, and I was relaxing. I realized that I was feeling in my arms and legs what I had always interpreted to be sort of a weight. But I said, that can't be. It must be that I never got down to the tone, you know, where you actually are maintaining tone, that causes stretch, and you get feedback, and that is how the human body works with the appropriate reception and feedback.

So I just concentrated a little bit harder and made it go away. All I had left was, I could see a little bit of movement with each heartbeat. Your central nervous nulls out the bounce of the body caused by your heartbeat. All of a sudden, you become acutely aware that your body is moving every time there is a heartbeat. There is a mass that is projected. In my case, it's five ounces that is projected from my heart every time it beats. That's not a lot, but if you are in a microgravity environment, you sense it.

So I got rid of the tone and basically got down to just being and seeing and thinking – in a sense, not having a body. I think when you don't have a body and you are still there, I guess, that's the soul, the intellect, or whatever it is.

There is something incredible about floating, or being weightless. Floating implies buoyancy, but there's no buoyancy, just weightlessness. And yet you're sitting there, you're breathing air, you're at 70 degrees, and you are just looking at the Earth go by. It is a pretty incredible feeling.

In terms of medical experiments, I think that men and women are pretty similar in space. The areas that one would expect differences, I think, are in terms of fluid shifts, simply because of vascular permeability. Women get edema, swelling, more than men do. On Earth, I think that there may well be some differences in how fluids are distributed within the body among men and women. Calcium metabolism is different. And I think there is clearly a concern in long duration flights in protecting women more aggressively than in men, but that, in part, has to do with body size. The petite, lightweight, female couch potato, or sedentary person, is much more likely to get osteoporosis than say, Shannon Lucid [a female American biochemist and astronaut, who flew on five space missions, including a 188-day mission to the Mir space station], who is much larger and athletic. She puts enough load on her skeleton that she should be relatively protected. Some of the other members of the astronaut corps, much smaller, may well be at more risk. But those are things that can be looked at. They are certainly not showstoppers.

Chapter 15
Eileen Collins

Eileen Collins earned her private pilot's license while she was in college. After earning a Bachelor's degree in mathematics and economics, she enlisted in the Air Force. Collins became a NASA astronaut in 1991. She flew four space shuttle missions, two as pilot and two as commander. In fact, she was the first woman pilot and first woman commander of a shuttle. She spent a total of more than 537 hours in space. Collins was inducted into the International Space Hall of Fame in 2001.

After I graduated from Elmira [New York] Free Academy High School in 1974, I went on to a community college for my first two years because was I was really needed at home to help out with the family. At Corning Community College, I got an Associate's degree in math and science in 1976. Then I transferred to Syracuse University and graduated in 1978 with a degree in math and economics. From there, I went on into the military. But as a lover of education, I continued with my schooling. The Air Force sent me to Stanford University, and I received a Master's degree in operations research in 1986 from the School of Engineering. And then in 1989, I received a Master of Arts degree in space systems management from Webster University. That degree, I earned by going to school in the evening for two years.

One of the reasons I selected Syracuse, other than the fact it was a very good school, was that they had a ROTC program. I was looking for that, because I knew during my sophomore year in college that I wanted to go into the military. I wasn't really sure what I specifically wanted to do.

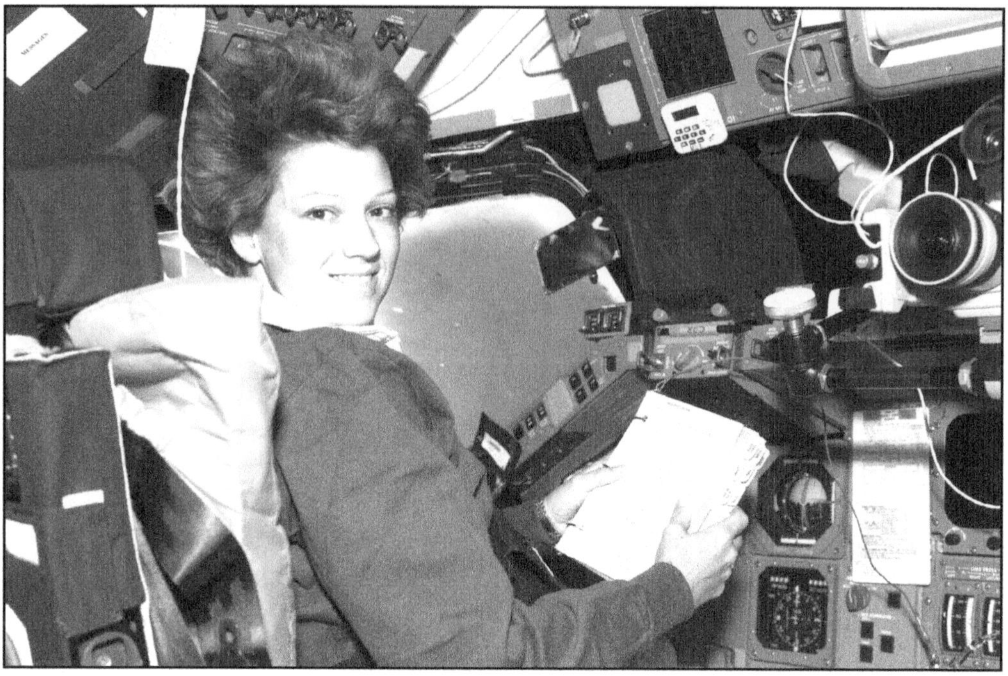

Eileen Collins in the commander's seat of the shuttle Columbia on the STS-93 mission in July 1999. Courtesy of NASA.

I just wanted to join the military, and the ROTC, at that time, was the way for me to go. The military academies were not accepting women back in those days, so a military academy was not an option for me. That pretty much left ROTC. I was very happy with the ROTC program. It offered exactly what I wanted.

The first [airplane] flight I ever took was on a commercial airliner with my mother when I was nineteen. I started flying the following summer at age twenty, a little Cessna 150. I took lessons, because I just said, "You know, this is what I want to do." So I saved up my money, and at the Corning Regional Airport in Elmira, I visited their flight operations school and signed up for classes. That is kind of how I got started. The Air Force called me two weeks early. I went off to the Air Force just one check ride short of getting my private pilot's license. But after I graduated from pilot's training, I took that check ride, and finally got my private FAA ticket.

I was sent to a screening program, which at the time was at Hondo [Air Force Base], which was a small airfield just outside of San Antonio, Texas. It was a very, very good training program in preparation for pilot training. I think we flew about fourteen flights in a Cessna 172. This was back in the days when we had Iranian students learning to fly here in the US. So we got to fly and learn almost side by side with the Iranian students. They were in a different "flight" [group]. My flight called ourselves "the Guards and the girls." We had four women and ten men. The four women were going on to active duty, but the ten men were all preparing to go through pilot training and then go back to their Guard units. There were two other "flights" with Iranian students.

After Hondo, this would be August or September 1978, I went to Vance Air Force Base in Enid, Oklahoma. I was one of the first four women to ever go through pilot training at Vance Air Force Base. I should say that there were at least three classes of women that had, in the previous two years, gone through training at Williams Air Force Base in Arizona. But the program was getting expanded. The women were not just going to attend Williams, but now they were going to attend all the training bases that the Air Force had. So I was one of the four women that started at Vance. It was quite interesting, because we attracted a lot of attention there at Vance Air Force Base. Nobody knew what to expect with these women flying. But I would have to say that we pretty much flew very similar to the guys. That was over twenty years ago now [Collins' interview was conducted in 2001], and women have been flying in the military ever since.

The Air Force, before going into pilot training, they have height requirements, both standing height and sitting height. I was right at the minimum on the sitting height, although I was several inches above minimum on the standing height. It was important that the pilot could hold the brake during the engine runup just prior to takeoff. And it was important that you could fit into the ejection seat and safely eject. I really didn't have any problems with any of those, as far as sitting in the airplane and flying the airplane. We did have some gals that were right at the minimum, and they did have problems holding the brakes. What you had to do was to discover a technique to pump up the brakes, maybe lift yourself up out of the seat and use your toes and press very hard. So, you could work around it.

I also found it was difficult to lower the canopy if you were facing into the wind. So I learned a little technique for putting my elbow up against my waist and pushing the lever that way. This was in a T-38. The T-37 canopy is electric, and you just push a switch. But the T-38 canopy is a lever that you use to manually lower it. It took awhile, but I learned little tricks for flying the airplane. I was fortunate that I was able to fly the T-38. Back in the early 1980s, women could not fly fighters, or bombers, or any aircraft that could be considered a combat aircraft.

I got my Air Force wings in September 1979 with Class 7908 at Vance Air Force Base. I am just very proud of the fact that I went through pilot training at Vance with a great group of people. There was a lot of camaraderie. It was in a relatively small town, so I found that people got to be, I would say more like a family, and got to know each other pretty well. That was a nice atmosphere to be in. I liked it so much that I volunteered to come back after graduation as an instructor at Vance. I was fortunate enough that I was assigned as a T-38 instructor pilot to the 25th Flight Training Squadron. I stayed there for almost three years as an instructor pilot and a check pilot.

Randolph Air Force Base is the Pilot Instructor Training, which we called Pit, PIT. That was about four months. My very first day at PIT was the day that the hostages were taken in Iran. It was in November 1979.

The three years I was at Vance, I was the only woman who was an instructor in the squadron. The month that I left permanently was the month that the second woman instructor came along. There just really weren't that many women, so you just kind of stood out. The years I was there, I was very careful that I did a good job so I could really leave a good impression for the women to follow. I didn't want the women to have a hard time of it. I wanted to make it as easy for them as I could.

When I left Vance in December 1982, I went to Altus [AFB in Oklahoma] for three months of training in the C-141, followed by an almost three-year tour at Travis Air Force Base in California. I just loved the C-141. I was a co-pilot, aircraft commander, and an instructor pilot in the C-141. It is interesting that it was not my first choice. I had asked to fly fighters. I wanted to fly the F-15 and the F-106, which was getting ready to be phased out at that point in time. Obviously I couldn't do it because of the restriction on women in combat, but I got sent to the C-141. It was a wonderful job. Looking back, it was one of the best jobs I've had, an opportunity to fly with a crew around the world. And with the missions, I have been to some very interesting places. I've got to see how people live in countries around the world. I was able to fly in part of the Granada operation in 1983 and returned some of the civilians that were medical students that were held hostage by the soldiers. We brought them back to

Charleston. That was the most important mission that I have flown in my career as a military pilot.

There was one woman who was an instructor prior to me in the T-38. She was at a different base. I was the second T-38 woman instructor in the Air Force, but the first at Vance Air Force Base. I only remember one occasion of resistance, but before I say what that was, let me say that the vast majority of the students were just great. I think once you get in there and you start actually doing your job and you start working, you show them that you love what you are doing, you really know your job very well, and the instructors' goal is to make their students the best possible pilots that they can be. You tell them that. "I want to make you the best pilot that you can possibly be." It becomes a team. That was the attitude that I had. I found that I had a very good relationship with my students, even the students that I just flew with on a one-time basis.

There was one time that I had resistance, and it was very obvious. When I was a check pilot, in my last year at Vance, the only thing I did was give check rides to students. We had a student from Saudi Arabia that was scheduled to fly with me on a check ride. I was going to give him a grade at the end of the check ride, which was a big deal. When I saw his name on the schedule, I told my scheduler, "Please don't fly me with him, because they prefer not to fly with women because of their culture." My scheduler said, "Eileen, this is the USA. We are going to do it this way, and I don't have anyone else I can fly him with." When the student, who was actually flight commander, saw his name on the schedule, he asked to be changed and it got changed. I have no hard feelings about this. By the way, the student did not do well on his check ride. He had to take it over again after he had flown with this other instructor. I guess that isn't important. I think what was important was that, you know, I am not overly sensitive to people's feelings about whom they fly with. I just focus on the mission. I do the best job that I can. And I have found that people will come around. They really will.

One time, I had a colonel who I just happened to be on a team with down at Randolph Air Force Base when I was a first lieutenant. After we were done working together that week, he said to me, "You can fly on my wing any time." I will never forget that he said that. I looked at him and I almost said, "Well, you can fly on my wing, too." But because he was a colonel, I didn't say that.

I will have to say that we were taking baby steps back then getting into flying. I didn't want to make a great big statement. I didn't want to go in there and change anything. I just wanted to fly. That's all I wanted to do. I was very mission oriented. I loved teaching. I loved being with people. And more than anything, I loved flying. I think I just did my job. In 1993, when this country finally lifted the ban on women flying combat aircraft, although I didn't have the opportunity to turn around and go fly those missions, I feel like I did something that helped that barrier come down. And I feel very thankful that I was able to help in regard. Now, that doesn't mean that every woman is going to be forced to go into combat. What it does mean is that women who want to fly and want to go into combat and are really good at it, can do it if they want. It is a very small number of women that are flying in combat aircraft right now, but at least they have the chance.

I very much wanted to go to test pilot school, and I was looking for another job that would help get me in a good position to go to test pilot school. I wanted to get a degree in engineering so I could better serve my job. I found the opportunity to teach at the Air Force Academy, so I applied for this position. The math department at the Air Force Academy hired me. They have a program where they send you to school to get your Master's degree. So they sent me out to Stanford for one year. I could have done it in eighteen months, but I did it in eleven months, because I was very anxious to get back to flying. So I doubled up my courses, got my degree at Stanford and taught at the Air Force Academy for three years. I taught all the calculus courses: differential calculus, integral calculus, multivariate statistics. I taught linear algebra, and I also taught one semester in the Aeronautical Engineering Department.

I flew the T-41, which is a Cessna 172, in the screening program at the Academy for cadets. That was three years. While I was at the Academy, I was applying to the Air Force Test Pilot School. Just as a side note, I had to apply three times to get in. The first time I applied, they wouldn't let me meet the board because I hadn't spent two years' time on station at the Academy. So I applied again the following year. They called and said I couldn't meet the board because I was on a three-year directed duty assignment. So finally, the very next year I said, "If I wait another year, I am going to be too old." Well, they gave me a waiver so we could go over ten years. I think it was after ten years and three months' time in service, you are too old. They gave me a waiver for that. I only needed like a one-month waiver.

I did get accepted into the test pilot school, and I started in June of 1989. I was a major. I became the class leader because I was the senior ranking officer. It was myself, twenty-three guys, and one other gal. She was a civilian engineer. In my opinion, it was a very big challenge, being the class leader with all these guys. I was kind of wondering how it was going to go, but it turned to be a great year. We had a great team. I thought we did a good job. Everybody graduated and went on to their jobs. I graduated in June 1990.

I had applied to the Air Force Test Pilot School and the astronaut program at the same time. This was in late 1988, early 1989. The acceptance process for the test pilot school went rather quickly, but the astronaut selection process takes a while. My second month in test pilot school, it was September 1989, NASA called me and said, "We want to interview you." They go through the interview process every two years. I had to go to my operations officer and my school commandant to ask permission to miss a week of school so I could go to NASA and interview for the astronaut program. They were wonderful. They let me go. I had to promise to make up all that work, which I did.

The first week in October 1989, I went to Johnson Space Center and interviewed for the astronaut class of 1990. Then on January 16 of 1990, NASA called me. I was out flying the A-37 on a spin ride when the call came in. I came back from my flight, and there was a note on the board to call NASA. I thought, "Well, this is it. They are going to tell me I'm in or I'm out." So I called NASA, and the secretary connected me with Duane Ross, who is the administrative lead for the astronaut board. He connected me with Don Puddy, who at the time was the director of flight crew operations, and he was the president of the board. Then Don Puddy connected me with John Young, who is, as you know, a famous astronaut. I thought, "Well this is leading to something." And John Young said to me "Well, Eileen, do you still want to come and work for us?" I said, "Yes, I would." And he said, "Well, we want you to come down in July. You are going to be in the class of 1990."

Of course, I don't remember anything he said to me after that. I thought, "Well, I finally made it." It was like a dream come true. But at the end of the phone call, he asked me if I had any questions. Now, I had applied both as a pilot and as a mission specialist. So I said, "Well, am I going to be a pilot or a mission specialist?" He said, "You are going to be a pilot. You are going to be our first woman pilot." So another challenge that I had to face. But I remember that was just a little bit past half-way of my test pilot school, and of course I was highly motivated to finish the course. We graduated on June 8 of 1990, and I reported in at NASA on July 16 of 1990. So the timing worked out. Throughout my career, my timing never worked. But that was the one time in my career where my timing was good.

The astronaut training program was twelve months; actually, it turned out to be about nine months, because we went a little bit quicker through it. There were twenty-three of us in the class. It was at Johnson Space Center. We started July 16 of 1990. We did space shuttle training. We flew the T-38. We did, obviously, classroom, simulators and travel to every NASA site to learn our mission. And we had what they called enrichment training, which included survival training, and a trip to New Mexico on a geology field trip. We had speakers come in to Johnson Space Center to talk to us about the history of the space program, former astronauts, flight directors. We had many, many classes on Earth observation, and just a variety of subjects.

We did full-time training for nine months, probably 90 percent of that was learning the space shuttle, and learning the mission of the space shuttle and the duties, the responsibilities in your particular position. At the nine month point, we picked up jobs. All the astronauts, if you are not training or flying your mission, you have a support job, where you help the other group get ready for their flight. So before your first flight and in between flights, we do that. We picked up our jobs at the nine month point, and we also became available to do speaking engagements. That is when we got very busy.

This may be a good point in time to tell you the various jobs that I have had in the office between my flights. My first job was in space shuttle engineering. I worked with the full-time engineers and flight controllers on the shuttle system from April of '91 until February of '92. Then in February of '92 I became a, well, the slang we use for this job is "Cape Crusader," where astronauts fly to the Cape and support the people down at the Cape getting the orbiters ready to fly. We do testing down there, and we also support the crews that are going up next. We check out the space suits for them. We strap them in, prelaunch. When they come back and land, we go in and get them out of the orbiter. It was just a super job because I got to work in the actual space shuttle itself. I did that job for fifteen months.

Then I became the CapCom. The CapCom is the astronaut in mission control who actually talks on the radio to the crew in space. I only did that job for about three to five months. In September of '93, I got assigned to my first flight. My first flight was STS-63, where our crew rendezvoused with the Russian space station Mir. The six of us crew, actually, five of us were Americans. We were the first Americans to see the Russian space station. We didn't dock, but we did a close approach rendezvous to 30 feet. Tested out the shuttle's navigation, communication, and flying systems to make sure that the subsequent flight, which would be the first docking, would be ready to go. I flew that mission in February 1995.

When I came back from that job, I went back to the CapCom job. I did that job until I was assigned to STS-84. That was the sixth shuttle mission to actually dock with the space station Mir. That mission flew in May of 1997. We were the shuttle crew that docked with Mir in between the two major accidents that they had. In March of '97 they had a fire, and in June of '97 they had a collision with a Progress vehicle. So we were up there in between those two. We got to be friends and co-workers with that Russian crew in space.

I came back from that flight and became the chief of the space hardware branch in the astronaut office, which included the space station and space shuttle. At this point in time, the space station was really growing in the development phase. The work was picking up, and we had many astronauts supporting all aspects of space station development. My branch was pretty much the hardware development. I did that job for ten months. Then in March in 1998, I was assigned to STS-93, which was a mission to take up the largest and most powerful X-ray telescope that had ever been launched in space. Initially, it was called AXAF, which is Advanced X-ray Astrophysics Facility. Later it was named Chandra, the Chandra X-ray Observatory. I was so proud to have the opportunity to be part of this mission. I was the commander. We had a crew of five, and our job was to safely deliver this telescope to space. The mission had many delays because it seems that whenever we found something that just wasn't perfect, the decision was made to fix it, which was the right decision. It turned out that the Chandra X-ray Observatory was almost one and a half billion dollars. We didn't want to make a mistake. Any potential problem that could have happened was fixed.

This mission actually flew in July of 1999. It was a tremendous amount of publicity around this mission, both before and after the mission, because of the first woman commander. After the flight, I traveled to places and met people that I never dreamed that I would have a chance to meet. That was my last flight, and that was two and a half years ago, now. Since then I had my second child. I didn't mention I had my first child in between my first and second flights. Then I had my second child after my third flight. So it has been a while since I have last flown, but we have a very, very large office right now. There is a much longer waiting time between flights. I am going up on my fourth flight. Right now, it is scheduled for November of 2002, although that mission most likely will delay a couple of months to early 2003. [Because of the space shuttle *Columbia* disaster in February 2003, Collins' flight STS-114 did not launch until July 2005.] That will be a docking with the International Space Station. We will take up the Expedition 7 crew, which is the seventh crew to live and work aboard the space station. And we will bring back Expedition 6 with all their logistics and their science operations.

My first mission, I was the first woman pilot of the shuttle. People often ask me what that was like. I would have to say that it was probably like flying as the pilot on any other mission, as far as the training and my day-to-day activities. I really didn't feel like I was singled out, or that there was anything different because I was going to be the first woman. But within my mind I knew I wanted to set a good precedent for the women that followed me, which was the way I had felt my entire career. It was a little bit easier at NASA. There were women specialists that flew in space before me. They are such professionals and they had all done such good jobs, they paved an easy road for me. So I would have to say that within NASA you do your day-to-day job; you don't go around thinking, "I am a woman, and I am different than the majority here." It is not that way. Everybody is so mission oriented that you are always focused on the mission. You don't care, what does this person look like that I am working with?

When I flew my second flight, there was[n't] much publicity, and much less requests to speak and travel after the flight. Now, on my third flight [STS-93 in July 1999], just to give an idea, I was the first woman commander. The amount of media attention around this was probably three times greater than when I flew as the first woman pilot. The mission was announced at the White House. I actually flew to the White House. I went into the Oval Office with my husband, and we met the President [Bill Clinton] and the First Lady. NASA Administrator Dan Goldin was with us. We had a meeting in the White House. Then we went into the

Roosevelt room, where the announcement was going to be made. The First Lady spoke, the administrator spoke, then I spoke, and the President spoke. The room was packed. There were reporters out there that I had seen on TV my whole life. There were famous women. Sally Ride was there.

I remember the feeling that I had when I peeked into that room, before I walked in to speak. I peeked into that room and I thought, "There is no way that I am going to be able to do this." And I thought, "These people are here for me?" Then I just stopped myself, and I said, "No, they are not here for you, Eileen. They are here for the first woman commander. Just go in there and be the first woman commander." I went and I made my speech, and everything was fine. Then I went home.

I am not the kind of person that seeks out publicity. But on the other hand, I don't shy away from it. I feel like I have a sense of duty to do some of these things. Well, after that ceremony, the next day I went back to Houston, and got into training. Training went just fine. The mission was extremely successful. The flight went off, basically, without a hitch. We had a few small things that broke on our mission. We had to fix a camera, and we had to fix an experiment that was overheating. But for the most part, the flight went really, really well.

We came back on the sixth day, and as we flew across the state of Texas at about 10 p.m., we left a streak across the sky. People still talk about this. There is a plasma that builds up around the orbiter as we slam back into the atmosphere, due to the air molecules that were hitting the shuttle. As heat is building up with the friction, and the orbiter jets are firing, it leaves a plasma which, if the conditions are just right, will light up the sky. It did. We came in and we landed.

Of course, I was very conscious of the fact that I was going to be the first woman that was going to land the shuttle. The women pilots across the country were always reminding me that they were really looking forward to me making a great landing. Nothing like pressure! But we came in, and I did what I was trained to do. We landed and I got out of the orbiter, and I had no idea what I was going to be facing. Probably four to five months after that, I did events like the Jay Leno show. We did Oprah Winfrey. We did the Regis and Kathy Lee show. We went to the World Series. I got to ring the opening bell on the New York Stock Exchange. I can't even think of half the events that I did. I went to Manhattan seven times for events that were unbelievable.

It was difficult for me having a child at home and doing all of this traveling. But I knew it was going to last for only a couple of months. And I knew it was an opportunity for me to talk about the space program, to get people refocused back on the space program and what we were doing, in the fact we were getting ready to build a space station. Eventually, someday, we are going to get people back to the Moon. Someday, we are going to get people on Mars. This was the message that I had. Also getting young people, both boys and girls, interested doing well in school, studying math and science, and preparing themselves to be part of the space program someday. Those were the messages that I wanted to take out. To me, it was a great opportunity to get those messages out to young people, and I really took advantage of that.

Then probably six months after that mission, I became pregnant with my second child, and I had to slow down and stay home. Most of my activity was working a full-time job in the astronaut office. I had to turn down all speaking requests, but I still, to this day, accept all of the requests for interviews on the Johnson Space Center, and even our telephone interviews. I still do all of those. Again, mainly, trying to reach young people and get the message out about how exciting the space program is and things they can do and have a career in the space program.

I have a five-year-old girl and an almost one-year-old boy that are just my pride and joy. They are my number one priority. I am a mom just like all other moms. I think being a mom is really great job. The last time I flew in space, my five year old was three and a half. I know she was very aware that I went up on the space shuttle, and she knows where I sit on the shuttle. She knew what clothes I wore and what food I ate. She knew I took her picture with me. But I think that her attitude was that everybody's mother flies on the space shuttle. She didn't realize that it was something different.

My next flight, she will be seven. She is going to have a much different perspective of the mission. Of course, my boy will be only two for the next flight. It is kind of a difficult thing to say goodbye to your kids. I remember saying goodbye to Bridget when she was three and a half. I knew I wasn't going to see her for thirteen days. I had seven days of quarantine and then six days on the flight. But because we had a three-day delay, it stretched out for sixteen days. It is tough being away from your kids for so long. But parents go through that all the time, being separated from the kids for whatever reason. I think at that time, it strengthened our relationship when I came back. It was just really neat being back with her again, and I took a little bit of time off and spent some time with her.

We have a training aircraft for the space shuttle. It is a Gulfstream II. It has been modified to dive at the ground with the same performance characteristics as the shuttle, and the flying qualities have been modified in the software. So what you feel as you fly, or make input to the control stick, is the same as what you feel as you fly the shuttle. We even fly the airplane suited up in our launch and entry suit. As you train for a flight, there are different rules for how many flights you need to have. Right now, I need to fly once a month, but when I get within six months of the launch, I will fly every two weeks. Then at three months from the launch, you have to fly once a week. In quarantine, you can fly every day if you want, three, or even four days before your launch.

[During a shuttle landing] the commander and the pilot each have the controls. The pilot is there to back up, normally. In fact, every shuttle flight to date, the commander has flown and made the landing manually, with the pilot backing up the commander. The pilot's job is to put the gear down and to deploy the drag chute, and various other things. If you are having problems, the pilot can work the speed brake. The pilot also makes the standard mandatory calls as you fly down the heading alignment, the final approach in landing. The commander's job is to put the "airplane" where it needs to be and land it.

The length of the deorbit burn depends on your altitude. Normally, it is somewhere between two and three minutes. You turn the shuttle around backward, and you light your OMS, Orbital Maneuvering System, engines. I have seen burns as low as two minutes. Our last flight, we were 150 miles [high], deploying the Chandra Observatory, or you can be up at 350 miles with the Hubble Space Telescope. In those cases, you can have a burn as long as three minutes.

After you do the burn, you flip the orbiter back around so the nose is into the flight path, and as you enter the atmosphere, you attain a 40-degree angle between the shuttle's longitudinal axis and the air stream. You fly down with a 40-degree angle of attack until you get very low in the atmosphere, and then you become an airplane. Then you start lowering that angle of attack. You manage your energy returns and the speed brake. When the orbiter goes subsonic, which is about 60,000 feet, the commander will take over manual control and fly around in a circle, which we call a heading alignment cone. It is not really a circle, but if you looked at it on a map it would appear to be a half circle of 180 degrees or a full circle of 360 degrees, or anywhere in between. We usually fly that thing to help us manage our energy. If you are too fast, you can fly a bigger circle, and if you are too slow, you can fly a smaller circle, to ensure that you hit the runway at the right spot.

I have had a lot of really neat experiences in space. They are not really strange, but your body is adapting to zero gravity. It starts immediately. When the main engines cut off at eight and one half minutes, things start floating. You can take your pencil out, hold it front of you, let go of it, and it's floating. You think, oh my gosh, there really is no gravity up here, like they told me my whole life. Then you look out the window and you watch the sun come up, and it is just breathtaking, the colors and looking down at the Earth. It is just an experience I can't describe in words.

Back to the human side of it, you feel your face is very fat. We call it fluid shift. Your legs get skinny. Eventually, and you are going to feel this immediately, your helmet is floating off, your suit is floating up on you. You get out of your seat and you float downstairs, you take your suit off and you put on your shirt-sleeve type clothes. The first thing you feel in space is clumsiness. You are going to bump into things. You are going to feel a little bit disoriented. You are going to see your crew members floating by you upside down. It takes a little while to get used to this. I found, on my first flight, it took me three days to get to the point to where I was able to zip around and work as efficiently as I can on Earth. It is almost like getting on roller skates and trying to do work around the house on roller skates. You are just having some problems until you get good at it.

The ceiling is still the ceiling, and where the floor was is still the floor. Although as you get more experienced up there in space, you can start using the volume in different ways. For example, I discovered that the best place for me to sleep was on the ceiling. I tried sleeping on the ceiling with my face facing the ceiling and with my face facing the floor, or laying on my side, so to speak. So I always tied my sleeping bag to the ceiling in the middeck and slept there. The shuttle cockpit, flight deck, and middeck look very small when you walk in the trainer on Earth, but really when you get up in space you have more volume than you think you would, because you can walk on the ceiling, for example.

When people ask me about being in space, I tell them the two things that really are the human side of space are just physically being in space, maneuvering and operating and working in zero gravity; and the second thing is the view out the window. As I look back and look at these mountains over here, to me they are beautiful. It is hard to describe that to someone that has never seen it. You could say they are brown, you can see the geological formations and the blue sky. It is difficult to explain the views as you look down on the Earth. It is, to me, beautiful, breathtaking.

The most stunning view I can remember is the Middle East. On my second flight, we had a pass over the Middle East and I actually had time off, where I could look out the window for about fifteen or twenty minutes. The sun came up over Greece. We were flying over Europe, and I looked down and you could see Europe was nighttime, and you could see the coastline because you could see all those city lights. The Mediterranean Sea was dark, and Africa was mostly all dark. You couldn't really make out the coastline of Africa. But as we flew over Greece, we were flying kind of towards the east, and towards the southeast you could see what we call the terminator, which was the line between the darkness and the light. You could see where sunrise was taking place on Earth.

Then we flew down over Israel, and the Sinai Peninsula, and Egypt. You could see the Red Sea, the Gulf of Aqaba, the Gulf of Suez, you could see the Suez Canal, the Persian Gulf, Iran, Iraq, just looking down at this area it was a stunning contrast between the tan desert and the very, very blue water. You think about all of the Biblical history and the wars that were fought and all of the people that lived there. It was just incredible.

Then we flew down over the Himalayan Mountains over India, down across the Indian Ocean, and over Australia before I had to go back to work. I remember that scene so well. I hope I get a chance to see that again, I would like to film it on one of my future flights. The other thing about looking at the Earth, there are little islands that are like little gems. Many, many of these little, small islands in the Pacific Ocean that are just so pretty, and the coral reefs. The Pacific Ocean is absolutely huge. I don't think you ever realize how big it was. You realize the Earth is very much a water planet. There are a lot of deserts out there, and there are a lot of mountains out there. Things you don't really think about in your everyday life.

People sometimes ask if you can see the pyramids from space, or you can see the Great Wall of China from space. I have never been able to see them, even looking for them. But if you get a camera with a large lens, something where you can just zoom in, or even a pair of binoculars, you may be able to see the pyramids if the sun is at a low angle and casting a shadow. Can you see manmade objects in space? In a way, yes. You can see agricultural patterns. Sometimes, for example the country of Israel, on the southern border, you can see it is darker, and there is a straight line that shows on the northern side. On the northern side you have agriculture, and on the southern side you have desert. You can actually see borders between countries.

You can see fires, looking down on Earth. These are probably area fires. We know there is burning going on in the rain forests and in various parts of the world for development. I will also say, the Island of Madagascar, which is off the east coast of Africa, I don't want to say it is totally cleared, but when you look down on it you see all kinds of erosion and the rivers are not clear water. From space, water will look kind of a grayish color, but if there is erosion it will be brown, maybe a light brown. The rivers in Madagascar, for example there is a Betsiboka River delta, which is very brown from space. You can see all of this sediment that is being brought down from the mountains when it rains, because the forests have been burned away.

On my second flight, I photographed the entire island on several passes we had made. My goal was to have these pictures available to scientists who were trying to document what was taking place in the environment there. You could see the same thing in South America, which obviously is much bigger. It would be hard to photograph the entire thing from close range. But if you take pictures of various regions from the Skylab program back in the 1970s and compare them to pictures of today, you could see there is so much jungle that has been cleared away. You could also, during the day, see the smoke from the fires. At night you could see the bright lights. Anytime you are up there, there are going to be fires. These are areas of concern.

What we can do, as astronauts, is we can document that. Before we fly, the Earth scientists, geologists, meteorologists, oceanographers will tell us here are areas we want you to photograph, and this in particular is what we are interested in. And they can show us slides of previous pictures. Then during the mission, we will get a message uplink. Every day we will have dozens of sites on Earth and the time we will be over them, and they ask us to take pictures of this particular [area], with what kind of camera and what kind of lens, and even with what light settings.

Oh, another thing I should mention. At night, at any time of the year, if you are over the equator, at night there will be thunderstorms, and they are just beautiful. Those lights flashing! You look down on those thunderstorms, and it just quite a lovely sight to see.

I know the shuttle-Mir program is going to hold a special place in history. The program started, really, before the first flight. The first flight started in 1995. In February of 1995, I was the pilot on that mission [STS-63] that did a close flyby to Mir. We rendezvoused at 30 feet. Our job was to test out all the equipment and the procedures. That paved the way for the first docking flight, which took place in June of '95. After the first docking flight, we flew several more missions. We took astronauts up to Mir that spent anywhere between three and six months on the Mir space station.

The reason we did this program was the Mir space station was up there, an opportunity for our astronauts. And NASA as an agency had to get experience at long duration spaceflights to help us get the International Space Station program started using real experience. I am so glad that we went to the Mir program because it made us so much better prepared for the International Space Station. Looking back, I have to say that the shuttle-Mir program was a tremendous success, despite the fact there was so much controversy going on. People outside of NASA were saying we should not send our astronauts to Mir because it was unsafe. There were leaks aboard Mir. The space station was old. They actually had a fire, and they had a Progress resupply vehicle that collided with the Mir space station. In that collision, there was depressurization. We completely lost the Specter module because all of the air leaked out of it. It took awhile for the astronauts and the cosmonauts to isolate the Specter to get the hatch up there to prevent the loss of all the air from Mir.

Two close calls and an old, ageing spaceship. It was easy for people to say, "Let's pull our Americans out of this program, because we are going to hurt somebody. We are going to kill somebody." But NASA looked very, very closely. We had several independent groups take a look at what we were doing with the Russians, what the Russians were doing with their Mir space station. Is it really technically safe to be there? Technically, the decision was made to continue the program, and we were able to do so safely. In that program, we built a working relationship with Russians which has lasted successfully to this date. Not without its bumps and bruises, but in the end it has been successful.

Let me say something about my flight, STS-63, which was on

Discovery and launched February 3 of '95. Since Apollo-Soyuz, we had really worked with the Russians in the space program. I think it was in '75 that we had Apollo-Soyuz. So it had been twenty years. After we had main engine cutoff, we had three jets fail. Two of the jets just failed off, but one of the jets failed because it was leaking oxidizer into space. That jet was on the right OMS [orbital maneuvering systems] pod and it was facing in an up direction. When we approached the Mir space station, we approached from the same direction that this jet was leaking. The Russians were afraid that these particles were going to attach to their space station, and even, worse, attach to the solar arrays and degrade the energy potential of their solar arrays. I don't blame them for being concerned about it. So after our first day in space, we got a message that there was a chance we wouldn't do our rendezvous. That was something we had been training for a year and working with the Russians for a year.

Well, work went on on the ground. We were up in space and talking to the cosmonauts on Mir on our radio. But on the ground, our managers were talking about the pluses and minuses that could be done. We had gone through several cycles on the jet. We were closing the manifold, aiming the jet toward the sun, trying to fake it out to get it to stop leaking. Well, by day four, which was our rendezvous, the leak had been slowed to a point where it was just minuscule, if any leak at all. The Russians agreed to let us go in, and do the close approach and rendezvous.

We actually found out early, because the cosmonauts on the Mir called us on the VHF radio and said their manager just told them we were now cleared to do the approach and rendezvous. And then a couple hours later, Houston called us and said, "Hey, you are now cleared to do the approach and rendezvous." "Yeah, well the Russians have already told us. So they beat you to it." I would say that was one of the great successes of my flight. Not the fact that we did the rendezvous and everything worked, but the fact we had a failure and we were able to work with the Russians to overcome this failure and make the mission happen anyway.

Another interesting thing that I should mention, which goes back a year before the launch, was when our management at NASA and the Russian management agreed to even do this mission. The original plan was that we would go to a thousand feet [of Mir], I can't remember how many meters. The Russians were talking in meters and we were talking in feet. But this gradually kept getting changed. We would negotiate with the Russians to go in closer. We were really glad that we agreed to a thousand feet, but we really were not going to learn that much by going to a thousand feet. We got to get in closer. So we negotiated for a long time.

I remember the talks in Mission Control with Victor Blagov, who was the flight director on the Russian side of the mission. Our commander, Jim Wetherbee, would be talking with him and there would be a translator. They wouldn't agree right away, but eventually, we talked them down to a hundred feet. We had a radar, we had a TCS, our trajectory control system, which is a laser. These are our navigation aids. We had a handheld laser. We wanted to test our navigation instruments and we really need to get in closer. We wanted to test our comm [communication] system. We need to get in closer. We need to spend more time. So we agreed to a hundred feet.

Then as the months went by, we really need to get in closer. Let's get somewhere closer than a hundred feet. And Russia would say, "No, no, no, we don't want to do that." Of course, they are concerned about their space station. Well, you know, if we are going to go in and do this docking mission, we really need to go in closer on the rendezvous mission. So we eventually talked them down to 30 feet, because we wanted to see if the shuttle handling qualities were going to be any different when you got in close. Of course, there really wasn't any difference between being at 100 feet or at 30 feet, but it gave us opportunity to practice the procedures up to 30 feet.

There was some question as to when the shuttle jets fire, could that cause any contamination on the Russian space station. So we went to a mode that flight control called low Z, which would not fire any jets in the Z direction, which happened to be a minus Z direction. But that would be straight at the Russian space station. So none of those jets were fired. So to slow down, we fired our X jets, which actually fire in a fore and aft direction from the shuttle. But they are canted just a little bit in the direction where you get some braking out of them. I would have to show a picture to really describe it, but we used a low Z mode on the jets. These procedures were developed on Mir, and then we came to this space station program and, heck, we pretty much used the same procedures on the station that we had used on our rendezvous with the Mir, with some differences. But we had fewer unknowns.

Chapter 16
Stars in Their Eyes

Stars and planets are the beacons that inspire space exploration. The two scientists represented in this chapter studied different aspects of astronomy, one to gauge the practicality of manned space expeditions, and the other to take advantage of an orbital observatory to study the Sun. Their recollections span a thirty-year period during which human spaceflight transitioned from a dream that might happen to an event that took place multiple times in a year. The United States flew nine space shuttle missions in 1985, the most of any year during the program's 1981-2011 period of operation.

> *In the 1950s, scientists were concerned about several factors that might have made manned spaceflight impractical. One of the questions was whether a belt of natural debris surrounded the Earth. If so, a collision between a spaceship and an orbiting rock could be catastrophic. Astronomer Clyde Tombaugh, who had worked on the V-2 missile experiments at White Sands Proving Ground (see Chapter 3), decided to investigate that possibility.*

In 1955, I came in to New Mexico State University to do a project looking for possible small, natural Earth satellites. Some of us were at Flagstaff [Arizona] for a while; they had a fast Schmidt camera there. Then we decided we wanted to look in closer in the plane of the equator. Well, we could not observe that from here because the curvature of the Earth would hide it. So we had to go to the equator and be right under the celestial equator to explore that zone. That's why we went to Ecuador, in South America. The university had the logistics set up for making such a foreign expedition. I sent two men down there with a telescope and fast cameras for a few years.

We didn't find any of those rocks going around, like in Saturn's rings, so we concluded then that man could go into space or go to the Moon with reasonable safety from possible collision on the way there. As you know, no one has had a collision. We took over 15,000 films and didn't find a single rock, so we figured the coast was clear. It was not a perfect coverage, but it was enough to sample it, and we felt confident that we knew what we were up against. That was just before the advent of the artificial satellites and multi-stage rockets. The technique that I developed for that was a high-angular tracking rate for cameras. This procedure was later used, even today, by those who are engaged in tracking our artificial satellites. I had forged a new technique here, a new type of instrumentation.

George Simon was born in Frankfort am Main, Germany, in 1934. His family, which was Jewish, escaped from Germany just before World War II started. By the time he was in high school, his family had settled in Iowa. He earned a Bachelor's degree in physics and mathematics from Grinnell College in Iowa in 1955. After earning his Ph.D. in physics from the California Institute of Technology (Cal Tech) in 1963, he worked at solar observatories in California and New Mexico. He became an expert in solar physics, which can significantly affect satellites and manned spacecraft. Dr. Simon became a NASA astronaut in 1978 and worked ground support for the Spacelab 2 mission. He was inducted into the International Space Hall of Fame in 2001.

I guess I got my interest in physics and mathematics mainly from my father, because he was a teacher all his life. He had taught in Germany before he came to the United States, and his main field was physics, mathematics, chemistry, pHi Losophy. So it just came naturally to me. It was my best subject anyway, so I just naturally gravitated in that direction.

My choice of a career in astrophysics happened pretty much by fate, as most things in life seem to occur. Things aren't usually planned, they just happen somehow. I majored in physics and math as an undergraduate at Grinnell College in Iowa, and then at Cal Tech I was working on a Ph.D. in physics. Every Thursday, either the Cal Tech professors or visiting scientists came in and gave a research seminar. They went on week after week, and I never found anything that really appealed to me. Most of the physics projects in those days, and very much still today, involved very large teams of people, especially in nuclear physics, where they have those huge accelerators. If you read a paper in that area, sometimes there are thirty or forty authors. And I thought, gee, I really want something I can do by myself or with just a few people.

One day, I heard a talk by one of the professors there, Robert Layton from the Cal Tech physics department, who had taken some experiments at Mount Wilson Observatory [near Los Angeles], which was run jointly by Cal Tech and by the Carnegie Institution of Washington. He had taken the solar telescope at Mt. Wilson, modified these instruments, and was doing some very interesting things and learning about the sun. Things that no one had seen before, magnetic fields on the sun, the motions of gases. I became fascinated by it and after the talk, I asked whether he needed another graduate student and he said yes. So that's how it all happened. I've been in the field ever since.

My degree was in the physics department, although it was in solar astronomy. I had a choice when I left, whether to go into the defense industry in California, which meant a lot of money and not a whole lot of job satisfaction, I felt. So, I ended up looking at the various opportunities for continuing my solar research. In the United States, if you want to be a solar observer, that mainly means to stay in the southwestern part of the United States. There are really only about three or four observatories. So I visited them all and decided that Sacramento Peak Observatory in Sunspot, New Mexico, was the most appealing. I came here in 1963 and I've been here ever since, now for some twenty-three and a half years.

I guess the people at the observatory were the main reason I chose that place, plus the facilities there. Sacramento Peak at that time, and still today, is the outstanding solar institution in the world. The director at that time, the first director who was there something like twenty-five years, Dr. John W. Evans, was a very inspiring sort of a director. He wrote me a letter at the time, when he knew I was considering different places. He said, "You really must come to Sunspot. It's really about the closest thing to heaven on earth." And I guess for all those reasons combined, we did decide to come here. It was quite a change, having lived in Los Angeles for five years among all those people and the traffic and all, a couple million population, to suddenly come to Sunspot, New Mexico, on top of a mountain with a population of 120, and that included the children and the dogs and cats. But we adapted to it very well and have enjoyed it ever since.

When I first came to Sac Peak, I continued exploring some of the questions that had been raised in my thesis research. And really, over the years, I have concentrated without a whole lot of deviation from these very fundamental problems in solar astronomy, which are the problems of convection in the sun. The heat from the sun comes to the surface as bubbles, hot bubbles of gas, and we call this convection. It's the same thing as when you're boiling soup on your stove. And this convection, the motions of the gases, interacts with the magnetic fields on the sun. The sun has very strong magnetic fields. The Earth's magnetic field is one-fifth of a gauss. On the sun, we have sun spots which have magnetic fields of 3,000 gauss. That's 15,000 times stronger than the Earth's magnetic field. Therefore, on the sun we have very violent electrical and magnetic storms, and the result of these storms are solar flares.

Solar flares are big explosions on the sun, much, much stronger than any thermonuclear bomb on the Earth would be. And these flares, these eruptions, send particles and radiation into space, a lot of which reaches the Earth. When it reaches the Earth, it has a lot of bad effects. It can affect weather, we believe now. That's still somewhat debatable, but the evidence leans more and more in that direction. Also, it causes electrical power outages, power surges, and it can completely disrupt satellites. This is one reason the military and also the communications people, satellite TV and such, are very concerned. These electronic devices in space can be destroyed by solar radiation. And for astronauts, if you get a bad enough flare, you can actually kill astronauts in space. So, we are very concerned about what the sun is doing.

My research has been directed toward trying to understand these phenomena, how they interact with each other in the sun to produce these storms, and to predict when we might expect to see these storms. If one knows ahead of time, one can take, in some cases, precautionary measures. So, it is a very important area of research, and scientifically it is very challenging and stimulating because the problems are very complex. We still don't have computers big enough to simulate well all the complex things which happen on the sun or in any star. The sun just happens to be the most interesting star, because it's close enough. It not only maintains life for us, but it's something we can study in detail. Essentially, all other stars look like a point, even with the biggest telescope, whereas we can actually resolve details on the sun's surface. So, it is a very important area of research. It's kept me going for some twenty-five years, and I still have more problems to tackle than I have time for.

I first got interested in the space program, well, from the first space launch. I have always found that a very exciting area. At first, NASA only accepted pilots, and very good pilots, as astronauts. But at some point later on, it was recognized that if NASA was going to accomplish the science that it was supposed to do, they had to have some scientists, some experts in these areas. So in 1965, they had the first competition for scientist-astronauts. I applied for that but was rejected because at the time you had to have a lot of credentials which I didn't have. You had to be under thirty-five, which I had passed by then. You had to already be a pilot with some experience, you couldn't wear glasses, and other things. It took me out of the competition. Two years later, they had eased up a bit on these restrictions so I applied again, and I got into the finals. I think they started with 900 people, and I was in the last sixty-nine that they took down to San Antonio to Brooks Air Force Base. They ran us through all of these horrible tests for a week, and I didn't make it. They finally picked eleven out of that original 900, so at that point I figured that was it for me in the space program.

I think I didn't answer some of the questions very well in that competition because at the end they had a board of astronauts who asked very probing questions such as, "Suppose we selected you to go to the Moon, but you have to work in geology. Would you like to do that?" I suppose the proper answer for a gung-ho astronaut is, "I'll go, I'll do whatever you want." But I said, "Well, I think in my case, I'm an astronomer. I'm not a geologist, and that would be a mistake to send someone with my training to the Moon." Maybe that answer wasn't too good. Of course, that was a good question to ask Jack Schmitt, who became a New Mexico astronaut and then a [US] senator, because that was his field. And it turned out that, of all those scientist-astronauts that had been selected, he was the only one who ever got to the Moon. Despite wanting to do research, NASA ended up sending only pilots to the Moon, except for Jack Schmitt, so he really was unique in that situation. They also asked me, “Suppose we tell you there’s only a fifty-fifty chance that you’re going to come back from this mission. Will you still like to do it?” Being a pragmatist, I thought, well, I’d really like it better if the odds were ninety-ten or something like that. So again, I don’t know how much those questions made a difference, but anyway they didn’t pick me so I went back to Sac Peak, doing research and forgot about it.

Then about ten years later, [preparation for] the Spacelab 2 mission came along, which involved four solar telescopes. One of them was being designed by people I had been working with all through the years, so I was asked to be a member of the scientific team. It was being built by the Lockheed Corporation. I joined the team, and by that time NASA no longer really had the scientist-astronaut category. They now had a mission specialist category, which were scientists who belonged to NASA. They also had to go through pilot training, but they didn’t have to be pilots to start with. So the scientist-astronaut had been converted to mission specialists. But the scientific community said there are some cases where the scientific experiments are so complex that you cannot train a mission specialist, one of your people, in the year or two that you have, to really do these experiments properly. So the scientific community wants to have its own experts fly on this kind of mission. NASA agreed and created the position of a payload specialist. Now, the payload specialist is not a true astronaut, he’s sort of a hitchhiker. He belongs to the scientific team, not to NASA.

For our mission, which was very complex, they decided that there would be two payload specialists. All the scientific team members were asked, “Do you want to apply to be one of these two payload specialists?” Well, of course, I already had shown the interest years before, and I said, of course. Out of the twenty or so scientists who applied, they ended up picking four in 1978. I was one of the four. Two of them were to fly on the first mission, called Spacelab 2. Everyone expected there would later be a reflight, and the two back-ups would not only be real back-ups in case something happened to one of the prime two, but they would fly on the second mission. So, I figured that NASA can’t possibly put so much money into one mission and not refly it if it’s successful, so that my chance would eventually come. I was happy to do that, and it has been a very interesting experience.

About 90 percent or more of the training was learning how to do the experiments correctly. We were experts on the solar experiments. That was one of the main criteria for choosing us in the first place, so it wasn't so difficult to understand the science that we were trying to do. But one had to learn every step of every experiment, and there were thirteen experiments. You couldn't do on-the-job training in space. When you figure it costs something like $20,000 an hour for our flight, if you waste an hour or two you have really spent a lot of money and lost a lot of good science, most importantly. So we learned and trained for about six or seven years on all the procedures, and we helped the scientists themselves develop the procedures. We learned, after awhile, how to run these experiments better than most of the scientists did, and got very good at it. So most of our training was actually planning the mission in great detail. We knew what we were going to do every minute of the mission that lasts something like 180 hours. And then 10 percent or so of the training was just learning how to live aboard the space shuttle – how you function on board, how you stow your gear, how you fix your food, where you sleep, and how to escape in an emergency. But that didn't really absorb very much of our time.

Spacelab 2 was finally launched in July of '85. It had been scheduled originally for 1981 and ended up being about four years late. During the mission, the role of the back-up payload specialist, of which I was one, was to work at the control center in Houston. Now, what you see on TV normally is the Mission Control Center. The scientific teams were in the payload operations control center, the POCC. We called it the "POK." We had trained together with the two payload specialists who did fly. We could have substituted up to the last minute for them. We knew exactly what they were doing, or were supposed to be doing. And if they had a question, we were probably the ones on the ground who could best answer the question. So we served as the communicator with the crew on board the shuttle, and we worked something like thirteen-hour shifts during this eight-day mission, talking to the crew and coordinating things. We were sort of the focal point between the communications from upstairs to all the scientists spread around the control center. We controlled the talking going back and forth, and tried to make the mission go as well as possible. It was a very exciting and a tiring time, and it was well worth it.

Our Spacelab 2 mission was really the most complex, by far, of any mission NASA had ever tried. Not only were there thirteen experiments, but they were each very complex, difficult to run, and took a lot of man interaction. They weren't experiments that could easily be done remotely without a man. There is debate about whether we should have manned or unmanned spaceflights. Clearly, sometimes you can have an unmanned one, and other times you have to have a man.

My experiment was called SOUP. SOUP stands for Solar Optical Universal Polarimeter. The proposal first went to NASA in 1976, so it took nine years to get this mission going. It isn't like your normal scientific experiment. To do a space mission takes roughly a decade from the time you first think about it until you get it done.

One of the things that NASA sprung on us in the last couple of weeks before we flew was that we had planned a competitive experiment between Coke and Pepsi. Coke and Pepsi wanted to demonstrate how one could drink carbonated beverages in space, so they designed a special container to be used in the shuttle in a zero-gravity atmosphere. Well, so here came this experiment called the X-ray Telescope from Birmingham University in England. They said forget Coke, forget Pepsi. Spacelab 2 put experiment 7-UP instead, because their experiment number was seven, you see. We always talk about British humor as a little hard to understand, but this one was easy enough and very clever.

Spacelab 2 got off to a very bad start because our flight lost one of the main engines about five minutes into the flight. It was the first time a main engine had failed. They were high enough so that they were able to get on up into orbit, but it wasn't the orbit that had been planned. We ended up being about 70 kilometers, about 40 miles, too low. Also, in order to get into orbit at all, they had to jettison fuel, so we were short of fuel when we finally got there. So we couldn't do all the maneuvers we wanted, and all our very carefully planned timelines had gotten completely fouled up because when you're in a lower orbit you go around the Earth faster. In fact, we went around about 100 seconds faster in every 90-minute orbit. Now, you may well say, that's only a minute and a half, so what? But in a day, that adds up to about twenty-seven minutes, and that's almost a third of an orbit. We were completely off on our timelines, so this minute-by-minute planning that we had done for so many years was just tossed out the window.

In real time, the people on the ground had to revise everything to make it fit into the new orbital schedule. That took a lot of work, and a lot of things were changed. Some experiments had needed to get up as high as possible because they were looking, for example, at X-rays and ultraviolet emission, which is absorbed by the Earth's atmosphere. So by being in a lower orbit, the light was coming through more atmosphere and was being absorbed. So some experiments didn't work as well as they could. On the other hand, some who preferred a lower orbit were getting more data than they had thought they would get. So it all balanced out, I guess, in the end. But it made a very difficult mission, and it turned out to be probably, according to NASA people I talked to, one of the most successful ever flown.

The *Challenger* disaster, after being a horrible, shocking experience for all of us, has had, I think, a terrible and a long-term effect on all of space science research in the United States. Out of something like 120 or 130 science missions which NASA had on the schedule board before the *Challenger* disaster, over eighty of them have been cancelled. We're down to fifty-some out of 130. Almost every mission, except several that were so far along that they were essentially ready to go at the time of the disaster, have been put on indefinite hold. Who knows what indefinite hold means? It most likely means never. I don't see how we're going to come out of this situation in the next decade. What few flights there still will be over the next five years are almost completely bought by the military now. Civilian, basic research is really on the back burner.

The reflight that I was sure I was going to get, which has been called the Sunlab 1 mission, is one of those on indefinite hold. Instead of flying this year as had been scheduled, we don't know whether it'll go at all, and if it goes, it looks like '91 or '92. By then I'm going to be fifty-eight or so, and who knows whether I'm going to be okay to go or whether I'll be completely senile at that time. We were also planning a much bigger solar telescope, because the small one we flew on Spacelab 2 was so successful. Now we knew we could plan a bigger telescope, where we could see much finer details on the sun and really get down to those tiny magnetic field regions that we're really interested in learning about. That is not in the '87 budget. It may come back in the '88 budget, but at the moment it isn't there at all. Now they're talking, if it flies at all again, '92. Or maybe they'll wait until the [International] space station is operating, and that may be '94 or '95.

So I think we may well be close to a decade without any solar experiments in space. Although I'm concerned with solar astronomy, the people in all other scientific disciplines who count on space are in the same boat, and it's a very tragic thing that we're facing. In fact, the next solar experiment that the US is sending up is going to be launched by the Japanese in '91. We have a difficult time ahead of us. The French, of course, with their Ariane rockets are going to get a lot of the business and probably do most of the science for the next decade. There are the Japanese. The Russians, of course, have an ongoing, very strong program, and now even the Chinese are planning to launch satellites for Americans, both commercial and scientific. It's sort of ironic that the country that has been the leader in space is going to have a hard time staying up there. So, the ramifications of the *Challenger* disaster are just really very serious. I don't know what will happen with science in the next decade. Very little, I'm afraid.

In recent years, more and more of my research has been done with solar data from rockets and satellites. It has become clear that there are many things about the sun we can't learn from the ground. There are three primary reasons for that. One is some of the radiation doesn't get through the Earth's atmosphere, and there's a lot of physics to be learned from the ultraviolet radiation. So if you want to see it and study it, you have to get up in a rocket or a spacecraft.

Secondly, if you've ever looked through a telescope yourself on the ground, you know that whatever you're looking at sort of shimmers and dances around. We call that "atmospheric seeing" and we talk about "bad seeing." That's the worst thing. It's a constant fact of life on the ground that whatever you're looking at shakes around. That means that a tiny object gets blurred, and we can never resolve it properly. Again, if we can get above the atmosphere which causes this problem, then we get perfectly steady images without any blurring and motion. That's what we got on Spacelab 2 and why we've made such tremendous progress. Things that we only had dreamed about, we've learned from our Spacelab 2 experiments. In fact, I'd been waiting twenty-five years since I started to see some things which I had expected to see but could never see from the ground. Now, in the last three months, in looking over our films, we have seen some of these things. You can imagine how exciting this is. Something that you gave up on twenty years ago because you couldn't see it no matter how hard you tried, and suddenly it's presented to you. So, that's the second reason.

The third reason is that on the ground we have the day-night cycle. It seems that every time we see an exciting event on the sun we hit sunset, and twelve hours later, at sunrise, the phenomenon has disappeared and we've lost it. Or we can tell something happened just before sunrise and, gee, it would have been nice to see it all. Well, from space, if you get in a high enough orbit, there is no night at all, and you can see the sun constantly. From the shuttle, you only have about a forty-minute night so the interruptions, instead of twelve hours, are only forty minutes, and therefore we have a much better chance to see the evolution of features on the sun. It is expensive to go into space, much cheaper to do experiments from the ground. But if you really need the data and there's no other way to get them, then you have to go into space.

Chapter 17
Anyone Else Out There?

Is the Earth the only planet in the universe that supports sentient life?

If advanced civilizations exist on other planets, have they developed interstellar or intergalactic travel?

Have extraterrestrial beings visited Earth?

These questions are intriguing and, as yet, unanswered. Many people, including respected scientists and pilots, have seen strange objects in the sky. Such "unidentified flying objects," or UFOs, have been investigated by government agencies, scientists, and private individuals. The topic stimulates people's imaginations to the extent that the term *UFO* has nearly become synonymous with *alien spacecraft* or *flying saucer*.

During the late 1940s and 1950s, the foundational era of manned spaceflight, UFO sightings were frequent and believable. Adults of that day had watched aviation develop from the first manned flight to the fighters, bombers, and jet planes of World War II. The war had ended with the most fearsome invention of all time, the atomic bomb. Now, scientists were talking seriously about human space exploration. The post-war period, rather than being a time of peace, was a time of Cold War fears and technological competition.

The following paragraph from the May 7, 1949, issue of *The Saturday Evening Post* sets the context for considering UFO reports from that era. It is from a two-part set of articles titled "What You Can Believe About Flying Saucers," by Sidney Shalett:

> It is a jittery age we live in, particularly since our scientists and military spokesmen have started talking about sending rockets to the moon and about

> experiments to by-pass the law of gravitation by creating a man-made planet that will streak off the earth at 25,000 miles per hour or so and start circling in our orbit. Though we have not yet produced the rocket-to-the-moon and the homemade satellite, it is small wonder that harassed humans, already suffering from atomic psychosis, have started seeing saucers and Martians.

Roswell or the "Roswell incident" are mentioned in many of the excerpts below. They refer to a field of debris found near Roswell, in southeastern New Mexico, in early July 1947. Mack [often incorrectly spelled *Mac*] Brazel, foreman on the Foster ranch, found the wreckage. He reported the find and took Roswell Army Air Field Intelligence Officer Jesse Marcel to the site. Marcel collected much of the debris and took it to Brigadier General Roger Ramey, commander of the Eighth Air Force at Fort Worth, Texas. A US Army spokesman at first described it as remnants of a flying saucer that had crashed. Ramey issued an official statement the following day saying, instead, that it was pieces of a high-altitude weather balloon and its radar targets. Federal investigations of the event eventually concluded that it was actually pieces of a research balloon that was highly classified at that time. Some people accept that explanation, while others still contend that it was, indeed, an extraterrestrial spacecraft. Some even believe that alien creatures were recovered by the military personnel who gathered the debris.

Flying saucer crashes and recovered alien bodies or living creatures have allegedly happened in other places in New Mexico as well. People have reported UFO sightings and found indications of extraterrestrial visitors all over the world, but ufologists (those who study UFOs) consider the Roswell incident and other New Mexico occurrences to be particularly convincing.

Like members of the general public, space pioneers include some who are convinced that aliens have visited our planet, some who believe it is possible but are uncertain whether it has actually happened, and some who believe each UFO can be explained by a human event or a natural phenomenon of the Earth and its atmosphere. The space pioneers featured in this chapter represent the range of those beliefs.

Fred Wilcoxson lived in Roswell when the Roswell incident occurred. He worked at Air Force Bases in Roswell and later in Alamogordo as an airplane engine mechanic, a skill he had learned as a civilian and in the Army Air Forces during World War II. Working on experimental aircraft development and advanced radar techniques, he held a Top Secret security clearance in the 1970s.

I was born in Bear, Arizona. It used to be a railroad section that doesn't exist anymore. I'm sure it drove the security people nuts trying to find out where it was.

After World War II ended, I decided to go down to Roswell. This was in the spring of '47. I moved my little trailer, my wife, and my young son down to Roswell and worked as an apprentice carpenter. That wasn't my line though, and eventually I left that. What happened in the meantime, in '47 of course, everybody now knows, the crash at Roswell. When that front page deal appeared on the paper and said "We have recovered a flying saucer," I was rather excited. Anybody would have been excited. Then the next day, they come out with the balloon story. But what most people don't know is that a day or so after this balloon whitewash, they came out and said we've got to check on this a little bit more. About a week and a half to two weeks later, they showed the picture on the front page of a rancher holding a radar kite, which was just basically a piece of foil, and said this is what he found. Well, I knew that this isn't what he found because that just doesn't make sense.

I filed an application with Walker Air Force Base in Roswell for aircraft engine buildup mechanic. I got an appointment to that job in the spring of '48. That proved to be very interesting because I started building a home and had our little 25-foot, M-system house trailer parked there. A friend was waiting on us to finish supper, and he says, "Look at this, look at this!" At east-northeast of Roswell, this real brilliant circular disk was floating gently, settling gently down to earth at a 45 degree angle, extremely brilliant. When it got just above the horizon, the green disappeared and some blips of light shot out of it like a Roman candle. At that time, that meant nothing to me, except it was pretty. But later years, I discovered what I had seen crash was a flying saucer. This was a little less than a year after the crash at Roswell, but I didn't put all these pieces together until about thirty years later. And the blips of light were men ejecting from this craft.

The fact that the green went out, they saw they couldn't make it. They shut the power off, and the green disappeared, which was an electrostatic glow. I found that out thirty years later. Put all these little bits of pieces of life together and finally, down the road somewhere, they started fitting together.

I was unhappy at Roswell, and the Korean War was starting. I thought I'd better find me a new home. I had read about Boeing building B-47s in Wichita, Kansas. So I thought, well that's a good home. I drove up there one wintery day and I hired on with Boeing. That was in January of 1951. In the fall of '56, some mysterious voice spoke to me, and I told my wife I was going to move her back to our home in Roswell. That was all right with her because she was glad to get out of Wichita and those hellacious storms they have there.

The same time I moved her to Roswell, I came over to Alamogordo to see how my application was doing, and I found out that I was the first on the list. So on the same trip, I had a job in Alamogordo. I went back to Boeing, turned my two weeks and started to work out at Holloman just before Thanksgiving in '56.

One extremely important thing in my life happened. My son was going to go to summer school over at Silver City. This was in 1966, and I convinced my wife that he and I should go over to Lake Powell and go fishing. After fishing one morning, we were parked out in the middle of nowhere, really. I heard this noise that sounded like an airplane or glider. I thought well, heavens, somebody has lost their engine, they're gliding to land. They had a landing strip just probably east of where I was, and I stood there looking, waiting for this airplane to land. Pretty soon the noise started going away, and the poor guy hadn't ever landed. I just happened to look up in the air, and there were two beautiful flying saucers, round with the saucer shape to them. They were in Air Force formation, one behind and to the side of the other, slightly below the other.

You hear all these stories about aliens and extraterrestrials, but, "Ah ha," I said, "They're United States Air Force." They were off-white. Evidently, when I first heard them, they were right overhead, probably within two or three thousand feet. I wish I had looked up. I wished I'd had the video camera. But I knew Roswell was a lie. I'd come to that conclusion a long time before that, and here was the best evidence you could ever see.

That was these two saucers, and they were rising up vertically and moving ahead. They weren't tilted any, and all of a sudden they disappeared. They must have really crammed the power to them. They were probably within a quarter of a mile or less to me.

One day, I was having lunch with my friend Michael and he said, "Did you live in Roswell when the crash at Roswell happened?" I said yes. Before that, I knew the things existed, and I knew the government was lying about them but kept it in the back of my mind. Then he gave me some literature, and I realized all they had found out and about all these books that had been written on the subject. Fifty percent of most books are farce. They have to speak on aliens and stuff or they couldn't even get the books printed. So I got to reading it more and more. I got interested. I told Michael one day, I got tired of hearing this "Oh, we copied this from aliens." I'm going to find out the truth about this. So I started scratching and I got acquainted with John Price of the [UFO] museum over in Roswell.

I started pecking around, and then one day we were over there and my wife bought this copy of the front page of the paper. Well heavens, I had read that. I didn't need that, but I got to looking at that paper and got to reading it. Everybody said when it passed over Roswell, the saucer was heading northwest. The paper says this saucer was heading towards Five Mile Hill. They did not say it was heading northwest. I bought *Roads and Trails of New Mexico* and I was looking at that, and lo and behold, Five Mile Hill is west of Roswell. That craft was not heading northwest, it was heading due west.

Donald Schmitt [a UFO researcher] was giving a lecture there one time. I stuck around after the lecture, and I got a couple of names. So I talked to this friend, and in 1960 he had gone elk hunting up at the Fort Stanton Mesa. Well, if you come from Roswell and you're going toward this Five Mile Hill and you keep going west, with a slight deviation you run into Fort Stanton Mesa. He had gone hunting up there with a Lockheed engineer in 1960. He said it had started raining and freezing. Their britches were wet, and their britches were frozen. They were miserable. They had separated for a while and they got back together, and this guy said he had seen a piece of a flying saucer. He said it was part of the flight control, but it was impacted into this ravine there.

I didn't believe that. I thought, "There's some more balloon." But then I got to realizing. Here we are, we're going west of Roswell and we've got an indentation that something happened. And the most popular part of the Roswell story is that two [crafts] had a midair collision. If they had a midair collision over Fort Stanton, they were both going back to home. Now, if you're heading west, where are you going? You can doggone bet that they're going to White Sands Missile Range.

Later on, my wife had a good friend she had known for many years. She was executive secretary at the officers' club out at the Base in '42 when it first opened. She said there was rampant talk about saucers all the time she was out there. And that was '42 to '45. *Rampant* was her word, not mine. So here we keep rolling the date back, and these things were brought into this area in '42.

Michael Shayne used to have a radio program here. One night, he had a UFO program, open mic. He did not get this guy's name, but the guy called from High Rolls [a small community about 80 miles southwest of Roswell]. He said, "Oh, we had these lights landing up here around High Rolls." Michael says, "What year was that?" 1935.

One other interesting thing I'd like to get on this tape is when I worked at Walker Air Force Base in Roswell in '48. I worked with people that had also worked there during the time the crash took place. Or what is known as the crash. There was nothing unusual happening at the Base at that time. Everything was just routine. Well, I made contact with a person that was my work leader all the time I was there. His name's Alfred E. Pearson. He used to always put the *E* in there. I've had several phone conversations with him, and I asked him, "Did you go over and see that wreckage?" I have to believe him, because I have never known Alfred to lie to me. He said, "No, I thought it was one of our airplanes." But the guy that was in charge of reclamation, George Hollyfield, went out and picked up the remains of that saucer. He had a flatbed [truck] and a fork lift. They loaded it up and brought it home and put it in the hangar.

I found this out later, but previous to that I had talked to his wife. She told me that George took some friends in to see the wreckage one morning. That was the long Fourth of July weekend, but he took some friends and showed it to them. And she said at the same time he showed them the wreckage, the bodies were under that tarp. I worked around George for all those years, and nothing was ever, ever said about what was known as the crash at Roswell. It wasn't that big of a deal. He said the bodies were there, what was probably left of the bodies. Evidently, they had laid out there for a while. Some people say a week. I say it could have been a period of months. All that was left was bones. They had laid out there and coyotes had eaten them up, and all you had was the bones.

Dennis Balthaser served in an engineering battalion in the Air Force and later worked for the Texas Highway Department for more than thirty years doing quality assurance and quality control of materials that were used for highway construction. He retired in 1996 at the age of fifty-four and soon decided he was too young to be retired. That was when he moved to Roswell.

I had an interest in UFOs, as a hobby more than anything, for several years. An engineering company in Roswell was looking for a technician. I applied and came to Roswell, was interviewed, and was hired. That worked out pretty well because with my interest in UFOs and a UFO museum being there. On weekends, I would go by the museum and volunteer. Started going by at noon and after work to pick up correspondence and correspond with people who had written to the museum. I did that from 1996 to 1998. Then I decided to do it as an independent researcher. Basically, for the last twelve to fifteen years, I have been doing it as an independent. I write editorials every other month for thirty websites and UFO magazines. I do lectures at different conferences, do a lot of radio and television interviews.

Probably about 1978, 1980, I had bought a book by Jacques Vallee, and then I started hearing some things about Roswell. [UFO researcher] Stanton Freedman interviewed Major Marcel, the [Army Air Forces] intelligence officer [who believed the Roswell debris was extraterrestrial], in '78, and Berlitz and Moore came out with a book [*The Roswell Incident*]. Pretty soon, I started hearing about other books, and Roswell became my primary interest in ufology. Since then, I have expanded to other areas: Area 51 up in Nevada, the secret base has been one of my areas of research and continues to be, underground bases. That probably is because of my engineering background. I was curious about the equipment that was used for underground facilities. How the facilities were done. How they were made. What their purpose was. Then in 2001, I was asked to join the Great Pyramids of Giza Advisory Board. Which has turned out to be fascinating research because the three pyramids of Giza are physical structures that we can see, that we don't know a whole lot about. There is very little information on how they were built or who built them. So to me, that is some fascinating research. And I have limited myself to those few areas. In ufology, there are areas like abductions, crop circles, and things like that you can get into. But I learned years ago that if you dedicate your research to a few items, you can do better research because there is so much information out there that you really become confused with all of the information that's available.

There are several of us that continue to do research with Roswell, Stanton Friedman being primary and being the original civilian researcher on Roswell. Frank Warren is doing a lot of work. David Rudiack has done work on the General Ramey memo. There are several of us, Schmitt and Randall and Tom Kerry are still doing Roswell research. A lot of people think Roswell's done. But as long as there are new witnesses, or witnesses that are still alive, that can be interviewed and information gathered, my feeling is that we have not been told the truth about Roswell.

Based on the witnesses that I've talked to and the information we have gathered over the last sixty-four years, I am of the opinion that it was a, I've heard the term of "craft of unknown origin." I believe it was probably an extraterrestrial craft because from the information we have, it definitely wasn't a weather balloon. And it wasn't anything that the Russians, the Germans, or the Japanese had. Nor do I believe that it was anything we had. I think after sixty-four years, it could be made public if it is in fact something we have.

One of the problems that I have as a researcher is I have spent a whole lot of time putting out fires. There is so much misinformation and disinformation out there. If you go to the Internet and look up Roswell, you are going to come up with over 2,000 websites on Roswell. Ninety percent of those people have never been to Roswell. Never interviewed a witness. Don't know where the crash sites are. It is second- and third-hand research.

Many times at lectures people will say, "Why don't they contact us?" My immediate answer is "Why would they want to? What do we have to offer?" I don't see where we have anything to offer. If they can get here from wherever they are from, they have technology and knowledge that we can't even comprehend. And I certainly don't go out into the back yard and talk to insects.

There are a couple of theories. The main one, of course, is the thunderstorm, which ranch foreman Mack Brazel talked about. He said he heard a crash louder than thunder, then went out the next day to check his sheep. His windmills were damaged, and he found a debris field that was several hundreds yard wide by three-quarters of a mile long. Sheep wouldn't go across it. He had to take them around it to get them to water. That is one theory. Another theory that I've heard is the use of microwave radar that could have messed up their propulsion or their guidance system in the craft. Maybe they didn't have thunderstorms or lightning where they were from, so this type of thing could play havoc with the guidance system or the propulsion system of a craft.

In Aztec, New Mexico, in 1948, ten months later, a 99-foot craft with thirteen bodies [was found]. For years, I didn't put any confidence in that story until I got to meet Scott Ramsey and his wife, Suzanne. They have been doing some very serious research on Aztec for years. In fact, they have a book coming out, probably the end of this year or early next year, about the Aztec crash [*The Aztec Incident: Recovery at Hart Canyon* was published in April 2012]. The more I am hearing about Aztec, the more I am believing that it was an actual event.

What has always amazed me about witnesses, and I have interviewed from a hundred to a hundred and fifty first- and secondhand witnesses, is that these people didn't know each other. So it wasn't some story that they made up in a kitchen or living room in 1947. For people that don't understand the Southwest, when you are talking about a neighbor on a ranch, you may be talking ten or twenty miles away. So seeing each other is not a daily occurrence. When they work sheep, yeah, they'll get together, and they'll eat together, and they work the cattle and stuff like that. But this was not a story that was made up by these people. Most of the people that we have interviewed, the firsthand witnesses, are extremely credible. The rancher, for example, knew what weather balloons were. He had recovered weather balloons, because they had a tag on them on where to turn them in for a small reward. As a ranch foreman in 1947, he didn't make much money. So if he had a chance to make a couple of dollars by picking up a weather balloon somewhere, he would do it. And he knew what they were. But he had no idea what the debris was he found on that morning after the thunderstorm.

It happened in July 1947. The story died three days later, after General Ramey came out with his cover-up story. It didn't go public again until

1978, when Stanton Friedman located Major Jesse Marcel in Huma, Louisiana, retired, and interviewed him. Basically, the research started again thirty years later.

My theory, and this strictly my theory on Roswell, the only thing recovered was debris and some bodies, apparently. Until our military gets the technological advantage by back engineering, they are not going to go public or admit it happened. Because our military has access to knowledge that would be critical worldwide. If a country had that type of knowledge on propulsion systems and guidance systems, and how they can travel in space and things like that, that would be of utmost value.

Back then, I think that would have been a reality. Today, I don't believe that one bit, because today people are more educated than they were in '47. Young people pretty much accept the fact that there are ETs because they have been brought up with it by TV and movies. So I don't look for panic to be a problem today. I think the problem is, who in our government is going to admit that they lied to us for sixty-four years? Because it involves a lot more than UFOs. If they open that can of worms, then they are going to have to open a lot of other ones also.

I'll give you two good examples of the type of research I do. Are you familiar with the General Ramey photo taken in his office with Major Marcel and with Colonel Dubose, Ramey's assistant? A couple of years ago, I got to looking at those pictures. In fact, I got copies from the University of Texas, Arlington library that has the originals. In one photograph, along the back wall, next to a chair is an unopened package. Many people wouldn't even see that. I don't look for pretty faces in pictures. I am looking for other stuff. Well, there is this package that has not been unwrapped, but on the floor is brown paper and a weather balloon. What's in that package back there on the floor? Could that be the material that Major Marcel actually brought from Roswell to Fort Worth, to General Ramey? More importantly than that, the brown paper that is laying on the floor, if you look at that closely, that is off of a brand-new roll of paper [not wrinkled]. That means nothing was wrapped in there. Those are two things in that picture that I have a problem with the validity of a weather balloon. And if it was a Mogul balloon [see below and Chapter 4], as they said later, why is none of the equipment in that picture? The sonobuoys, reflectors, and things like that. All you have there is weather balloon. That is what's on the floor, a weather balloon. But the Mogul balloon had equipment on it that should have been there. The debunkers and critics look at [Mogul] Flight 4 out of Alamogordo as being the one that caused the Roswell crash. We've disproven that. Flight 4 was cancelled because of weather. The balloon was filled with helium and probably released, but there was no equipment on it because they had to cancel the flight.

Charles Moore was a chemical engineer and meteorologist who became heavily involved in using high-altitude balloons for long-range detection of nuclear devices in the late 1940s. He helped develop Project Mogul, a top-secret project intended to detect Soviet atomic bomb tests. Helium-filled balloons carried microphones to altitudes in the range of 30,000 to 40,000 feet to detect sound waves generated by the explosions. The project was operated by Watson Laboratories of the Army Air Forces' Air Materiel Command. New York University was contracted to develop the Mogul balloons. Initially, the microphones were carried by a string of rubber balloons up to 600 feet long attached to a chain of kite-like structures covered with aluminum foil to permit radar tracking. Before long, the string of rubber balloons was replaced by a single, large polyethylene balloon. As Moore mentioned, a B-17 airplane was usually used to track each balloon visually.

I suspect that the New York University balloon flights probably are responsible for the Roswell "incident." During June and early July of '47, there was quite a bit of excitement in the United States. A pilot named Arnold, in maybe the last week of June of '47, had seen some unidentified objects over the northwestern United States, Washington state. This report of his sighting touched off a flood of other reports. Every time that we flew balloons in late June and early July, the local radio station in Alamogordo would carry reports of flying saucers being seen over the Tularosa Valley [between Alamogordo and White Sands Proving Ground, about 100 miles southwest of Roswell]. And from the description over the radio, it was very clear that people were seeing the balloons that we were launching in the early morning. Alamogordo Army Air Field was pretty well isolated. There was nothing between the town and the Base. We usually put our flights up in early morning, so that after sunrise people would see the balloons already in the air. No one was in position to ever see them launched. They would just appear in the air. So we were quite sure that the unidentified object reports, I guess they were flying saucer reports in the Tularosa Valley, were ones that we were generating.

On the fourth of June 1947, we put up a sonobuoy [an expendable sonar system] on meteorological balloons. I have a memory of that flight, and I think that's the one that caused the Roswell incident, because it was tracked by the B-17 to the Capitan Mountains [about 50 miles northwest of Roswell (all such references in this chapter are measured "as the crow flies")]. There were two little towns in that part of New Mexico. One's called Bluewater and the other's Arabela [which is about 17 miles from the debris field Brazel found on the Foster ranch]. Those were exotic names to me, and they locked into my memory. The crew of the B-17 was unhappy with the signal they are receiving and stopped tracking when the balloon was between Capitan Peak and Arabela.

For that early flight with equipment being sent down by a B-17, we were not allowed to carry our radiosonde [receiver for atmospheric data transmitted from the balloon]. The radiosonde ground station weighed around 500 pounds or more and would not have easily fit in the bomb bay of the B-17. So the person running the program, Dr. James Peoples, said, "We've got a radar down in the North Area at Alamogordo, and we'll just track your balloons with radar." So we ended up carrying radar targets on most flights. They were more like an XYZ coordinate system turned at an angle. They weren't really tetrahedral.

My best reconstruction was [Mogul] Flight 4 was flown on June 4, and then we had a formal flight in which we actually put some radiosonde equipment on June 5. That was Flight 5. That one went east of Alamogordo, about twenty or thirty miles east of the Bottomless Lakes. That's the one that landed near an oil rig there. When the balloon came down and started dragging across the ground, the people from the oil rig started chasing it. The pilot of the B-17 buzzed them to scare them off. We had weapons carriers and drove from Alamogordo and retrieved that.

Then there was a pause for about a day while we prepared another flight. We had another flight on, I think, the seventh of June. I think that one landed in the mountains around High Rolls. These were reasonably successful flights, as far as carrying the Watson Laboratory microphone [an expendable microphone capable of detecting, at long range, low-frequency sound transmissions generated by explosions and missiles].

We then flew back on the B-17 to Newark and we were waiting for the delivery. All this was done while the first polyethylene balloons were being fabricated. We got our first delivery about the first of July. The balloons for New York University Flight 8 were ten General Mills, 7-foot-diameter balloons that we launched at three o'clock in the morning in anticipation of a V-2 that was to have been fired from the [White Sands] Proving Ground. The firing of the V-2 was postponed after the balloons were in the air. Part of the Project Mogul effort, in addition to monitoring nuclear explosions, was also to monitor Russian missiles, and so there was great interest in getting the sound signatures from the German V-2 rockets that were being fired. So we requested to put up another flight, but we had used all of the General Mills balloons that had been delivered. So we put up a cluster of meteorological balloons.

Albert Krairie prepared those flights. He had transferred to Alamogordo Army Air Field in December of '46, and he was making surface firings of explosives from the Tularosa Valley. He was also responsible for having 500-pound bombs dropped off Bermuda and off the Jersey coast. He was trying to pick up the signals from those explosions in Alamogordo. He had two mathematicians, two lady data analysts. One, Vivian Bushnell, died in October of this year [1995], and a younger lady, Eileen Ulrich, that we recently located and talk to. She's alive and very active.

Later that day, we flew Flight 9, and as I remember, there was an accident over at White Sands. Somebody spilled propellant or oxidizer, and the flight that was to have gone on the third of July around 19:00 [seven o'clock in the evening] was scrubbed because of the accident. We already had the polyethylene balloon and the neoprene backup balloons inflated, and there was no way we could put the helium back in the bottles. And no way we could save the balloons in the hangar to use another day because the neoprene balloons, just like pliofilm, degrade after exposure to sunlight. After the film stretched and they're exposed to light, even on the ground, they'll burst in about twenty-four hours or so. So rather than rip the balloons open, since we're flight oriented, we took the microphones and, I guess, the radar target off, and flew it with a dummy load, probably a sand bag. The radar crew would have been on duty from three clock in the morning that day, and now it was around 19:00, so I'm pretty sure that the Watson Laboratory radar crew didn't stay around to track a flight that was carrying no instruments. If they weren't going to be around track it, there was no point in our having a radar target. So I think it's probable that we took the radar target off and just flew a sand bag. That was Flight 9.

We made another flight, Flight 10, I think on the fifth of July. Over the Fourth of July weekend, we received some more polyethylene balloons, and on the seventh of July we flew one of H.A. Smith's 15-foot-diameter balloons, and some more of the little General Mills 7-foot balloons in a cluster. That was a very successful, controlled altitude flight. It went up to about 17,000 feet. The balloons weren't very big, so it didn't go as high as we wanted, but it floated at constant level. We had a fancy control that would drop ballast after the balloon reached some maximum altitude and started down again. We had a little liquid switch that would close when the pressure increased, and that would seal a diaphragm off and lock us to that altitude. Then if the balloon descended further, the diaphragm would close, and that would open a needle valve. It was all a mechanical and electrical thing. Then we'd drop ballast until the balloon went up, and a diaphragm expanded and sealed the ballast's flow off. It was liquid ballast. We used a grade of kerosene that wouldn't freeze. We used a liquid because you could control it.

The first flight we had, we did use sand and just had a little tube with a sieve in the bottom. When we wanted to drop ballast, we'd drop one chunk of sand. You could meter liquid. The same difference between solid rockets and liquid fuel rockets: one's controllable, and the other's not. In any event, this flight was exceedingly successful, and among other things, it gave us scientific mystery, because as it went at constant level at about 17,000 feet over the Sacramento Mountains [about 20 miles northeast of Alamogordo], it curved to the right, anticyclonically. It turns out that there was a very lovely meteorological theorem about the flow of air over a mountain barrier and how that generates vortisity [a measure of the rotation of air in a horizontal plane]. And because of this, a scientific paper got written up on this flight. It's called Flight 11-A because we also flew Flights 11-B, C, and D.

This flight landed about 19 miles west of Roswell Army Air Field and was tracked all the way by a C-54. The C-54 pilot stuck with it until it hit the ground, so he spotted where it was on the ground. We had radiosonde on it back at Alamogordo until it got down below the level of the mountain. That was on the seventh of July, the same day that the people from Roswell Army Air Field went out to pick up the debris from a ranch that's between Corona and Roswell, the Foster ranch, where the debris was found. The next morning, we'd used up all of our balloons. There was twenty-three of us from Watson Laboratory and from the NYU balloon group. We got onto the C-54 and flew east, leaving about 10:30 in the morning on July 8, the same day that the press release was made from the public information officer over in Roswell that Roswell had come into possession of a flying saucer. We went to Wright Field [in Dayton, Ohio] that night. We landed at Wright-Patterson and the twenty-one men in the party plus the pilot, co-pilot, and engineer, went off to the BOQ [Bachelor Officers' Quarters]. There were no accommodations for the ladies, so they got fed into Dayton to the YWCA. Ellen, the young lady mathematician, remembers this very vividly, because she said the beds were alive with bed bugs. She and Vivian Bushnell sat up all night in a chair. There's no question she remembers it.

I have a vague memory of what Ellen says happened next. She said the next morning, as we were trying to get on the flight [to Newark], we were pursued by the reporters who thought that our plane had brought the debris from Roswell into Wright Field. Our program, even the name of our program, was top secret, real top secret. Dr. Peoples wanted no part of talking to reporters, and so he evaded the reporters. I have a vague memory of someone trying to talk to us, and we stood around the Base Ops [Base Operations Center] looking like, "not us."

The pilot dropped us off at Newark, and somewhere in there, we learned about the flying saucer flap. I was completely convinced it was our Flight 11-A that had been picked up. Nobody would recognize it because it was the first real flight of a 15-foot-diameter polyethylene balloon that had ever been made. No balloons like that had been made in the world before. It's easy to understand why nobody could identify it. So for many years, I was quite convinced that it was either that, or it may have been Flight 8, because Flight 8 was seen dragging over the desert, going toward the north. I never knew where that stuff had been recovered, but if it was west of Roswell, it was Flight 11. If it was north of Roswell, it could've been Flight 8. So for years, I was convinced that it was that.

I have a large file of documents that were obtained under the Freedom of Information Act. I take strong issue with people that say the Air Force lied about this. What we were doing was not for meteorological purpose, but we were using meteorological equipment. It needs to be said that those radar targets were a new design that was finalized in March of 1945 and procured later that year. I'm sure they never got widespread distribution. The ML-306B targets that we had were new, and there was no way that the rancher, no way that the intelligence operator, Marcel, or anyone, except maybe the weather officer who never got called in at Roswell, would have recognized the target. In any event, in 1979 the people thought that there had been a cover-up, and I think there had been a cover-up because of Project Mogul. At least, people in the military didn't know what it was, but they knew that they weren't supposed to talk about it. And that got transferred over to the idea that it was a flying disk being covered up.

A fellow named William Moore came to talk to me in '79, and I was quite convinced then that one of our polyethylene balloons had caused it. He told me that, no, it couldn't have been a balloon, because there was some sort of heavy craft that landed. It plowed big furrows in the ground. It shed part of the craft, then it took off again and landed over on the plains of San Agustin, about 110 miles west of the Corona site, and had a final crash landing there. People came and saw aliens. He asked me, "Could your balloon have done that?" I was careful to condition [my answer]. I said, "Based on the description you gave me, I think that that could not have been one of our balloons."

Two years ago this investigator, [Robert] Todd, asked me about Mogul and sent me some evidence that it was unclassified. I told him a good bit of what we did and that it had been my opinion for years that one of our balloons did it. Then he sent me the interview from the Roswell *Daily Record* on July 8, that famous day when the press release was made and the day that we flew back to Wright Field. The rancher described a bunch of sticks, tin foil, smoky rubber, and some tape that had flowers on it. In all, the sticks were about 36 inches long, and the whole package gathered up together would weigh about five pounds. Boy, there was no mention in there of polyethylene at all. So I was quite convinced it's not one of our polyethylene balloons, but it's very much like the flights that we put up in early June of that year.

Since then, Todd sent me more and more information, and I think it's really Flight 4 that did it, the one that went to Arabela and then Bluewater. There's some more stuff coming on that I can't talk about right now, that will handle the matter of the dummies, or the bodies, or whatever they are. Turns out that there's quite another thing that will be of interest to you when it comes out. We're not in position to talk about that end of it, but there's a rational explanation of those too. The first part of it, what the rancher described, most people think that he would have had a cover-up. But if you're familiar with military communication or civilian communication, there's no way that within an hour of the thing being identified in Fort Worth to a rancher being interviewed on a radio station in Roswell, someone could have gotten the word to have a cover-up story that fits so closely what we were doing, which was classified and nobody knew about it. There's no way that one of those targets could have been substituted for real debris in General Ramey's office because we had the only targets in New Mexico. Those targets were not used widely meteorologically. They were really developed for the field artillery; however, we were using them to track. It's quite understandable why new generation people who weren't there wouldn't know anything about it. However, it's now a quite profitable commercial enterprise. It's a tourist attraction for Roswell, and there are people with a vested interest in keeping this alive. So I suspect that it's not all at rest, but I can assure you that I think the chances of it being one of our balloons are a lot greater than it being from another civilization outside the Earth.

Robert Montgomery worked for Land-Air, a civilian contractor that did work for both Holloman Air Force Base and White Sands Proving Ground/Missile Range. He arrived in New Mexico in 1948, and quickly became aware of several UFO sightings. Around that time, the government was concerned about a series of reports of strange "green fireballs" being seen near sensitive facilities such as White Sands Proving Ground and Los Alamos National Laboratory in northern New Mexico. In 1950, Land-Air was hired to conduct a study, Project Twinkle, to try to find an explanation for the phenomena. Montgomery participated in that study.

Back in the very early '50s, there were numerous UFO sightings out here at Holloman. They would be chasing with the Askanias or some of the other cameras, and they'd see some little white deals going around the missile, or wherever they were looking at.

At that point in time, I was in charge of timing and communications out there at Building 1103. We had tape recorders that were going all the time, recording the intercom discussions that were going on during the missile firings. Some of these fellows would see these so-called flying saucers out here and there while they were tracking the bird [missile], and they would get excited and start hollering about them. We recorded all this stuff. After the program was over on firings where we had recorded discussions like this, the Air Force would come in and confiscate our tapes real quick. And they classified it.

We don't think they were ever pinned down and photographed to where you could identify them, but there was more activity going on in other areas, up at Vaughan, New Mexico, and Datil, New Mexico. The Air Force had some activity going on there, and people were seeing flying saucers. One story was that some sergeant was driving a jeep up there when he saw this hovering light moving. Chased the thing around and fired his .45 at it. Never did plummet to earth. There's another story about a sheep herder up there near Vaughn that came in and told a story of a long, cigar-shaped object that hovered over the plateau up there while he was herding sheep. They went up to find evidence of it. They found an empty Eastman Kodak film package, or something. That was the only evidence they ever found of that one.

A few of the sightings that were going on there, and the secrecy involved in all this stuff, the Air Force decided they'd better get serious about it. They awarded Land-Air a contract for something that they call Project Twinkle. I was involved in going up to set up the equipment at Vaughn on Project Twinkle. They gave us a little guardhouse-type building that they had moved up there to get in out of the cold. You couldn't lie down and sleep in it, because there wasn't room for that. You could at least get warm in there. We had an Askania set up outside. We had a portable timing generator that could run that Askania and another Askania over at Datil so to triangulate on any flying saucers that were identified. We had a thirty-power, huge telescope on a pedestal that we can search the skies with at Vaughn.

Another fellow and I went up and set up this equipment at Vaughn. We manned that thing for a week or ten days, I don't recall how long now. We had fleeced-lined flying suits, because that was the only heat we really had. We'd look through the telescope, and we'd scan the skies. And we used the telescope for looking downtown to see what was going on downtown, because we were up on a high plateau over the town there. When looking at the skies, we saw an awfully bright object a lot of times that we finally decided was Venus, but we weren't too sure it wasn't something else at the time.

An incident happened, though, when I was coming back from there. My wife came to pick me up. Coming back, we were coming between Socorro and Tularosa. All of a sudden, here's a flying saucer coming right at us, right over the highway. Scared us to death. We pulled off the side. The thing came down right over us. It turned out to be Mr. Ryan, who owned a ranch there north of Tularosa. He had a private landing strip out there, and that was his landing light on his plane coming in.

At that time, we were on edge about flying saucers.

Meteorologist Duke Gildenberg worked with Charles Moore at Holloman Air Force Base on Project Mogul. Subsequently, Gildenberg worked at Wright-Patterson Air Force Base in Dayton, Ohio, and at Holloman on Project Moby Dick. That was a program to develop equipment and techniques for a secret surveillance operation that would carry cameras above the Soviet Union with high-altitude balloons.

We wound up with great UFO stories there, because nobody in Ohio had ever seen a poly balloon before. Nobody had ever seen anything like that before because it was before rocket-launched satellites. We had the first things to go up and stay up. In the stratosphere, just like today, the reason you see a lot of satellites right after sunset is because they are still up in the sun, and the Earth will be dark. So you see a maximum number. I know because I am an amateur astronomer. You see a number right after sunset. Our Skyhook balloons [plastic balloons developed in the late 1940s for high-altitude research] were the first ones to do that. It wasn't much of a lag, but the Earth already was in twilight, and we [i.e., our balloon] would be up there in the stratosphere. When you get sunset in the stratosphere, you get changing colors in the sunset on the plastic balloon. You get all kinds of reports. It was purple, it was white, it was red, for the same balloon. And they're up there where the line of sight is 400 miles. So you get reports from all over the countryside on them.

We used to track the flights with UFO reports. You call up the police station at Truth or Consequences [a town about 75 miles northwest of Alamogordo], for instance. "You have any UFO reports today?" "Yeah, we got one." I said, "Which direction?" "South." "Okay, keep an eye on it." (Laughs) I was a part of this. Our Blue Book office was located in our Balloon Branch office. That is the other thing I worked on when I first got here. I immediately started working on Project Blue Book [the US Air Force effort to investigate UFOs]. But that wasn't full time. Just when a report came in, we would analyze it. We had so many reports with Moby Dick that we didn't even register them after a while. We weren't in the main office. It was a reporting station. For the average base, it was in the Intelligence Office. They didn't analyze them. They just gathered information and sent them to Wright Field. The difference was that in our office, we analyzed them. We handled all the local stuff, anyway. The reason was because so many of them were our Skyhooks that they said, "Being as they were coming from you, we might as well put the office here."

We also handled Project Twinkle [which investigated reports of green fireballs]. The tracking sites were using standard optical sites on the [Holloman] Range. Major Duty showed me some, that's all that I can remember. I didn't do the official analysis on them. I did a couple special shots [films], since I was an amateur astronomer in addition to being a meteorologist. He showed me just one that I can remember. It looked like a meteor, except it changed course. So, he says, "What do you think about this?" I say, "Let me think." I really had no idea until about six months later. I was out with my telescope from sunset to sunrise on the weekends, and I saw one just like it. All it was, was a meteor that had broken up, and one piece went shooting off this way, and the front piece died out. So, it looked like it was one piece that changed course. If you took a picture of it, it looked like it went, not exactly 90 degrees, but was a pretty good turn.

We started on Excelsior [see Chapter 4], well, it was first called Project High Dive. That started with the dummies in '54. We did dummies for three years, just getting the parachute testing. They designed the parachute system, then redesigned it and redesigned it until they finally got it down right. Umpteen tests with that. It was written up and case closed. Then the Roswell affair, you know, the target area was mostly the fifty-mile impact area. When we had an abort, we would keep on trying, to the last minute. By the time we got through, we couldn't get the dummies out [retrieve them]. It was too late to get them on the Range, or even near Alamogordo.

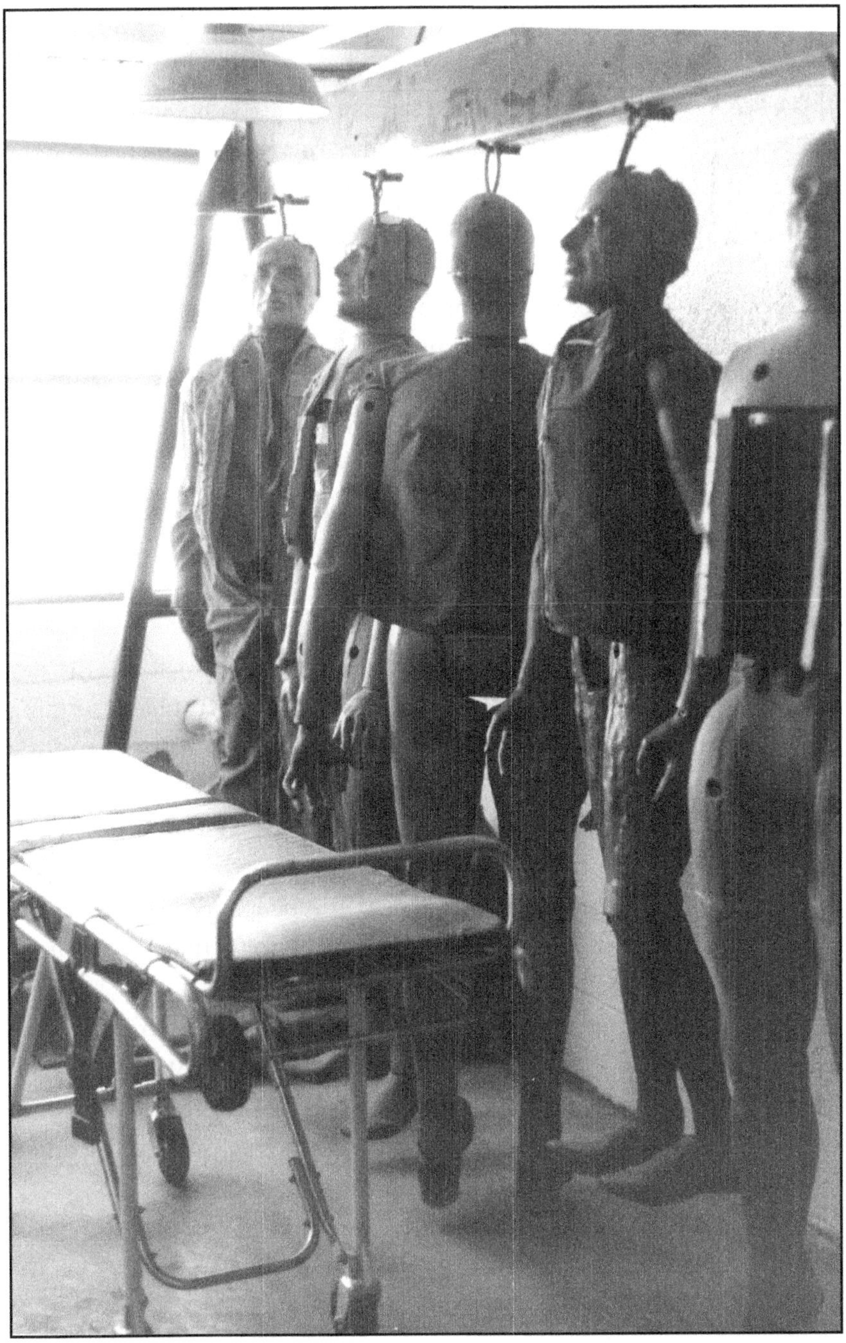

The dummies used at Holloman were surprisingly lifelike. Courtesy of US Air Force.

We overshot the Sacs [Sacramento Mountains and] there was Roswell. So we dumped a bunch of them in that area, one or three or four anyway, that we can remember.

There was one that Mac [Air Force historian James McAndrew] didn't write up, where one [of the dummies] got buried in a snowdrift. This woman was driving by and just saw the legs sticking out, and started to scream. This was over north of Roswell. It was in December. The snow was enough to bury the dummies. She saw the legs sticking out, and she screamed. Our recovery guys went over there and calmed her down. "Hey, it wasn't really a human after all." We had one mission where we cancelled out about midnight because of some kind of electronic problem. By that time, it was one o'clock in the morning, so some of the crew took one of the dummies and put a rope around its neck and hung it on the end of the hallway. We were all coming in late, because we had been working until two o'clock in the morning. The secretaries came in, looked down the long hallway, and thought someone had just committed suicide. They started to scream. Finally, one of the other troops arrives and says, "Okay, calm down, that's a dummy." Boy, did they ream us out when we got there. (Laughs)

We had another case where we were out at the launch site the day before the launch. I forecast it would be no rain, no problem keeping them overnight. So we left the dummies out at the launch site. A little wind kicked up, and they were kind of swinging around in the wind just a little bit. An MP came driving by and he took a shot at one of them. (Laughs) You know, it was another proof that they really did look like something alive, at least from a distance. The people didn't realize that any dummy looks real if you see it from a distance or at night. He actually took a shot at the damn thing.

As time goes on, they become four feet and green. Actually, the heights they gave in all those stories kind of bracketed it. At Wright Field, they had them like seven feet. So they went all the way from four feet to seven feet. One of the principal Roswell witnesses said about five and a half, which was the same as the female dummies.

Jack Glasco, a photographer working for a civilian contractor at Holloman Air Force Base in the 1950s and '60s, was involved in balloon projects including Mogul and Moby Dick. In this excerpt, he talked about a piece of equipment designed to test a parachute system for a soft landing on Mars. It was a 15-foot-diameter aeroshell that looked somewhat like a flying saucer. It was carried aloft by a balloon launched at Roswell and released over White Sands Missile Range. Then its cluster of rocket engines were ignited, carrying it to an altitude of 160,000 feet, where the atmosphere's density was similar to that of Mars and the parachute system could be tested. The Voyager program, for which the parachute system was originally tested, was later replaced by the Viking program. Viking 1 and Viking 2 both landed successfully on Mars in 1976.

Periodically, White Sands would bring journalists on the White Sands and give them a tour, just for publicity purposes and to let the world know what the government was doing at White Sands. Many of these journalists were looking for something sensational. And after this Roswell incident became public, people were reading about it, and so many people believed that an alien space ship landed over there.

We had a tour scheduled for White Sands for the journalists. This was after the Voyager test had been sent aloft and had come back down, and we recovered it. It was designed for a soft landing on Mars. It had already been hauled into the boneyard at White Sands. The boneyard is where we put stuff that has been used, and they just store it until they decide what to do with it. This busload of journalists was driving along, and they look out there. There is a chain link fence, and inside the boneyard is this Voyager [aeroshell]. This young journalist said, "Ah ha, you do have flying saucers. There is one of them right there." So we stopped the bus and went out there, and we told everybody what it was. It was this Voyager Project that was done to test the design of a vehicle that would soft land on Mars.

Okay. I am sitting in my living room, watching TV one day, and this picture comes on TV. It originated in Roswell somewhere, and they were presenting it as an alien spaceship that landed. I started laughing, and I couldn't stop laughing. I shot these pictures. I was on the whole project. There is nothing alien or out of this world about it.

A PEPP aeroshell being prepared for launch. Courtesy of US Air Force.

James Winker studied aeronautical engineering at the University of Minnesota between 1946 and 1952, where he became friends with the Piccards, a famous ballooning family (see Chapter 4). The Japanese bombing balloons he mentioned in this excerpt were also known as Fu-Go balloons. In late 1944 and early 1945, the Japanese released more than 9,000 hydrogen-filled balloons carrying bombs or incendiary devices. The jet stream carried the balloons toward the United States, where the Japanese hoped they would cause damage when the balloons finally lost altitude and crashed to the ground. Perhaps as many as 1,000 Fu-Go balloons reached the United States, but they inflicted very little damage. The only fatalities occurred when a group of people found a balloon on the ground in southern Oregon. While they were examining it, its bomb exploded and killed all six of them. Winker's UFO/alien experience did not relate to the Roswell incident, but to the strange markings on the Plains of Nazca in Peru.

I was the fourth of seven children. My older brothers were my greatest influence, and they were all aviation nuts. That was a time in history when aviation was really coming out and becoming very common. It was shortly after [Charles] Lindbergh's trans-Atlantic flight, and there was a lot of air racing in the '30s. I'm sure those things all had an influence. We lived only about 10 miles from Charles Lindbergh's boyhood home, so he was quite a local hero.

My three brothers before me had all gone into aeronautical engineering. It was almost automatic that I would do it, then. My specialization developed as I was in the course of study. For one thing, one of my professors was Doctor Jean Piccard. He taught what was called a seminar course. I guess it was designed to just make you think. He was good at that. I enjoyed studying under him. He asked for term papers, so we had to do experiments and write them up. That was of interest.

The first year I was at the University of Minnesota, I was already in the ROTC program. Come along February of 1947, a young fellow named Donald Piccard, son of Jean Piccard, was going to make a balloon flight. He was in the ROTC also, and he wanted to have a crew to help launch him. Boy, I was probably first in line. This was one of the Japanese bombing balloons that had been recovered. He had patched it up and put on a different load line system and a basket to stand in. The balloon was not very large, only about 30 feet in diameter, maybe even 25. He flew this out of downtown Minneapolis and was in the air an hour or two. I was on his launch crew, and we chased and then helped restrain the balloon as he disembarked.

This is also the same time when General Mills was doing the early pioneering work on scientific ballooning. Probably the clincher was, I got a summer job working for them one year. I continued part time through the rest of my years at the University and then went full-time with them when I graduated in 1952. I started out in data analysis. Looking back, I think that that was probably a very fortunate thing, because we analyzed the instruments that told how the balloon performed and how the equipment on the balloon performed. I got a really good grounding in how balloons operate. Already, they were using polyethylene. The scientific balloon did not become practical until polyethylene was invented and was developed enough so that it would perform this kind of service.

[How about this balloon that you specially built for the test down in the Plains of Nazca?]

What's at Nazca is commonly known as the Nazca lines. These are paths taken out of the high plains. They moved rocks and dirt, and when you move this you get a different coloration in the soil. They have tremendous, big figures. There are animal figures, and there are just straight lines, and there are tapered lines. The Nazca plains are very high, very flat, extend for miles and very, very low rainfall, so there is no erosion. These lines were made many centuries ago; historians estimate 1,500 to 2,500 years ago. And the lines are still there. Nobody knew about it until about 1920, when the first airplane flew over the area. They were just flabbergasted. Where did these things come from?

There was an organization called the International Explorer's Society. Their purpose was to find places where something significant – you might call it primitive technology – occurred at some time or another, and they would develop tourist trips to go to those places and do those things. On one of their trips to Peru, they were going to set up this idea of tours going into Nazca. There has always been speculation about the pictures there. Why were they put there? One line of thought is somebody must have been able to see them from above, or there would have been no point in doing it. How would they get up to see them? The Society came across information which made it appear that there might have been ancient balloons developed, built, used centuries earlier. So this developed into a program to do what they call experimental archeology – to try and do something in the way it could have been done, or would have been done in primitive times and just to show that maybe this idea isn't flaky, that maybe it really could have happened.

Of course, there is another line of thought, that aliens have done this. The scenario that the Explorer's Society used was that they probably had balloons. They could have flown balloons, and they could have seen the big, big figures on the ground with that. Not that they needed the balloons to construct them, just to see them. Maybe they were religious or something like that. That's where the idea generated. We were given the small contract to build a balloon like what they had seen in some primitive graphics. It would have been a hot air balloon, of course, and probably inflated in the style of the early smoke hot air balloons, where you just put it over a fire pit, and let the fire build up and warm it up. It's restrained by people around the edge. Then when it gets enough lift, you let it go and it carries the gondola up with people in it. That's the pathway we looked at and successfully accomplished it. We flew the balloon in 1975. The two people that flew with it said that they no sooner got off the ground than there was this array of figures around them. It was beautiful.

We used a cotton material, tightly woven. They have actual samples of textiles from those early days, so we were able to somewhat duplicate what we knew would have been available at that time. That type of balloon could carry just two people. It was an odd shape. It was a tetrahedron. That would have been unusual, except that's what the figures on the old pottery looked like. A tetrahedron has four triangular sides, the top one and three [on the sides, coming together at a point below]. I'd say from point to point on the triangle it maybe would have been about 30, 35 feet, something like that.

They would lose altitude by leakage and cooling. They don't carry any heat with them, so it's just the heat that they have generated while it's still held down. In this case, you'd measure it in minutes. I suppose they reached their peak height of four, five, six hundred feet in just a few minutes, and then they gradually round off. I don't think they were in the air more than ten minutes, fifteen at the most.

> *In the late 1960s, Thelma "Dee" Swope worked for Radiation, Inc., a civilian contractor at Holloman Air Force Base (see Chapter 2). Since the late 1940s, there had been reports of various kinds of unexplained phenomena in the region around White Sands Missile Range and Holloman Air Force Base. Swope helped create one of them.*

The emerald laser project was the fun one. At the time, I was married to Captain Don Pearson, who was in Foreign Technology Development (FTD) out at Holloman. He decided he wanted to cross-train. He wasn't getting anywhere in Foreign Technology, and it was a little boring. He had his aeronautical engineering degree, so he wanted to get out of FTD and into something that was a little more hands-on. So he transferred into the laser research project, which had just begun. They had just started working with the ruby and the emerald lasers. When Don got his transfer, when he cross-trained into that department, I was working with Radiation, near where Don's offices were. They didn't have enough personnel, and according to Don, this was an Air Force thing. They didn't want a lot of civilians involved in it.

So Don asked me, since we were married and since I was familiar with some of the basics that have to be done, if I would go out in the evenings and help him out at the [high speed test] track to do some distance, long-range testing with the beam. The way it works in my mind is they were trying to develop a plate that the laser would go through the emerald and bounce off of this plate. He used convex and then concave, and he tried different styles. He would have the different plates ground to change the focusing of the beam. My job, after I helped him set up the transmitter in the blockhouse, was I would run out to the truck on the side of the track next to the roadway. On the back of that truck, we had this huge target, and my job was to help Don for range testing. I had a little walkie-talkie, and I would be driving this truck down, and he would see how far down the road I could get before we would lose the beam. So he would adjust the plate. For three, four, or five hours we would be out together in this little blockhouse, and I would be driving the truck down the track. He would tell me, "Out of range." I would drive the truck back.

Go in the blockhouse, help him convert the plate, and we would change [the position of] the emerald. It was a huge emerald. It was the biggest stone I had ever seen in my life. Of course, it wasn't cut like you would see in a diamond ring. It was the raw stone.

It was a real emerald. The artificial came later. But that was another thing that was sort of exciting. Don was in charge of the emerald. It was almost like Fort Knox. He put it in a little box with all of this special material around it, and then the little box went into a metal case. Then the metal case had to go into a shoe box, and the shoe box went into a regular store box. His idea was that if he controlled this, he was responsible for it. He didn't want anyone to know that he had anything valuable. So he would disguise it after he got it into its protective case. So I was running around in my car with this box that people thought was just clothes or whatever. And I had this expensive emerald. I don't know how much it was worth.

We had to do all the testing after nine o'clock at night. I guess it was so the earth would cool or the atmosphere would settle. I really didn't understand why, I just knew it was exciting. It was revolutionary at the time.

One of the interesting sidelines to working with the emerald laser was one time, it had come a really rainy season. It had rained and rained for a week, and so we hadn't been out at the track doing much work with the laser. So finally Don said to me, "We really need to get out there," because we had gotten to the point where we could get the beam to travel 500 feet, which was quite a feat at that time. And he wanted to press it and see if he could get 600 out of it. Well, like I said, it had rained, and rained, and rained, and we had been delayed. Finally, the rains had quit. Supposedly, it was not going to rain this night. We went out, and this was in July of '69, if I am not mistaken. We went out, set up the laser, did all of our testing, and we did increase the range another 20 to 30 feet. Which was great. We were both really excited.

But while we were out there, the frogs were driving us nuts! They were *reep, reep, rip, rip*. It sounded like a million of them. They were so loud you could hardly hear yourself talk. Over the radio, I would yell, "Can you hear me?" "No, I can hear all of the frogs singing to us." They were all over the place. It sounded so weird. Well, we had been really stretching the range of the laser. And it was a beautiful, clear, clear night. The stars were out, no clouds, full moon, beautiful, just beautiful. It was almost ideal for what we were working on. The next evening, I picked up the Alamogordo *Daily News*. The front-page headline was something like, "Green Alien Light Seen in Desert." I read those headlines, and I thought, "Oh my gosh!" Because it was a top-secret thing we were working on. I am thinking "Oh, no. I don't think I want to do this." Fortunately I wasn't on the payroll, so they couldn't hang me too big. I continued reading through the article, and the article stated this strange sound. There were two men working on some kind of pipeline or something outside of Holloman Air Force Base. They had heard the strange noises. Well, that was the frogs! That was ongoing, and it would come and go. They would chirp for a while, and then they would get real quiet, and chirp some more. And the men said they had seen this green light that was running along the road out there. So they had an alien sighting.

It was written like a UFO sighting. I showed that to my husband, and I said, "You know what, we have been discovered." We had to lay low. We had to stop the project for about a month. We could not afford to do any more testing while the public awareness was that great. But to this day, I think, "Well, I have the explanation for a UFO sighting that happened here in Alamogordo. I know where that one came from."

While working at White Sands Proving Ground, Clyde Tombaugh (see Chapter 3) saw several phenomena he could not explain, including many green fireballs. In about 1955, he wrote to UFO researcher Leonard Stringfield, "I have seen three objects within the past seven years which defied any explanation of known phenomena such as Venus, atmospheric optics, meteors, or planes. I am a professional, highly skilled observing astronomer." These three objects were "green fireballs which were unusual in behavior from scores of normal green fireballs." In his 1986 oral history interview, however, he demurred from discussing such sightings.

[Have you ever been involved in any UFO investigations?] Not officially. I've seen a few strange things, but I don't know what they are.

[Would you relate your experiences with them?] Well, I prefer not, because when I do that, I get a flock of mail demanding a story on that.

I want to keep a low profile, so I'd rather not. That gets to be a touchy problem, because it's generally misunderstood. Then I'm blamed for a lot of nonsense and so on. I've even been threatened. So, I don't talk about what I've seen.

Chapter 18

Pioneers of the Future

Space pioneers came from many backgrounds, some highly educated and some not. But all contributed to man's achievement of visiting the Moon and living in long-duration Earth-orbital missions. Some flew on those missions, some coordinated the flights in Mission Control, and some worked behind the scenes as software developers, researchers, welders, or weather forecasters.

As they reflected on grand achievements, some stressful episodes, and a few horrific failures, some pioneers naturally turned their thoughts to the future of manned spaceflight. What comes next? What are the government and the society of the United States willing to envision and pursue in space exploration? Why should we strive for long-duration missions such as a visit to Mars or establishing a manned outpost on the Moon?

Many of the pioneers expressed frustration that NASA has not sent a manned mission beyond low-Earth orbit in more than forty years. That same frustration has been the motivating force for a new generation of space pioneers who are developing a commercial spaceflight industry. Some outstanding examples are Peter Diamandis, who established the X-Prize Foundation that led to the first privately funded, manned, suborbital spaceflight in 2004; Richard Branson, founder of Virgin Galactic, the world's first commercial spaceline for tourism and research flights; and Elon Musk, founder of Space Exploration Technologies Corporation (SpaceX), which delivered the first cargo on a commercial vehicle to the International Space Station in 2012.

The new generation of commercial space pioneers – in fact, all of the American people – can benefit from the experience and vision expressed by those who led the way during the twentieth century:

Duke Gildenberg was a space pioneer involved in the early manned spaceflight simulations, including the Manhigh balloon program (see Chapter 4). His characterization of those projects is a reminder of the challenge of pioneering a new human capability.

We learned as we went along. I went to visit Dave Simons [pilot of the second Manhigh flight] after he became a doctor with the FAA in California. I hadn't seen him for about eight years. I was on a business trip there and dropped in to say hello. He said something like "Did we really know what we were doing back then?" And I said, "Well, that was the purpose, Dave. If we knew what we were doing, there was no point in doing it."

In his 1998 oral history interview, James Gimbel looked to the future from the perspective of a person who worked in the Apollo and space shuttle programs (see Chapters 9 and 14).

I believe now that our technology is such that the lunar surface would be the better option [than an Earth-orbiting space station] for the long range. In my personal opinion, the information we would retrieve from the space station itself would be Earth resources information. You've got a [natural] satellite, and the distance between low-Earth orbit and the lunar surface is negligible. I mean, you can park reboost tanks now in orbit. I envision a used car dealer up there with a whole bunch of reboosts, selling them to passing-by spacecraft. The technology that puts that into space was not there, even fifteen years ago. All of a sudden, we find ourselves pushed so fast that our concepts are outweighed by the progression of the science.

So we've got a dilemma there. Whether we want to spend money and go at a very slow pace, which makes a lot of sense. But mankind having such a short life span in cosmic time, so to speak, is always anxious because he has this innate feeling of manifest destiny that is always there. That's in our genetic makeup. We are always impatient to see great things done, great movements in science. And there is a little conflict between your budgetary allotments for manifest destiny and actually doing this in a time frame that matters to someone like me. I was born in the science. I have been thirty years in the science. But I am very anxious. I would like to live to see something even greater coming. I have lived through the lunar lander, which was, to me when I was a child, completely out of scope. All of it was Buck Rogers. Science is marching so fast, we forget the point we started from.

Travel is pretty cheap, if you developed a vehicle to go to the Moon and return, and put a complex similar to the space station on the lunar surface, or under the lunar surface. With a lot of scientific complexity, you can generate water from the lunar surface if you are in the right place. Building a pipeline and a process plant may be out of scope. I would think that we have come to the point of time that we can reuse water effectively and replace it with special supply ships. Certainly, the shuttle is not going to do it. But there would be something similar to the shuttle that could return, launch, return, and land with enough cargo to support a small colony of scientists on the lunar surface.

You can begin to start an assembly point on the lunar surface. Doing it in space, the GNC (the guidance, navigation, and control) is outlandish. The cost of the GNC of parking hangars in space to assemble long-term space vehicles is going to be another political hurdle that no one really wants to get into. The arguments are going to be that you've got terra firma [on the Moon]. It's one-sixth gravity, but things still fall down. If you drop a wrench, you can pick it up. If you drop a wrench in orbit, about two orbits later you are going to get a wrench for dinner. It's going to come at you at 24,000 miles per hour, and it's going to blow holes in things.

It's so complex to do work in outer space. The EVA technologies and the suits have to be redesigned. We can no longer depend on the limited time that we have in the present suit. It gets pretty complex because the kinesis [movement] of the shuttle suit is not going to fulfill the requirements of a space station suit. The suit has to be more dexterous. It has to allow you to get up off the floor. You know, you are a turtle [in the current suits].

My Master's thesis was on the physiological and psychological effect of long-term space missions, i.e., Mars landing. I was in a good position to get data, and the research led me to believe that there are certain points of reality that we refuse to face. If I track them back far enough, I find budget restraints always at the helm of this disaster ship that we are riding. Cooler heads are going to prevail. But there are certain limitations that the human body has. You will not travel on a long-term space mission unless you take your gravity with you! Period! Design for it. It is easy. The technology to design gravity in your long-term space missions is a monetary decision, based on budget restraints. If you want to send a person out in zero gravity, don't expect him to live past three or four years. Or if he does, you don't ever bring him back to Earth. He lives a terrible life, the rest of his life in zero gravity somewhere, until his heart completely dissolves. If you don't need the organ, it will certainly go away. That is our human makeup.

In my research for my Master's, I used a Mars mission, only because people understood the complexity of a Mars mission. If we go to Mars, eighteen months with a solid propulsion system. Good. We go there. We pick up some rocks, and we come back. That's within the scope of our human capabilities. If we go to the nearest star system, forget it! You either take gravity, or you don't go. I have no reason to believe that we can go without something less than one g in a long-term space mission. The technology to get one g or one-half g is probably a year away. You can do it with a small, centrifuge-type of arrangement, where the sleeping couch will give you eight hours of one g inside this huge spacecraft.

We know that if we are going on a mission of long term, those people are gone. It might be that technology surpasses them, that we pass them on their way to their target destination. If you send a spacecraft with 300 people out for a 200-year voyage to a star system, back on Earth, technology is still evolving. They are stuck in their technology because what we designed for that trip is there. Unless they take engineers and manufacturing plants and low-bid contractors with them, they are going to be stable. They have the resources only to survive. They are flying along at maybe some speed, take a pick, half of light speed or nine-tenths light speed. Back on Earth, technologies are evolving, and we are starting to warp space and time. We build these ships that work off the gravitational pulls of these different planets. We zoom past them. We are sitting there fifty years before they get there, and we already have got a factory plant and are selling used cars before they even arrive.

Do we wait till the technology evolves? It will always be evolving. We might as well just dream, rather than spend the money to design and build. But the urge for manifest destiny is always going to be with us. It is another given. We can't take that out of us. We came out of a cave, and we are going to the stars. There is nothing going to stop that, except a cataclysmic event that puts us all back in the cave. We are going to the stars. It is a matter of 100 years, 150 years. It could be even 20 years. The research going on now, magnetic propulsion engines and ion engines look very promising. There are some givens. No, we will not freeze people. That technology problem is a thousand years away from us, if they were funded. There is nobody funding freeze technology.

Can we send a twelfth grade high school student into space and let him finish his education on the flight? Is it humane? Is it humane to do these things that we must do for that type of a mission? We have a whole spread. We have even got to send older people in order to get that full complexity of society. If we are going to take our society with us, we don't want that society to be some weird bunch of human look-alikes. We need to take our society with us. We need to take our morals, our ethics, our religions all with us because the whole point of the mission will be ruined if we don't. That's going to be complex and exciting. And that is where human factoring gets mixed in.

So, especially our age group, we need to leave enough historical information. You compile it in a way it is understandable by the normal person, where he can educate himself to, at least, know what these programs are about. They are still cloaked in a lot of secrecy. No sooner does NASA let out some information than it's marching so fast that some MIT Ph.D. says, "No, you are all wrong." And he is right. An organization as complex as NASA, and as weighted down by political interests as NASA is now, hesitates to put out real good information. It's very easy to put out this generally known information, and say, "Yeah, our selection committee is unbiased and unprejudiced, and we run this happy ship." That's bull.

That is part of the system that is not pertinent to what our ultimate goal and our ultimate mission is going to be, and that is inhabiting other star systems. There is no place on Earth you can go now and start a colony. But we still have this urge. The Earth as we know it is no longer going to facilitate that priority in our lives. And this needs to be understood by the guy that is painting a wall, by the guy that is running a bank, by the guy that is selling used cars.

So, how do you build hardware such as this? How do you train people to use the hardware on the longer missions, when each one is significantly longer? How to build engines and aircrafts that will take that? The shuttle is now approaching twenty years old. We've got metal fatigue in the shuttle. We are talking about materials that were heated in ovens and poured and cast. Aircraft don't last long. They are like us. They get old and their joints get stiff. In space, you are dealing with an environment which changes from 300 or 400 degrees below zero to 300 degrees above, depending on which way you are facing. If you are facing toward the sun, you're hot on one side and cold on the other. Look at the stress it is causing. It is dangerous. Space is still dangerous. We know from hard experience that we have to accept a lot of loss in space.

Flying the astronauts, you don't want to kill a flight crewman. This is ingrained into the ground crew: Don't hurt these astronauts. These guys, pound for pound, cost more than rib eye steak. There is a lot of training in these guys. You're talking about five or six million dollars' worth of training to get there. You don't want to hurt these people. But they are willing, by the nature of their personality, to walk out on the edge. They live to be there.

Some mistakes, especially in a dangerous field, are there. They are inherent to this system. We need to understand that. We are telling high schoolers how great space is. A little dose of reality saying, "Hey, we killed a lot of a people, from flying into a goose in a T-38 to not understanding the dynamics of solid boosters." We gloss over things. We don't tell the children that are going on these missions. We are not going on these missions. From our wheelchairs, we will be waving at them. They are going on the missions. They need to know reality, not PR hype!

Luckily, NASA is going under a great change now. My age group is leaving. They are retiring. I really enjoy watching the new people coming aboard, even if we waste some time reinventing the wheel. The old people seem to be more set: "It has worked this way forever and we are going to keep this way." Which was all right for building cattle fences and stocking fields, but it is not all right with the sciences. The sciences have to be free to expand, and we have to be forgiving of mistakes. One of the things that I have always been impressed with NASA is, if someone made a mistake and they raised their hand and said, "Hey, let's don't go forward here. I blew it!" Everybody gathers around like a protective duck and they solve the problem, and he is not ostracized for raising his hand. He is actually considered a hero for saying, "Hey, stop, I blew it!" In industry, a guy can get fired for making a mistake. He is out the door, or he is looked down upon by his peers.

Joe Kittinger has described himself as "the first man in space." In his 1957 Manhigh I flight, he gazed into the darkness of space at an altitude of 96,000 feet (just over 18 miles). Three years later, during his Excelsior III flight, he set several records that endured for fifty-two years until Felix Baumgartner broke all but one of them in 2012 in the Stratos project sponsored by the energy drink maker Red Bull (see Chapter 4).

I think it's wonderful, when you think about it, that the Stratos project was financed by private industry. Just like the new SpaceX is private industry, Richard Branson is private industry. Private industry can do it better and faster and cheaper. They're in a profit-making business, and so they don't waste money or time or end up with a massive bureaucracy like NASA is. I think it's wonderful that private industry is getting involved in space research, because it'll end up with a better product for a lot less money.

I think we're more risk averse today. I think we've lost sight of our sense of adventure, or sense of proportion, when it comes to doing those types of programs. Nowadays, the Air Force would never, ever think about approving me to go up to that altitude that had never been done before. We had a heck of a time getting permission as it was. But I think that the whole psyche of research and using a man to further knowledge has changed significantly in the last fifty years. I don't believe we're as adventurous a nation as we were then.

[President John] Kennedy threw out the challenge that we were going to go to the Moon. What he did was phenomenal. The only reason we went to the Moon was because of his challenge and diverting the money to finance it. What a phenomenal move that was, to say we're going to go to the Moon. And by gum, we did it.

I just don't think we have the same leadership, we don't have the same psyche. We have people that don't have the sense of adventure – they won't take any chances whatsoever. In all of the experiments that Dr. Stapp did (see Chapter 6), he never had a fatality. Because we did everything very carefully, and we monitored what we were doing and how we were doing it. We did it in a safe manner, and there was never anybody killed on any of Dr. Stapp's experiments. Because it was all done as safe as he could make it.

I personally wish we had been working on going to Mars a long time ago. It's going to be an extremely difficult task. It's going to be very, very hazardous. But it has to be done. It's human nature. We have to extend our boundaries. To me, Mars is the next great challenge, and it's going to take a significant effort on everybody's part. It should be a worldwide effort to make it happen, because it's going to be extremely expensive and very dangerous. But to me, that should be the next goal, that we should have been working on a long time ago.

I feel that we're in a very difficult situation right now, because of the attitude that we're taking toward research. I'm encouraged that private industry is getting involved in space research, because they can do it better and cheaper and without the bureaucracy that NASA has. I look forward to the future because of the fact that private industry is getting involved and replacing NASA as the main entity for research.

[Will private industry be able to give to young people today the sense of excitement and adventure that the Mercury and Apollo programs did?] I hope that they will be motivated to think about the opportunities that are available for research and extending the boundaries of man's knowledge. I hope that happens, because that's the only way it *will* happen. I think that one of the great challenges that we have as a nation is to get back to the thought that we have to develop engineers to work on the future. We're way behind in that respect. The STEM [science, technology, engineering, and mathematics] programs that are being extended and pushed, to me, is the salvation of the scientific endeavor, because we've fallen behind in our educational system when it comes to developing engineers.

Andrew Gaffney based his projections about a manned Mars mission on his nine-day experience on Spacelab 1 in 1991 (see Chapters 13 and 14).

I think there is a manned Mars mission that can be done in about thirteen or fourteen months. You could get there in about five months, spend a month on the surface, or at least a month in orbit, some of that time on the Martian surface, then spend about seven months on return. That was one option we studied.

There was a lot of debate when we were doing these planning sessions where, I think, some of the political, non-medical people wanted to do the two-year mission, where you would stay on the surface a year and then come back. I thought, quite frankly, they were crazy. I think the technical complexity, the medical complexity, of doing that kind of a mission [would be problematic]. If you can get there for a month, do some exploring, get samples, send the right people, do it right, and get them back, that is a pretty good start. And that's where you ought to start.

Norman Thagard flew four short-duration space shuttle missions before spending four months on the Russian space station Mir in 1995 (see Chapters 13 and 14).

I won't pretend to be a seer. I won't pretend to have a crystal ball. But I would hope that humans never lose that desire to explore and discover. And certainly, space is an area that we don't have wide access or easy access to. There are still a lot of things we could learn and do. We could go back to the Moon. We could go to Mars. I really hope that we do those things. And I think that we will do those things. It's hard for me to believe that humans won't do things which they are capable of doing, and certainly this business of sending humans off to Mars excites a lot of people. It is a difficult thing to do, and as Kennedy said, we do things not necessarily because they are easy, but because they are difficult. Besides, the opportunity is there to learn about the origins of the universe and the origins of humans, for instance. So I think those are things that we are going to want to do and things that we will do.

Because of his participation in Mission Control throughout the Mercury, Gemini, and Apollo programs, Gene Kranz understands manned spaceflight as well as anyone (see Chapter 12). In May 2008, Kranz told a Senate hearing on space exploration, "The ultimate goal is to return to the Moon and to establish a lasting human presence on its surface. The Moon remains relatively unexplored and also presents new and interesting scientific prospects, whether it's greater research into its unique geology or use as a fixed platform without atmospheric interference for a new generation of space telescopes. It is an exciting and dynamic initiative."

The reason we were able to get to the Moon in ten years was because, intelligently, we addressed the issue of risk. We recognized that chances were small, risk high, but it was a worthwhile risk to take. And we developed into damned good risk managers. Today, we are so risk averse. I know you watched John Glenn [America's first manned orbital flight, in 1962]. Two of the previous five rockets had blown up. Now, can you think about putting a man on a rocket today, where you've only had five test flights and two failed? You would never do it! In some ways, it is always easy to look in the rear view mirror. But I lived in the days of John Paul Stapp (see Chapter 6) and Joey Kittinger (see Chapter 4) and jet aces, et cetera. These were my heroes. Not celebrities that appear on television and they are there for an instant. I did a talk last night at our dinner table, and I talked about the process of growing up. I didn't know at that time who my mentors were. I talked about my mother. I talked about my high school history teacher. I talked about the guy who taught me to fly. I talked about my first bosses. Each one of these lessons I got from these people without knowing it. Then all of a sudden, I find myself marvelously prepared to go to the Moon, marvelously prepared to assume responsibilities for the shuttle program. I had never managed a program in my life. I had to pick up the responsibilities for the shuttle flight software. Scariest thing I had ever done. Because one bit of error in that thing was a bad day for the program. Basically, it is just at that point where you have the ability to select the people and the confidence to turn them loose.

Constellation [NASA's manned space exploration program that was cancelled in 2010] was my dreams and my hopes that I would see an American on the Moon in my lifetime. Okay, I am now concerned that maybe my children will never see an American on the Moon. But then I worry about the future of the nation that, basically, is so service oriented that it does not build things anymore. It does not conceive anything, except for the purpose of amusement. To me, the most revolutionary work that is going on today is in medicine and, to some extent, energy. But where are the other technologies that are going to keep the economic engine of our country going? What happens to the day when kids no longer want to be astronauts?

I have pretty much eliminated colleges from my speaking catalogue. The best times I have had is doing the fifth grade on up. I love to work with fifth graders, because they've still got the attention. I like my pre-K group. But we are losing them somewhere about the time they get into the eighth grade and high school area. We have to do something if we are going to address where we are going as a nation from the standpoint of leadership. The one hope I have is I'm starting to see military leaders come back from Iraq and Afghanistan, like this Allen West down in Florida. Hopefully, that might be the real benefit of the war in Iraq and in Afghanistan. Not that these countries are free, but because we have got some leaders who are capable of making "go/no go" decisions.

Walter Cunningham spent eleven days in Earth orbit on his 1978 Apollo 7 mission (see Chapter 11).

Supposedly, we've got a new mission for NASA. I am one of those that believe you don't have a mission until you have financing for that mission, and I don't see a commitment for that kind of financing. But we talk about going to Mars. People talk about, well, maybe we will have another race with the Chinese. It's not going to happen. The Chinese will take their own sweet time of doing what they want, and they will accomplish pretty much what they want. It won't be spectacular, but they will do it. And it's a whole lot easier today. The Chinese can talk about going to the Moon, and they probably won't do it as fast as you read in the paper. But they can talk about doing it, because they don't have to reproduce all of the activities that we did to get rid of the unknowns. It's all known now. Anybody can go out and do it.

That's the way I work in my life. If there is something challenging and it's tough to do, and I see that somebody else has done it, I say, well, what the hell, I can go do it. So the Chinese are going to go on to the Moon. All of the decent industrial powers of the world are committed to do certain things in space. And here in America we keep taking agonizing reappraisals of it. We are in a position that is the envy of all these other countries because of what we have and what we could do, if we could just gear our will to do it.

When we talk about going to Mars, we have the resources, we have the technology. That doesn't mean we have all the answers, but we have the technology to solve those problems. I don't know if we have the will to do it. We don't seem to be willing to pay the price. It's going to cost lives. It's going to cost a lot of money. We may not have the farsightedness to understand that it is a worthwhile investment, just like a lot of people were against the Apollo program in the '60s. But people are going to go to Mars. It's only a question of when do they go, who will make up the crew, what kind of culture they will have. Who, in fact, will get the recognition for doing this? I would like it to be America.

About the Author

Loretta Hall is a nonfiction author and speaker. Her 2011 book *Out of this World: New Mexico's Contributions to Space Travel* won six state, regional, and national awards. She speaks frequently about space history and is a Space Ambassador for the National Space Society. She is a member of the Southwest Oral History Association and the Historical Society of New Mexico's Speakers Bureau.

Loretta graduated from the University of Washington with a B.S. degree in mathematics. She is a member of Phi Beta Kappa. In 1977, she moved to Albuquerque, where she and her husband, Jerry, raised their three daughters.

In addition to her books on space history, Loretta wrote *Underground Buildings: More than Meets the Eye*. She is certified as a Green Building Technical Professional. Her website *SubsurfaceBuildings.com* provides information about civic and commercial earth-sheltered buildings in the United States.

Loretta welcomes you to visit her space history websites, *NMSpaceHistory.com* and *SpacePioneerWords.com*, and send her your comments.

Index

O

P

R

S

CPSIA information can be obtained at www.ICGtesting.com
Printed in the USA
LVOW03s1917050614

388736LV00002B/2/P

9 781936 744275